Student Solutions Man

MW00535427

Elementary Linear Algebra

SEVENTH EDITION

Ron Larson
The Pennsylvania State University/The Behrend College

BROOKS/COLE
CENGAGE Learning™

Australia • Brazil • Japan • Korea • Mexico • Singapore • Spain • United Kingdom • United States

BROOKS/COLE
CENGAGE Learning™

© 2013 Brooks/Cole, Cengage Learning

ALL RIGHTS RESERVED. No part of this work covered by the copyright herein may be reproduced, transmitted, stored or used in any form or by any means graphic, electronic, or mechanical, including but not limited to photocopying, recording, scanning, digitizing, taping, Web distribution, information networks, or information storage and retrieval systems, except as permitted under Section 107 or 108 of the 1976 United States Copyright Act, without the prior written permission of the publisher.

For product information and technology assistance, contact us at
Cengage Learning Customer & Sales Support,
1-800-354-9706

For permission to use material from this text or product,
submit all requests online at **www.cengage.com/permissions**
Further permissions questions can be emailed to
permissionrequest@cengage.com

ISBN-13: 978-1-133-11132-0

ISBN-10: 1-133-11132-7

Brooks/Cole
20 Davis Drive
Belmont, CA 94002
USA

Cengage Learning is a leading provider of customized learning solutions with office locations around the globe, including Singapore, the United Kingdom, Australia, Mexico, Brazil, and Japan. Locate your local office at **www.cengage.com/global**

Cengage Learning products are represented in Canada by Nelson Education, Ltd.

To learn more about Brooks/Cole, visit
www.cengage.com/brookscole

Purchase any of our products at your local college store or at our preferred online store **www.cengagebrain.com**

Printed in the United States of America
2 3 4 5 6 18 17 16 15 14

CONTENTS

© 2013 Cengage Learning. All Rights Reserved. May not be scanned, copied or duplicated, or posted to a publicly accessible website, in whole or in part.

Preface

This *Student Solutions Manual* is designed as a supplement to *Elementary Linear Algebra*, Seventh Edition, by Ron Larson. All references to chapters, theorems, and exercises relate to the main text. Solutions to every odd-numbered exercise in the text are given with all essential algebraic steps included. Although this supplement is not a substitute for good study habits, it can be valuable when incorporated into a well planned course of study.

We have made every effort to see that the solutions are correct. However, we would appreciate hearing about any errors or other suggestions for improvement. Good luck with your study of elementary linear algebra.

Ron Larson
Larson Texts, Inc.

© 2013 Cengage Learning. All Rights Reserved. May not be scanned, copied or duplicated, or posted to a publicly accessible website, in whole or in part.

CHAPTER 1
Systems of Linear Equations

© 2013 Cengage Learning. All Rights Reserved. May not be scanned, copied or duplicated, or posted to a publicly accessible website, in whole or in part.

CHAPTER 1
Systems of Linear Equations

Section 1.1 Introduction to Systems of Linear Equations

1. Because the equation is in the form $a_1x + a_2y = b$, it *is* linear in the variables x and y.

3. Because the equation cannot be written in the form $a_1x + a_2y = b$, it is *not* linear in the variables x and y.

5. Because the equation cannot be written in the form $a_1x + a_2y = b$, it is *not* linear in the variables x and y.

7. Choosing y as the free variable, let $y = t$ and obtain

$$2x - 4t = 0$$
$$2x = 4t$$
$$x = 2t.$$

So, you can describe the solution set as $x = 2t$ and $y = t$, where t is any real number.

9. Choosing y and z as the free variables, let $y = s$ and $z = t$, and obtain $x + s + t = 1$ or $x = 1 - s - t$. So, you can describe the solution set as $x = 1 - s - t$, $y = s$, and $z = t$, where s and t are any real numbers.

11. $2x + y = 4$

$x - y = 2$

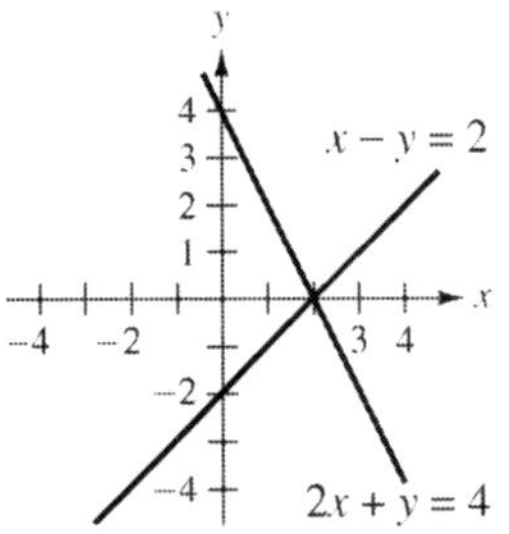

Adding the first equation to the second equation produces a new second equation, $3x = 6$ or $x = 2$. So, $y = 0$, and the solution is $x = 2$, $y = 0$. This is the point where the two lines intersect.

13. $x - y = 1$

$-2x + 2y = 5$

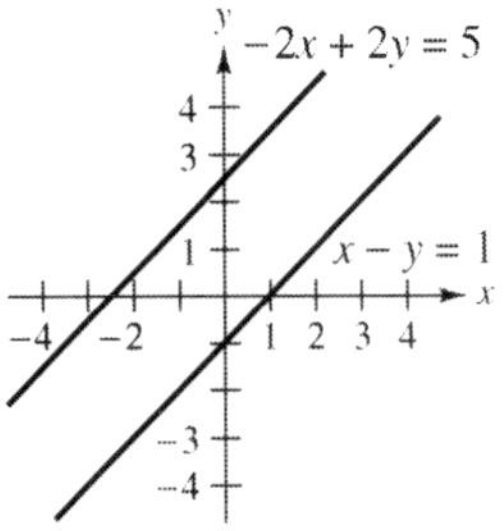

Adding 2 times the first equation to the second equation produces a new second equation.

$x - y = 1$

$0 = 7$

Because the second equation is a false statement, you can conclude that the original system of equations has no solution. Geometrically, the two lines are parallel.

15. $3x - 5y = 7$

$2x + y = 9$

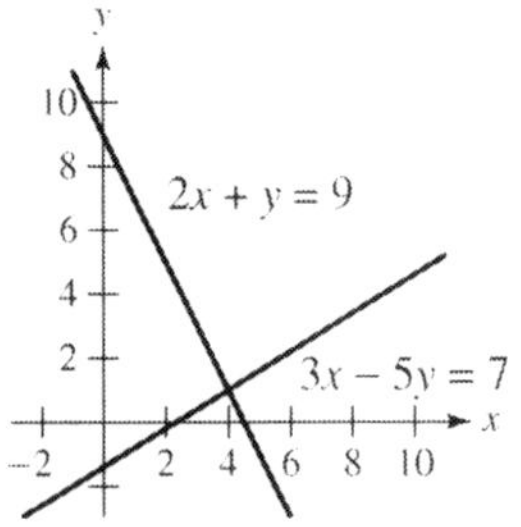

Adding the first equation to 5 times the second equation produces a new second equation, $13x = 52$ or $x = 4$. So, $2(4) + y = 9$, or $y = 1$, and the solution is: $x = 4$, $y = 1$. This is the point where the two lines intersect.

17. $2x - y = 5$

$5x - y = 11$

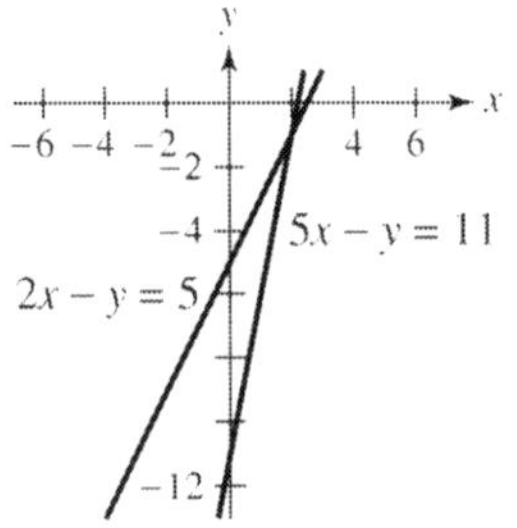

Subtracting the first equation from the second equation produces a new second equation, $3x = 6$ or $x = 2$. So, $2(2) - y = 5$ or $y = -1$, and the solution is: $x = 2$, $y = -1$. This is the point where the two lines intersect.

 © 2013 Cengage Learning. All Rights Reserved. May not be scanned, copied or duplicated, or posted to a publicly accessible website, in whole or in part.

19. $\dfrac{x+3}{4} + \dfrac{y-1}{3} = 1$

$2x - y = 12$

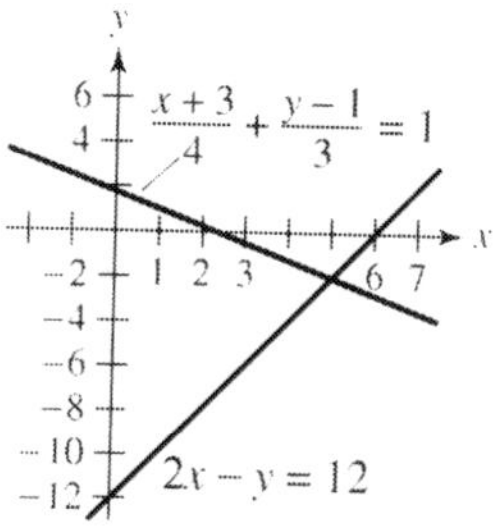

Multiplying the first equation by 12 produces a new first equation.

$3x + 4y = 7$

$2x - y = 12$

Adding the first equation to 4 times the second equation produces a new second equation, $11x = 55$ or $x = 5$. So, $2(5) - y = 12$ or $y = -2$, and the solution is: $x = 5$, $y = -2$. This is the point where the two lines intersect.

21. $0.05x - 0.03y = 0.07$

$0.07x + 0.02y = 0.16$

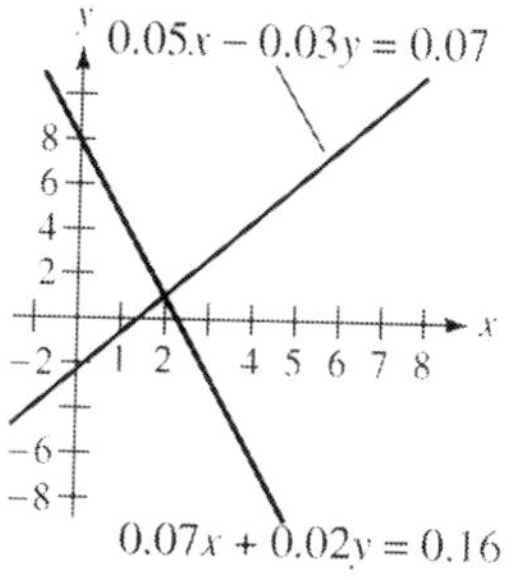

Multiplying the first equation by 200 and the second equation by 300 produces new equations.

$10x - 6y = 14$

$21x + 6y = 48$

Adding the first equation to the second equation produces a new second equation, $31x = 62$ or $x = 2$. So, $10(2) - 6y = 14$ or $y = 1$, and the solution is: $x = 2$, $y = 1$. This is the point where the two lines intersect.

23. $\dfrac{x}{4} + \dfrac{y}{6} = 1$

$x - y = 3$

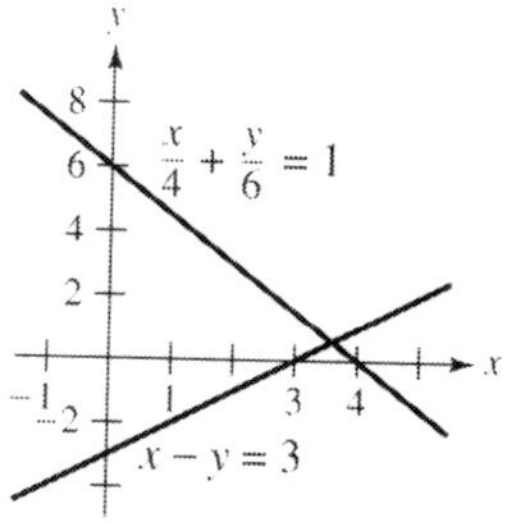

Adding 6 times the first equation to the second equation produces a new second equation, $\frac{5}{2}x = 9$ or $x = \frac{18}{5}$. So, $\frac{18}{5} - y = 3$ or $y = \frac{3}{5}$, and the solution is: $x = \frac{18}{5}$, $y = \frac{3}{5}$. This is the point where the two lines intersect.

25. From Equation 2 you have $x_2 = 3$. Substituting this value into Equation 1 produces $x_1 - 3 = 2$ or $x_1 = 5$. So, the system has exactly one solution: $x_1 = 5$ and $x_2 = 3$.

27. From Equation 3 you can conclude that $z = 0$. Substituting this value into Equation 2 produces

$2y + 0 = 3$

$y = \frac{3}{2}$.

Finally, by substituting $y = \frac{3}{2}$ and $z = 0$ into Equation 1, you obtain

$-x + \frac{3}{2} - 0 = 0$

$x = \frac{3}{2}$.

So, the system has exactly one solution: $x = \frac{3}{2}$, $y = \frac{3}{2}$, and $z = 0$.

29. Begin by rewriting the system in row-echelon form. The equations are interchanged.

$2x_1 + x_2 = 0$

$5x_1 + 2x_2 + x_3 = 0$

The first equation is multiplied by $\frac{1}{2}$.

$x_1 + \frac{1}{2}x_2 = 0$

$5x_1 + 2x_2 + x_3 = 0$

Adding -5 times the first equation to the second equation produces a new second equation.

$x_1 + \frac{1}{2}x_2 = 0$

$-\frac{1}{2}x_2 + x_3 = 0$

The second equation is multiplied by -2.

$x_1 + \frac{1}{2}x_2 = 0$

$x_2 - 2x_3 = 0$

To represent the solutions, choose x_3 to be the free variable and represent it by the parameter t. Because $x_2 = 2x_3$ and $x_1 = -\frac{1}{2}x_2$, you can describe the solution set as

$x_1 = -t$, $x_2 = 2t$, $x_3 = t$, where t is any real number.

31. (a)

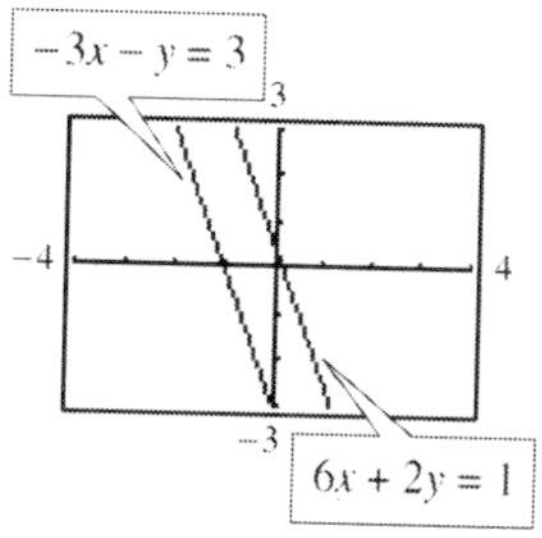

(b) The system is inconsistent.

© 2013 Cengage Learning. All Rights Reserved. May not be scanned, copied or duplicated, or posted to a publicly accessible website, in whole or in part.

33. (a)

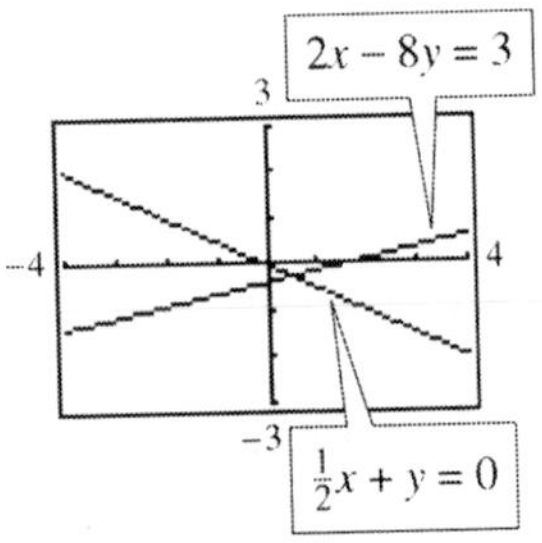

(b) The system is consistent.

(c) The solution is approximately $x = \frac{1}{2}$, $y = -\frac{1}{4}$.

(d) Adding $-\frac{1}{4}$ times the first equation to the second equation produces the new second equation $3y = -\frac{3}{4}$ or $y = -\frac{1}{4}$. So, $x = \frac{1}{2}$, and the solution is $x = \frac{1}{2}$, $y = -\frac{1}{4}$.

(e) The solutions in (c) and (d) are the same.

35. (a)

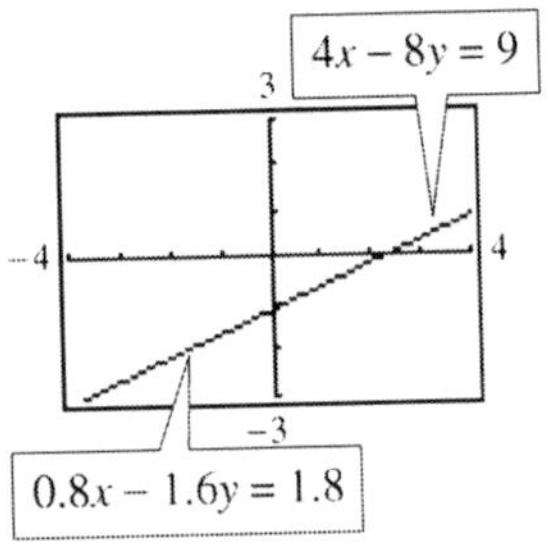

(b) The system is consistent.

(c) There are infinite solutions.

(d) The second equation is the result of multiplying both sides of the first equation by 0.2. A parametric representation of the solution set is given by $x = \frac{9}{4} + 2t$, $y = t$, where t is any real number.

(e) The solutions in (c) and (d) are consistent.

37. Adding −3 times the first equation to the second equation produces a new second equation.

$$x_1 - x_2 = 0$$
$$x_2 = -1$$

Now, using back-substitution you can conclude that the system has exactly one solution: $x_1 = -1$ and $x_2 = -1$.

39. Interchanging the two equations produces the system

$$u + 2v = 120$$
$$2u + v = 120.$$

Adding −2 times the first equation to the second equation produces a new second equation.

$$u + 2v = 120$$
$$-3v = -120$$

Solving the second equation you have $v = 40$. Substituting this value into the first equation gives $u + 80 = 120$ or $u = 40$. So, the system has exactly one solution: $u = 40$ and $v = 40$.

41. Dividing the first equation by 9 produces a new first equation.

$$x - \frac{1}{3}y = -\frac{1}{9}$$
$$\frac{1}{5}x + \frac{2}{5}y = -\frac{1}{3}$$

Adding $-\frac{1}{5}$ times the first equation to the second equation produces a new second equation.

$$x - \frac{1}{3}y = -\frac{1}{9}$$
$$\frac{7}{15}y = -\frac{14}{45}$$

Multiplying the second equation by $\frac{15}{7}$ produces a new second equation.

$$x - \frac{1}{3}y = -\frac{1}{9}$$
$$y = -\frac{2}{3}$$

Now, using back-substitution you can substitute $y = -\frac{2}{3}$ into the first equation to obtain $x + \frac{2}{9} = -\frac{1}{9}$ or $x = -\frac{1}{3}$. So, you can conclude that the system has exactly one solution: $x = -\frac{1}{3}$ and $y = -\frac{2}{3}$.

43. To begin, change the form of the first equation.

$$\frac{1}{4}x + \frac{1}{3}y = \frac{17}{6}$$
$$x - 3y = 20$$

Multiplying the first equation by 4 yields a new first equation.

$$x + \frac{4}{3}y = \frac{34}{3}$$
$$x - 3y = 20$$

Subtracting the first equation from the second equation yields a new second equation.

$$x + \frac{4}{3}y = \frac{34}{3}$$
$$\frac{13}{3}y = -\frac{26}{3}$$

Dividing the second equation by $\frac{13}{3}$ yields a new second equation.

$$x + \frac{4}{3}y = \frac{34}{3}$$
$$y = -2$$

Now, using back-substitution you can conclude that the system has exactly one solution: $x = 14$ and $y = -2$.

45. Multiplying the first equation by 50 and the second equation by 100 produces a new system.

$$x_1 - 2.5x_2 = -9.5$$
$$3x_1 + 4x_2 = 52$$

Adding −3 times the first equation to the second equation produces a new second equation.

$$x_1 - 2.5x_2 = -9.5$$
$$11.5x_2 = 80.5$$

Now, using back-substitution you can conclude that the system has exactly one solution: $x_1 = 8$ and $x_2 = 7$.

© 2013 Cengage Learning. All Rights Reserved. May not be scanned, copied or duplicated, or posted to a publicly accessible website, in whole or in part.

47. Adding −2 times the first equation to the second equation yields a new second equation.

$$\begin{aligned} x + y + z &= 6 \\ -3y - z &= -9 \\ 3x - z &= 0 \end{aligned}$$

Adding −3 times the first equation to the third equation yields a new third equation.

$$\begin{aligned} x + y + z &= 6 \\ -3y - z &= -9 \\ -3y - 4z &= -18 \end{aligned}$$

Dividing the second equation by −3 yields a new second equation.

$$\begin{aligned} x + y + z &= 6 \\ y + \tfrac{1}{3}z &= 3 \\ -3y - 4z &= -18 \end{aligned}$$

Adding 3 times the second equation to the third equation yields a new third equation.

$$\begin{aligned} x + y + z &= 6 \\ y + \tfrac{1}{3}z &= 3 \\ -3z &= -9 \end{aligned}$$

Dividing the third equation by −3 yields a new third equation.

$$\begin{aligned} x + y + z &= 6 \\ y + \tfrac{1}{3}z &= 3 \\ z &= 3 \end{aligned}$$

Now, using back-substitution you can conclude that the system has exactly one solution:
$x = 1$, $y = 2$, and $z = 3$.

49. Dividing the first equation by 3 yields a new first equation.

$$\begin{aligned} x_1 - \tfrac{2}{3}x_2 + \tfrac{4}{3}x_3 &= \tfrac{1}{3} \\ x_1 + x_2 - 2x_3 &= 3 \\ 2x_1 - 3x_2 + 6x_3 &= 8 \end{aligned}$$

Subtracting the first equation from the second equation yields a new second equation.

$$\begin{aligned} x_1 - \tfrac{2}{3}x_2 + \tfrac{4}{3}x_3 &= \tfrac{1}{3} \\ \tfrac{5}{3}x_2 - \tfrac{10}{3}x_3 &= \tfrac{8}{3} \\ 2x_1 - 3x_2 + 6x_3 &= 8 \end{aligned}$$

Adding −2 times the first equation to the third equation yields a new third equation.

$$\begin{aligned} x_1 - \tfrac{2}{3}x_2 + \tfrac{4}{3}x_3 &= \tfrac{1}{3} \\ \tfrac{5}{3}x_2 - \tfrac{10}{3}x_3 &= \tfrac{8}{3} \\ -\tfrac{5}{3}x_2 + \tfrac{10}{3}x_3 &= \tfrac{22}{3} \end{aligned}$$

At this point you should recognize that Equations 2 and 3 cannot both be satisfied. So, the original system of equations has no solution.

51. Dividing the first equation by 2 yields a new first equation.

$$\begin{aligned} x_1 + \tfrac{1}{2}x_2 - \tfrac{3}{2}x_3 &= 2 \\ 4x_1 + 2x_3 &= 10 \\ -2x_1 + 3x_2 - 13x_3 &= -8 \end{aligned}$$

Adding −4 times the first equation to the second equation produces a new second equation.

$$\begin{aligned} x_1 + \tfrac{1}{2}x_2 - \tfrac{3}{2}x_3 &= 2 \\ -2x_2 + 8x_3 &= 2 \\ -2x_1 + 3x_2 - 13x_3 &= -8 \end{aligned}$$

Adding 2 times the first equation to the third equation produces a new third equation.

$$\begin{aligned} x_1 + \tfrac{1}{2}x_2 - \tfrac{3}{2}x_3 &= 2 \\ -2x_2 + 8x_3 &= 2 \\ 4x_2 - 16x_3 &= -4 \end{aligned}$$

Dividing the second equation by −2 yields a new second equation.

$$\begin{aligned} x_1 + \tfrac{1}{2}x_2 - \tfrac{3}{2}x_3 &= 2 \\ x_2 - 4x_3 &= -1 \\ 4x_2 - 16x_3 &= -4 \end{aligned}$$

Adding −4 times the second equation to the third equation produces a new third equation.

$$\begin{aligned} x_1 + \tfrac{1}{2}x_2 - \tfrac{3}{2}x_3 &= 2 \\ x_2 - 4x_3 &= -1 \\ 0 &= 0 \end{aligned}$$

Adding $-\frac{1}{2}$ times the second equation to the first equation produces a new first equation.

$$\begin{aligned} x_1 + \tfrac{1}{2}x_3 &= \tfrac{5}{2} \\ x_2 - 4x_3 &= -1 \end{aligned}$$

Choosing $x_3 = t$ as the free variable, you can describe the solution as $x_1 = \frac{5}{2} - \frac{1}{2}t$, $x_2 = 4t - 1$, and $x_3 = t$, where t is any real number.

53. Adding −5 times the first equation to the second equation yields a new second equation.

$$\begin{aligned} x - 3y + 2z &= 18 \\ 0 &= -72 \end{aligned}$$

Because the second equation is a false statement, you can conclude that the original system of equations has no solution.

© 2013 Cengage Learning. All Rights Reserved. May not be scanned, copied or duplicated, or posted to a publicly accessible website, in whole or in part.

55. Adding -2 times the first equation to the second, 3 times the first equation to the third, and -1 times the first equation to the fourth, produces

$$\begin{aligned} x + y + z + w &= 6 \\ y - 2z - 3w &= -12 \\ 7y + 4z + 5w &= 22 \\ y - 2z &= -6. \end{aligned}$$

Adding -7 times the second equation to the third and -1 times the second equation to the fourth produces

$$\begin{aligned} x + y + z + w &= 6 \\ y - 2z - 3w &= -12 \\ 18z + 26w &= 106 \\ 3w &= 6. \end{aligned}$$

Using back-substitution, you find the original system has exactly one solution: $x = 1$, $y = 0$, $z = 3$, and $w = 2$.

Answers may vary slightly for Exercises 57 and 59.

57. Using a computer software program or graphing utility, you obtain

$x_1 = -15$, $x_2 = 40$, $x_3 = 45$, $x_4 = -75$.

59. Using a computer software program or graphing utility, you obtain

$x_1 = \frac{1}{5}$, $x_2 = -\frac{4}{5}$, $x_3 = \frac{1}{2}$.

61. $x = y = z = 0$ is clearly a solution.

Dividing the first equation by 4 yields a new first equation.

$$\begin{aligned} x + \tfrac{3}{4}y + \tfrac{17}{4}z &= 0 \\ 5x + 4y + 22z &= 0 \\ 4x + 2y + 19z &= 0 \end{aligned}$$

Adding -5 times the first equation to the second equation yields a new second equation.

$$\begin{aligned} x + \tfrac{3}{4}y + \tfrac{17}{4}z &= 0 \\ \tfrac{1}{4}y + \tfrac{3}{4}z &= 0 \\ 4x + 2y + 19z &= 0 \end{aligned}$$

Adding -4 times the first equation to the third equation yields a new third equation.

$$\begin{aligned} x + \tfrac{3}{4}y + \tfrac{17}{4}z &= 0 \\ \tfrac{1}{4}y + \tfrac{3}{4}z &= 0 \\ -y + 2z &= 0 \end{aligned}$$

Multiplying the second equation by 4 yields a new second equation.

$$\begin{aligned} x + \tfrac{3}{4}y + \tfrac{17}{4}z &= 0 \\ y + 3z &= 0 \\ -y + 2z &= 0 \end{aligned}$$

Adding the second equation to the third equation yields a new third equation.

$$\begin{aligned} x + \tfrac{3}{4}y + \tfrac{17}{4}z &= 0 \\ y + 3z &= 0 \\ 5z &= 0 \end{aligned}$$

Dividing the third equation by 5 yields a new third equation.

$$\begin{aligned} x + \tfrac{3}{4}y + \tfrac{17}{4}z &= 0 \\ y + 3z &= 0 \\ z &= 0 \end{aligned}$$

Now, using back-substitution you can conclude that the system has exactly one solution: $x = 0$, $y = 0$, and $z = 0$.

© 2013 Cengage Learning. All Rights Reserved. May not be scanned, copied or duplicated, or posted to a publicly accessible website, in whole or in part.

63. $x = y = z = 0$ is clearly a solution.

Dividing the first equation by 5 yields a new first equation.

$$\begin{aligned} x + y - \tfrac{1}{5}z &= 0 \\ 10x + 5y + 2z &= 0 \\ 5x + 15y - 9z &= 0 \end{aligned}$$

Adding -10 times the first equation to the second equation yields a new second equation.

$$\begin{aligned} x + y - \tfrac{1}{5}z &= 0 \\ -5y + 4z &= 0 \\ 5x + 15y - 9z &= 0 \end{aligned}$$

Adding -5 times the first equation to the third equation yields a new third equation.

$$\begin{aligned} x + y - \tfrac{1}{5}z &= 0 \\ -5y + 4z &= 0 \\ 10y - 8z &= 0 \end{aligned}$$

Dividing the second equation by -5 yields a new second equation.

$$\begin{aligned} x + y - \tfrac{1}{5}z &= 0 \\ y - \tfrac{4}{5}z &= 0 \\ 10y - 8z &= 0 \end{aligned}$$

Adding -10 times the second equation to the third equation yields a new third equation.

$$\begin{aligned} x + y - \tfrac{1}{5}z &= 0 \\ y - \tfrac{4}{5}z &= 0 \\ 0 &= 0 \end{aligned}$$

Adding -1 times the second equation to the first equation yields a new first equation.

$$\begin{aligned} x + \tfrac{3}{5}z &= 0 \\ y - \tfrac{4}{5}z &= 0 \end{aligned}$$

Choosing $z = t$ as the free variable you find the solution to be $x = -\frac{3}{5}t$, $y = \frac{4}{5}t$, and $z = t$, where t is any real number.

65. Let x = number of milligrams in 8-ounce glass of apple juice.

Let y = number of milligrams in 8-ounce glass of orange juice.

$x + y = 177.4$ Equation 1

$2x + 3y = 436.7$ Equation 2

Multiply Equation 1 by -2: $-2x - 2y = -354.8$

Add this to Equation 2 to eliminate x:

$$\begin{aligned} -2x - 2y &= -354.8 \\ 2x + 3y &= 436.7 \\ \hline y &= 81.9 \text{ mg} \end{aligned}$$

Back-substitute $y = 81.9$ into Equation 1: $x + 81.9 = 177.4$

$x = 95.5$ mg

Apple juice contains 95.5 milligrams of Vitamin C and orange juice contains 81.9 milligrams.

© 2013 Cengage Learning. All Rights Reserved. May not be scanned, copied or duplicated, or posted to a publicly accessible website, in whole or in part.

67. (a) True. You can describe the entire solution set using parametric representation.

$ax + by = c$

Choosing $y = t$ as the free variable, the solution is

$x = \frac{c}{a} - \frac{b}{a}t,\ y = t$, where t is any real number.

(b) False. For example, consider the system

$$x_1 + x_2 + x_3 = 1$$
$$x_1 + x_2 + x_3 = 2$$

which is an inconsistent system.

(c) False. A consistent system may have only one solution.

69. Because $x_1 = t$ and $x_2 = 3t - 4 = 3x_1 - 4$, one answer is the system

$$3x_1 - x_2 = 4$$
$$-3x_1 + x_2 = -4.$$

Letting $x_2 = t$, you get $x_1 = \frac{4 + t}{3} = \frac{4}{3} + \frac{t}{3}$.

71. Substituting $A = 1/x$ and $B = 1/y$ into the original system yields

$$12A - 12B = 7$$
$$3A + 4B = 0.$$

Reduce this system to row-echelon form. Dividing the first equation by 12 yields a new first equation.

$$A - B = \tfrac{7}{12}$$
$$3A + 4B = 0$$

Adding -3 times the first equation to the second equation yields a new second equation.

$$A - B = \tfrac{7}{12}$$
$$7B = -\tfrac{7}{4}$$

Dividing the second equation by 7 yields a new second equation.

$$A - B = \tfrac{7}{12}$$
$$B = -\tfrac{1}{4}$$

So, $A = -1/4$ and $B = 1/3$. Because $A = 1/x$ and $B = 1/y$, the solution of the original system of equations is: $x = 3$ and $y = -4$.

73. Substituting $A = 1/x$, $B = 1/y$, and $C = 1/z$, into the original system yields

$$2A + B - 3C = 4$$
$$4A + 2C = 10$$
$$-2A + 3B - 13C = -8.$$

Reduce this system to row-echelon form.

$$A + \tfrac{1}{2}B - \tfrac{3}{2}C = 2$$
$$-2B + 8C = 2$$
$$4B - 16C = -4$$

$$A + \tfrac{1}{2}B - \tfrac{3}{2}C = 2$$
$$B - 4C = -1$$

Letting $t = C$ be the free variable, you have

$C = t$, $B = 4t - 1$, and $A = \frac{(-t + 5)}{2}$. So, the solution to the original problem is

$x = \frac{2}{5 - t},\ y = \frac{1}{4t - 1},\ z = \frac{1}{t}$, where $t \neq 5, \frac{1}{4}, 0$.

75. Reduce the system to row-echelon form. Dividing the first equation by $\cos\theta$ yields a new second equation.

$$x + \left(\frac{\sin\theta}{\cos\theta}\right)y = \frac{1}{\cos\theta}$$
$$(-\sin\theta)x + (\cos\theta)y = 0$$

Multiplying the first equation by $\sin\theta$ and adding to the second equation yields a new second equation.

$$x + \left(\frac{\sin\theta}{\cos\theta}\right)y = \frac{1}{\cos\theta}$$
$$\left(\frac{1}{\cos\theta}\right)y = \frac{\sin\theta}{\cos\theta}$$

$\left(\text{Because } \frac{\sin^2\theta}{\cos\theta} + \cos\theta = \frac{\sin^2\theta + \cos^2\theta}{\cos\theta} = \frac{1}{\cos\theta}\right)$

Multiplying the second equation by $\cos\theta$ yields a new second equation.

$$x + \left(\frac{\sin\theta}{\cos\theta}\right)y = \frac{1}{\cos\theta}$$
$$y = \sin\theta$$

Substituting $y = \sin\theta$ into the first equation yields

$$x + \left(\frac{\sin\theta}{\cos\theta}\right)\sin\theta = \frac{1}{\cos\theta}$$
$$x = \frac{1 - \sin^2\theta}{\cos\theta} = \frac{\cos^2\theta}{\cos\theta} = \cos\theta.$$

So, the solution of the original system of equations is: $x = \cos\theta$ and $y = \sin\theta$.

© 2013 Cengage Learning. All Rights Reserved. May not be scanned, copied or duplicated, or posted to a publicly accessible website, in whole or in part.

77. For this system to have an infinite number of solutions, both equations need to be equivalent.

Multiply the second equation by −2.

$$\begin{aligned} 4x + ky &= 6 \\ -2kx - 2y &= 6 \end{aligned}$$

So, when $k = -2$, the system will have an infinite number of solutions.

79. Reduce the system to row-echelon form.

$$\begin{aligned} x + \quad ky &= 0 \\ \left(1 - k^2\right)y &= 0 \end{aligned}$$

$$\begin{aligned} x + ky &= 0 \\ y &= 0,\ 1 - k^2 \neq 0 \end{aligned}$$

$$\begin{aligned} x &= 0 \\ y &= 0,\ 1 - k^2 \neq 0 \end{aligned}$$

If $1 - k^2 \neq 0$, that is if $k \neq \pm 1$, the system will have exactly one solution.

81. To begin, reduce the system to row-echelon form.

$$\begin{aligned} x + 2y + kz &= 6 \\ (8 - 3k)z &= -14 \end{aligned}$$

This system will have no solution if $8 - 3k = 0$, that is, $k = \frac{8}{3}$.

83. Reducing the system to row-echelon form, you have

$$\begin{aligned} x + \quad y + \quad kz &= 3 \\ (k - 1)y + (1 - k)z &= -1 \\ (1 - k)y + \left(1 - k^2\right)z &= 1 - 3k \end{aligned}$$

$$\begin{aligned} x + \quad y + \quad kz &= 3 \\ (k - 1)y + \quad (1 - k)z &= -1 \\ \left(-k^2 - k + 2\right)z &= -3k. \end{aligned}$$

If $-k^2 - k + 2 = 0$, then there is no solution. So, if $k = 1$ or $k = -2$, there is not a unique solution.

85. (a) All three of the lines will intersect in exactly one point (corresponding to the solution point).

(b) All three of the lines will coincide (every point on these lines is a solution point).

(c) The three lines have no common point.

87. Answers will vary. (*Hint:* Choose three different values for x and solve the resulting system of linear equations in the variables a, b, and c.)

89.

$$\begin{aligned} x - 4y &= -3 \\ 5x - 6y &= 13 \end{aligned}$$

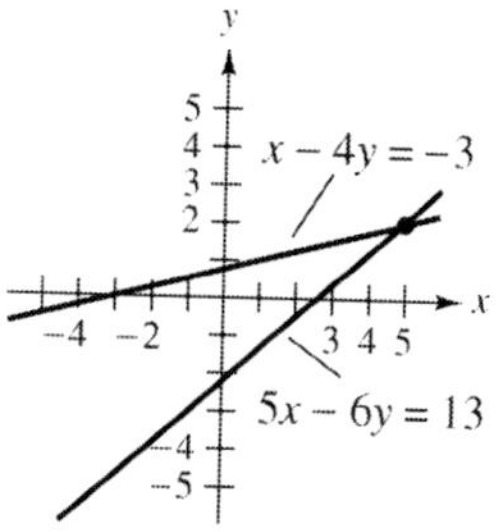

$$\begin{aligned} x - 4y &= -3 \\ 14y &= 28 \end{aligned}$$

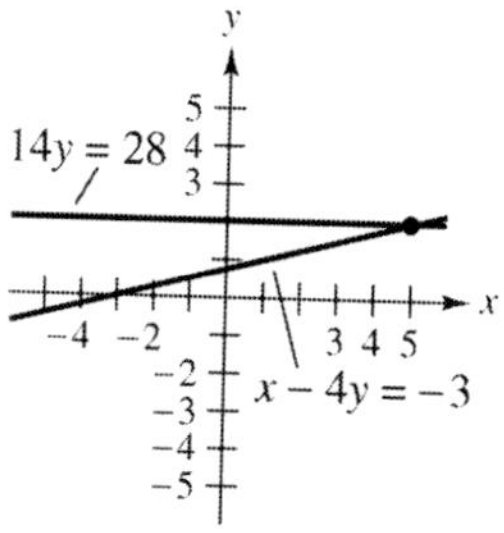

$$\begin{aligned} x - 4y &= -3 \\ y &= 2 \end{aligned}$$

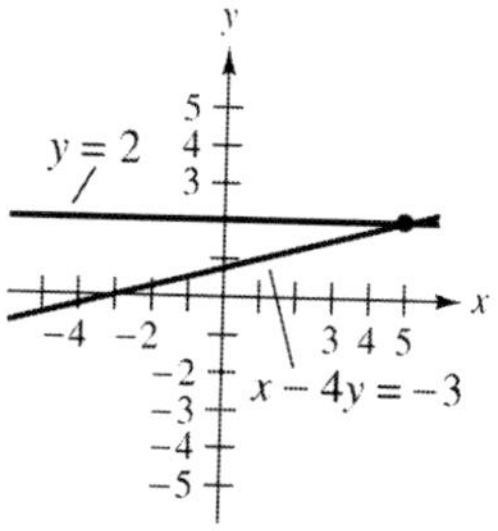

$$\begin{aligned} x &= 5 \\ y &= 2 \end{aligned}$$

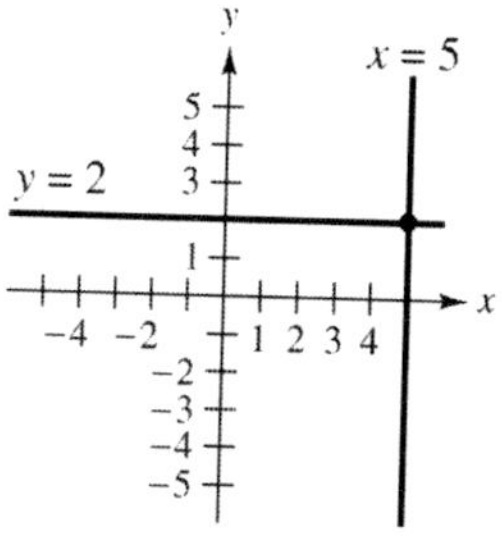

At each step, the lines always intersect at $(5, 2)$, which is the solution to the system of equations.

91. Solve each equation for y.

$$y = \frac{1}{100}x + 2$$

$$y = \frac{1}{99}x - 2$$

The graphs are misleading because, while they appear parallel, when the equations are solved for y they have slightly different slopes.

Section 1.2 Gaussian Elimination and Gauss-Jordan Elimination

1. Because the matrix has 3 rows and 3 columns, it has size 3×3.

3. Because the matrix has 2 rows and 4 columns, it has size 2×4.

5. Because the matrix has 4 rows and 5 columns, it has size 4×5.

7. $\begin{bmatrix} -2 & 5 & 1 \\ 3 & -1 & -8 \end{bmatrix} \Rightarrow \begin{bmatrix} 13 & 0 & -39 \\ 3 & -1 & -8 \end{bmatrix}$

Add 5 times Row 2 to Row 1.

9. $\begin{bmatrix} 0 & -1 & -5 & 5 \\ -1 & 3 & -7 & 6 \\ 4 & -5 & 1 & 3 \end{bmatrix} \Rightarrow \begin{bmatrix} -1 & 3 & -7 & 6 \\ 0 & -1 & -5 & 5 \\ 0 & 7 & -27 & 27 \end{bmatrix}$

Interchange Row 1 and Row 2. Then add 4 times the new Row 1 to Row 3.

© 2013 Cengage Learning. All Rights Reserved. May not be scanned, copied or duplicated, or posted to a publicly accessible website, in whole or in part.

11. Because the matrix is in reduced row-echelon form, convert back to a system of linear equations.

$x_1 = 0$

$x_2 = 2$

So, the solution is: $x_1 = 0$ and $x_2 = 2$.

13. Because the matrix is in row-echelon form, convert back to a system of linear equations.

$$\begin{aligned} x_1 - x_2 \quad &= 3 \\ x_2 - 2x_3 &= 1 \\ x_3 &= -1 \end{aligned}$$

Solve this system by back-substitution.

$$\begin{aligned} x_2 - 2(-1) &= 1 \\ x_2 &= -1 \end{aligned}$$

Substituting $x_2 = -1$ into Equation 1,

$$\begin{aligned} x_1 - (-1) &= 3 \\ x_1 &= 2. \end{aligned}$$

So, the solution is: $x_1 = 2$, $x_2 = -1$, and $x_3 = -1$.

15. Interchange the first and second rows.

$$\begin{bmatrix} 1 & -1 & 1 & 0 \\ 2 & 1 & -1 & 3 \\ 0 & 1 & 2 & 1 \end{bmatrix}$$

Interchange the second and third rows.

$$\begin{bmatrix} 1 & -1 & 1 & 0 \\ 0 & 1 & 2 & 1 \\ 2 & 1 & -1 & 3 \end{bmatrix}$$

Add -2 times the first row to the third row to produce a new third row.

$$\begin{bmatrix} 1 & -1 & 1 & 0 \\ 0 & 1 & 2 & 1 \\ 0 & 3 & -3 & 3 \end{bmatrix}$$

Add -3 times the second row to the third row to produce a new third row.

$$\begin{bmatrix} 1 & -1 & 1 & 0 \\ 0 & 1 & 2 & 1 \\ 0 & 0 & -9 & 0 \end{bmatrix}$$

Divide the third row by -9 to produce a new third row.

$$\begin{bmatrix} 1 & -1 & 1 & 0 \\ 0 & 1 & 2 & 1 \\ 0 & 0 & 1 & 0 \end{bmatrix}$$

Convert back to a system of linear equations.

$$\begin{aligned} x_1 - x_2 + x_3 &= 0 \\ x_2 + 2x_3 &= 1 \\ x_3 &= 0 \end{aligned}$$

Solve this system by back-substitution.

$x_2 = 1 - 2x_3 = 1 - 2(0) = 1$

$x_1 = x_2 - x_3 = 1 - 0 = 1$

So, the solution is: $x_1 = 1$, $x_2 = 1$, and $x_3 = 0$.

17. Because the matrix is in row-echelon form, convert back to a system of linear equations.

$$\begin{aligned} x_1 + 2x_2 \qquad + x_4 &= 4 \\ x_2 + 2x_3 + x_4 &= 3 \\ x_3 + 2x_4 &= 1 \\ x_4 &= 4 \end{aligned}$$

Solve this system by back-substitution.

$x_3 = 1 - 2x_4 = 1 - 2(4) = -7$

$x_2 = 3 - 2x_3 - x_4 = 3 - 2(-7) - 4 = 13$

$x_1 = 4 - 2x_2 - x_4 = 4 - 2(13) - 4 = -26$

So, the solution is: $x_1 = -26$, $x_2 = 13$, $x_3 = -7$, and $x_4 = 4$.

19. The matrix satisfies all three conditions in the definition of row-echelon form. Moreover, because each column that has a leading 1 (columns one and two) has zeros elsewhere, the matrix *is* in reduced row-echelon form.

21. Because the matrix has two non-zero rows without leading 1's, it is *not* in row-echelon form.

23. Because the matrix has a non-zero row without a leading 1, it is *not* in row-echelon form.

25. The augmented matrix for this system is

$$\begin{bmatrix} 1 & 2 & 7 \\ 2 & 1 & 8 \end{bmatrix}.$$

Adding -2 times the first row to the second row yields a new second row.

$$\begin{bmatrix} 1 & 2 & 7 \\ 0 & -3 & -6 \end{bmatrix}$$

Dividing the second row by -3 yields a new second row.

$$\begin{bmatrix} 1 & 2 & 7 \\ 0 & 1 & 2 \end{bmatrix}$$

Converting back to a system of linear equations produces

$$\begin{aligned} x + 2y &= 7 \\ y &= 2. \end{aligned}$$

Finally, using back-substitution you find that $x = 3$ and $y = 2$.

27. The augmented matrix for this system is

$$\begin{bmatrix} -1 & 2 & 1.5 \\ 2 & -4 & 3 \end{bmatrix}.$$

Gaussian elimination produces the following.

$$\begin{bmatrix} -1 & 2 & 1.5 \\ 2 & -4 & 3 \end{bmatrix} \Rightarrow \begin{bmatrix} 1 & -2 & -\frac{3}{2} \\ 2 & -4 & 3 \end{bmatrix} \Rightarrow \begin{bmatrix} 1 & -2 & -\frac{3}{2} \\ 0 & 0 & 6 \end{bmatrix}$$

Because the second row of this matrix corresponds to the equation $0 = 6$, you can conclude that the original system has no solution.

© 2013 Cengage Learning. All Rights Reserved. May not be scanned, copied or duplicated, or posted to a publicly accessible website, in whole or in part.

29. The augmented matrix for this system is

$$\begin{bmatrix} -3 & 5 & -22 \\ 3 & 4 & 4 \\ 4 & -8 & 32 \end{bmatrix}.$$

Dividing the first row by -3 yields a new first row.

$$\begin{bmatrix} 1 & -\frac{5}{3} & \frac{22}{3} \\ 3 & 4 & 4 \\ 4 & -8 & 32 \end{bmatrix}$$

Adding -3 times the first row to the second row yields a new second row.

$$\begin{bmatrix} 1 & -\frac{5}{3} & \frac{22}{3} \\ 0 & 9 & -18 \\ 4 & -8 & 32 \end{bmatrix}$$

Adding -4 times the first row to the third row yields a new third row.

$$\begin{bmatrix} 1 & -\frac{5}{3} & \frac{22}{3} \\ 0 & 9 & -18 \\ 0 & -\frac{4}{3} & \frac{8}{3} \end{bmatrix}$$

Dividing the second row by 9 yields a new second row.

$$\begin{bmatrix} 1 & -\frac{5}{3} & \frac{22}{3} \\ 0 & 1 & -2 \\ 0 & -\frac{4}{3} & \frac{8}{3} \end{bmatrix}$$

Adding $\frac{4}{3}$ times the second row to the third row yields a new third row.

$$\begin{bmatrix} 1 & -\frac{5}{3} & \frac{22}{3} \\ 0 & 1 & -2 \\ 0 & 0 & 0 \end{bmatrix}$$

Converting back to a system of linear equations produces

$$\begin{aligned} x - \tfrac{5}{3}y &= \tfrac{22}{3} \\ y &= -2. \end{aligned}$$

Finally, using back-substitution you find that the solution is: $x = 4$ and $y = -2$.

31. The augmented matrix for this system is

$$\begin{bmatrix} 1 & 0 & -3 & -2 \\ 3 & 1 & -2 & 5 \\ 2 & 2 & 1 & 4 \end{bmatrix}.$$

Gaussian elimination produces the following.

$$\begin{bmatrix} 1 & 0 & -3 & -2 \\ 3 & 1 & -2 & 5 \\ 2 & 2 & 1 & 4 \end{bmatrix} \Rightarrow \begin{bmatrix} 1 & 0 & -3 & -2 \\ 0 & 1 & 7 & 11 \\ 0 & 2 & 7 & 8 \end{bmatrix} \Rightarrow \begin{bmatrix} 1 & 0 & -3 & -2 \\ 0 & 1 & 7 & 11 \\ 0 & 0 & -7 & -14 \end{bmatrix}$$

Back substitution now yields

$$\begin{aligned} x_3 &= 2 \\ x_2 &= 11 - 7x_3 = 11 - (7)2 = -3 \\ x_1 &= -2 + 3x_3 = -2 + 3(2) = 4. \end{aligned}$$

So, the solution is: $x_1 = 4$, $x_2 = -3$, and $x_3 = 2$.

33. The augmented matrix for this system is

$$\begin{bmatrix} 2 & 0 & 3 & 3 \\ 4 & -3 & 7 & 5 \\ 8 & -9 & 15 & 10 \end{bmatrix}.$$

Gaussian elimination produces the following.

$$\begin{bmatrix} 2 & 0 & 3 & 3 \\ 4 & -3 & 7 & 5 \\ 8 & -9 & 15 & 10 \end{bmatrix} \Rightarrow \begin{bmatrix} 1 & 0 & \frac{3}{2} & \frac{3}{2} \\ 4 & -3 & 7 & 5 \\ 8 & -9 & 15 & 10 \end{bmatrix}$$

$$\Rightarrow \begin{bmatrix} 1 & 0 & \frac{3}{2} & \frac{3}{2} \\ 0 & -3 & 1 & -1 \\ 0 & -9 & 3 & -2 \end{bmatrix} \Rightarrow \begin{bmatrix} 1 & 0 & \frac{3}{2} & \frac{3}{2} \\ 0 & 1 & -\frac{1}{3} & \frac{1}{3} \\ 0 & 0 & 0 & 1 \end{bmatrix}$$

Because the third row corresponds to the equation $0 = 1$, there is no solution to the original system.

35. The augmented matrix for this system is

$$\begin{bmatrix} 4 & 12 & -7 & -20 & 22 \\ 3 & 9 & -5 & -28 & 30 \end{bmatrix}.$$

Dividing the first row by 4 yields a new first row.

$$\begin{bmatrix} 1 & 3 & -\frac{7}{4} & -5 & \frac{11}{2} \\ 3 & 9 & -5 & -28 & 30 \end{bmatrix}$$

Adding -3 times the first row to the second row yields a new second row.

$$\begin{bmatrix} 1 & 3 & -\frac{7}{4} & -5 & \frac{11}{2} \\ 0 & 0 & \frac{1}{4} & -13 & \frac{27}{2} \end{bmatrix}$$

Multiplying the second row by 4 yields a new second row.

$$\begin{bmatrix} 1 & 3 & -\frac{7}{4} & -5 & \frac{11}{2} \\ 0 & 0 & 1 & -52 & 54 \end{bmatrix}$$

Converting back to a system of linear equations produces

$$\begin{aligned} x + 3y - \tfrac{7}{4}z - 5w &= \tfrac{11}{2} \\ z - 52w &= 54. \end{aligned}$$

Choosing $y = s$ and $w = t$ as the free variables, you can describe the solution as $x = 100 - 3s + 96t$, $y = s$, $z = 54 + 52t$, and $w = t$, where s and t are any real numbers.

© 2013 Cengage Learning. All Rights Reserved. May not be scanned, copied or duplicated, or posted to a publicly accessible website, in whole or in part.

37. The augmented matrix for this system is

$$\begin{bmatrix} 3 & 3 & 12 & 6 \\ 1 & 1 & 4 & 2 \\ 2 & 5 & 20 & 10 \\ -1 & 2 & 8 & 4 \end{bmatrix}.$$

Gaussian elimination produces the following.

$$\begin{bmatrix} 3 & 3 & 12 & 6 \\ 1 & 1 & 4 & 2 \\ 2 & 5 & 20 & 10 \\ -1 & 2 & 8 & 4 \end{bmatrix} \Rightarrow \begin{bmatrix} 1 & 1 & 4 & 2 \\ 1 & 1 & 4 & 2 \\ 2 & 5 & 20 & 10 \\ -1 & 2 & 8 & 4 \end{bmatrix}$$

$$\Rightarrow \begin{bmatrix} 1 & 1 & 4 & 2 \\ 0 & 0 & 0 & 0 \\ 0 & 3 & 12 & 6 \\ 0 & 3 & 12 & 6 \end{bmatrix} \Rightarrow \begin{bmatrix} 1 & 1 & 4 & 2 \\ 0 & 3 & 12 & 6 \\ 0 & 0 & 0 & 0 \\ 0 & 0 & 0 & 0 \end{bmatrix}$$

$$\Rightarrow \begin{bmatrix} 1 & 1 & 4 & 2 \\ 0 & 1 & 4 & 2 \\ 0 & 0 & 0 & 0 \\ 0 & 0 & 0 & 0 \end{bmatrix} \Rightarrow \begin{bmatrix} 1 & 0 & 0 & 0 \\ 0 & 1 & 4 & 2 \\ 0 & 0 & 0 & 0 \\ 0 & 0 & 0 & 0 \end{bmatrix}$$

Letting $z = t$ be the free variable, the solution is: $x = 0$, $y = 2 - 4t$, and $z = t$, where t is any real number.

39. Using a computer software program or graphing utility, the augmented matrix reduces to

$$\begin{bmatrix} 1 & 0 & 0 & 0 & 0 & 2 \\ 0 & 1 & 0 & 0 & 0 & -2 \\ 0 & 0 & 1 & 0 & 0 & 3 \\ 0 & 0 & 0 & 1 & 0 & -5 \\ 0 & 0 & 0 & 0 & 1 & 1 \end{bmatrix}.$$

So, the solution is: $x_1 = 2$, $x_2 = -2$, $x_3 = 3$, $x_4 = -5$, and $x_5 = 1$.

41. The corresponding system of equations is

$$\begin{aligned} x_1 \qquad\quad &= 0 \\ x_2 + x_3 &= 0 \\ 0 &= 0. \end{aligned}$$

Letting $x_3 = t$ be the free variable, the solution is: $x_1 = 0$, $x_2 = -t$, and $x_3 = t$, where t is any real number.

43. The corresponding system of equations is

$$\begin{aligned} x_1 \qquad + x_4 &= 0 \\ x_3 \qquad &= 0 \\ 0 &= 0. \end{aligned}$$

Letting $x_4 = t$ and $x_3 = s$ be the free variables, the solution is: $x_1 = -t$, $x_2 = s$, $x_3 = 0$, and $x_4 = t$, where t and s are any real numbers.

45. x = amount at 9%

y = amount at 10%

z = amount at 12%

$$\begin{aligned} x + y + z &= 500{,}000 \\ 0.09x + 0.10y + 0.12z &= 52{,}000 \\ 2.5x - y \qquad &= 0 \end{aligned}$$

$$\left[\begin{array}{ccc|c} 1 & 1 & 1 & 500{,}000 \\ 0.09 & 0.10 & 0.12 & 52{,}000 \\ 2.5 & -1 & 0 & 0 \end{array}\right]$$

$$\begin{array}{r} \\ -0.09R_1 + R_2 \rightarrow \\ -2.5R_1 + R_3 \rightarrow \end{array} \left[\begin{array}{ccc|c} 1 & 1 & 1 & 500{,}000 \\ 0 & 0.01 & 0.03 & 7{,}000 \\ 0 & -3.5 & -2.5 & -1{,}250{,}000 \end{array}\right]$$

$$\begin{array}{r} \\ 100R_2 \rightarrow \\ 2R_3 \rightarrow \end{array} \left[\begin{array}{ccc|c} 1 & 1 & 1 & 500{,}000 \\ 0 & 1 & 3 & 700{,}000 \\ 0 & -7 & -5 & -2{,}500{,}000 \end{array}\right]$$

$$\begin{array}{r} -R_2 + R_1 \rightarrow \\ \\ 7R_2 + R_3 \rightarrow \end{array} \left[\begin{array}{ccc|c} 1 & 0 & -2 & -200{,}000 \\ 0 & 1 & 3 & 700{,}000 \\ 0 & 0 & 16 & 2{,}400{,}000 \end{array}\right]$$

$$\begin{array}{r} \\ \\ \frac{1}{16}R_3 \rightarrow \end{array} \left[\begin{array}{ccc|c} 1 & 0 & -2 & -200{,}000 \\ 0 & 1 & 3 & 700{,}000 \\ 0 & 0 & 1 & 150{,}000 \end{array}\right]$$

$$\begin{aligned} x - 2z &= -200{,}000 \\ y + 3z &= 700{,}000 \\ z &= 150{,}000 \end{aligned}$$

$$y + 3(150{,}000) = 700{,}000 \Rightarrow y = 250{,}000$$

$$x - 2(150{,}000) = -200{,}000 \Rightarrow x = 100{,}000$$

Solution: \$100,000 at 9%, \$250,000 at 10%, \$150,000 at 12%

© 2013 Cengage Learning. All Rights Reserved. May not be scanned, copied or duplicated, or posted to a publicly accessible website, in whole or in part.

47. (a) Because there are two rows and three columns, there are two equations and two variables.

(b) Gaussian elimination produces the following.

$$\begin{bmatrix} 1 & k & 2 \\ -3 & 4 & 1 \end{bmatrix} \Rightarrow \begin{bmatrix} 1 & k & 2 \\ 0 & 4+3k & 7 \end{bmatrix} \Rightarrow \begin{bmatrix} 1 & k & 2 \\ 0 & 1 & \dfrac{7}{4+3k} \end{bmatrix}$$

The system is consistent if $4 + 3k \neq 0$, or if $k \neq -\frac{4}{3}$.

Because there are two rows and three columns, there are two equations and three variables.
Gaussian elimination produces the following.

$$\begin{bmatrix} 1 & k & 2 & 0 \\ -3 & 4 & 1 & 0 \end{bmatrix} \Rightarrow \begin{bmatrix} 1 & k & 2 & 0 \\ 0 & 4+3k & 7 & 0 \end{bmatrix} \Rightarrow \begin{bmatrix} 1 & k & 2 & 0 \\ 0 & 1 & \dfrac{7}{4+3k} & 0 \end{bmatrix}$$

Notice that $4 + 3k \neq 0$, or $k \neq -\frac{4}{3}$. But if $-\frac{4}{3}$ is substituted for k in the original matrix, Gaussian elimination produces the following:

$$\begin{bmatrix} 1 & -\frac{4}{3} & 2 & 0 \\ -3 & 4 & 1 & 0 \end{bmatrix} \Rightarrow \begin{bmatrix} 1 & -\frac{4}{3} & 2 & 0 \\ 0 & 0 & 7 & 0 \end{bmatrix}$$

The system is consistent. So, the original system is consistent for all real k.

49. Begin by forming the augmented matrix for the system

$$\begin{bmatrix} 1 & 1 & 0 & 2 \\ 0 & 1 & 1 & 2 \\ 1 & 0 & 1 & 2 \\ a & b & c & 0 \end{bmatrix}.$$

Then use Gauss-Jordan elimination as follows.

$$\begin{bmatrix} 1 & 1 & 0 & 2 \\ 0 & 1 & 1 & 2 \\ 0 & -1 & 1 & 0 \\ a & b & c & 0 \end{bmatrix} \Rightarrow \begin{bmatrix} 1 & 1 & 0 & 2 \\ 0 & 1 & 1 & 2 \\ 0 & -1 & 1 & 0 \\ 0 & b-a & c & -2a \end{bmatrix}$$

$$\Rightarrow \begin{bmatrix} 1 & 1 & 0 & 2 \\ 0 & 1 & 1 & 2 \\ 0 & 0 & 2 & 2 \\ 0 & b-a & c & -2a \end{bmatrix} \Rightarrow \begin{bmatrix} 1 & 1 & 0 & 2 \\ 0 & 1 & 1 & 2 \\ 0 & 0 & 2 & 2 \\ 0 & 0 & a-b+c & -2b \end{bmatrix}$$

$$\Rightarrow \begin{bmatrix} 1 & 1 & 0 & 2 \\ 0 & 1 & 1 & 2 \\ 0 & 0 & 1 & 1 \\ 0 & 0 & a-b+c & -2b \end{bmatrix} \Rightarrow \begin{bmatrix} 1 & 1 & 0 & 2 \\ 0 & 1 & 1 & 2 \\ 0 & 0 & 1 & 1 \\ 0 & 0 & 0 & a+b+c \end{bmatrix}$$

$$\Rightarrow \begin{bmatrix} 1 & 1 & 0 & 2 \\ 0 & 1 & 0 & 1 \\ 0 & 0 & 1 & 1 \\ 0 & 0 & 0 & a+b+c \end{bmatrix} \Rightarrow \begin{bmatrix} 1 & 0 & 0 & 1 \\ 0 & 1 & 0 & 1 \\ 0 & 0 & 1 & 1 \\ 0 & 0 & 0 & a+b+c \end{bmatrix}$$

Converting back to a system of linear equations produces

$x = 1$

$y = 1$

$z = 1$

$0 = a + b + c.$

The system

(a) will have a unique solution if $a + b + c = 0$.

(b) will have no solution if $a + b + c \neq 0$.

(c) cannot have an infinite number of solutions.

© 2013 Cengage Learning. All Rights Reserved. May not be scanned, copied or duplicated, or posted to a publicly accessible website, in whole or in part.

51. Solve each pair of equations by Gaussian elimination as follows.

(a) Equations 1 and 2:

$$\begin{bmatrix} 4 & -2 & 5 & 16 \\ 1 & 1 & 0 & 0 \end{bmatrix} \Rightarrow \begin{bmatrix} 1 & 0 & \frac{5}{6} & \frac{8}{3} \\ 0 & 1 & -\frac{5}{6} & -\frac{8}{3} \end{bmatrix} \Rightarrow \begin{aligned} x &= \tfrac{8}{3} - \tfrac{5}{6}t, \\ y &= -\tfrac{8}{3} + \tfrac{5}{6}t, \\ z &= t \end{aligned}$$

(b) Equations 1 and 3:

$$\begin{bmatrix} 4 & -2 & 5 & 16 \\ -1 & -3 & 2 & 6 \end{bmatrix} \Rightarrow \begin{bmatrix} 1 & 0 & \frac{11}{14} & \frac{36}{14} \\ 0 & 1 & -\frac{13}{14} & -\frac{40}{14} \end{bmatrix} \Rightarrow \begin{aligned} x &= \tfrac{18}{7} - \tfrac{11}{14}t, \\ y &= -\tfrac{20}{7} + \tfrac{13}{14}t, \\ z &= t \end{aligned}$$

(c) Equations 2 and 3:

$$\begin{bmatrix} 1 & 1 & 0 & 0 \\ -1 & -3 & 2 & 6 \end{bmatrix} \Rightarrow \begin{bmatrix} 1 & 0 & 1 & 3 \\ 0 & 1 & -1 & -3 \end{bmatrix} \Rightarrow \begin{aligned} x &= 3 - t, \\ y &= -3 + t, \\ z &= t \end{aligned}$$

(d) Each of these systems has an infinite number of solutions.

53. Use Gauss-Jordan elimination as follows.

$$\begin{bmatrix} 1 & 2 \\ -1 & 2 \end{bmatrix} \Rightarrow \begin{bmatrix} 1 & 2 \\ 0 & 4 \end{bmatrix} \Rightarrow \begin{bmatrix} 1 & 2 \\ 0 & 1 \end{bmatrix} \Rightarrow \begin{bmatrix} 1 & 0 \\ 0 & 1 \end{bmatrix}$$

55. Begin by finding all possible first rows.

$[0 \quad 0], [0 \quad 1], [1 \quad 0], [1 \quad k]$

For each of these, examine the possible second rows.

$$\begin{bmatrix} 0 & 0 \\ 0 & 0 \end{bmatrix}, \begin{bmatrix} 0 & 1 \\ 0 & 0 \end{bmatrix}, \begin{bmatrix} 1 & 0 \\ 0 & 1 \end{bmatrix}, \begin{bmatrix} 1 & k \\ 0 & 0 \end{bmatrix}$$

These represent all possible 2×2 reduced row-echelon matrices.

57. (a) True. In the notation $m \times n$, m is the number of rows of the matrix. So, a 6×3 matrix has six rows.

(b) True. On page 16, after Example 4, the sentence reads, "Every matrix is row-equivalent to a matrix in row-echelon form."

(c) False. Consider the row-echelon form

$$\begin{bmatrix} 1 & 0 & 0 & 0 & 0 \\ 0 & 1 & 0 & 0 & 1 \\ 0 & 0 & 1 & 0 & 2 \\ 0 & 0 & 0 & 1 & 3 \end{bmatrix}$$

which gives the solution $x_1 = 0$, $x_2 = 1$, $x_3 = 2$, and $x_4 = 3$.

(d) True. Theorem 1.1 states that if a homogeneous system has fewer equations than variables, then it must have an infinite number of solutions.

59. To show that it is possible, you only need to give one example, such as

$$x_1 + x_2 + x_3 = 0$$
$$x_1 + x_2 + x_3 = 1$$

which has fewer equations than variables and obviously has no solution.

61. First, a and c cannot both be zero. So, assume $a \neq 0$, and use row reduction as follows.

$$\begin{bmatrix} a & b \\ c & d \end{bmatrix} \Rightarrow \begin{bmatrix} a & b \\ 0 & \dfrac{-cb}{a} + d \end{bmatrix} \Rightarrow \begin{bmatrix} a & b \\ 0 & ad - bc \end{bmatrix}$$

So, $ad - bc \neq 0$. Similarly, if $c \neq 0$, interchange rows and proceed as above. So the original matrix is row equivalent to the identity if and only if $ad - bc \neq 0$.

63. Form the augmented matrix for this system.

$$\begin{bmatrix} \lambda - 2 & 1 & 0 \\ 1 & \lambda - 2 & 0 \end{bmatrix}$$

Reduce the system using elementary row operations.

$$\begin{bmatrix} 1 & \lambda - 2 & 0 \\ \lambda - 2 & 1 & 0 \end{bmatrix} \Rightarrow \begin{bmatrix} 1 & \lambda - 2 & 0 \\ 0 & \lambda^2 - 4\lambda + 3 & 0 \end{bmatrix}$$

To have a nontrivial solution you must have

$$\lambda^2 - 4\lambda + 3 = 0$$
$$(\lambda - 1)(\lambda - 3) = 0.$$

So, if $\lambda = 1$ or $\lambda = 3$, the system will have nontrivial solutions.

© 2013 Cengage Learning. All Rights Reserved. May not be scanned, copied or duplicated, or posted to a publicly accessible website, in whole or in part.

65. $\begin{bmatrix} a & b \\ c & d \end{bmatrix} \Rightarrow \begin{bmatrix} a-c & b-d \\ c & d \end{bmatrix} \Rightarrow \begin{bmatrix} a-c & b-d \\ a & b \end{bmatrix} \Rightarrow \begin{bmatrix} -c & -d \\ a & b \end{bmatrix} \Rightarrow \begin{bmatrix} c & d \\ a & b \end{bmatrix}$

The rows have been interchanged. In general, the second and third elementary row operations can be used in this manner to interchange two rows of a matrix. So, the first elementary row operation is, in fact, redundant.

67. (a) When a system of linear equations is inconsistent, the row-echelon form of the corresponding augmented matrix will have a row that is all zeros except for the last entry.

(b) When a system of linear equations has infinitely many solutions, the row-echelon form of the corresponding augmented matrix will have a row that consists entirely of zeros or more than one column with no leading 1's. The last column will not contain a leading 1.

Section 1.3 Applications of Systems of Linear Equations

1. (a) Because there are three points, choose a second-degree polynomial, $p(x) = a_0 + a_1x + a_2x^2$. Then substitute $x = 2, 3,$ and 4 into $p(x)$ and equate the results to $y = 5, 2,$ and 5, respectively.

$$a_0 + a_1(2) + a_2(2)^2 = a_0 + 2a_1 + 4a_2 = 5$$
$$a_0 + a_1(3) + a_2(3)^2 = a_0 + 3a_1 + 9a_2 = 2$$
$$a_0 + a_1(4) + a_2(4)^2 = a_0 + 4a_1 + 16a_2 = 5$$

Form the augmented matrix

$$\begin{bmatrix} 1 & 2 & 4 & 5 \\ 1 & 3 & 9 & 2 \\ 1 & 4 & 16 & 5 \end{bmatrix}$$

and use Gauss-Jordan elimination to obtain the equivalent reduced row-echelon matrix

$$\begin{bmatrix} 1 & 0 & 0 & 29 \\ 0 & 1 & 0 & -18 \\ 0 & 0 & 1 & 3 \end{bmatrix}.$$

So, $p(x) = 29 - 18x + 3x^2$.

(b)

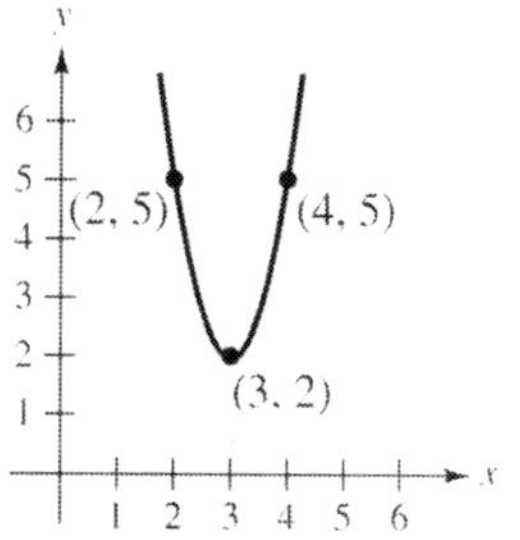

3. (a) Because there are three points, choose a second-degree polynomial, $p(x) = a_0 + a_1x + a_2x^2$. Then substitute $x = 2, 3,$ and 5 into $p(x)$ and equate the results to $y = 4, 6,$ and 10, respectively.

$$a_0 + a_1(2) + a_2(2)^2 = a_0 + 2a_1 + 4a_2 = 4$$
$$a_0 + a_1(3) + a_2(3)^2 = a_0 + 3a_1 + 9a_2 = 6$$
$$a_0 + a_1(5) + a_2(5)^2 = a_0 + 5a_1 + 25a_2 = 10$$

Use Gauss-Jordan elimination on the augmented matrix for this system.

$$\begin{bmatrix} 1 & 2 & 4 & 4 \\ 1 & 3 & 9 & 6 \\ 1 & 5 & 25 & 10 \end{bmatrix} \Rightarrow \begin{bmatrix} 1 & 0 & 0 & 0 \\ 0 & 1 & 0 & 2 \\ 0 & 0 & 1 & 0 \end{bmatrix}$$

So, $p(x) = 2x$.

(b)

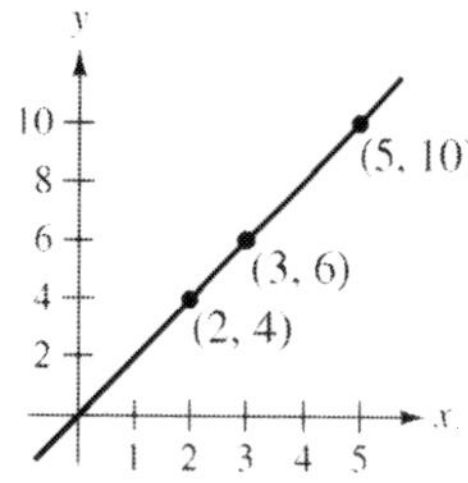

© 2013 Cengage Learning. All Rights Reserved. May not be scanned, copied or duplicated, or posted to a publicly accessible website, in whole or in part.

5. (a) Because there are four points, choose a third-degree polynomial, $p(x) = a_0 + a_1x + a_2x^2 + a_3x^3$. Then substitute $x = -1, 0, 1$, and 4 into $p(x)$ and equate the results to $y = 3, 0, 1$, and 58, respectively.

$$\begin{aligned} a_0 + a_1(-1) + a_2(-1)^2 + a_3(-1)^3 &= a_0 - a_1 + a_2 - a_3 = 3 \\ a_0 + a_1(0) + a_2(0)^2 + a_3(0)^3 &= a_0 = 0 \\ a_0 + a_1(1) + a_2(1)^2 + a_3(1)^3 &= a_0 + a_1 + a_2 + a_3 = 1 \\ a_0 + a_1(4) + a_2(4)^2 + a_3(4)^3 &= a_0 + 4a_1 + 16a_2 + 64a_3 = 58 \end{aligned}$$

Use Gauss-Jordan elimination on the augmented matrix for this system.

$$\begin{bmatrix} 1 & -1 & 1 & -1 & 3 \\ 1 & 0 & 0 & 0 & 0 \\ 1 & 1 & 1 & 1 & 1 \\ 1 & 4 & 16 & 64 & 58 \end{bmatrix} \Rightarrow \begin{bmatrix} 1 & 0 & 0 & 0 & 0 \\ 0 & 1 & 0 & 0 & -\frac{3}{2} \\ 0 & 0 & 1 & 0 & 2 \\ 0 & 0 & 0 & 1 & \frac{1}{2} \end{bmatrix}$$

So, $p(x) = -\frac{3}{2}x + 2x^2 + \frac{1}{2}x^3$.

(b)

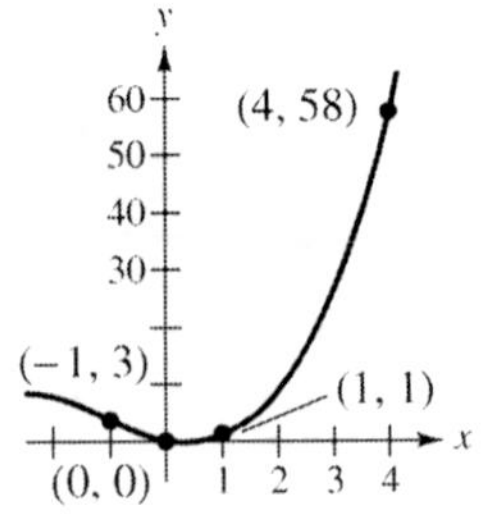

7. (a) Because there are five points, choose a fourth-degree polynomial, $p(x) = a_0 + a_1x + a_2x^2 + a_3x^3 + a_4x^4$. Then substitute $x = -2, -1, 0, 1$, and 2 into $p(x)$ and equate the results to $y = 28, 0, -6, -8$, and 0, respectively.

$$\begin{aligned} a_0 + a_1(-2) + a_2(-2)^2 + a_3(-2)^3 + a_4(-2)^4 &= a_0 - 2a_1 + 4a_2 - 8a_3 + 16a_4 = 28 \\ a_0 + a_1(-1) + a_2(-1)^2 + a_3(-1)^3 + a_4(-1)^4 &= a_0 - a_1 + a_2 - a_3 + a_4 = 0 \\ a_0 + a_1(0) + a_2(0)^2 + a_3(0)^3 + a_4(0)^4 &= a_0 = -6 \\ a_0 + a_1(1) + a_2(1)^2 + a_3(1)^3 + a_4(1)^4 &= a_0 + a_1 + a_2 + a_3 + a_4 = -8 \\ a_0 + a_1(2) + a_2(2)^2 + a_3(2)^3 + a_4(2)^4 &= a_0 + 2a_1 + 4a_2 + 8a_3 + 16a_4 = 0 \end{aligned}$$

Form the augmented matrix

$$\begin{bmatrix} 1 & -2 & 4 & -8 & 16 & 28 \\ 1 & -1 & 1 & -1 & 1 & 0 \\ 1 & 0 & 0 & 0 & 0 & -6 \\ 1 & 1 & 1 & 1 & 1 & -8 \\ 1 & 2 & 4 & 8 & 16 & 0 \end{bmatrix}$$

and use Gauss-Jordan elimination to obtain the equivalent reduced row-echelon matrix

$$\begin{bmatrix} 1 & 0 & 0 & 0 & 0 & -6 \\ 0 & 1 & 0 & 0 & 0 & -3 \\ 0 & 0 & 1 & 0 & 0 & 1 \\ 0 & 0 & 0 & 1 & 0 & -1 \\ 0 & 0 & 0 & 0 & 1 & 1 \end{bmatrix}.$$

So, $p(x) = -6 - 3x + x^2 - x^3 + x^4$.

(b)

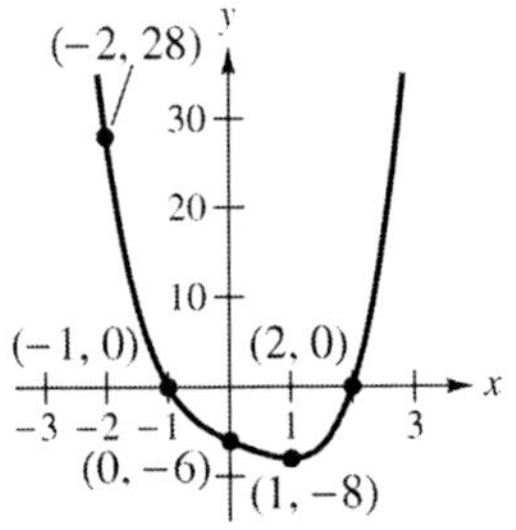

© 2013 Cengage Learning. All Rights Reserved. May not be scanned, copied or duplicated, or posted to a publicly accessible website, in whole or in part.

9. (a) Because there are three points, choose the second-degree polynomial, $p(x) = a_0 + a_1x + a_2x^2$. Then substitute $x = 2006, 2007$, and 2008 into $p(x)$ and equate the results to $y = 5, 7$, and 12, respectively.

$$a_0 + a_1(2006) + a_2(2006)^2 = a_0 + 2006a_1 + 4{,}024{,}036a_2 = 5$$
$$a_0 + a_1(2007) + a_2(2007)^2 = a_0 + 2007a_1 + 4{,}028{,}049a_2 = 7$$
$$a_0 + a_1(2008) + a_2(2008)^2 = a_0 + 2008a_1 + 4{,}032{,}064a_2 = 12$$

Use Gauss-Jordan elimination on the augmented matrix for this system.

$$\begin{bmatrix} 1 & 2006 & 4{,}024{,}036 & 5 \\ 1 & 2007 & 4{,}028{,}049 & 7 \\ 1 & 2008 & 4{,}032{,}064 & 12 \end{bmatrix} \Rightarrow \begin{bmatrix} 1 & 0 & 0 & 6{,}035{,}056 \\ 0 & 1 & 0 & -6017.5 \\ 0 & 0 & 1 & 1.5 \end{bmatrix}$$

So, $p(x) = 6{,}035{,}056 - 6017.5x + 1.5x^2$.

(b)

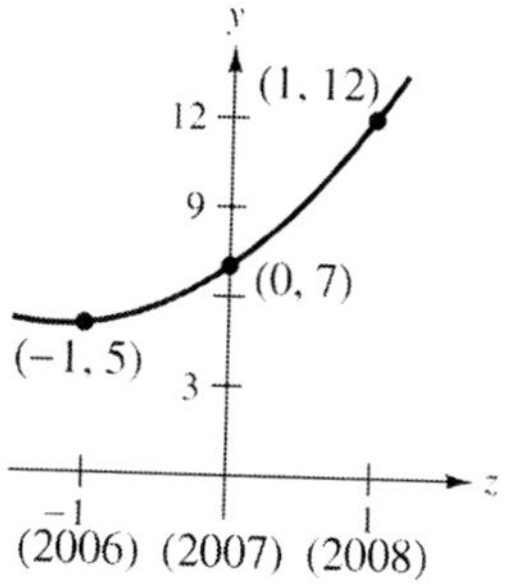

11. (a) Because there are three points, choose the second-degree polynomial, $p(x) = a_0 + a_1x + a_2x^2$. Then substitute $x = 0.072, 0.120$, and 0.148 into $p(x)$ and equate the results to $y = 0.203, 0.238$, and 0.284, respectively.

$$a_0 + a_1(0.072) + a_2(0.072)^2 = a_0 + 0.072a_1 + 0.005184a_2 = 0.203$$
$$a_0 + a_1(0.120) + a_2(0.120)^2 = a_0 + 0.12a_1 + 0.0144a_2 = 0.238$$
$$a_0 + a_1(0.148) + a_2(0.148)^2 = a_0 + 0.148a_1 + 0.021904a_2 = 0.284$$

Use Gauss-Jordan elimination on the augmented matrix for this system.

$$\begin{bmatrix} 1 & 0.072 & 0.005184 & 0.203 \\ 1 & 0.12 & 0.0144 & 0.238 \\ 1 & 0.148 & 0.021904 & 0.284 \end{bmatrix} \Rightarrow \begin{bmatrix} 1 & 0 & 0 & 0.25437 \\ 0 & 1 & 0 & -1.57910 \\ 0 & 0 & 1 & 12.02224 \end{bmatrix}$$

So, $p(x) \approx 0.254 - 1.579x + 12.022x^2$.

(b)

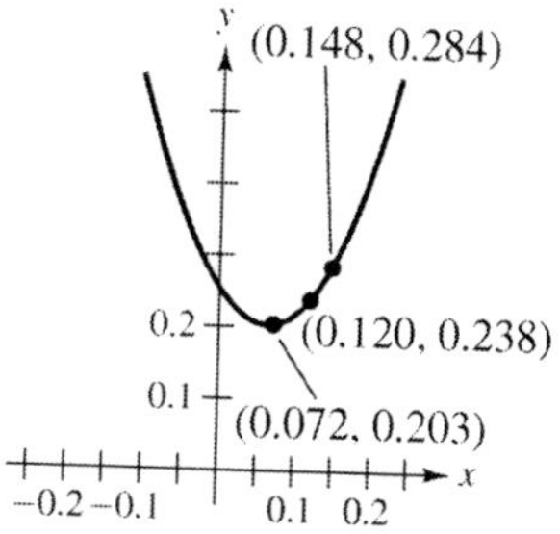

© 2013 Cengage Learning. All Rights Reserved. May not be scanned, copied or duplicated, or posted to a publicly accessible website, in whole or in part.

13. Choosing a second-degree polynomial approximation, $p(x) = a_0 + a_1x + a_2x^2$, substitute $x = 0, \frac{\pi}{2}$, and π into $p(x)$ and equate the results to $y = 0, 1$, and 0, respectively.

$$\begin{aligned} a_0 \qquad\qquad\qquad &= 0 \\ a_0 + \frac{\pi}{2}a_1 + \frac{\pi^2}{4}a^2 &= 1 \\ a_0 + \pi a_1 + \pi^2 a_2 &= 0 \end{aligned}$$

Then form the augmented matrix

$$\begin{bmatrix} 1 & 0 & 0 & 0 \\ 1 & \frac{\pi}{2} & \frac{\pi^2}{4} & 1 \\ 1 & \pi & \pi^2 & 0 \end{bmatrix}$$

and use Gauss-Jordan elimination to obtain the equivalent reduced row-echelon matrix

$$\begin{bmatrix} 1 & 0 & 0 & 0 \\ 0 & 1 & 0 & \frac{4}{\pi} \\ 0 & 0 & 1 & -\frac{4}{\pi^2} \end{bmatrix}.$$

So, $p(x) = \frac{4}{\pi}x - \frac{4}{\pi^2}x^2 = \frac{4}{\pi^2}\left(\pi x - x^2\right)$.

Furthermore,

$$\begin{aligned} \sin\frac{\pi}{3} \approx p\left(\frac{\pi}{3}\right) &= \frac{4}{\pi^2}\left[\pi\left(\frac{\pi}{3}\right) - \left(\frac{\pi}{3}\right)^2\right] \\ &= \frac{4}{\pi^2}\left[\frac{2\pi^2}{9}\right] = \frac{8}{9} \approx 0.889. \end{aligned}$$

Note that $\sin \pi/3 = 0.866$ to three significant digits.

15. Assume that the equation of the circle is $x^2 + ax + y^2 + by - c = 0$. Because each of the given points lies on the circle, you have the following linear equations.

$$\begin{aligned} (1)^2 + a(1) + (3)^2 + b(3) - c &= a + 3b - c + 10 = 0 \\ (-2)^2 + a(-2) + (6)^2 + b(6) - c &= -2a + 6b - c + 40 = 0 \\ (4)^2 + a(4) + (2)^2 + b(2) - c &= 4a + 2b - c + 20 = 0 \end{aligned}$$

Use Gauss-Jordan elimination on the system.

$$\begin{bmatrix} 1 & 3 & -1 & -10 \\ -2 & 6 & -1 & -40 \\ 4 & 2 & -1 & -20 \end{bmatrix} \Rightarrow \begin{bmatrix} 1 & 0 & 0 & -10 \\ 0 & 1 & 0 & -20 \\ 0 & 0 & 1 & -60 \end{bmatrix}$$

So, the equation of the circle is $x^2 - 10x + y^2 - 20y + 60 = 0$ or $(x - 5)^2 + (y - 10)^2 = 65$.

© 2013 Cengage Learning. All Rights Reserved. May not be scanned, copied or duplicated, or posted to a publicly accessible website, in whole or in part.

17. Because you are given three points, choose a second-degree polynomial, $p(x) = a_0 + a_1x + a_2x^2$. Because the x-values are large, use the translation $z = x - 2000$ to obtain $(-10, 249)$, $(0, 281)$, and $(10, 309)$. Substituting the given points into $p(x)$ produces the following system of linear equations.

$$\begin{aligned} a_0 + a_1(-10) + a_2(-10)^2 &= a_0 - 10a_1 + 100a_2 = 249 \\ a_0 + a_1(0) + a_2(0)^2 &= a_0 = 281 \\ a_0 + a_1(10) + a_2(10)^2 &= a_0 + 10a_1 + 100a_2 = 309 \end{aligned}$$

Form the augmented matrix

$$\begin{bmatrix} 1 & -10 & 100 & 249 \\ 1 & 0 & 0 & 281 \\ 1 & 10 & 100 & 309 \end{bmatrix}$$

and use Gauss-Jordan elimination to obtain the equivalent reduced row-echelon matrix

$$\begin{bmatrix} 1 & 0 & 0 & 281 \\ 0 & 1 & 0 & 3 \\ 0 & 0 & 1 & -0.02 \end{bmatrix}.$$

So, $p(z) = 281 + 3z - 0.02z^2$ and $p(x) = 281 + 3(x - 2000) - 0.02(x - 2000)^2$. To predict the population in 2020 and 2030, substitute these values into $p(x)$.

$$p(2020) = 281 + 3(20) - 0.02(20)^2 = 333 \text{ million}$$

$$p(2030) = 281 + 3(30) - 0.02(30)^2 = 353 \text{ million}$$

19. (a) Letting $z = x - 2000$, the four points are $(3, 10, 526)$, $(4, 11, 330)$, $(5, 12, 715)$, and $(6, 12, 599)$.

Let $p(z) = a_0 + a_1z + a_1z^2 + a_3z^3$.

$$\begin{aligned} a_0 + a_1(3) + a_2(3)^2 + a_3(3)^3 &= a_0 + 3a_1 + 9a_2 + 27a_3 = 10,526 \\ a_0 + a_1(4) + a_2(4)^2 + a_3(4)^3 &= a_0 + 4a_1 + 16a_2 + 64a_3 = 11,330 \\ a_0 + a_1(5) + a_2(5)^2 + a_3(5)^3 &= a_0 + 5a_1 + 25a_2 + 125a_3 = 12,715 \\ a_0 + a_1(6) + a_2(6)^2 + a_3(6)^3 &= a_0 + 6a_1 + 36a_2 + 216a_3 = 12,599 \end{aligned}$$

(b) Use Gauss-Jordan elimination to solve the system.

$$\begin{bmatrix} 1 & 3 & 9 & 27 & 10,526 \\ 1 & 4 & 16 & 64 & 11,330 \\ 1 & 5 & 25 & 125 & 12,715 \\ 1 & 6 & 36 & 216 & 12,599 \end{bmatrix} \Rightarrow \begin{bmatrix} 1 & 0 & 0 & 0 & 32,420 \\ 0 & 1 & 0 & 0 & -17,538.5 \\ 0 & 0 & 1 & 0 & 4454.5 \\ 0 & 0 & 0 & 1 & -347 \end{bmatrix}$$

So, $p(z) = 32,420 - 17,538.5z + 4454.5z^2 - 347z^3$, and

$p(x) = 32,420 - 17,538.5(x - 2000) + 4454.5(x - 2000)^2 - 347(x - 2000)^3$.

To determine the reasonableness of the model for years after 2006, compare the predicted values in 2007–2010 to the actual values.

x	2007	2008	2009	2010
$p(x)$	8900	–464	–17,575	–44,515
Actual	14,065	17,681	14,569	18,700

The model does not produce reasonable outcomes after 2006.

© 2013 Cengage Learning. All Rights Reserved. May not be scanned, copied or duplicated, or posted to a publicly accessible website, in whole or in part.

21. (i) $p(-1) = a_0 + (-1)a_1 + (-1)^2 a_2 = a_0 - a_1 + a_2$

$p(0) = a_0 + (0)a_1 + (0)^2 a_2 = a_0$

$p(1) = a_0 + (1)a_1 + (1)^2 a_2 = a_0 + a_1 + a_2$

(ii) $a_0 - a_1 + a_2 = 0$

$a_0 = 0$

$a_0 + a_1 + a_2 = 0$

(iii) Form the augmented matrix

$$\begin{bmatrix} 1 & -1 & 1 & 0 \\ 1 & 0 & 0 & 0 \\ 1 & 1 & 1 & 0 \end{bmatrix}$$

and use Gauss-Jordan elimination to obtain the equivalent reduced row-echelon matrix

$$\begin{bmatrix} 1 & 0 & 0 & 0 \\ 0 & 1 & 0 & 0 \\ 0 & 0 & 1 & 0 \end{bmatrix}.$$

So, $a_0 = 0$, $a_1 = 0$, and $a_2 = 0$.

23. To begin, substitute $x = -1$ and $x = 1$ into $p(x) = a_0 + a_1x + a_2x^2 + a_3x^3$ and equate the results to $y = 2$ and $y = -2$, respectively.

$a_0 - a_1 + a_2 - a_3 = 2$

$a_0 + a_1 + a_2 + a_3 = -2$

Then, differentiate p, yielding $p'(x) = a_1 + 2a_2x + 3a_3x^2$. Substitute $x = -1$ and $x = 1$ into $p'(x)$ and equate the results to 0.

$a_1 - 2a_2 + 3a_3 = 0$

$a_1 + 2a_2 + 3a_3 = 0$

Combining these four equations into one system and forming the augmented matrix, you obtain

$$\begin{bmatrix} 1 & -1 & 1 & -1 & 2 \\ 1 & 1 & 1 & 1 & -2 \\ 0 & 1 & -2 & 3 & 0 \\ 0 & 1 & 2 & 3 & 0 \end{bmatrix}.$$

Use Gauss-Jordan elimination to find the equivalent reduced row-echelon matrix

$$\begin{bmatrix} 1 & 0 & 0 & 0 & 0 \\ 0 & 1 & 0 & 0 & -3 \\ 0 & 0 & 1 & 0 & 0 \\ 0 & 0 & 0 & 1 & 1 \end{bmatrix}.$$

So, $p(x) = -3x + x^3$.

The graph of $y = p(x)$ is shown at the right.

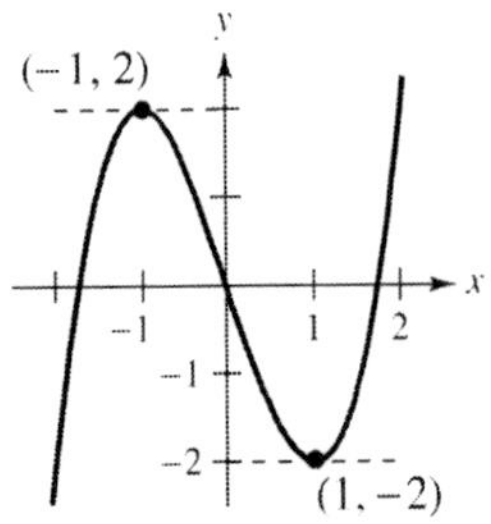

25. (a) Letting $p(x) = a_0 + a_1x + a_2x^2$, substitute $x = 0, 2,$ and 4 into $p(x)$ and equate the results to $y = 1, \frac{1}{3}$, and $\frac{1}{5}$, respectively.

$a_0 + a_1(0) + a_2(0)^2 = a_0 = 1$

$a_0 + a_1(2) + a_2(2)^2 = a_0 + 2a_1 + 4a_2 = \frac{1}{3}$

$a_0 + a_1(4) + a_2(4)^2 = a_0 + 4a_1 + 16a_2 = \frac{1}{5}$

Use Gauss-Jordan elimination on the augmented matrix for this system.

$$\begin{bmatrix} 1 & 0 & 0 & 1 \\ 1 & 2 & 4 & \frac{1}{3} \\ 1 & 4 & 16 & \frac{1}{5} \end{bmatrix} \Rightarrow \begin{bmatrix} 1 & 0 & 0 & 1 \\ 0 & 1 & 0 & -\frac{7}{15} \\ 0 & 0 & 1 & \frac{1}{15} \end{bmatrix}$$

So, $p(x) = 1 - \frac{7}{15}x + \frac{1}{15}x^2$.

(b) Letting $p(x) = a_0 + a_1x + a_2x^2$, substitute $x = 0, 2,$ and 4 into $p(x)$ and equate the results to $y = 1, 3$, and 5, respectively.

$a_0 + a_1(0) + a_2(0)^2 = a_0 = 1$

$a_0 + a_1(2) + a_2(2)^2 = a_0 + 2a_1 + 4a_2 = 3$

$a_0 + a_1(4) + a_2(4)^2 = a_0 + 4a_1 + 16a_2 = 5$

Use Gauss-Jordan elimination on the augmented matrix for this system.

$$\begin{bmatrix} 1 & 0 & 0 & 1 \\ 1 & 2 & 4 & 3 \\ 1 & 4 & 16 & 5 \end{bmatrix} \Rightarrow \begin{bmatrix} 1 & 0 & 0 & 1 \\ 0 & 1 & 0 & 1 \\ 0 & 0 & 1 & 0 \end{bmatrix}$$

So, $p(x) = 1 + x$.

The graphs of $y = 1/p(x) = 1/(1 + x)$ and that of the function $y = 1 - \frac{7}{15}x + \frac{1}{15}x^2$ are shown below.

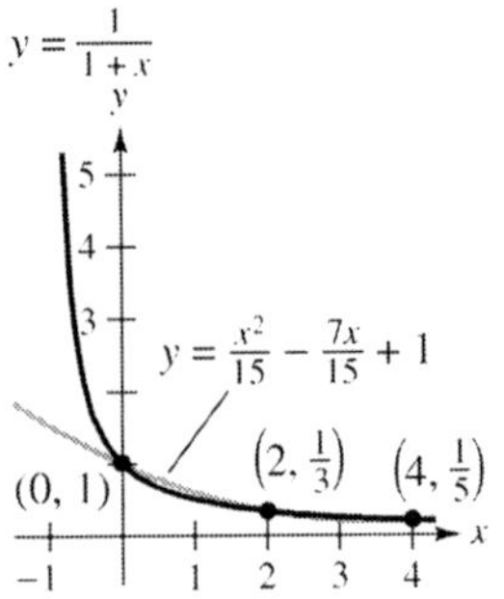

© 2013 Cengage Learning. All Rights Reserved. May not be scanned, copied or duplicated, or posted to a publicly accessible website, in whole or in part.

27. (a) Each of the network's six junctions gives rise to a linear equation as shown below.

input = output

$$600 = x_1 + x_3$$
$$x_1 = x_2 + x_4$$
$$x_2 + x_5 = 500$$
$$x_3 + x_6 = 600$$
$$x_4 + x_7 = x_6$$
$$500 = x_5 + x_7$$

Rearrange these equations, form the augmented matrix, and use Gauss-Jordan elimination.

$$\begin{bmatrix} 1 & 0 & 1 & 0 & 0 & 0 & 0 & 600 \\ 1 & -1 & 0 & -1 & 0 & 0 & 0 & 0 \\ 0 & 1 & 0 & 0 & 1 & 0 & 0 & 500 \\ 0 & 0 & 1 & 0 & 0 & 1 & 0 & 600 \\ 0 & 0 & 0 & 1 & 0 & -1 & 1 & 0 \\ 0 & 0 & 0 & 0 & 1 & 0 & 1 & 500 \end{bmatrix} \Rightarrow \begin{bmatrix} 1 & 0 & 0 & 0 & 0 & -1 & 0 & 0 \\ 0 & 1 & 0 & 0 & 0 & 0 & -1 & 0 \\ 0 & 0 & 1 & 0 & 0 & 1 & 0 & 600 \\ 0 & 0 & 0 & 1 & 0 & -1 & 1 & 0 \\ 0 & 0 & 0 & 0 & 1 & 0 & 1 & 500 \\ 0 & 0 & 0 & 0 & 0 & 0 & 0 & 0 \end{bmatrix}$$

Letting $x_7 = t$ and $x_6 = s$ be the free variables, you have

$x_1 = s$

$x_2 = t$

$x_3 = 600 - s$

$x_4 = s - t$

$x_5 = 500 - t$

$x_6 = s$

$x_7 = t$, where s and t are any real numbers.

(b) If $x_6 = x_7 = 0$, then the solution is $x_1 = 0$, $x_2 = 0$, $x_3 = 600$, $x_4 = 0$, $x_5 = 500$, $x_6 = 0$, and $x_7 = 0$.

(c) If $x_5 = 1000$ and $x_6 = 0$, then the solution is $x_1 = 0$, $x_2 = -500$, $x_3 = 600$, $x_4 = 500$, $x_5 = 1000$, $x_6 = 0$, and $x_7 = -500$.

29. (a) Each of the network's four junctions gives rise to a linear equation, as shown below.

input = output

$$200 + x_2 = x_1$$
$$x_4 = x_2 + 100$$
$$x_3 = x_4 + 200$$
$$x_1 + 100 = x_3$$

Rearranging these equations and forming the augmented matrix, you obtain

$$\begin{bmatrix} 1 & -1 & 0 & 0 & 200 \\ 0 & 1 & 0 & -1 & -100 \\ 0 & 0 & 1 & -1 & 200 \\ 1 & 0 & -1 & 0 & -100 \end{bmatrix}.$$

Gauss-Jordan elimination produces the matrix

$$\begin{bmatrix} 1 & 0 & 0 & -1 & 100 \\ 0 & 1 & 0 & -1 & -100 \\ 0 & 0 & 1 & -1 & 200 \\ 0 & 0 & 0 & 0 & 0 \end{bmatrix}.$$

Letting $x_4 = t$, you have $x_1 = 100 + t$, $x_2 = -100 + t$, $x_3 = 200 + t$, and $x_4 = t$, where t is a real number.

(b) When $x_4 = t = 0$, then $x_1 = 100$, $x_2 = -100$, and $x_3 = 200$.

(c) When $x_4 = t = 100$, then $x_1 = 200$, $x_2 = 0$, and $x_3 = 300$.

© 2013 Cengage Learning. All Rights Reserved. May not be scanned, copied or duplicated, or posted to a publicly accessible website, in whole or in part.

31. Applying Kirchoff's first law to either junction produces

$$I_1 + I_3 = I_2$$

and applying the second law to the two paths produces

$$R_1I_1 + R_2I_2 = 4I_1 + 3I_2 = 3$$
$$R_2I_2 + R_3I_3 = 3I_2 + I_3 = 4.$$

Rearrange these equations, form the augmented matrix, and use Gauss-Jordan elimination.

$$\begin{bmatrix} 1 & -1 & 1 & 0 \\ 4 & 3 & 0 & 3 \\ 0 & 3 & 1 & 4 \end{bmatrix} \Rightarrow \begin{bmatrix} 1 & 0 & 0 & 0 \\ 0 & 1 & 0 & 1 \\ 0 & 0 & 1 & 1 \end{bmatrix}$$

So, $I_1 = 0$, $I_2 = 1$, and $I_3 = 1$.

33. (a) To find the general solution, let A have a volts and B have b volts. Applying Kirchoff's first law to either junction produces

$$I_1 + I_3 = I_2$$

and applying the second law to the two paths produces

$$R_1I_1 + R_2I_2 = I_1 + 2I_2 = a$$
$$R_2I_2 + R_3I_3 = 2_2 + 3I_3 = b.$$

Rearrange these three equations and form the augmented matrix.

$$\begin{bmatrix} 1 & -1 & 1 & 0 \\ 1 & 2 & 0 & a \\ 0 & 2 & 4 & b \end{bmatrix}$$

Gauss-Jordan elimination produces the matrix

$$\begin{bmatrix} 1 & 0 & 0 & (3a - b)/7 \\ 0 & 1 & 0 & (4a + b)/14 \\ 0 & 0 & 1 & (3b - 2a)/14 \end{bmatrix}.$$

When $a = 5$ and $b = 8$, then $I_1 = 1$, $I_2 = 2$, and $I_3 = 1$.

(b) When $a = 2$ and $b = 6$, then $I_1 = 0$, $I_2 = 1$, and $I_3 = 1$.

35.
$$\begin{aligned} T_1 &= \frac{60 + 20 + T_2 + T_3}{4} \\ T_2 &= \frac{60 + 50 + T_1 + T_4}{4} \\ T_3 &= \frac{20 + 10 + T_1 + T_4}{4} \\ T_4 &= \frac{50 + 10 + T_2 + T_3}{4} \end{aligned} \Rightarrow \begin{aligned} 4T_1 - T_2 - T_3 &= 80 \\ -T_1 + 4T_2 - T_4 &= 110 \\ -T_1 + 4T_3 - T_4 &= 30 \\ - T_2 - T_3 + 4T_4 &= 60 \end{aligned}$$

Use Gauss-Jordan elimination to solve this system.

$$\begin{bmatrix} 4 & -1 & -1 & 0 & 80 \\ -1 & 4 & 0 & -1 & 110 \\ -1 & 0 & 4 & -1 & 30 \\ 0 & -1 & -1 & 4 & 60 \end{bmatrix} \Rightarrow \begin{bmatrix} 1 & 0 & 0 & 0 & 37.5 \\ 0 & 1 & 0 & 0 & 45 \\ 0 & 0 & 1 & 0 & 25 \\ 0 & 0 & 0 & 1 & 32.5 \end{bmatrix}$$

So, $T_1 = 37.5°C$, $T_2 = 45°C$, $T_3 = 25°C$, and $T_4 = 32.5°C$.

© 2013 Cengage Learning. All Rights Reserved. May not be scanned, copied or duplicated, or posted to a publicly accessible website, in whole or in part.

37. $\dfrac{4x^2}{(x+1)^2(x-1)} = \dfrac{A}{x-1} + \dfrac{B}{x+1} + \dfrac{C}{(x+1)^2}$

$$4x^2 = A(x+1)^2 + B(x+1)(x-1) + C(x-1)$$
$$4x^2 = Ax^2 + 2Ax + A + Bx^2 - B + Cx - C$$
$$4x^2 = (A+B)x^2 + (2A+C)x + A - B - C$$

So,
$$\begin{aligned} A + B \quad\quad &= 4 \\ 2A \quad\quad + C &= 0 \\ A - B - C &= 0. \end{aligned}$$

Use Gauss-Jordan elimination to solve the system.

$$\begin{bmatrix} 1 & 1 & 0 & 4 \\ 2 & 0 & 1 & 0 \\ 1 & -1 & -1 & 0 \end{bmatrix} \Rightarrow \begin{bmatrix} 1 & 0 & 0 & 1 \\ 0 & 1 & 0 & 3 \\ 0 & 0 & 1 & -2 \end{bmatrix}$$

The solution is: $A = 1$, $B = 3$, and $C = -2$.

So, $\dfrac{4x^2}{(x+1)^2(x-1)} = \dfrac{1}{x-1} + \dfrac{3}{x+1} - \dfrac{2}{(x+1)^2}$.

39. Use Gauss-Jordan elimination to solve the system.

$$\begin{bmatrix} 2 & 0 & 1 & 0 \\ 0 & 2 & 1 & 0 \\ 1 & 1 & 0 & 4 \end{bmatrix} \Rightarrow \begin{bmatrix} 1 & 0 & 0 & 2 \\ 0 & 1 & 0 & 2 \\ 0 & 0 & 1 & -4 \end{bmatrix}$$

So, $x = 2$, $y = 2$, and $\lambda = -4$.

Review Exercises for Chapter 1

1. Because the equation cannot be written in the form $a_1x + a_2y = b$, it is *not* linear in the variables x and y.

3. Because the equation is in the form $a_1x + a_2y = b$, it *is* linear in the variables x and y.

5. Because the equation cannot be written in the form $a_1x + a_2y = b$, it is *not* linear in the variables x and y.

7. Choosing y and z as the free variables and letting $y = s$ and $z = t$, you have

$$-4x + 2s - 6t = 1$$
$$-4x = 1 - 2s + 6t$$
$$x = -\tfrac{1}{4} + \tfrac{1}{2}s - \tfrac{3}{2}t.$$

So, the solution set can be described as $x = -\frac{1}{4} + \frac{1}{2}s - \frac{3}{2}t$, $y = s$, and $z = t$, where s and t are real numbers.

9. Row reduce the augmented matrix for this system.

$$\begin{bmatrix} 1 & 1 & 2 \\ 3 & -1 & 0 \end{bmatrix} \Rightarrow \begin{bmatrix} 1 & 1 & 2 \\ 0 & -4 & -6 \end{bmatrix} \Rightarrow \begin{bmatrix} 1 & 1 & 2 \\ 0 & 1 & \frac{3}{2} \end{bmatrix} \Rightarrow \begin{bmatrix} 1 & 0 & \frac{1}{2} \\ 0 & 1 & \frac{3}{2} \end{bmatrix}$$

Converting back to a linear system, the solution is: $x = \frac{1}{2}$ and $y = \frac{3}{2}$.

11. Rearrange the equations as shown below.

$$x - y = -4$$
$$2x - 3y = 0$$

Row reduce the augmented matrix for this system.

$$\begin{bmatrix} 1 & -1 & -4 \\ 2 & -3 & 0 \end{bmatrix} \Rightarrow \begin{bmatrix} 1 & -1 & -4 \\ 0 & -1 & 8 \end{bmatrix} \Rightarrow \begin{bmatrix} 1 & -1 & -4 \\ 0 & 1 & -8 \end{bmatrix} \Rightarrow \begin{bmatrix} 1 & 0 & -12 \\ 0 & 1 & -8 \end{bmatrix}$$

Converting back to a linear system, the solution is: $x = -12$ and $y = -8$.

13. Row reduce the augmented matrix for this system.

$$\begin{bmatrix} 1 & 1 & 0 \\ 2 & 1 & 0 \end{bmatrix} \Rightarrow \begin{bmatrix} 1 & 1 & 0 \\ 0 & -1 & 0 \end{bmatrix} \Rightarrow \begin{bmatrix} 1 & 1 & 0 \\ 0 & 1 & 0 \end{bmatrix} \Rightarrow \begin{bmatrix} 1 & 0 & 0 \\ 0 & 1 & 0 \end{bmatrix}$$

Converting back to a linear system, the solution is: $x = 0$ and $y = 0$.

© 2013 Cengage Learning. All Rights Reserved. May not be scanned, copied or duplicated, or posted to a publicly accessible website, in whole or in part.

15. The augmented matrix for this system is

$$\begin{bmatrix} 1 & -1 & 9 \\ -1 & 1 & 1 \end{bmatrix}$$

which is equivalent to the reduced row-echelon matrix

$$\begin{bmatrix} 1 & -1 & 0 \\ 0 & 0 & 1 \end{bmatrix}.$$

Because the second row corresponds to $0 = 1$, which is a false statement, you can conclude that the system has no solution.

17. Expanding the second equation, $3x + 2y = 0$, the augmented matrix for this system is

$$\begin{bmatrix} \frac{1}{2} & -\frac{1}{3} & 0 \\ 3 & 2 & 0 \end{bmatrix}$$

which is equivalent to the reduced row-echelon matrix

$$\begin{bmatrix} 1 & 0 & 0 \\ 0 & 1 & 0 \end{bmatrix}.$$

So, the solution is: $x = 0$ and $y = 0$.

19. Multiplying both equations by 100 and forming the augmented matrix produces

$$\begin{bmatrix} 20 & 30 & 14 \\ 40 & 50 & 20 \end{bmatrix}.$$

Use Gauss-Jordan elimination as shown below.

$$\begin{bmatrix} 1 & \frac{3}{2} & \frac{7}{10} \\ 40 & 50 & 20 \end{bmatrix} \Rightarrow \begin{bmatrix} 1 & \frac{3}{2} & \frac{7}{10} \\ 0 & -10 & -8 \end{bmatrix} \Rightarrow \begin{bmatrix} 1 & \frac{3}{2} & \frac{7}{10} \\ 0 & 1 & \frac{4}{5} \end{bmatrix} \Rightarrow \begin{bmatrix} 1 & 0 & -\frac{1}{2} \\ 0 & 1 & \frac{4}{5} \end{bmatrix}$$

So, the solution is: $x_1 = -\frac{1}{2}$ and $x_2 = \frac{4}{5}$.

21. Because the matrix has 2 rows and 3 columns, it has size 2×3.

23. This matrix corresponds to the system

$$x_1 + 2x_2 = 0$$
$$x_3 = 0.$$

Choosing $x_2 = t$ as the free variable, you can describe the solution as $x_1 = -2t$, $x_2 = t$, and $x_3 = 0$, where t is a real number.

25. This matrix has the characteristic stair step pattern of leading 1's so it is in row-echelon form. However, the leading 1 in row three of column four has 1's above it, so the matrix is *not* in reduced row-echelon form.

27. Because the first row begins with -1, this matrix is not in row-echelon form.

29. The augmented matrix for this system is

$$\begin{bmatrix} -1 & 1 & 2 & 1 \\ 2 & 3 & 1 & -2 \\ 5 & 4 & 2 & 4 \end{bmatrix}$$

which is equivalent to the reduced row-echelon matrix

$$\begin{bmatrix} 1 & 0 & 0 & 2 \\ 0 & 1 & 0 & -3 \\ 0 & 0 & 1 & 3 \end{bmatrix}.$$

So, the solution is: $x = 2$, $y = -3$, and $z = 3$.

31. Use Gauss-Jordan elimination on the augmented matrix.

$$\begin{bmatrix} 2 & 3 & 3 & 3 \\ 6 & 6 & 12 & 13 \\ 12 & 9 & -1 & 2 \end{bmatrix} \Rightarrow \begin{bmatrix} 1 & 0 & 0 & \frac{1}{2} \\ 0 & 1 & 0 & -\frac{1}{3} \\ 0 & 0 & 1 & 1 \end{bmatrix}$$

So, $x = \frac{1}{2}$, $y = -\frac{1}{3}$, and $z = 1$.

33. The augmented matrix for this system is

$$\begin{bmatrix} 1 & -2 & 1 & -6 \\ 2 & -3 & 0 & -7 \\ -1 & 3 & -3 & 11 \end{bmatrix}$$

which is equivalent to the reduced row-echelon matrix

$$\begin{bmatrix} 1 & 0 & -3 & 4 \\ 0 & 1 & -2 & 5 \\ 0 & 0 & 0 & 0 \end{bmatrix}.$$

Choosing $z = t$ as the free variable, you find that the solution set can be described by $x = 4 + 3t$, $y = 5 + 2t$, and $z = t$, where t is a real number.

35. Use Gauss-Jordan elimination on the augmented matrix.

$$\begin{bmatrix} 1 & 2 & 6 & 1 \\ 2 & 5 & 15 & 4 \\ 3 & 1 & 3 & -6 \end{bmatrix} \Rightarrow \begin{bmatrix} 1 & 2 & 6 & 1 \\ 0 & 1 & 3 & 2 \\ 0 & 0 & 0 & 1 \end{bmatrix}$$

Because the third row corresponds to the false statement $0 = 1$, there is no solution.

© 2013 Cengage Learning. All Rights Reserved. May not be scanned, copied or duplicated, or posted to a publicly accessible website, in whole or in part.

37. The augmented matrix for this system is

$$\begin{bmatrix} 2 & 1 & 1 & 2 & -1 \\ 5 & -2 & 1 & -3 & 0 \\ -1 & 3 & 2 & 2 & 1 \\ 3 & 2 & 3 & -5 & 12 \end{bmatrix}$$

which is equivalent to the reduced row-echelon matrix

$$\begin{bmatrix} 1 & 0 & 0 & 0 & 1 \\ 0 & 1 & 0 & 0 & 4 \\ 0 & 0 & 1 & 0 & -3 \\ 0 & 0 & 0 & 1 & -2 \end{bmatrix}.$$

So, the solution is: $x_1 = 1$, $x_2 = 4$, $x_3 = -3$, and $x_4 = -2$.

39. Using a graphing utility, the augmented matrix reduces to

$$\begin{bmatrix} 1 & 0 & 0 & 0 \\ 0 & 1 & 4 & 2 \\ 0 & 0 & 0 & 0 \\ 0 & 0 & 0 & 0 \end{bmatrix}.$$

Choosing $z = t$ as the free variable, you find that the solution set can be described by $x = 0$, $y = z - 4t$, and $z = t$, where t is a real number.

41. Using a graphing utility, the augmented matrix reduces to

$$\begin{bmatrix} 1 & 0 & 0 & 0 & 1 \\ 0 & 1 & 0 & 0 & 0 \\ 0 & 0 & 1 & 0 & 4 \\ 0 & 0 & 0 & 1 & -2 \end{bmatrix}$$

So, the solution is: $x = 1$, $y = 0$, $z = 4$, and $w = -2$.

43. Use Gauss-Jordan elimination on the augmented matrix.

$$\begin{bmatrix} 1 & -2 & -8 & 0 \\ 3 & 2 & 0 & 0 \end{bmatrix} \Rightarrow \begin{bmatrix} 1 & 0 & -2 & 0 \\ 0 & 1 & 3 & 0 \end{bmatrix}$$

Choosing $x_3 = t$ as the free variable, you find that the solution set can be described by $x_1 = 2t$, $x_2 = -3t$, and $x_3 = t$, where t is a real number.

45. The augmented matrix for this system is

$$\begin{bmatrix} 2 & -8 & 4 & 0 \\ 3 & -10 & 7 & 0 \\ 0 & 10 & 5 & 0 \end{bmatrix}$$

which is equivalent to the reduced row-echelon matrix

$$\begin{bmatrix} 1 & 0 & 4 & 0 \\ 0 & 1 & \frac{1}{2} & 0 \\ 0 & 0 & 0 & 0 \end{bmatrix}.$$

Choosing $x_3 = t$ as the free variable, you find that the solution set can be described by $x_1 = -4t$, $x_2 = -\frac{1}{2}t$, and $x_3 = t$, where t is a real number.

47. Forming the augmented matrix

$$\begin{bmatrix} k & 1 & 0 \\ 1 & k & 1 \end{bmatrix}$$

and using Gauss-Jordan elimination, you obtain

$$\begin{bmatrix} 1 & k & 1 \\ k & 1 & 0 \end{bmatrix} \Rightarrow \begin{bmatrix} 1 & k & 1 \\ 0 & 1-k^2 & -k \end{bmatrix} \Rightarrow \begin{bmatrix} 1 & k & 1 \\ 0 & 1 & \dfrac{k}{k^2-1} \end{bmatrix} \Rightarrow \begin{bmatrix} 1 & 0 & \dfrac{-1}{k^2-1} \\ 0 & 1 & \dfrac{k}{k^2-1} \end{bmatrix}, \; k^2 - 1 \neq 0.$$

So, the system is inconsistent if $k = \pm 1$.

49. Row reduce the augmented matrix.

$$\begin{bmatrix} 1 & 2 & 3 \\ a & b & -9 \end{bmatrix} \Rightarrow \begin{bmatrix} 1 & 2 & 3 \\ 0 & (b-2a) & (-9-3a) \end{bmatrix}$$

(a) There will be no solution if $b - 2a = 0$ and $-9 - 3a \neq 0$. That is, if $b = 2a$ and $a = -3$.

(b) There will be exactly one solution if $b \neq 2a$.

(c) There will be an infinite number of solutions if $b = 2a$ and $a = -3$. That is, if $a = -3$ and $b = -6$.

51. You can show that two matrices of the same size are row equivalent if they both row reduce to the same matrix. The two given matrices are row equivalent because each is row equivalent to the identity matrix.

© 2013 Cengage Learning. All Rights Reserved. May not be scanned, copied or duplicated, or posted to a publicly accessible website, in whole or in part.

53. Adding a multiple of Row 1 to each row yields the following matrix.

$$\begin{bmatrix} 1 & 2 & 3 & \cdots & n \\ 0 & -n & -2n & \cdots & -(n-1)n \\ 0 & -2n & -4n & \cdots & -2(n-1)n \\ \vdots & \vdots & \vdots & & \vdots \\ 0 & -(n-1)n & -2(n-1)n & \cdots & -(n-1)(n-1)n \end{bmatrix}$$

Every row below Row 2 is a multiple of Row 2. Therefore, reduce these rows to zeros.

$$\begin{bmatrix} 1 & 2 & 3 & \cdots & n \\ 0 & -n & -2n & \cdots & -(n-1)n \\ 0 & 0 & 0 & \cdots & 0 \\ \vdots & \vdots & \vdots & \cdots & \vdots \\ 0 & 0 & 0 & & 0 \end{bmatrix}$$

Dividing Row 2 by $-n$ yields a new second row.

$$\begin{bmatrix} 1 & 2 & 3 & \cdots & n \\ 0 & 1 & 2 & \cdots & n-1 \\ 0 & 0 & 0 & \cdots & 0 \\ \vdots & \vdots & \vdots & & \vdots \\ 0 & 0 & 0 & \cdots & 0 \end{bmatrix}$$

Adding -2 times Row 2 to Row 1 yields a new first row.

$$\begin{bmatrix} 1 & 0 & -1 & \cdots & 2-n \\ 0 & 1 & 2 & \cdots & n-1 \\ 0 & 0 & 0 & \cdots & 0 \\ \vdots & \vdots & \vdots & & \vdots \\ 0 & 0 & 0 & \cdots & 0 \end{bmatrix}$$

This matrix is in reduced row-echelon form.

55. (a) False. See page 3, following Example 2.

(b) True. See page 5, Example 4(b).

57. Let x = number of touchdowns, y = number of extra-point kicks, and z = number of field goals.

$$\begin{aligned} 6x + y + 3z &= 45 \\ x - y \qquad &= 0 \\ x \qquad - 6z &= 0 \end{aligned}$$

Use Gauss-Jordan elimination to solve the system.

$$\begin{bmatrix} 6 & 1 & 3 & 45 \\ 1 & -1 & 0 & 0 \\ 1 & 0 & -6 & 0 \end{bmatrix} \Rightarrow \begin{bmatrix} 1 & 0 & 0 & 6 \\ 0 & 1 & 0 & 6 \\ 0 & 0 & 1 & 1 \end{bmatrix}$$

Because $x = 6$, $y = 6$, and $z = 1$, there were 6 touchdowns, 6 extra-point kicks, and 1 field goal.

59. $$\frac{8x^2}{(x-1)^2(x+1)} = \frac{A}{x+1} + \frac{B}{x-1} + \frac{C}{(x-1)^2}$$

$$\begin{aligned} 8x^2 &= A(x-1)^2 + B(x-1)(x+1) + C(x+1) \\ 8x^2 &= Ax^2 - 2Ax + A + Bx^2 - B + Cx + C \\ 8x^2 &= (A+B)x^2 + (-2A+C)x + A - B + C \end{aligned}$$

So, $$\begin{aligned} A + B \qquad &= 8 \\ -2A \qquad + C &= 0 \\ A - B + C &= 0. \end{aligned}$$

Use Gauss-Jordan elimination to solve the system.

$$\begin{bmatrix} 1 & 1 & 0 & 8 \\ -2 & 0 & 1 & 0 \\ 1 & -1 & 1 & 0 \end{bmatrix} \Rightarrow \begin{bmatrix} 1 & 0 & 0 & 2 \\ 0 & 1 & 0 & 6 \\ 0 & 0 & 1 & 4 \end{bmatrix}$$

The solution is: $A = 2$, $B = 6$, and $C = 4$.

So, $$\frac{8x^2}{(x-1)^2(x+1)} = \frac{2}{x+1} + \frac{6}{x-1} + \frac{4}{(x-1)^2}.$$

© 2013 Cengage Learning. All Rights Reserved. May not be scanned, copied or duplicated, or posted to a publicly accessible website, in whole or in part.

61. (a) Because there are three points, choose a second-degree polynomial, $p(x) = a_0 + a_1x + a_2x^2$. By substituting the values at each point into this equation you obtain the system

$$a_0 + 2a_1 + 4a_2 = 5$$
$$a_0 + 3a_1 + 9a_2 = 0$$
$$a_0 + 4a_1 + 16a_2 = 20.$$

Forming the augmented matrix

$$\begin{bmatrix} 1 & 2 & 4 & 5 \\ 1 & 3 & 9 & 0 \\ 1 & 4 & 16 & 20 \end{bmatrix}$$

and using Gauss-Jordan elimination you obtain

$$\begin{bmatrix} 1 & 0 & 0 & 90 \\ 0 & 1 & 0 & -\frac{135}{2} \\ 0 & 0 & 1 & \frac{25}{2} \end{bmatrix}.$$

So, $p(x) = 90 - \frac{135}{2}x + \frac{25}{2}x^2$.

(b)

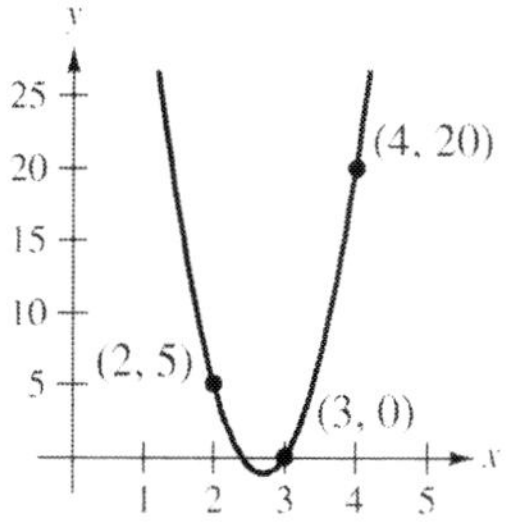

63. Establish the first year as $x = 0$ and substitute the values at each point into $p(x) = a_0 + a_1x + a_2x^2$ to obtain the system

$$a_0 = 50$$
$$a_0 + a_1 + a_2 = 60$$
$$a_0 + 2a_1 + 4a_2 = 75.$$

Forming the augmented matrix

$$\begin{bmatrix} 1 & 0 & 0 & 50 \\ 1 & 1 & 1 & 60 \\ 1 & 2 & 4 & 75 \end{bmatrix}$$

and using Gauss-Jordan elimination, you obtain

$$\begin{bmatrix} 1 & 0 & 0 & 50 \\ 0 & 1 & 0 & \frac{15}{2} \\ 0 & 0 & 1 & \frac{5}{2} \end{bmatrix}.$$

So, $p(x) = 50 + \frac{15}{2}x + \frac{5}{2}x^2$. To predict the sales in the fourth year, evaluate $p(x)$ when $x = 3$.

$$p(3) = 50 + \tfrac{15}{2}(3) + \tfrac{5}{2}(3)^2 = \$95$$

65. (a) There are three points: (0, 80), (4, 68), and (80, 30). Because you are given three points, choose a second-degree polynomial, $p(x) = a_0 + a_1x + a_2x^2$. Substituting the given points into $p(x)$ produces the following system of linear equations.

$$a_0 + (0)a_1 + (0)^2a_2 = a_0 = 80$$
$$a_0 + (4)a_1 + (4)^2a_2 = a_0 + 4a_1 + 16a_2 = 68$$
$$a_0 + (80)a_1 + (80)^2a_2 = a_0 + 80a_1 + 6400a_2 = 30$$

(b) Form the augmented matrix

$$\begin{bmatrix} 1 & 0 & 0 & 80 \\ 1 & 4 & 16 & 68 \\ 1 & 80 & 6400 & 30 \end{bmatrix}$$

and use Gauss-Jordan elimination to obtain the equivalent reduced row-echelon matrix

$$\begin{bmatrix} 1 & 0 & 0 & 80 \\ 0 & 1 & 0 & -\frac{25}{8} \\ 0 & 0 & 1 & \frac{1}{32} \end{bmatrix}.$$

So, $p(x) = 80 - \frac{25}{8}x + \frac{1}{32}x^2$.

(c) The graphing utility gives $a_0 = 80$, $a_1 = -\frac{25}{8}$, and $a_2 = \frac{1}{32}$. In other words, $p(x) = 80 - \frac{25}{8}x + \frac{1}{32}x^2$.

(d) The results of (b) and (c) are the same.

(e) There is precisely one polynomial function of degree $n - 1$ (or less) that fits n distinct points.

67. Applying Kirchoff's first law to either junction produces $I_1 + I_3 = I_2$ and applying the second law to the two paths produces

$$R_1I_1 + R_2I_2 = 3I_1 + 4I_2 = 3$$
$$R_2I_2 + R_3I_3 = 4I_2 + 2I_3 = 2.$$

Rearrange these equations, form the augmented matrix, and use Gauss-Jordan elimination.

$$\begin{bmatrix} 1 & -1 & 1 & 0 \\ 3 & 4 & 0 & 3 \\ 0 & 4 & 2 & 2 \end{bmatrix} \Rightarrow \begin{bmatrix} 1 & 0 & 0 & \frac{5}{13} \\ 0 & 1 & 0 & \frac{6}{13} \\ 0 & 0 & 1 & \frac{1}{13} \end{bmatrix}$$

So, the solution is $I_1 = \frac{5}{13}$, $I_2 = \frac{6}{13}$, and $I_3 = \frac{1}{13}$.

© 2013 Cengage Learning. All Rights Reserved. May not be scanned, copied or duplicated, or posted to a publicly accessible website, in whole or in part.

CHAPTER 2
Matrices

© 2013 Cengage Learning. All Rights Reserved. May not be scanned, copied or duplicated, or posted to a publicly accessible website, in whole or in part.

CHAPTER 2
Matrices

Section 2.1 Operations with Matrices

1. $x = -4,\ y = 22$

3. $2x + 1 = 5 \qquad 3y - 5 = 4$

$2x = 4 \qquad 3y = 9$

$x = 2 \qquad y = 3$

5. (a) $A + B = \begin{bmatrix} 1 & -1 \\ 2 & -1 \end{bmatrix} + \begin{bmatrix} 2 & -1 \\ -1 & 8 \end{bmatrix} = \begin{bmatrix} 1+2 & -1-1 \\ 2-1 & -1+8 \end{bmatrix} = \begin{bmatrix} 3 & -2 \\ 1 & 7 \end{bmatrix}$

(b) $A - B = \begin{bmatrix} 1 & -1 \\ 2 & -1 \end{bmatrix} - \begin{bmatrix} 2 & -1 \\ -1 & 8 \end{bmatrix} = \begin{bmatrix} 1-2 & -1+1 \\ 2+1 & -1-8 \end{bmatrix} = \begin{bmatrix} -1 & 0 \\ 3 & -9 \end{bmatrix}$

(c) $2A = 2\begin{bmatrix} 1 & -1 \\ 2 & -1 \end{bmatrix} = \begin{bmatrix} 2(1) & 2(-1) \\ 2(2) & 2(-1) \end{bmatrix} = \begin{bmatrix} 2 & -2 \\ 4 & -2 \end{bmatrix}$

(d) $2A - B = \begin{bmatrix} 2 & -2 \\ 4 & -2 \end{bmatrix} - \begin{bmatrix} 2 & -1 \\ -1 & 8 \end{bmatrix} = \begin{bmatrix} 0 & -1 \\ 5 & -10 \end{bmatrix}$

(e) $B + \frac{1}{2}A = \begin{bmatrix} 2 & -1 \\ -1 & 8 \end{bmatrix} + \frac{1}{2}\begin{bmatrix} 1 & -1 \\ 2 & -1 \end{bmatrix} = \begin{bmatrix} 2 & -1 \\ -1 & 8 \end{bmatrix} + \begin{bmatrix} \frac{1}{2} & -\frac{1}{2} \\ 1 & -\frac{1}{2} \end{bmatrix} = \begin{bmatrix} \frac{5}{2} & -\frac{3}{2} \\ 0 & \frac{15}{2} \end{bmatrix}$

7. (a) $A + B = \begin{bmatrix} 6 & -1 \\ 2 & 4 \\ -3 & 5 \end{bmatrix} + \begin{bmatrix} 1 & 4 \\ -1 & 5 \\ 1 & 10 \end{bmatrix} = \begin{bmatrix} 6+1 & -1+4 \\ 2+(-1) & 4+5 \\ -3+1 & 5+10 \end{bmatrix} = \begin{bmatrix} 7 & 3 \\ 1 & 9 \\ -2 & 15 \end{bmatrix}$

(b) $A - B = \begin{bmatrix} 6 & -1 \\ 2 & 4 \\ -3 & 5 \end{bmatrix} - \begin{bmatrix} 1 & 4 \\ -1 & 5 \\ 1 & 10 \end{bmatrix} = \begin{bmatrix} 6-1 & -1-4 \\ 2-(-1) & 4-5 \\ -3-1 & 5-10 \end{bmatrix} = \begin{bmatrix} 5 & -5 \\ 3 & -1 \\ -4 & -5 \end{bmatrix}$

(c) $2A = 2\begin{bmatrix} 6 & -1 \\ 2 & 4 \\ -3 & 5 \end{bmatrix} = \begin{bmatrix} 2(6) & 2(-1) \\ 2(2) & 2(4) \\ 2(-3) & 2(5) \end{bmatrix} = \begin{bmatrix} 12 & -2 \\ 4 & 8 \\ -6 & 10 \end{bmatrix}$

(d) $2A - B = \begin{bmatrix} 12 & -2 \\ 4 & 8 \\ -6 & 10 \end{bmatrix} - \begin{bmatrix} 1 & 4 \\ -1 & 5 \\ 1 & 10 \end{bmatrix} = \begin{bmatrix} 12-1 & -2-4 \\ 4-(-1) & 8-5 \\ -6-1 & 10-10 \end{bmatrix} = \begin{bmatrix} 11 & -6 \\ 5 & 3 \\ -7 & 0 \end{bmatrix}$

(e) $B + \frac{1}{2}A = \begin{bmatrix} 1 & 4 \\ -1 & 5 \\ 1 & 10 \end{bmatrix} + \frac{1}{2}\begin{bmatrix} 6 & -1 \\ 2 & 4 \\ -3 & 5 \end{bmatrix} = \begin{bmatrix} 1 & 4 \\ -1 & 5 \\ 1 & 10 \end{bmatrix} + \begin{bmatrix} 3 & -\frac{1}{2} \\ 1 & 2 \\ -\frac{3}{2} & \frac{5}{2} \end{bmatrix} = \begin{bmatrix} 4 & \frac{7}{2} \\ 0 & 7 \\ -\frac{1}{2} & \frac{25}{2} \end{bmatrix}$

9. (a) $A + B = \begin{bmatrix} 3 & 2 & -1 \\ 2 & 4 & 5 \\ 0 & 1 & 2 \end{bmatrix} + \begin{bmatrix} 0 & 2 & 1 \\ 5 & 4 & 2 \\ 2 & 1 & 0 \end{bmatrix} = \begin{bmatrix} 3+0 & 2+2 & -1+1 \\ 2+5 & 4+4 & 5+2 \\ 0+2 & 1+1 & 2+0 \end{bmatrix} = \begin{bmatrix} 3 & 4 & 0 \\ 7 & 8 & 7 \\ 2 & 2 & 2 \end{bmatrix}$

(b) $A - B = \begin{bmatrix} 3 & 2 & -1 \\ 2 & 4 & 5 \\ 0 & 1 & 2 \end{bmatrix} - \begin{bmatrix} 0 & 2 & 1 \\ 5 & 4 & 2 \\ 2 & 1 & 0 \end{bmatrix} = \begin{bmatrix} 3-0 & 2-2 & -1-1 \\ 2-5 & 4-4 & 5-2 \\ 0-2 & 1-1 & 2-0 \end{bmatrix} = \begin{bmatrix} 3 & 0 & -2 \\ -3 & 0 & 3 \\ -2 & 0 & 2 \end{bmatrix}$

© 2013 Cengage Learning. All Rights Reserved. May not be scanned, copied or duplicated, or posted to a publicly accessible website, in whole or in part.

(c) $2A = 2\begin{bmatrix} 3 & 2 & -1 \\ 2 & 4 & 5 \\ 0 & 1 & 2 \end{bmatrix} = \begin{bmatrix} 2(3) & 2(2) & 2(-1) \\ 2(2) & 2(4) & 2(5) \\ 2(0) & 2(1) & 2(2) \end{bmatrix} = \begin{bmatrix} 6 & 4 & -2 \\ 4 & 8 & 10 \\ 0 & 2 & 4 \end{bmatrix}$

(d) $2A - B = 2\begin{bmatrix} 3 & 2 & -1 \\ 2 & 4 & 5 \\ 0 & 1 & 2 \end{bmatrix} - \begin{bmatrix} 0 & 2 & 1 \\ 5 & 4 & 2 \\ 2 & 1 & 0 \end{bmatrix} = \begin{bmatrix} 6 & 4 & -2 \\ 4 & 8 & 10 \\ 0 & 2 & 4 \end{bmatrix} - \begin{bmatrix} 0 & 2 & 1 \\ 5 & 4 & 2 \\ 2 & 1 & 0 \end{bmatrix} = \begin{bmatrix} 6 & 2 & -3 \\ -1 & 4 & 8 \\ -2 & 1 & 4 \end{bmatrix}$

(e) $B + \frac{1}{2}A = \begin{bmatrix} 0 & 2 & 1 \\ 5 & 4 & 2 \\ 2 & 1 & 0 \end{bmatrix} + \frac{1}{2}\begin{bmatrix} 3 & 2 & -1 \\ 2 & 4 & 5 \\ 0 & 1 & 2 \end{bmatrix} = \begin{bmatrix} 0 & 2 & 1 \\ 5 & 4 & 2 \\ 2 & 1 & 0 \end{bmatrix} + \begin{bmatrix} \frac{3}{2} & 1 & -\frac{1}{2} \\ 1 & 2 & \frac{5}{2} \\ 0 & \frac{1}{2} & 1 \end{bmatrix} = \begin{bmatrix} \frac{3}{2} & 3 & \frac{1}{2} \\ 6 & 6 & \frac{9}{2} \\ 2 & \frac{3}{2} & 1 \end{bmatrix}$

11. $A = \begin{bmatrix} 6 & 0 & 3 \\ -1 & -4 & 0 \end{bmatrix}, B = \begin{bmatrix} 8 & -1 \\ 4 & -3 \end{bmatrix}$

(a) $A + B$ is not possible. A and B have different sizes.

(b) $A - B$ is not possible. A and B have different sizes.

(c) $2A = \begin{bmatrix} 12 & 0 & 6 \\ -2 & -8 & 0 \end{bmatrix}$

(d) $2A - B$ is not possible. A and B have different sizes.

(e) $B + \frac{1}{2}A$ is not possible. A and B have different sizes.

13. (a) $c_{21} = 2a_{21} - 3b_{21} = 2(-3) - 3(0) = -6$

(b) $c_{13} = 2a_{13} - 3b_{13} = 2(4) - 3(-7) = 29$

15. Expanding both sides of the equation produces

$$\begin{bmatrix} 4x & 4y \\ 4z & -4 \end{bmatrix} = \begin{bmatrix} 2y + 8 & 2z + 2x \\ -2x + 10 & 2 - 2x \end{bmatrix}.$$

By setting corresponding entries equal to each other, you obtain four equations.

$$\begin{aligned} 4x &= 2y + 8 &\Rightarrow\quad 4x - 2y &= 8 \\ 4y &= 2z + 2x &\Rightarrow\quad 2x - 4y + 2z &= 0 \\ 4z &= -2x + 10 &\Rightarrow\quad 2x + 4z &= 10 \\ -4 &= 2 - 2x &\Rightarrow\quad 2x &= 6 \end{aligned}$$

Gauss-Jordan elimination produces $x = 3$, $y = 2$, and $z = 1$.

17. (a) $AB = \begin{bmatrix} 1 & 2 \\ 4 & 2 \end{bmatrix}\begin{bmatrix} 2 & -1 \\ -1 & 8 \end{bmatrix} = \begin{bmatrix} 1(2) + 2(-1) & 1(-1) + 2(8) \\ 4(2) + 2(-1) & 4(-1) + 2(8) \end{bmatrix} = \begin{bmatrix} 0 & 15 \\ 6 & 12 \end{bmatrix}$

(b) $BA = \begin{bmatrix} 2 & -1 \\ -1 & 8 \end{bmatrix}\begin{bmatrix} 1 & 2 \\ 4 & 2 \end{bmatrix} = \begin{bmatrix} 2(1) + (-1)(4) & 2(2) + (-1)(2) \\ -1(1) + 8(4) & -1(2) + 8(2) \end{bmatrix} = \begin{bmatrix} -2 & 2 \\ 31 & 14 \end{bmatrix}$

19. (a) $AB = \begin{bmatrix} 2 & -1 & 3 \\ 5 & 1 & -2 \\ 2 & 2 & 3 \end{bmatrix}\begin{bmatrix} 0 & 1 & 2 \\ -4 & 1 & 3 \\ -4 & -1 & -2 \end{bmatrix} = \begin{bmatrix} 2(0) + (-1)(-4) + 3(-4) & 2(1) + (-1)(1) + 3(-1) & 2(2) + (-1)(3) + 3(-2) \\ 5(0) + 1(-4) + (-2)(-4) & 5(1) + 1(1) + (-2)(-1) & 5(2) + 1(3) + (-2)(-2) \\ 2(0) + 2(-4) + 3(-4) & 2(1) + 2(1) + 3(-1) & 2(2) + 2(3) + 3(-2) \end{bmatrix}$

$$= \begin{bmatrix} -8 & -2 & -5 \\ 4 & 8 & 17 \\ -20 & 1 & 4 \end{bmatrix}$$

(b) $BA = \begin{bmatrix} 0 & 1 & 2 \\ -4 & 1 & 3 \\ -4 & -1 & -2 \end{bmatrix}\begin{bmatrix} 2 & -1 & 3 \\ 5 & 1 & -2 \\ 2 & 2 & 3 \end{bmatrix}$

$$= \begin{bmatrix} 0(2) + 1(5) + 2(2) & 0(-1) + 1(1) + 2(2) & 0(3) + 1(-2) + 2(3) \\ -4(2) + 1(5) + 3(2) & -4(-1) + 1(1) + 3(2) & -4(3) + 1(-2) + 3(3) \\ -4(2) + (-1)(5) + (-2)(2) & -4(-1) + (-1)(1) + (-2)(2) & -4(3) + (-1)(-2) + (-2)(3) \end{bmatrix}$$

$$= \begin{bmatrix} 9 & 5 & 4 \\ 3 & 11 & -5 \\ -17 & -1 & -16 \end{bmatrix}$$

© 2013 Cengage Learning. All Rights Reserved. May not be scanned, copied or duplicated, or posted to a publicly accessible website, in whole or in part.

21. (a) AB is not defined because A is 3×2 and B is 3×3.

(b) $$BA = \begin{bmatrix} 0 & -1 & 0 \\ 4 & 0 & 2 \\ 8 & -1 & 7 \end{bmatrix}\begin{bmatrix} 2 & 1 \\ -3 & 4 \\ 1 & 6 \end{bmatrix} = \begin{bmatrix} 0(2)+(-1)(-3)+0(1) & 0(1)+(-1)(4)+0(6) \\ 4(2)+0(-3)+2(1) & 4(1)+0(4)+2(6) \\ 8(2)+(-1)(-3)+7(1) & 8(1)+(-1)(4)+7(6) \end{bmatrix} = \begin{bmatrix} 3 & -4 \\ 10 & 16 \\ 26 & 46 \end{bmatrix}$$

23. (a) $$AB = \begin{bmatrix} 3 & 2 & 1 \end{bmatrix}\begin{bmatrix} 2 \\ 3 \\ 0 \end{bmatrix} = \begin{bmatrix} 3(2)+2(3)+1(0) \end{bmatrix} = \begin{bmatrix} 12 \end{bmatrix}$$

(b) $$BA = \begin{bmatrix} 2 \\ 3 \\ 0 \end{bmatrix}\begin{bmatrix} 3 & 2 & 1 \end{bmatrix} = \begin{bmatrix} 2(3) & 2(2) & 2(1) \\ 3(3) & 3(2) & 3(1) \\ 0(3) & 0(2) & 0(1) \end{bmatrix} = \begin{bmatrix} 6 & 4 & 2 \\ 9 & 6 & 3 \\ 0 & 0 & 0 \end{bmatrix}$$

25. (a) $$AB = \begin{bmatrix} -1 & 3 \\ 4 & -5 \\ 0 & 2 \end{bmatrix}\begin{bmatrix} 1 & 2 \\ 0 & 7 \end{bmatrix} = \begin{bmatrix} -1(1)+3(0) & -1(2)+3(7) \\ 4(1)+(-5)(0) & 4(2)+(-5)(7) \\ 0(1)+2(0) & 0(2)+2(7) \end{bmatrix} = \begin{bmatrix} -1 & 19 \\ 4 & -27 \\ 0 & 14 \end{bmatrix}$$

(b) BA is not defined because B is a 2×2 matrix and A is a 3×2 matrix.

27. (a) $$AB = \begin{bmatrix} 0 & -1 & 0 \\ 4 & 0 & 2 \\ 8 & -1 & 7 \end{bmatrix}\begin{bmatrix} 2 \\ -3 \\ 1 \end{bmatrix} = \begin{bmatrix} 0(2)+(-1)(-3)+0(1) \\ 4(2)+0(-3)+2(1) \\ 8(2)+(-1)(-3)+7(1) \end{bmatrix} = \begin{bmatrix} 3 \\ 10 \\ 26 \end{bmatrix}$$

(b) BA is not defined because B is 3×1 and A is 3×3.

29. (a) $$AB = \begin{bmatrix} 6 \\ -2 \\ 1 \\ 6 \end{bmatrix}\begin{bmatrix} 10 & 12 \end{bmatrix} = \begin{bmatrix} 6(10) & 6(12) \\ -2(10) & -2(12) \\ 1(10) & 1(12) \\ 6(10) & 6(12) \end{bmatrix} = \begin{bmatrix} 60 & 72 \\ -20 & -24 \\ 10 & 12 \\ 60 & 72 \end{bmatrix}$$

(b) BA is not defined because B is a 1×2 matrix and A is a 4×1 matrix.

31. $A + B$ is defined and has size 3×4 because A and B have size 3×4.

33. $\frac{1}{2}D$ is defined and has size 4×2 because D has size 4×2.

35. AC is defined. Because A has size 3×4 and C has size 4×2, the size of AC is 3×2.

37. $E - 2A$ is not defined because E and $2A$ have different sizes.

39. As a system of linear equations, $A\mathbf{x} = \mathbf{0}$ is

$$\begin{aligned} 2x_1 - x_2 - x_3 &= 0 \\ x_1 - 2x_2 + 2x_3 &= 0 \end{aligned}.$$

Use Gauss-Jordan elimination on the augmented matrix for this system.

$$\begin{bmatrix} 2 & -1 & -1 & 0 \\ 1 & -2 & 2 & 0 \end{bmatrix} \Rightarrow \begin{bmatrix} 1 & 0 & -\frac{4}{3} & 0 \\ 0 & 1 & -\frac{5}{3} & 0 \end{bmatrix}$$

Choosing $x_3 = 3t$, the solution is $x_1 = 4t$, $x_2 = 5t$, and $x_3 = 3t$, where t is any real number.

41. In matrix form $A\mathbf{x} = \mathbf{b}$, the system is

$$\begin{bmatrix} -1 & 1 \\ -2 & 1 \end{bmatrix}\begin{bmatrix} x_1 \\ x_2 \end{bmatrix} = \begin{bmatrix} 4 \\ 0 \end{bmatrix}.$$

Use Gauss-Jordan elimination on the augmented matrix.

$$\begin{bmatrix} -1 & 1 & 4 \\ -2 & 1 & 0 \end{bmatrix} \Rightarrow \begin{bmatrix} 1 & 0 & 4 \\ 0 & 1 & 8 \end{bmatrix}$$

So, the solution is $\begin{bmatrix} x_1 \\ x_2 \end{bmatrix} = \begin{bmatrix} 4 \\ 8 \end{bmatrix}$.

43. In matrix form $A\mathbf{x} = \mathbf{b}$, the system is

$$\begin{bmatrix} -2 & -3 \\ 6 & 1 \end{bmatrix}\begin{bmatrix} x_1 \\ x_2 \end{bmatrix} = \begin{bmatrix} -4 \\ -36 \end{bmatrix}.$$

Use Gauss-Jordan elimination on the augmented matrix.

$$\begin{bmatrix} -2 & -3 & -4 \\ 6 & 1 & -36 \end{bmatrix} \Rightarrow \begin{bmatrix} 1 & 0 & -7 \\ 0 & 1 & 6 \end{bmatrix}$$

So, the solution is $\begin{bmatrix} x_1 \\ x_2 \end{bmatrix} = \begin{bmatrix} -7 \\ 6 \end{bmatrix}$.

© 2013 Cengage Learning. All Rights Reserved. May not be scanned, copied or duplicated, or posted to a publicly accessible website, in whole or in part.

45. In matrix form $A\mathbf{x} = \mathbf{b}$, the system is

$$\begin{bmatrix} 1 & -2 & 3 \\ -1 & 3 & -1 \\ 2 & -5 & 5 \end{bmatrix}\begin{bmatrix} x_1 \\ x_2 \\ x_3 \end{bmatrix} = \begin{bmatrix} 9 \\ -6 \\ 17 \end{bmatrix}.$$

Use Gauss-Jordan elimination on the augmented matrix.

$$\begin{bmatrix} 1 & -2 & 3 & 9 \\ -1 & 3 & -1 & -6 \\ 2 & -5 & 5 & 17 \end{bmatrix} \Rightarrow \begin{bmatrix} 1 & 0 & 0 & 1 \\ 0 & 1 & 0 & -1 \\ 0 & 0 & 1 & 2 \end{bmatrix}$$

So, the solution is $\begin{bmatrix} x_1 \\ x_2 \\ x_3 \end{bmatrix} = \begin{bmatrix} 1 \\ -1 \\ 2 \end{bmatrix}.$

47. In matrix form $A\mathbf{x} = \mathbf{b}$, the system is

$$\begin{bmatrix} 1 & -5 & 2 \\ -3 & 1 & -1 \\ 0 & -2 & 5 \end{bmatrix}\begin{bmatrix} x_1 \\ x_2 \\ x_3 \end{bmatrix} = \begin{bmatrix} -20 \\ 8 \\ -16 \end{bmatrix}.$$

Use Gauss-Jordan elimination on the augmented matrix.

$$\begin{bmatrix} 1 & -5 & 2 & -20 \\ -3 & 1 & -1 & 8 \\ 0 & -2 & 5 & -16 \end{bmatrix} \Rightarrow \begin{bmatrix} 1 & 0 & 0 & -1 \\ 0 & 1 & 0 & 3 \\ 0 & 0 & 1 & -2 \end{bmatrix}$$

So, the solution is $\begin{bmatrix} x_1 \\ x_2 \\ x_3 \end{bmatrix} = \begin{bmatrix} -1 \\ 3 \\ -2 \end{bmatrix}.$

49. The augmented matrix row reduces as follows.

$$\begin{bmatrix} 1 & -1 & 2 & -1 \\ 3 & -3 & 1 & 7 \end{bmatrix} \Rightarrow \begin{bmatrix} 1 & -1 & 0 & 3 \\ 0 & 0 & 1 & -2 \end{bmatrix}$$

There are an infinite number of solutions.

Answers will vary. *Sample answer*: $x_3 = -2$, $x_2 = 0$, $x_1 = 3$.

So,

$$\mathbf{b} = \begin{bmatrix} -1 \\ 7 \end{bmatrix} = 3\begin{bmatrix} 1 \\ 3 \end{bmatrix} + 0\begin{bmatrix} -1 \\ -3 \end{bmatrix} - 2\begin{bmatrix} 2 \\ 1 \end{bmatrix}.$$

51. The augmented matrix row reduces as follows.

$$\begin{bmatrix} 1 & 1 & -5 & 3 \\ 1 & 0 & -1 & 1 \\ 2 & -1 & -1 & 0 \end{bmatrix} \Rightarrow \begin{bmatrix} 1 & 0 & 0 & 1 \\ 0 & 1 & 0 & 2 \\ 0 & 0 & 1 & 0 \end{bmatrix}$$

So, $\mathbf{b} = \begin{bmatrix} 3 \\ 1 \\ 0 \end{bmatrix} = 1\begin{bmatrix} 1 \\ 1 \\ 2 \end{bmatrix} + 2\begin{bmatrix} 1 \\ 0 \\ -1 \end{bmatrix} + 0\begin{bmatrix} -5 \\ -1 \\ -1 \end{bmatrix}.$

53. Expanding the left side of the equation produces

$$\begin{bmatrix} 1 & 2 \\ 3 & 5 \end{bmatrix}A = \begin{bmatrix} 1 & 2 \\ 3 & 5 \end{bmatrix}\begin{bmatrix} a_{11} & a_{12} \\ a_{21} & a_{22} \end{bmatrix}$$

$$= \begin{bmatrix} a_{11} + 2a_{21} & a_{12} + 2a_{22} \\ 3a_{11} + 5a_{21} & 3a_{12} + 5a_{22} \end{bmatrix} = \begin{bmatrix} 1 & 0 \\ 0 & 1 \end{bmatrix}$$

from which you obtain the system

$$\begin{aligned} a_{11} \quad + 2a_{21} \quad\quad &= 1 \\ a_{12} \quad\quad + 2a_{22} &= 0 \\ 3a_{11} \quad + 5a_{21} \quad\quad &= 0 \\ 3a_{12} \quad\quad + 5a_{22} &= 1. \end{aligned}$$

Solving by Gauss-Jordan elimination yields $a_{11} = -5$, $a_{12} = 2$, $a_{21} = 3$, and $a_{22} = -1$. So, you have

$$A = \begin{bmatrix} -5 & 2 \\ 3 & -1 \end{bmatrix}.$$

55. Expand the left side of the matrix equation.

$$\begin{bmatrix} 1 & 2 \\ 3 & 4 \end{bmatrix}\begin{bmatrix} a & b \\ c & d \end{bmatrix} = \begin{bmatrix} 6 & 3 \\ 19 & 2 \end{bmatrix}$$

$$\begin{bmatrix} a + 2c & b + 2d \\ 3a + 4c & 3b + 4d \end{bmatrix} = \begin{bmatrix} 6 & 3 \\ 19 & 2 \end{bmatrix}$$

By setting corresponding entries equal to each other, obtain four equations.

$$\begin{aligned} a \quad + 2c \quad\quad &= 6 \\ b \quad\quad + 2d &= 3 \\ 3a \quad + 4c \quad\quad &= 19 \\ 3b \quad\quad + 4d &= 2 \end{aligned}$$

Gauss-Jordan elimination produces $a = 7$, $b = -4$, $c = -\frac{1}{2}$, and $d = \frac{7}{2}$.

57. $$AA = \begin{bmatrix} -1 & 0 & 0 \\ 0 & 2 & 0 \\ 0 & 0 & 3 \end{bmatrix}\begin{bmatrix} -1 & 0 & 0 \\ 0 & 2 & 0 \\ 0 & 0 & 3 \end{bmatrix} = \begin{bmatrix} -1(-1) + 0(0) + 0(0) & -1(0) + 0(2) + 0(0) & -1(0) + 0(0) + 0(3) \\ 0(-1) + 2(0) + 0(0) & 0(0) + 2(2) + 0(0) & 0(0) + 2(0) + 0(3) \\ 0(-1) + 0(0) + 3(0) & 0(0) + 0(2) + 3(0) & 0(0) + 0(0) + 3(3) \end{bmatrix} = \begin{bmatrix} 1 & 0 & 0 \\ 0 & 4 & 0 \\ 0 & 0 & 9 \end{bmatrix}$$

59. $$AB = \begin{bmatrix} 2 & 0 \\ 0 & -3 \end{bmatrix}\begin{bmatrix} -5 & 0 \\ 0 & 4 \end{bmatrix} = \begin{bmatrix} -10 & 0 \\ 0 & -12 \end{bmatrix}$$

$$BA = \begin{bmatrix} -5 & 0 \\ 0 & 4 \end{bmatrix}\begin{bmatrix} 2 & 0 \\ 0 & -3 \end{bmatrix} = \begin{bmatrix} -10 & 0 \\ 0 & -12 \end{bmatrix}$$

© 2013 Cengage Learning. All Rights Reserved. May not be scanned, copied or duplicated, or posted to a publicly accessible website, in whole or in part.

61. Let A and B be diagonal matrices of sizes $n \times n$.

Then, $AB = [c_{ij}] = \left[\sum_{k=1}^{n} a_{ik}b_{kj}\right]$

where $c_{ij} = 0$ if $i \neq j$, and $c_{ii} = a_{ii}b_{ii}$ otherwise.

The entries of BA are exactly the same.

63. The trace is the sum of the elements on the main diagonal.

$1 + (-2) + 3 = 2$

65. The trace is the sum of the elements on the main diagonal.

$1 + 1 + 1 + 1 = 4$

67. (a)
$$\begin{aligned}\operatorname{Tr}(A + B) &= \operatorname{Tr}([a_{ij} + b_{ij}]) \\ &= \sum_{i=1}^{n}(a_{ii} + b_{ii}) \\ &= \sum_{i=1}^{n} a_{ii} + \sum_{i=1}^{n} b_{ii} \\ &= \operatorname{Tr}(A) + \operatorname{Tr}(B)\end{aligned}$$

(b)
$$\begin{aligned}\operatorname{TR}(cA) &= \operatorname{Tr}([ca_{ij}]) = \sum_{i=1}^{n} ca_{ii} = c\sum_{i=1}^{n} a_{ii} \\ &= c\operatorname{Tr}(A)\end{aligned}$$

69. Expand $AB = BA$ as follows.

$$\begin{bmatrix} w & x \\ y & z \end{bmatrix}\begin{bmatrix} 1 & 1 \\ -1 & 1 \end{bmatrix} = \begin{bmatrix} 1 & 1 \\ -1 & 1 \end{bmatrix}\begin{bmatrix} w & x \\ y & z \end{bmatrix}$$

$$\begin{bmatrix} w - x & w + x \\ y - z & y + z \end{bmatrix} = \begin{bmatrix} w + y & x + z \\ -w + y & -x + z \end{bmatrix}$$

This yields the system of equations

$$\begin{aligned} -x - y \quad &= 0 \\ w \quad -z &= 0 \\ w \quad -z &= 0 \\ x + y \quad &= 0. \end{aligned}$$

Using Gauss-Jordan elimination, you can solve this system to obtain $w = t$, $x = -s$, $y = s$, and $z = t$, where s and t are any real numbers. So, $w = z$ and $x = -y$.

71. Let $A = \begin{bmatrix} a_{11} & a_{12} \\ a_{21} & a_{22} \end{bmatrix}$,

then the given matrix equation expands to

$$\begin{bmatrix} a_{11} + a_{21} & a_{12} + a_{22} \\ a_{11} + a_{21} & a_{12} + a_{22} \end{bmatrix} = \begin{bmatrix} 1 & 0 \\ 0 & 1 \end{bmatrix}.$$

Because $a_{11} + a_{21} = 1$ and $a_{11} + a_{21} = 0$ cannot both be true, conclude that there is no solution.

73. (a)
$$A^2 = \begin{bmatrix} i & 0 \\ 0 & i \end{bmatrix}\begin{bmatrix} i & 0 \\ 0 & i \end{bmatrix} = \begin{bmatrix} i^2 & 0 \\ 0 & i^2 \end{bmatrix} = \begin{bmatrix} -1 & 0 \\ 0 & -1 \end{bmatrix}$$

$$A^3 = A^2A = \begin{bmatrix} -1 & 0 \\ 0 & -1 \end{bmatrix}\begin{bmatrix} i & 0 \\ 0 & i \end{bmatrix} = \begin{bmatrix} -i & 0 \\ 0 & -i \end{bmatrix}$$

$$A^4 = A^3A = \begin{bmatrix} -i & 0 \\ 0 & -i \end{bmatrix}\begin{bmatrix} i & 0 \\ 0 & i \end{bmatrix} = \begin{bmatrix} -i^2 & 0 \\ 0 & -i^2 \end{bmatrix} = \begin{bmatrix} 1 & 0 \\ 0 & 1 \end{bmatrix}$$

A^2, A^3, and A^4 are diagonal matrices with diagonal entries i^2, i^3, and i^4, respectively.

(b)
$$B^2 = \begin{bmatrix} 0 & -i \\ i & 0 \end{bmatrix}\begin{bmatrix} 0 & -i \\ i & 0 \end{bmatrix} = \begin{bmatrix} 1 & 0 \\ 0 & 1 \end{bmatrix} = A^4 = I$$

75. Assume that A is an $m \times n$ matrix and B is a $p \times q$ matrix.

Because AB is defined, you have that $n = p$ and AB is $m \times q$.

Because BA is defined, you have that $m = q$ and so AB is an $m \times m$ square matrix.

Likewise, because BA is defined, $m = q$ and so BA is $p \times n$.

Because AB is defined, you have $n = p$. Therefore, BA is an $n \times n$ square matrix.

77. Let $AB = [c_{ij}]$, where $c_{ij} = \sum_{k=1}^{n} a_{ik}b_{kj}$. If the ith row of A has all zero entries, then $a_{ik} = 0$ for $k = 1, \ldots, n$. So, $c_{ij} = 0$ for all $j = 1, \ldots, n$, and the ith row of AB has all zero entries. To show the converse is not true, consider $AB = \begin{bmatrix} 2 & 1 \\ -2 & -1 \end{bmatrix}\begin{bmatrix} 1 & 1 \\ -2 & -2 \end{bmatrix} = \begin{bmatrix} 0 & 0 \\ 0 & 0 \end{bmatrix}$.

79.
$$\begin{aligned} BA &= \begin{bmatrix} 3.50 & 6.00 \end{bmatrix}\begin{bmatrix} 125 & 100 & 75 \\ 100 & 175 & 125 \end{bmatrix} \\ &= \begin{bmatrix} \$1037.5 & \$1400 & \$1012.5 \end{bmatrix}\end{aligned}$$

The entries of BA represent the profit for both crops at each of the three outlets.

© 2013 Cengage Learning. All Rights Reserved. May not be scanned, copied or duplicated, or posted to a publicly accessible website, in whole or in part.

81. $P^2 = \begin{bmatrix} 0.6 & 0.1 & 0.1 \\ 0.2 & 0.7 & 0.1 \\ 0.2 & 0.2 & 0.8 \end{bmatrix}\begin{bmatrix} 0.6 & 0.1 & 0.1 \\ 0.2 & 0.7 & 0.1 \\ 0.2 & 0.2 & 0.8 \end{bmatrix} = \begin{bmatrix} 0.40 & 0.15 & 0.15 \\ 0.28 & 0.53 & 0.17 \\ 0.32 & 0.32 & 0.68 \end{bmatrix}$

This product represents the changes in party affiliation after *two* elections.

83. $AB = \left[\begin{array}{cc|cc} 1 & 2 & 0 & 0 \\ 0 & 1 & 0 & 0 \\ \hline 0 & 0 & 2 & 1 \end{array}\right]\left[\begin{array}{cc|c} 1 & 2 & 0 \\ -1 & 1 & 0 \\ \hline 0 & 0 & 1 \\ 0 & 0 & 3 \end{array}\right] = \left[\begin{array}{cc|c} -1 & 4 & 0 \\ -1 & 1 & 0 \\ \hline 0 & 0 & 5 \end{array}\right]$

85. (a) True. On page 43, "... for the product of two matrices to be defined, the number of columns of the first matrix must equal the number of rows of the second matrix."

(b) True. On page 46, "... the system $A\mathbf{x} = \mathbf{b}$ is consistent if and only if $\mathbf{b}$ can be expressed as ... a linear combination, where the coefficients of the linear combination are a solution of the system."

87. (a) $AT = \begin{bmatrix} 0 & -1 \\ 1 & 0 \end{bmatrix}\begin{bmatrix} 1 & 2 & 3 \\ 1 & 4 & 2 \end{bmatrix} = \begin{bmatrix} -1 & -4 & -2 \\ 1 & 2 & 3 \end{bmatrix}$

$AAT = \begin{bmatrix} 0 & -1 \\ 1 & 0 \end{bmatrix}\begin{bmatrix} -1 & -4 & -2 \\ 1 & 2 & 3 \end{bmatrix} = \begin{bmatrix} -1 & -2 & -3 \\ -1 & -4 & -2 \end{bmatrix}$

Triangle associated with T

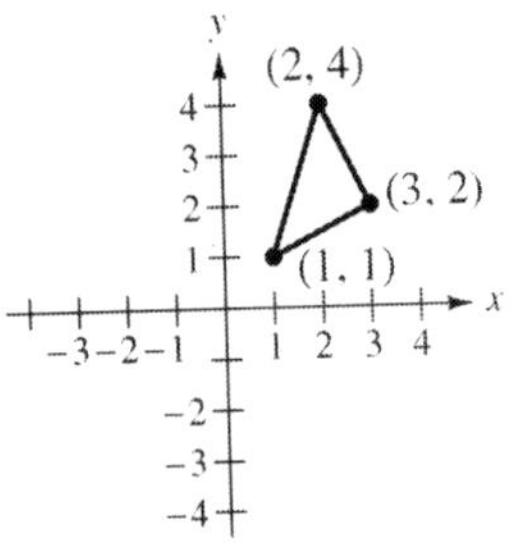

Triangle associated with AT

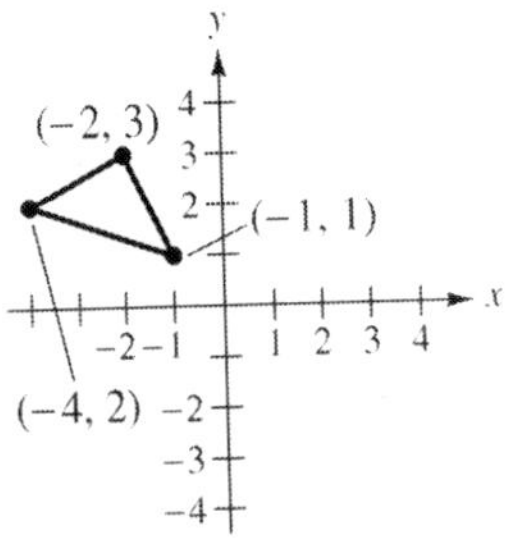

Triangle associated with AAT

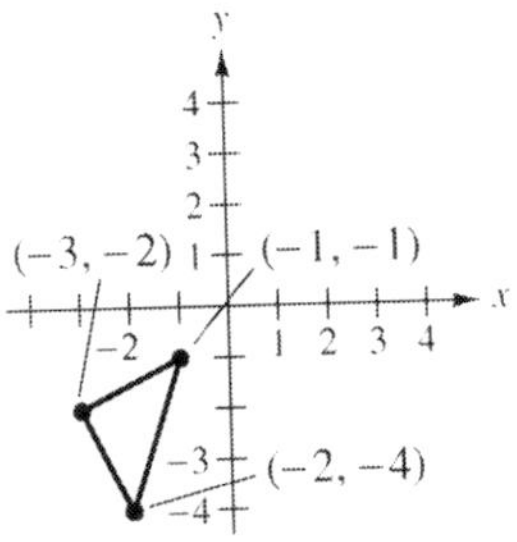

The transformation matrix A rotates the triangle about the origin in a counterclockwise direction through $90°$.

(b) Given the triangle associated with AAT, the transformation that would produce the triangle associated with AT would be a rotation about the origin of $90°$ in a clockwise direction. Another such rotation would produce the triangle associated with T.

Section 2.2 Properties of Matrix Operations

1. $\begin{bmatrix} -5 & 0 \\ 3 & -6 \end{bmatrix} + \begin{bmatrix} 7 & 1 \\ -2 & -1 \end{bmatrix} + \begin{bmatrix} -10 & -8 \\ 14 & 6 \end{bmatrix} = \begin{bmatrix} -5 + 7 + (-10) & 0 + 1 + (-8) \\ 3 + (-2) + 14 & -6 + (-1) + 6 \end{bmatrix} = \begin{bmatrix} -8 & -7 \\ 15 & -1 \end{bmatrix}$

3. $4\left(\begin{bmatrix} -4 & 0 & 1 \\ 0 & 2 & 3 \end{bmatrix} - \begin{bmatrix} 2 & 1 & -2 \\ 3 & -6 & 0 \end{bmatrix}\right) = 4\begin{bmatrix} -6 & -1 & 3 \\ -3 & 8 & 3 \end{bmatrix} = \begin{bmatrix} -24 & -4 & 12 \\ -12 & 32 & 12 \end{bmatrix}$

5. $-3\left(\begin{bmatrix} 0 & -3 \\ 7 & 2 \end{bmatrix} + \begin{bmatrix} -6 & 3 \\ 8 & 1 \end{bmatrix}\right) - 2\begin{bmatrix} 4 & -4 \\ 7 & -9 \end{bmatrix} = -3\begin{bmatrix} -6 & 0 \\ 15 & 3 \end{bmatrix} - \begin{bmatrix} 8 & -8 \\ 14 & -18 \end{bmatrix} = \begin{bmatrix} 18 & 0 \\ -45 & -9 \end{bmatrix} - \begin{bmatrix} 8 & -8 \\ 14 & -18 \end{bmatrix} = \begin{bmatrix} 10 & 8 \\ -59 & 9 \end{bmatrix}$

7. $3\begin{bmatrix} 1 & 2 \\ 3 & 4 \end{bmatrix} + (-4)\begin{bmatrix} 0 & 1 \\ -1 & 2 \end{bmatrix} = \begin{bmatrix} 3 & 6 \\ 9 & 12 \end{bmatrix} + \begin{bmatrix} 0 & -4 \\ 4 & -8 \end{bmatrix} = \begin{bmatrix} 3 & 2 \\ 13 & 4 \end{bmatrix}$

9. $ab(B) = (3)(-4)\begin{bmatrix} 0 & 1 \\ -1 & 2 \end{bmatrix} = (-12)\begin{bmatrix} 0 & 1 \\ -1 & 2 \end{bmatrix} = \begin{bmatrix} 0 & -12 \\ 12 & -24 \end{bmatrix}$

© 2013 Cengage Learning. All Rights Reserved. May not be scanned, copied or duplicated, or posted to a publicly accessible website, in whole or in part.

11. $[3-(-4)]\left(\begin{bmatrix}1&2\\3&4\end{bmatrix}-\begin{bmatrix}0&1\\-1&2\end{bmatrix}\right)=7\begin{bmatrix}1&1\\4&2\end{bmatrix}=\begin{bmatrix}7&7\\28&14\end{bmatrix}$

13. (a)

$$3X+2A=B$$
$$3X+\begin{bmatrix}-8&0\\2&-10\\-6&4\end{bmatrix}=\begin{bmatrix}1&2\\-2&1\\4&4\end{bmatrix}$$
$$3X=\begin{bmatrix}9&2\\-4&11\\10&0\end{bmatrix}$$
$$X=\begin{bmatrix}3&\frac{2}{3}\\-\frac{4}{3}&\frac{11}{3}\\\frac{10}{3}&0\end{bmatrix}$$

(b)

$$2A-5B=3X$$
$$\begin{bmatrix}-8&0\\2&-10\\-6&4\end{bmatrix}-\begin{bmatrix}5&10\\-10&5\\20&20\end{bmatrix}=3X$$
$$\begin{bmatrix}-13&-10\\12&-15\\-26&-16\end{bmatrix}=3X$$
$$\begin{bmatrix}-\frac{13}{3}&-\frac{10}{3}\\4&-5\\-\frac{26}{3}&-\frac{16}{3}\end{bmatrix}=X$$

(c)

$$X-3A+2B=O$$
$$X=3A-2B$$
$$=\begin{bmatrix}-12&0\\3&-15\\-9&6\end{bmatrix}-\begin{bmatrix}2&4\\-4&2\\8&8\end{bmatrix}$$
$$=\begin{bmatrix}-14&-4\\7&-17\\-17&-2\end{bmatrix}$$

(d)

$$6X-4A-3B=0$$
$$6X=4A+3B$$
$$6X=\begin{bmatrix}-16&0\\4&-20\\-12&8\end{bmatrix}+\begin{bmatrix}3&6\\-6&3\\12&12\end{bmatrix}$$
$$6X=\begin{bmatrix}-13&6\\-2&-17\\0&20\end{bmatrix}$$
$$X=\begin{bmatrix}-\frac{13}{6}&1\\-\frac{1}{3}&-\frac{17}{6}\\0&\frac{10}{3}\end{bmatrix}$$

15. $B(CA)=\begin{bmatrix}1&3\\-1&2\end{bmatrix}\left(\begin{bmatrix}0&1\\-1&0\end{bmatrix}\begin{bmatrix}1&2&3\\0&1&-1\end{bmatrix}\right)=\begin{bmatrix}1&3\\-1&2\end{bmatrix}\begin{bmatrix}0&1&-1\\-1&-2&-3\end{bmatrix}=\begin{bmatrix}-3&-5&-10\\-2&-5&-5\end{bmatrix}$

17. $\left(\begin{bmatrix}1&3\\-1&2\end{bmatrix}+\begin{bmatrix}0&1\\-1&0\end{bmatrix}\right)\begin{bmatrix}1&2&3\\0&1&-1\end{bmatrix}=\begin{bmatrix}1&4\\-2&2\end{bmatrix}\begin{bmatrix}1&2&3\\0&1&-1\end{bmatrix}=\begin{bmatrix}1&6&-1\\-2&-2&-8\end{bmatrix}$

19. $\left(-2\begin{bmatrix}1&3\\-1&2\end{bmatrix}\right)\left(\begin{bmatrix}0&1\\-1&0\end{bmatrix}+\begin{bmatrix}0&1\\-1&0\end{bmatrix}\right)=\begin{bmatrix}-2&-6\\2&-4\end{bmatrix}\begin{bmatrix}0&2\\-2&0\end{bmatrix}=\begin{bmatrix}12&-4\\8&4\end{bmatrix}$

21. (a) $(AB)C=\left(\begin{bmatrix}1&2\\3&4\end{bmatrix}\begin{bmatrix}0&1\\2&3\end{bmatrix}\right)\begin{bmatrix}3&0\\0&1\end{bmatrix}=\begin{bmatrix}4&7\\8&15\end{bmatrix}\begin{bmatrix}3&0\\0&1\end{bmatrix}=\begin{bmatrix}12&7\\24&15\end{bmatrix}$

(b) $A(BC)=\begin{bmatrix}1&2\\3&4\end{bmatrix}\left(\begin{bmatrix}0&1\\2&3\end{bmatrix}\begin{bmatrix}3&0\\0&1\end{bmatrix}\right)=\begin{bmatrix}1&2\\3&4\end{bmatrix}\begin{bmatrix}0&1\\6&3\end{bmatrix}=\begin{bmatrix}12&7\\24&15\end{bmatrix}$

23. (a) $AB=\begin{bmatrix}-2&1\\0&3\end{bmatrix}\begin{bmatrix}4&0\\-1&2\end{bmatrix}=\begin{bmatrix}-9&2\\-3&6\end{bmatrix}$

$BA=\begin{bmatrix}4&0\\-1&2\end{bmatrix}\begin{bmatrix}-2&1\\0&3\end{bmatrix}=\begin{bmatrix}-8&4\\2&5\end{bmatrix}\neq AB$

25. $AC=\begin{bmatrix}0&1\\0&1\end{bmatrix}\begin{bmatrix}2&3\\2&3\end{bmatrix}=\begin{bmatrix}2&3\\2&3\end{bmatrix}=\begin{bmatrix}1&0\\1&0\end{bmatrix}\begin{bmatrix}2&3\\2&3\end{bmatrix}=BC$, but $A\neq B$.

© 2013 Cengage Learning. All Rights Reserved. May not be scanned, copied or duplicated, or posted to a publicly accessible website, in whole or in part.

27. $AB = \begin{bmatrix} 3 & 3 \\ 4 & 4 \end{bmatrix}\begin{bmatrix} 1 & -1 \\ -1 & 1 \end{bmatrix} = \begin{bmatrix} 0 & 0 \\ 0 & 0 \end{bmatrix}$, but $A \neq 0$ and $B \neq 0$.

29. $IA = \begin{bmatrix} 1 & 0 \\ 0 & 1 \end{bmatrix}\begin{bmatrix} 1 & 2 \\ 0 & -1 \end{bmatrix} = \begin{bmatrix} 1 & 2 \\ 0 & -1 \end{bmatrix}$

31. $$\begin{bmatrix} 1 & 2 \\ 0 & -1 \end{bmatrix}\left(\begin{bmatrix} 1 & 0 \\ 0 & 1 \end{bmatrix} + \begin{bmatrix} 1 & 2 \\ 0 & -1 \end{bmatrix}\right) = \begin{bmatrix} 1 & 2 \\ 0 & -1 \end{bmatrix}\begin{bmatrix} 2 & 2 \\ 0 & 0 \end{bmatrix} = \begin{bmatrix} 2 & 2 \\ 0 & 0 \end{bmatrix}$$

33. $A^2 = \begin{bmatrix} 1 & 2 \\ 0 & -1 \end{bmatrix}\begin{bmatrix} 1 & 2 \\ 0 & -1 \end{bmatrix} = \begin{bmatrix} 1 & 0 \\ 0 & 1 \end{bmatrix} = I$

35. In general, $AB \neq BA$ for matrices. So,

$$(A + B)(A - B) = A^2 + BA - AB - B^2 \neq A^2 - B^2.$$

37. $D^T = \begin{bmatrix} 1 & -2 \\ -3 & 4 \\ 5 & -1 \end{bmatrix}^T = \begin{bmatrix} 1 & -3 & 5 \\ -2 & 4 & -1 \end{bmatrix}$

39. $$(AB)^T = \left(\begin{bmatrix} -1 & 1 & -2 \\ 2 & 0 & 1 \end{bmatrix}\begin{bmatrix} -3 & 0 \\ 1 & 2 \\ 1 & -1 \end{bmatrix}\right)^T = \begin{bmatrix} 2 & 4 \\ -5 & -1 \end{bmatrix}^T = \begin{bmatrix} 2 & -5 \\ 4 & -1 \end{bmatrix}$$

$$B^T A^T = \begin{bmatrix} -3 & 0 \\ 1 & 2 \\ 1 & -1 \end{bmatrix}^T \begin{bmatrix} -1 & 1 & -2 \\ 2 & 0 & 1 \end{bmatrix}^T = \begin{bmatrix} -3 & 1 & 1 \\ 0 & 2 & -1 \end{bmatrix}\begin{bmatrix} -1 & 2 \\ 1 & 0 \\ -2 & 1 \end{bmatrix} = \begin{bmatrix} 2 & -5 \\ 4 & -1 \end{bmatrix}$$

41. $$(AB)^T = \left(\begin{bmatrix} 2 & 1 \\ 0 & 1 \\ -2 & 1 \end{bmatrix}\begin{bmatrix} 2 & 3 & 1 \\ 0 & 4 & -1 \end{bmatrix}\right)^T = \begin{bmatrix} 4 & 10 & 1 \\ 0 & 4 & -1 \\ -4 & -2 & -3 \end{bmatrix}^T = \begin{bmatrix} 4 & 0 & -4 \\ 10 & 4 & -2 \\ 1 & -1 & -3 \end{bmatrix}$$

$$B^T A^T = \begin{bmatrix} 2 & 3 & 1 \\ 0 & 4 & -1 \end{bmatrix}^T \begin{bmatrix} 2 & 1 \\ 0 & 1 \\ -2 & 1 \end{bmatrix}^T = \begin{bmatrix} 2 & 0 \\ 3 & 4 \\ 1 & -1 \end{bmatrix}\begin{bmatrix} 2 & 0 & -2 \\ 1 & 1 & 1 \end{bmatrix} = \begin{bmatrix} 4 & 0 & -4 \\ 10 & 4 & -2 \\ 1 & -1 & -3 \end{bmatrix}$$

43. (a) $$A^T A = \begin{bmatrix} 4 & 0 \\ 2 & 2 \\ 1 & -1 \end{bmatrix}\begin{bmatrix} 4 & 2 & 1 \\ 0 & 2 & -1 \end{bmatrix} = \begin{bmatrix} 16 & 8 & 4 \\ 8 & 8 & 0 \\ 4 & 0 & 2 \end{bmatrix}$$

(b) $$AA^T = \begin{bmatrix} 4 & 2 & 1 \\ 0 & 2 & -1 \end{bmatrix}\begin{bmatrix} 4 & 0 \\ 2 & 2 \\ 1 & -1 \end{bmatrix} = \begin{bmatrix} 21 & 3 \\ 3 & 5 \end{bmatrix}$$

45. (a) $$A^T A = \begin{bmatrix} 0 & 8 & -2 & 0 \\ -4 & 4 & 3 & 0 \\ 3 & 0 & 5 & -3 \\ 2 & 1 & 1 & 2 \end{bmatrix}\begin{bmatrix} 0 & -4 & 3 & 2 \\ 8 & 4 & 0 & 1 \\ -2 & 3 & 5 & 1 \\ 0 & 0 & -3 & 2 \end{bmatrix} = \begin{bmatrix} 68 & 26 & -10 & 6 \\ 26 & 41 & 3 & -1 \\ -10 & 3 & 43 & 5 \\ 6 & -1 & 5 & 10 \end{bmatrix}$$

(b) $$AA^T = \begin{bmatrix} 0 & -4 & 3 & 2 \\ 8 & 4 & 0 & 1 \\ -2 & 3 & 5 & 1 \\ 0 & 0 & -3 & 2 \end{bmatrix}\begin{bmatrix} 0 & 8 & -2 & 0 \\ -4 & 4 & 3 & 0 \\ 3 & 0 & 5 & -3 \\ 2 & 1 & 1 & 2 \end{bmatrix} = \begin{bmatrix} 29 & -14 & 5 & -5 \\ -14 & 81 & -3 & 2 \\ 5 & -3 & 39 & -13 \\ -5 & 2 & -13 & 13 \end{bmatrix}$$

47. (a) True. See Theorem 2.1, part 1.

(b) True. See Theorem 2.3, part 1.

(c) False. See Theorem 2.6, part 4, or Example 9.

(d) True. See Example 10.

© 2013 Cengage Learning. All Rights Reserved. May not be scanned, copied or duplicated, or posted to a publicly accessible website, in whole or in part.

49. (a) $aX + bY = a\begin{bmatrix}1\\0\\1\end{bmatrix} + b\begin{bmatrix}1\\1\\0\end{bmatrix} = \begin{bmatrix}2\\-1\\3\end{bmatrix}$

This matrix equation yields the linear system

$$\begin{aligned} a + b &= 2 \\ b &= -1 \\ a &= 3. \end{aligned}$$

The only solution to this system is: $a = 3$ and $b = -1$.

(b) $aX + bY = a\begin{bmatrix}1\\0\\1\end{bmatrix} + b\begin{bmatrix}1\\1\\0\end{bmatrix} = \begin{bmatrix}1\\1\\1\end{bmatrix}$

This matrix equation yields the linear system

$$\begin{aligned} a + b &= 1 \\ b &= 1 \\ a &= 1. \end{aligned}$$

The system is inconsistent. So, no values of a and b will satisfy the equation.

(c) $aX + bY + cW = a\begin{bmatrix}1\\0\\1\end{bmatrix} + b\begin{bmatrix}1\\1\\0\end{bmatrix} + c\begin{bmatrix}1\\1\\1\end{bmatrix} = \begin{bmatrix}0\\0\\0\end{bmatrix}$

This matrix equation yields the linear system

$$\begin{aligned} a + b + c &= 0 \\ b + c &= 0 \\ a + c &= 0. \end{aligned}$$

Then $a = -c$, so $b = 0$. Then $c = 0$, so $a = b = c = 0$.

(d) $aX + bY + cZ = a\begin{bmatrix}1\\0\\1\end{bmatrix} + b\begin{bmatrix}1\\1\\0\end{bmatrix} + c\begin{bmatrix}2\\-1\\3\end{bmatrix} = \begin{bmatrix}0\\0\\0\end{bmatrix}$

This matrix equation yields the linear system

$$\begin{aligned} a + b + 2c &= 0 \\ b - c &= 0 \\ a + 3c &= 0. \end{aligned}$$

Using Gauss-Jordan elimination, the solution is $a = -3t$, $b = t$, and $c = t$, where t is any real number.

If $t = 1$, then $a = -3$, $b = 1$, and $c = 1$.

51. $A^{19} = \begin{bmatrix}1^{19} & 0 & 0\\0 & (-1)^{19} & 0\\0 & 0 & 1^{19}\end{bmatrix} = \begin{bmatrix}1 & 0 & 0\\0 & -1 & 0\\0 & 0 & 1\end{bmatrix}$

53. There are four possibilities, such that $A^2 = B$, namely

$$A = \begin{bmatrix}3 & 0\\0 & 2\end{bmatrix}, A = \begin{bmatrix}-3 & 0\\0 & 2\end{bmatrix}, A = \begin{bmatrix}3 & 0\\0 & -2\end{bmatrix}, A = \begin{bmatrix}-3 & 0\\0 & -2\end{bmatrix}.$$

55.
$$\begin{aligned} f(A) &= \begin{bmatrix}2 & 0\\4 & 5\end{bmatrix}^2 - 5\begin{bmatrix}2 & 0\\4 & 5\end{bmatrix} + 2\begin{bmatrix}1 & 0\\0 & 1\end{bmatrix} \\ &= \begin{bmatrix}2 & 0\\4 & 5\end{bmatrix}\begin{bmatrix}2 & 0\\4 & 5\end{bmatrix} - \begin{bmatrix}10 & 0\\20 & 25\end{bmatrix} + \begin{bmatrix}2 & 0\\0 & 2\end{bmatrix} \\ &= \begin{bmatrix}4 & 0\\28 & 25\end{bmatrix} + \begin{bmatrix}-8 & 0\\-20 & -23\end{bmatrix} \\ &= \begin{bmatrix}-4 & 0\\8 & 2\end{bmatrix} \end{aligned}$$

57.
$$\begin{aligned} A + (B + C) &= [a_{ij}] + ([b_{ij}] + [c_{ij}]) \\ &= ([a_{ij}] + [b_{ij}]) + [c_{ij}] \\ &= (A + B) + C \end{aligned}$$

59. $1A = 1[a_{ij}] = [1a_{ij}] = [a_{ij}] = A$

61. (1) $A + O_{mn} = [a_{ij}] + [0] = [a_{ij} + 0] = [a_{ij}] = A$

(2)
$$\begin{aligned} A + (-A) &= [a_{ij}] + [-a_{ij}] \\ &= [a_{ij} + (-a_{ij})] \\ &= [0] \\ &= O_{mn} \end{aligned}$$

(3) Let $cA = O_{mn}$ and suppose $c \neq 0$. Then,

$O_{mn} = cA \Rightarrow c^{-1}O_{mn} = c^{-1}(cA) = (c^{-1}c)A = A$

and so, $A = O_{mn}$. If $c = 0$, then

$OA = [0 \cdot a_{cj}] = [0] = O_{mn}$.

63. (1) The entry in the ith row and jth column of AI_n is

$a_{i1}0 + \cdots + a_{ij}1 + \cdots + a_{in}0 = a_{ij}$.

(2) The entry in the ith row and jth column of I_mA is

$0a_{1j} + \cdots + 1a_{ij} + \cdots + 0a_{nj} = a_{ij}$.

65. $(AA^T)^T = (A^T)^T A^T = AA^T$, which implies that AA^T is symmetric.

Similarly, $(A^TA)^T = A^T(A^T)^T = A^TA$, which implies that A^TA is symmetric.

67. Because $A^T = \begin{bmatrix}0 & -2\\2 & 0\end{bmatrix} = -\begin{bmatrix}0 & 2\\-2 & 0\end{bmatrix} = -A$, the matrix is skew-symmetric.

© 2013 Cengage Learning. All Rights Reserved. May not be scanned, copied or duplicated, or posted to a publicly accessible website, in whole or in part.

69. Because $A^T = \begin{bmatrix} 0 & 2 & 1 \\ 2 & 0 & 3 \\ 1 & 3 & 0 \end{bmatrix} = A$, the matrix is symmetric.

71. Because $A^T = -A$, the diagonal element a_{ii} satisfies $a_{ii} = -a_{ii}$, or $a_{ii} = 0$.

73. (a) If A is a square matrix of order n, then A^T is a square matrix of order n. Let

$$A = \begin{bmatrix} a_{11} & a_{12} & a_{13} & \cdots & a_{1n} \\ a_{21} & a_{22} & a_{23} & \cdots & a_{2n} \\ a_{31} & a_{32} & a_{33} & \cdots & a_{3n} \\ \vdots & \vdots & \vdots & \vdots & \vdots \\ a_{n1} & a_{n2} & a_{n3} & \cdots & a_{nn} \end{bmatrix}. \text{ Then } A^T = \begin{bmatrix} a_{11} & a_{21} & a_{31} & \cdots & a_{n1} \\ a_{12} & a_{22} & a_{32} & \cdots & a_{n2} \\ a_{13} & a_{23} & a_{33} & \cdots & a_{n3} \\ \vdots & \vdots & \vdots & & \vdots \\ a_{1n} & a_{2n} & a_{3n} & \cdots & a_{nn} \end{bmatrix}.$$

Now form the sum and scalar multiple, $\frac{1}{2}(A + A^T)$.

$$\tfrac{1}{2}(A + A^T) = \tfrac{1}{2}\begin{bmatrix} a_{11} + a_{11} & a_{12} + a_{21} & a_{13} + a_{31} & \cdots & a_{1n} + a_{n1} \\ a_{21} + a_{12} & a_{22} + a_{22} & a_{23} + a_{32} & \cdots & a_{2n} + a_{n2} \\ a_{31} + a_{13} & a_{32} + a_{23} & a_{33} + a_{33} & \cdots & a_{3n} + a_{n3} \\ \vdots & \vdots & \vdots & & \vdots \\ a_{n1} + a_{1n} & a_{n2} + a_{2n} & a_{n3} + a_{3n} & \cdots & a_{nn} + a_{nn} \end{bmatrix}$$

Note that, for the matrix $A + A^T$, the ijth entry is equal to the jith entry for all $i \neq j$. So, $\frac{1}{2}(A + A^T)$ is symmetric.

(b) Use the matrices in part (a).

$$\tfrac{1}{2}(A - A^T) = \tfrac{1}{2}\begin{bmatrix} a_{11} - a_{11} & a_{12} - a_{21} & a_{13} - a_{31} & \cdots & a_{1n} - a_{n1} \\ a_{21} - a_{12} & a_{22} - a_{22} & a_{23} - a_{32} & \cdots & a_{2n} - a_{n2} \\ a_{31} - a_{13} & a_{32} - a_{23} & a_{33} - a_{33} & \cdots & a_{3n} - a_{n3} \\ \vdots & \vdots & \vdots & & \vdots \\ a_{n1} - a_{1n} & a_{n2} - a_{2n} & a_{n3} - a_{3n} & \cdots & a_{nn} - a_{nn} \end{bmatrix}$$

Note that, for the matrix $A - A^T$, the ijth entry is the negative of the jith entry for all $i \neq j$. So, $\frac{1}{2}(A - A^T)$ is skew-symmetric.

(c) For any square matrix A of order n, let $B = \frac{1}{2}(A + A^T)$ and $C = \frac{1}{2}(A - A^T)$. By part (a), B is symmetric. By part (b), C is skew-symmetric. And $A = \frac{1}{2}(A + A^T) + \frac{1}{2}(A - A^T) = B + C$ as desired.

(d) For the given A, $A^T = \begin{bmatrix} 2 & -3 & 4 \\ 5 & 6 & 1 \\ 3 & 0 & 1 \end{bmatrix}$. Using the notation of part (c),

$$B + C = \tfrac{1}{2}(A + A^T) + \tfrac{1}{2}(A - A^T) = \begin{bmatrix} 2 & 1 & \frac{7}{2} \\ 1 & 6 & \frac{1}{2} \\ \frac{7}{2} & \frac{1}{2} & 1 \end{bmatrix} + \begin{bmatrix} 0 & 4 & -\frac{1}{2} \\ -4 & 0 & -\frac{1}{2} \\ \frac{1}{2} & \frac{1}{2} & 0 \end{bmatrix} = \begin{bmatrix} 2 & 5 & 3 \\ -3 & 6 & 0 \\ 4 & 1 & 1 \end{bmatrix} = A.$$

75. (a) An example of a 2×2 matrix of the given form is $A_2 = \begin{bmatrix} 0 & 1 \\ 0 & 0 \end{bmatrix}$.

An example of a 3×3 matrix of the given form is $A_3 = \begin{bmatrix} 0 & 1 & 2 \\ 0 & 0 & 3 \\ 0 & 0 & 0 \end{bmatrix}$.

© 2013 Cengage Learning. All Rights Reserved. May not be scanned, copied or duplicated, or posted to a publicly accessible website, in whole or in part.

(b) $A_2^2 = \begin{bmatrix} 0 & 0 \\ 0 & 0 \end{bmatrix}$

$A_3^2 = \begin{bmatrix} 0 & 0 & 3 \\ 0 & 0 & 0 \\ 0 & 0 & 0 \end{bmatrix}$ and $A_3^3 = \begin{bmatrix} 0 & 0 & 0 \\ 0 & 0 & 0 \\ 0 & 0 & 0 \end{bmatrix}$

(c) The conjecture is that if A is a 4×4 matrix of the given form, then A^4 is the 4×4 zero matrix. A graphing utility shows this to be true.

(d) If A is an $n \times n$ matrix of the given form, then A^n is the $n \times n$ zero matrix.

Section 2.3 The Inverse of a Matrix

1. $AB = \begin{bmatrix} 2 & 1 \\ 5 & 3 \end{bmatrix}\begin{bmatrix} 3 & -1 \\ -5 & 2 \end{bmatrix} = \begin{bmatrix} 6-5 & -2+2 \\ 15-15 & -5+6 \end{bmatrix} = \begin{bmatrix} 1 & 0 \\ 0 & 1 \end{bmatrix}$

$BA = \begin{bmatrix} 3 & -1 \\ -5 & 2 \end{bmatrix}\begin{bmatrix} 2 & 1 \\ 5 & 3 \end{bmatrix} = \begin{bmatrix} 6-5 & 3-3 \\ -10+10 & -5+6 \end{bmatrix} = \begin{bmatrix} 1 & 0 \\ 0 & 1 \end{bmatrix}$

3. $AB = \begin{bmatrix} 1 & 2 \\ 3 & 4 \end{bmatrix}\begin{bmatrix} -2 & 1 \\ \frac{3}{2} & -\frac{1}{2} \end{bmatrix} = \begin{bmatrix} 1 & 0 \\ 0 & 1 \end{bmatrix}$

$BA = \begin{bmatrix} -2 & 1 \\ \frac{3}{2} & -\frac{1}{2} \end{bmatrix}\begin{bmatrix} 1 & 2 \\ 3 & 4 \end{bmatrix} = \begin{bmatrix} 1 & 0 \\ 0 & 1 \end{bmatrix}$

5. $AB = \begin{bmatrix} -2 & 2 & 3 \\ 1 & -1 & 0 \\ 0 & 1 & 4 \end{bmatrix}\left(\frac{1}{3}\right)\begin{bmatrix} -4 & -5 & 3 \\ -4 & -8 & 3 \\ 1 & 2 & 0 \end{bmatrix} = \left(\frac{1}{3}\right)\begin{bmatrix} 3 & 0 & 0 \\ 0 & 3 & 0 \\ 0 & 0 & 3 \end{bmatrix} = \begin{bmatrix} 1 & 0 & 0 \\ 0 & 1 & 0 \\ 0 & 0 & 1 \end{bmatrix}$

$BA = \left(\frac{1}{3}\right)\begin{bmatrix} -4 & -5 & 3 \\ -4 & -8 & 3 \\ 1 & 2 & 0 \end{bmatrix}\begin{bmatrix} -2 & 2 & 3 \\ 1 & -1 & 0 \\ 0 & 1 & 4 \end{bmatrix} = \left(\frac{1}{3}\right)\begin{bmatrix} 3 & 0 & 0 \\ 0 & 3 & 0 \\ 0 & 0 & 3 \end{bmatrix} = \begin{bmatrix} 1 & 0 & 0 \\ 0 & 1 & 0 \\ 0 & 0 & 1 \end{bmatrix}$

7. $\begin{bmatrix} A & I \end{bmatrix} = \begin{bmatrix} 2 & 0 & 1 & 0 \\ 0 & 3 & 0 & 1 \end{bmatrix}$

$\begin{matrix} \frac{1}{2}R_1 \rightarrow \\ \frac{1}{3}R_2 \rightarrow \end{matrix} \begin{bmatrix} 1 & 0 & \frac{1}{2} & 0 \\ 0 & 1 & 0 & \frac{1}{3} \end{bmatrix} = \begin{bmatrix} I & A^{-1} \end{bmatrix}$

$A^{-1} = \begin{bmatrix} \frac{1}{2} & 0 \\ 0 & \frac{1}{3} \end{bmatrix}$

9. Use the formula $A^{-1} = \dfrac{1}{ad - bc}\begin{bmatrix} d & -b \\ -c & a \end{bmatrix}$, where

$A = \begin{bmatrix} a & b \\ c & d \end{bmatrix} = \begin{bmatrix} 1 & 2 \\ 3 & 7 \end{bmatrix}$. So, the inverse is

$A^{-1} = \dfrac{1}{(1)(7) - (2)(3)}\begin{bmatrix} 7 & -2 \\ -3 & 1 \end{bmatrix} = \begin{bmatrix} 7 & -2 \\ -3 & 1 \end{bmatrix}$.

11. Use the formula $A^{-1} = \dfrac{1}{ad - bc}\begin{bmatrix} d & -b \\ -c & a \end{bmatrix}$, where

$A = \begin{bmatrix} a & b \\ c & d \end{bmatrix} = \begin{bmatrix} -7 & 33 \\ 4 & -19 \end{bmatrix}$. So, the inverse is

$A^{-1} = \dfrac{1}{(-7)(-19) - 33(4)}\begin{bmatrix} -19 & -33 \\ -4 & -7 \end{bmatrix} = \begin{bmatrix} -19 & -33 \\ -4 & -7 \end{bmatrix}$.

13. Adjoin the identity matrix to form

$\begin{bmatrix} A & I \end{bmatrix} = \begin{bmatrix} 1 & 1 & 1 & 1 & 0 & 0 \\ 3 & 5 & 4 & 0 & 1 & 0 \\ 3 & 6 & 5 & 0 & 0 & 1 \end{bmatrix}$.

Using elementary row operations, rewrite this matrix in reduced row-echelon form.

$\begin{bmatrix} I & A^{-1} \end{bmatrix} = \begin{bmatrix} 1 & 0 & 0 & 1 & 1 & -1 \\ 0 & 1 & 0 & -3 & 2 & -1 \\ 0 & 0 & 1 & 3 & -3 & 2 \end{bmatrix}$

Therefore, the inverse is $A^{-1} = \begin{bmatrix} 1 & 1 & -1 \\ -3 & 2 & -1 \\ 3 & -3 & 2 \end{bmatrix}$.

© 2013 Cengage Learning. All Rights Reserved. May not be scanned, copied or duplicated, or posted to a publicly accessible website, in whole or in part.

15. Adjoin the identity matrix to form

$$[A \; I] = \begin{bmatrix} 1 & 2 & -1 & 1 & 0 & 0 \\ 3 & 7 & -10 & 0 & 1 & 0 \\ 7 & 16 & -21 & 0 & 0 & 1 \end{bmatrix}.$$

Using elementary row operations, you cannot form the identity matrix on the left side.

$$\begin{bmatrix} 1 & 0 & 13 & 0 & -16 & 7 \\ 0 & 1 & -7 & 0 & 7 & -3 \\ 0 & 0 & 0 & 1 & 2 & -1 \end{bmatrix}.$$

Therefore, the matrix is singular and has no inverse.

17. Adjoin the identity matrix to form

$$[A \; I] = \begin{bmatrix} 1 & 1 & 2 & 1 & 0 & 0 \\ 3 & 1 & 0 & 0 & 1 & 0 \\ -2 & 0 & 3 & 0 & 0 & 1 \end{bmatrix}.$$

Using elementary row operations, rewrite this matrix in reduced row-echelon form.

$$[I \; A^{-1}] = \begin{bmatrix} 1 & 0 & 0 & -\frac{3}{2} & \frac{3}{2} & 1 \\ 0 & 1 & 0 & \frac{9}{2} & -\frac{7}{2} & -3 \\ 0 & 0 & 1 & -1 & 1 & 1 \end{bmatrix}$$

Therefore, the inverse is

$$A^{-1} = \begin{bmatrix} -\frac{3}{2} & \frac{3}{2} & 1 \\ \frac{9}{2} & -\frac{7}{2} & -3 \\ -1 & 1 & 1 \end{bmatrix}.$$

19. Adjoin the identity matrix to form

$$[I \; A^{-1}] = \begin{bmatrix} 2 & 0 & 0 & 1 & 0 & 0 \\ 0 & 3 & 0 & 0 & 1 & 0 \\ 0 & 0 & 5 & 0 & 0 & 1 \end{bmatrix}.$$

Using elementary row operations, reduce the matrix as follows.

$$[A \; I] = \begin{bmatrix} 1 & 0 & 0 & \frac{1}{2} & 0 & 0 \\ 0 & 1 & 0 & 0 & \frac{1}{3} & 0 \\ 0 & 0 & 1 & 0 & 0 & \frac{1}{5} \end{bmatrix}$$

Therefore, the inverse is

$$A^{-1} = \begin{bmatrix} \frac{1}{2} & 0 & 0 \\ 0 & \frac{1}{3} & 0 \\ 0 & 0 & \frac{1}{5} \end{bmatrix}.$$

21. Adjoin the identity matrix to form

$$[I \; A^{-1}] = \begin{bmatrix} 0.6 & 0 & -0.3 & 1 & 0 & 0 \\ 0.7 & -1 & 0.2 & 0 & 1 & 0 \\ 1 & 0 & -0.9 & 0 & 0 & 1 \end{bmatrix}.$$

Using elementary row operations, reduce the matrix as follows.

$$[A \; I] = \begin{bmatrix} 1 & 0 & 0 & 3.75 & 0 & -1.25 \\ 0 & 1 & 0 & 3.458\overline{3} & -1 & -1.375 \\ 0 & 0 & 1 & 4.1\overline{6} & 0 & -2.5 \end{bmatrix}$$

Therefore, the inverse is $A^{-1} = \begin{bmatrix} 3.75 & 0 & -1.25 \\ 3.458\overline{3} & -1 & -1.375 \\ 4.1\overline{6} & 0 & -2.5 \end{bmatrix}.$

23. Adjoin the identity matrix to form

$$[A \; I] = \begin{bmatrix} 1 & 0 & 0 & 1 & 0 & 0 \\ 3 & 4 & 0 & 0 & 1 & 0 \\ 2 & 5 & 5 & 0 & 0 & 1 \end{bmatrix}.$$

Using elementary row operations, rewrite this matrix in reduced row-echelon form.

$$[I \; A^{-1}] = \begin{bmatrix} 1 & 0 & 0 & 1 & 0 & 0 \\ 0 & 1 & 0 & -\frac{3}{4} & \frac{1}{4} & 0 \\ 0 & 0 & 1 & \frac{7}{20} & -\frac{1}{4} & \frac{1}{5} \end{bmatrix}$$

Therefore, the inverse is $A^{-1} = \begin{bmatrix} 1 & 0 & 0 \\ -\frac{3}{4} & \frac{1}{4} & 0 \\ \frac{7}{20} & -\frac{1}{4} & \frac{1}{5} \end{bmatrix}.$

25. Adjoin the identity matrix to form

$$[A \; I] = \begin{bmatrix} -8 & 0 & 0 & 0 & 1 & 0 & 0 & 0 \\ 0 & 1 & 0 & 0 & 0 & 1 & 0 & 0 \\ 0 & 0 & 0 & 0 & 0 & 0 & 1 & 0 \\ 0 & 0 & 0 & -5 & 0 & 0 & 0 & 1 \end{bmatrix}.$$

Using elementary row operations, you cannot form the identity matrix on the left side. Therefore, the matrix A is singular and has no inverse.

27. Use a graphing utility or a computer software program.

The inverse is

$$A^{-1} = \begin{bmatrix} -24 & 7 & 1 & -2 \\ -10 & 3 & 0 & -1 \\ -29 & 7 & 3 & -2 \\ 12 & -3 & -1 & 1 \end{bmatrix}.$$

29. Using a graphing utility or a computer software program, you find that the matrix is singular and therefore has no inverse.

© 2013 Cengage Learning. All Rights Reserved. May not be scanned, copied or duplicated, or posted to a publicly accessible website, in whole or in part.

31. $A = \begin{bmatrix} 2 & 3 \\ -1 & 5 \end{bmatrix}$

$ad - bc = (2)(5) - (3)(-1) = 13$

$A^{-1} = \frac{1}{13}\begin{bmatrix} 5 & -3 \\ 1 & 2 \end{bmatrix} = \begin{bmatrix} \frac{5}{13} & -\frac{3}{13} \\ \frac{1}{13} & \frac{2}{13} \end{bmatrix}$

33. $A = \begin{bmatrix} -4 & -6 \\ 2 & 3 \end{bmatrix}$

$ad - bc = (-4)(3) - (-2)(-6) = 0$

Because $ad - bc = 0$, A^{-1} does not exist.

35. $A = \begin{bmatrix} \frac{7}{2} & -\frac{3}{4} \\ \frac{1}{5} & \frac{4}{5} \end{bmatrix}$

$ad - bc = \left(\frac{7}{2}\right)\left(\frac{4}{5}\right) - \left(-\frac{3}{4}\right)\left(\frac{1}{5}\right) = \frac{28}{10} + \frac{3}{20} = \frac{59}{20}$

$A^{-1} = \frac{20}{59}\begin{bmatrix} \frac{4}{5} & \frac{3}{4} \\ -\frac{1}{5} & \frac{7}{2} \end{bmatrix} = \begin{bmatrix} \frac{16}{59} & \frac{15}{59} \\ -\frac{4}{59} & \frac{70}{59} \end{bmatrix}$

37. $A^{-2} = \left(A^{-1}\right)^2 = \left(-\frac{1}{2}\begin{bmatrix} 3 & 2 \\ 1 & 0 \end{bmatrix}\right)^2 = \frac{1}{4}\begin{bmatrix} 11 & 6 \\ 3 & 2 \end{bmatrix}$

$A^{-2} = \left(A^2\right)^{-1} = \left(\begin{bmatrix} 2 & -6 \\ -3 & 11 \end{bmatrix}\right)^{-1} = \frac{1}{4}\begin{bmatrix} 11 & 6 \\ 3 & 2 \end{bmatrix}$

The results are equal.

39. $A^{-2} = \left(A^{-1}\right)^2 = \left(\begin{bmatrix} -\frac{1}{2} & 0 & 0 \\ 0 & 1 & 0 \\ 0 & 0 & \frac{1}{3} \end{bmatrix}\right)^2 = \begin{bmatrix} \frac{1}{4} & 0 & 0 \\ 0 & 1 & 0 \\ 0 & 0 & \frac{1}{9} \end{bmatrix}$

$A^{-2} = \left(A^2\right)^{-1} = \left(\begin{bmatrix} 4 & 0 & 0 \\ 0 & 1 & 0 \\ 0 & 0 & 9 \end{bmatrix}\right)^{-1} = \begin{bmatrix} \frac{1}{4} & 0 & 0 \\ 0 & 1 & 0 \\ 0 & 0 & \frac{1}{9} \end{bmatrix}$

The results are equal.

41. (a) $(AB)^{-1} = B^{-1}A^{-1} = \begin{bmatrix} 7 & -3 \\ 2 & 0 \end{bmatrix}\begin{bmatrix} 2 & 5 \\ -7 & 6 \end{bmatrix} = \begin{bmatrix} 35 & 17 \\ 4 & 10 \end{bmatrix}$

(b) $\left(A^T\right)^{-1} = \left(A^{-1}\right)^T = \begin{bmatrix} 2 & 5 \\ -7 & 6 \end{bmatrix}^T = \begin{bmatrix} 2 & -7 \\ 5 & 6 \end{bmatrix}$

(c) $(2A)^{-1} = \frac{1}{2}A^{-1} = \frac{1}{2}\begin{bmatrix} 2 & 5 \\ -7 & 6 \end{bmatrix} = \begin{bmatrix} 1 & \frac{5}{2} \\ -\frac{7}{2} & 3 \end{bmatrix}$

43. (a) $(AB)^{-1} = B^{-1}A^{-1} = \begin{bmatrix} 2 & 4 & \frac{5}{2} \\ -\frac{3}{4} & 2 & \frac{1}{4} \\ \frac{1}{4} & \frac{1}{2} & 2 \end{bmatrix}\begin{bmatrix} 1 & -\frac{1}{2} & \frac{3}{4} \\ \frac{3}{2} & \frac{1}{2} & -2 \\ \frac{1}{4} & 1 & \frac{1}{2} \end{bmatrix} = \begin{bmatrix} \frac{69}{8} & \frac{7}{2} & -\frac{21}{4} \\ \frac{37}{16} & \frac{13}{8} & -\frac{71}{16} \\ \frac{3}{2} & \frac{17}{8} & \frac{3}{16} \end{bmatrix} = \frac{1}{16}\begin{bmatrix} 138 & 56 & -84 \\ 37 & 26 & -71 \\ 24 & 34 & 3 \end{bmatrix}$

(b) $\left(A^T\right)^{-1} = \left(A^{-1}\right)^T = \begin{bmatrix} 1 & \frac{3}{2} & \frac{1}{4} \\ -\frac{1}{2} & \frac{1}{2} & 1 \\ \frac{3}{4} & -2 & \frac{1}{2} \end{bmatrix} = \frac{1}{4}\begin{bmatrix} 4 & 6 & 1 \\ -2 & 2 & 4 \\ 3 & -8 & 2 \end{bmatrix}$

(c) $(2A)^{-1} = \frac{1}{2}A^{-1} = \begin{bmatrix} \frac{1}{2} & -\frac{1}{4} & \frac{3}{8} \\ \frac{3}{4} & \frac{1}{4} & -1 \\ \frac{1}{8} & \frac{1}{2} & \frac{1}{4} \end{bmatrix} = \frac{1}{8}\begin{bmatrix} 4 & -2 & 3 \\ 6 & 2 & -8 \\ 1 & 4 & 2 \end{bmatrix}$

45. The coefficient matrix for each system is $A = \begin{bmatrix} 1 & 2 \\ 1 & -2 \end{bmatrix}$ and the formula for the inverse of a 2×2 matrix produces

$A^{-1} = \frac{1}{-2-2}\begin{bmatrix} -2 & -2 \\ -1 & 1 \end{bmatrix} = \begin{bmatrix} \frac{1}{2} & \frac{1}{2} \\ \frac{1}{4} & -\frac{1}{4} \end{bmatrix}.$

(a) $\mathbf{x} = A^{-1}\mathbf{b} = \begin{bmatrix} \frac{1}{2} & \frac{1}{2} \\ \frac{1}{4} & -\frac{1}{4} \end{bmatrix}\begin{bmatrix} -1 \\ 3 \end{bmatrix} = \begin{bmatrix} 1 \\ -1 \end{bmatrix}$

The solution is: $x = 1$ and $y = -1$.

(b) $\mathbf{x} = A^{-1}\mathbf{b} = \begin{bmatrix} \frac{1}{2} & \frac{1}{2} \\ \frac{1}{4} & -\frac{1}{4} \end{bmatrix}\begin{bmatrix} 10 \\ -6 \end{bmatrix} = \begin{bmatrix} 2 \\ 4 \end{bmatrix}$

The solution is: $x = 2$ and $y = 4$.

© 2013 Cengage Learning. All Rights Reserved. May not be scanned, copied or duplicated, or posted to a publicly accessible website, in whole or in part.

47. The coefficient matrix for each system is

$$A = \begin{bmatrix} 1 & 2 & 1 \\ 1 & 2 & -1 \\ 1 & -2 & 1 \end{bmatrix}.$$

Using the algorithm to invert a matrix, you find that the inverse is

$$A^{-1} = \begin{bmatrix} 0 & \frac{1}{2} & \frac{1}{2} \\ \frac{1}{4} & 0 & -\frac{1}{4} \\ \frac{1}{2} & -\frac{1}{2} & 0 \end{bmatrix}$$

(a) $\mathbf{x} = A^{-1}\mathbf{b} = \begin{bmatrix} 0 & \frac{1}{2} & \frac{1}{2} \\ \frac{1}{4} & 0 & -\frac{1}{4} \\ \frac{1}{2} & -\frac{1}{2} & 0 \end{bmatrix}\begin{bmatrix} 2 \\ 4 \\ -2 \end{bmatrix} = \begin{bmatrix} 1 \\ 1 \\ -1 \end{bmatrix}$

The solution is: $x_1 = 1$, $x_2 = 1$, and $x_3 = -1$.

(b) $\mathbf{x} = A^{-1}\mathbf{b} = \begin{bmatrix} 0 & \frac{1}{2} & \frac{1}{2} \\ \frac{1}{4} & 0 & -\frac{1}{4} \\ \frac{1}{2} & -\frac{1}{2} & 0 \end{bmatrix}\begin{bmatrix} 1 \\ 3 \\ -3 \end{bmatrix} = \begin{bmatrix} 0 \\ 1 \\ -1 \end{bmatrix}$

The solution is: $x_1 = 0$, $x_2 = 1$, and $x_3 = -1$.

49. Using a graphing utility or a computer software program, you have

$A\mathbf{x} = \mathbf{b}$

$$\mathbf{x} = A^{-1}\mathbf{b} = \begin{bmatrix} 0 \\ 1 \\ 2 \\ -1 \\ 0 \end{bmatrix} \text{ where}$$

$$A = \begin{bmatrix} 1 & 2 & -1 & 3 & -1 \\ 1 & -3 & 1 & 2 & -1 \\ 2 & 1 & 1 & -3 & 1 \\ 1 & -1 & 2 & 1 & -1 \\ 2 & 1 & -1 & 2 & 1 \end{bmatrix},\ \mathbf{x} = \begin{bmatrix} x_1 \\ x_2 \\ x_3 \\ x_4 \\ x_5 \end{bmatrix},\ \text{and } \mathbf{b} = \begin{bmatrix} -3 \\ -3 \\ 6 \\ 2 \\ -3 \end{bmatrix}.$$

The solution is: $x_1 = 0$, $x_2 = 1$, $x_3 = 2$, $x_4 = -1$, and $x_5 = 0$.

51. Using a graphing utility or a computer software program, you have

$A\mathbf{x} = \mathbf{b}$

$$\mathbf{x} = A^{-1}\mathbf{b} = \begin{bmatrix} 1 \\ -2 \\ 3 \\ 0 \\ 1 \\ -2 \end{bmatrix} \text{ where}$$

$$A = \begin{bmatrix} 2 & -3 & 1 & -2 & 1 & -4 \\ 3 & 1 & -4 & 1 & -1 & 2 \\ 4 & 1 & -3 & 4 & -1 & 2 \\ -5 & -1 & 4 & 2 & -5 & 3 \\ 1 & 1 & -3 & 4 & -3 & 1 \\ 3 & -1 & 2 & -3 & 2 & -6 \end{bmatrix},$$

$$\mathbf{x} = \begin{bmatrix} x_1 \\ x_2 \\ x_3 \\ x_4 \\ x_5 \\ x_6 \end{bmatrix},\ \text{and } \mathbf{b} = \begin{bmatrix} 20 \\ -16 \\ -12 \\ -2 \\ -15 \\ 25 \end{bmatrix}.$$

The solution is: $x_1 = 1$, $x_2 = -2$, $x_3 = 3$, $x_4 = 0$, $x_5 = 1$, and $x_6 = -2$.

53. Using the formula for the inverse of a 2×2 matrix, you have

$$A^{-1} = \frac{1}{2x - 9}\begin{bmatrix} -3 & -x \\ 2 & 3 \end{bmatrix}.$$

Letting $A^{-1} = A$, you find that $1/(2x - 9) = -1$.

So, $x = 4$.

55. The matrix $\begin{bmatrix} 4 & x \\ -2 & -3 \end{bmatrix}$ will be singular if $ad - bc = (4)(-3) - (x)(-2) = 0$, which implies that $2x = 12$ or $x = 6$.

57. First find $2A$.

$$2A = \left[(2A)^{-1}\right]^{-1} = \frac{1}{4 - 6}\begin{bmatrix} 4 & -2 \\ -3 & 1 \end{bmatrix} = \begin{bmatrix} -2 & 1 \\ \frac{3}{2} & -\frac{1}{2} \end{bmatrix}$$

Then divide by 2 to obtain

$$A = \tfrac{1}{2}(2A) = \frac{1}{2}\begin{bmatrix} -2 & 1 \\ \frac{3}{2} & -\frac{1}{2} \end{bmatrix} = \begin{bmatrix} -1 & \frac{1}{2} \\ \frac{3}{4} & -\frac{1}{4} \end{bmatrix}.$$

59. Using the formula for the inverse of a 2×2 matrix, you have

$$A^{-1} = \frac{1}{ad - bc}\begin{bmatrix} d & -b \\ -c & a \end{bmatrix} = \frac{1}{\sin^2\theta + \cos^2\theta}\begin{bmatrix} \sin\theta & -\cos\theta \\ \cos\theta & \sin\theta \end{bmatrix} = \begin{bmatrix} \sin\theta & -\cos\theta \\ \cos\theta & \sin\theta \end{bmatrix}.$$

© 2013 Cengage Learning. All Rights Reserved. May not be scanned, copied or duplicated, or posted to a publicly accessible website, in whole or in part.

61. Adjoin the identity matrix to form

$$[F \quad I] = \begin{bmatrix} 0.008 & 0.004 & 0.003 & 1 & 0 & 0 \\ 0.004 & 0.006 & 0.004 & 0 & 1 & 0 \\ 0.003 & 0.004 & 0.008 & 0 & 0 & 1 \end{bmatrix}.$$

Using elementary row operations, reduce the matrix as follows.

$$[I \quad F^{-1}] = \begin{bmatrix} 1 & 0 & 0 & 188.24 & -117.65 & -11.76 \\ 0 & 1 & 0 & -117.65 & 323.53 & -117.65 \\ 0 & 0 & 1 & -11.76 & -117.65 & 188.24 \end{bmatrix}$$

So, $F^{-1} = \begin{bmatrix} 188.24 & -117.65 & -11.76 \\ -117.65 & 323.53 & -117.65 \\ -11.76 & -117.65 & 188.24 \end{bmatrix}$ and

$$\mathbf{w} = F^{-1}\mathbf{d} = \begin{bmatrix} 188.24 & -117.65 & -11.76 \\ -117.65 & 323.53 & -117.65 \\ -11.76 & -117.65 & 188.24 \end{bmatrix}\begin{bmatrix} 0.585 \\ 0.640 \\ 0.835 \end{bmatrix} = \begin{bmatrix} 25 \\ 40 \\ 75 \end{bmatrix}.$$

63. (a) True. See Theorem 2.10, part 1.

(b) False. See Theorem 2.9.

(c) True. See "Finding the Inverse of a Matrix by Gauss-Jordan Elimination," part 2, on page 64.

65. Use mathematical induction. The property is clearly true if $k = 1$. Suppose the property is true for $k = n$, and consider the case for $k = n + 1$.

$$(A^{n+1})^{-1} = (AA^n)^{-1}$$

$$= (A^n)^{-1}A^{-1} = \underbrace{(A^{-1} \cdots A^{-1})}_{n \text{ times}}A^{-1} = \underbrace{A^{-1} \cdots A^{-1}}_{n+1 \text{ times}}$$

67. Let A be symmetric and nonsingular. Then $A^T = A$ and $(A^{-1})^T = (A^T)^{-1} = A^{-1}$. Therefore, A^{-1} is symmetric.

69.
$$\begin{aligned}(I - 2A)(I - 2A) &= I^2 - 2IA - 2AI + 4A^2 \\ &= I - 4A + 4A^2 \\ &= I - 4A + 4A \quad (\text{because } A = A^2) \\ &= I\end{aligned}$$

So, $(I - 2A)^{-1} = I - 2A$.

71. Because A is invertible, you can multiply both sides of the equation $AB = O$ by A^{-1} to obtain the following.

$$A^{-1}(AB) = A^{-1}O$$
$$(A^{-1}A)B = O$$
$$B = O$$

73. No. For instance, $\begin{bmatrix} 1 & 0 \\ 0 & 1 \end{bmatrix} + \begin{bmatrix} -1 & 0 \\ 0 & -1 \end{bmatrix} = \begin{bmatrix} 0 & 0 \\ 0 & 0 \end{bmatrix}$.

75. To find the inverses, take the reciprocals of the diagonal entries.

(a) $A^{-1} = \begin{bmatrix} -1 & 0 & 0 \\ 0 & \frac{1}{3} & 0 \\ 0 & 0 & \frac{1}{2} \end{bmatrix}$

(b) $A^{-1} = \begin{bmatrix} 2 & 0 & 0 \\ 0 & 3 & 0 \\ 0 & 0 & 4 \end{bmatrix}$

77. (a) Let $H = I - 2\mathbf{u}\mathbf{u}^T$, where $\mathbf{u}^T\mathbf{u} = 1$. Then

$$H^T = (I - 2\mathbf{u}\mathbf{u}^T)^T = I^T - 2(\mathbf{u}\mathbf{u}^T)^T = I - 2\left[(\mathbf{u}^T)^T\mathbf{u}^T\right] = I - 2\mathbf{u}\mathbf{u}^T = H.$$

So, H is symmetric. Furthermore,

$$HH = (I - 2\mathbf{u}\mathbf{u}^T)(I - 2\mathbf{u}\mathbf{u}^T) = I^2 - 4\mathbf{u}\mathbf{u}^T + 4(\mathbf{u}\mathbf{u}^T)^2 = I - 4\mathbf{u}\mathbf{u}^T + 4\mathbf{u}\mathbf{u}^T = I.$$

So, H is nonsingular.

© 2013 Cengage Learning. All Rights Reserved. May not be scanned, copied or duplicated, or posted to a publicly accessible website, in whole or in part.

(b) $\mathbf{u}^T\mathbf{u} = \begin{bmatrix} \frac{\sqrt{2}}{2} & \frac{\sqrt{2}}{2} & 0 \end{bmatrix}\begin{bmatrix} \frac{\sqrt{2}}{2} \\ \frac{\sqrt{2}}{2} \\ 0 \end{bmatrix} = \begin{bmatrix} \frac{2}{4} + \frac{2}{4} + 0 \end{bmatrix} = 1.$

$$H = I_n - 2\mathbf{u}\mathbf{u}^T = \begin{bmatrix} 1 & 0 & 0 \\ 0 & 1 & 0 \\ 0 & 0 & 1 \end{bmatrix} - 2\begin{bmatrix} \frac{\sqrt{2}}{2} \\ \frac{\sqrt{2}}{2} \\ 0 \end{bmatrix}\begin{bmatrix} \frac{\sqrt{2}}{2} & \frac{\sqrt{2}}{2} & 0 \end{bmatrix} = \begin{bmatrix} 1 & 0 & 0 \\ 0 & 1 & 0 \\ 0 & 0 & 1 \end{bmatrix} - 2\begin{bmatrix} \frac{1}{2} & \frac{1}{2} & 0 \\ \frac{1}{2} & \frac{1}{2} & 0 \\ 0 & 0 & 0 \end{bmatrix} = \begin{bmatrix} 0 & -1 & 0 \\ -1 & 0 & 0 \\ 0 & 0 & 1 \end{bmatrix}$$

79. If P is nonsingular, then P^{-1} exists. Use matrix multiplication to solve for A. Because all of the matrices are $n \times n$ matrices

$$AP = PD$$
$$APP^{-1} = PDP^{-1}$$
$$A = PDP^{-1}.$$

No, it is not necessarily true that $A = D$.

81. To find the inverse of a matrix, first the matrix must be square. Next, adjoin the $n \times n$ identity matrix to the given matrix A to obtain $[A \quad I]$. Use Gauss-Jordan elimination to row reduce the matrix to $[I \quad A^{-1}]$. If you cannot row reduce, then A is noninvertible.

83. Write the linear equations so that the variable terms are on the left and the constant term is on the right. Then you can write this as a matrix equation $AX = B$. For matrix A, each row represents the variable terms and each column represents a variable. For matrix X, write the variables vertically, so that the first column of matrix A represents the variable put in the first row, and so on. For matrix B, each row represents the constant term from an equation.

To solve the matrix equation $AX = B$, multiply each side by the inverse of A, A^{-1}, to obtain $X = A^{-1}B$. The product $A^{-1}B$ will be a matrix that will have the same dimensions as matrix X and will represent the solution of the system of linear equations.

Section 2.4 Elementary Matrices

1. The matrix *is* elementary. It can be obtained by multiplying the second row of I_2 by 2.

3. The matrix *is* elementary. Two times the first row was added to the second row.

5. The matrix is *not* elementary. The first row was multiplied by 2 and the second and third rows were interchanged.

7. The matrix *is* elementary. It can be obtained by multiplying the second row of I_4 by -5, and adding the result to the third row.

9. B is obtained by interchanging the first and third rows of A. So,

$$E = \begin{bmatrix} 0 & 0 & 1 \\ 0 & 1 & 0 \\ 1 & 0 & 0 \end{bmatrix}.$$

11. A is obtained by interchanging the first and third rows of B. So,

$$E = \begin{bmatrix} 0 & 0 & 1 \\ 0 & 1 & 0 \\ 1 & 0 & 0 \end{bmatrix}.$$

13.

Matrix	Elementary Row Operation	Elementary Matrix
$\begin{bmatrix} 5 & 10 & -5 \\ 0 & 1 & 7 \end{bmatrix}$	$R_1 \leftrightarrow R_2$	$\begin{bmatrix} 0 & 1 \\ 1 & 0 \end{bmatrix}$
$\begin{bmatrix} 1 & 2 & -1 \\ 0 & 1 & 7 \end{bmatrix}$	$\left(\frac{1}{5}\right)R_1 \rightarrow R_1$	$\begin{bmatrix} \frac{1}{5} & 0 \\ 0 & 1 \end{bmatrix}$

So, $\begin{bmatrix} \frac{1}{5} & 0 \\ 0 & 1 \end{bmatrix}\begin{bmatrix} 0 & 1 \\ 1 & 0 \end{bmatrix}\begin{bmatrix} 0 & 1 & 7 \\ 5 & 10 & -5 \end{bmatrix} = \begin{bmatrix} 1 & 2 & -1 \\ 0 & 1 & 7 \end{bmatrix}.$

© 2013 Cengage Learning. All Rights Reserved. May not be scanned, copied or duplicated, or posted to a publicly accessible website, in whole or in part.

15. Matrix / Elementary Row Operation / Elementary Matrix

$\begin{bmatrix} 1 & -2 & -1 & 0 \\ 0 & 4 & 8 & -4 \\ 0 & 0 & 2 & 1 \end{bmatrix}$ $\quad R_3 + bR_1 \rightarrow R_3$ $\quad \begin{bmatrix} 1 & 0 & 0 \\ 0 & 1 & 0 \\ 6 & 0 & 1 \end{bmatrix}$

$\begin{bmatrix} 1 & -2 & -1 & 0 \\ 0 & 1 & 2 & -1 \\ 0 & 0 & 2 & 1 \end{bmatrix}$ $\quad \left(\frac{1}{4}\right)R_2 \rightarrow R_2$ $\quad \begin{bmatrix} 1 & 0 & 0 \\ 0 & \frac{1}{4} & 0 \\ 0 & 0 & 1 \end{bmatrix}$

$\begin{bmatrix} 1 & -2 & -1 & 0 \\ 0 & 1 & 2 & -1 \\ 0 & 0 & 1 & \frac{1}{2} \end{bmatrix}$ $\quad \left(\frac{1}{2}\right)R_3 \rightarrow R_3$ $\quad \begin{bmatrix} 1 & 0 & 0 \\ 0 & 1 & 0 \\ 0 & 0 & \frac{1}{2} \end{bmatrix}$

So, $\begin{bmatrix} 1 & 0 & 0 \\ 0 & 1 & 0 \\ 0 & 0 & \frac{1}{2} \end{bmatrix}\begin{bmatrix} 1 & 0 & 0 \\ 0 & \frac{1}{4} & 0 \\ 0 & 0 & 1 \end{bmatrix}\begin{bmatrix} 1 & 0 & 0 \\ 0 & 1 & 0 \\ 6 & 0 & 1 \end{bmatrix}\begin{bmatrix} 1 & -2 & -1 & 0 \\ 0 & 4 & 8 & -4 \\ -6 & 12 & 8 & 1 \end{bmatrix} = \begin{bmatrix} 1 & -2 & -1 & 0 \\ 0 & 1 & 2 & -1 \\ 0 & 0 & 1 & \frac{1}{2} \end{bmatrix}.$

17. To obtain the inverse matrix, reverse the elementary row operation that produced it. So, interchange the first and second rows of I_2 to obtain

$$E^{-1} = \begin{bmatrix} 0 & 1 \\ 1 & 0 \end{bmatrix}.$$

19. To obtain the inverse matrix, reverse the elementary row operation that produced it. So, interchange the first and third rows to obtain

$$E^{-1} = \begin{bmatrix} 0 & 0 & 1 \\ 0 & 1 & 0 \\ 1 & 0 & 0 \end{bmatrix}.$$

21. To obtain the inverse matrix, reverse the elementary row operation that produced it. So, divide the first row of I_3 by k to obtain

$$E^{-1} = \begin{bmatrix} \frac{1}{k} & 0 & 0 \\ 0 & 1 & 0 \\ 0 & 0 & 1 \end{bmatrix}, \; k \neq 0.$$

23. Find a sequence of elementary row operations that can be used to rewrite A in reduced row-echelon form.

$\begin{bmatrix} 1 & 0 \\ 3 & -2 \end{bmatrix}$ $\quad R_1 \leftrightarrow R_2$ $\quad E_1 = \begin{bmatrix} 0 & 1 \\ 1 & 0 \end{bmatrix}$

$\begin{bmatrix} 1 & 0 \\ 0 & -2 \end{bmatrix}$ $\quad R_2 - 3R_1 \rightarrow R_2$ $\quad E_2 = \begin{bmatrix} 1 & 0 \\ -3 & 1 \end{bmatrix}$

$\begin{bmatrix} 1 & 0 \\ 0 & 1 \end{bmatrix}$ $\quad \left(-\frac{1}{2}\right)R_2 \rightarrow R_2$ $\quad E_3 = \begin{bmatrix} 1 & 0 \\ 0 & -\frac{1}{2} \end{bmatrix}$

Use the elementary matrices to find the inverse.

$$A^{-1} = E_3E_2E_1$$
$$= \begin{bmatrix} 1 & 0 \\ 0 & -\frac{1}{2} \end{bmatrix}\begin{bmatrix} 1 & 0 \\ -3 & 1 \end{bmatrix}\begin{bmatrix} 0 & 1 \\ 1 & 0 \end{bmatrix}$$
$$= \begin{bmatrix} 1 & 0 \\ \frac{3}{2} & -\frac{1}{2} \end{bmatrix}\begin{bmatrix} 0 & 1 \\ 1 & 0 \end{bmatrix} = \begin{bmatrix} 0 & 1 \\ -\frac{1}{2} & \frac{3}{2} \end{bmatrix}$$

25. Find a sequence of elementary row operations that can be used to rewrite A in reduced row-echelon form.

$\begin{bmatrix} 1 & 0 & -1 \\ 0 & 1 & -\frac{1}{6} \\ 0 & 0 & 4 \end{bmatrix}$ $\quad \left(\frac{1}{6}\right)R_2 \rightarrow R_2$ $\quad E_1 = \begin{bmatrix} 1 & 0 & 0 \\ 0 & \frac{1}{6} & 0 \\ 0 & 0 & 1 \end{bmatrix}$

$\begin{bmatrix} 1 & 0 & -1 \\ 0 & 1 & -\frac{1}{6} \\ 0 & 0 & 1 \end{bmatrix}$ $\quad \left(\frac{1}{4}\right)R_3 \rightarrow R_3$ $\quad E_2 = \begin{bmatrix} 1 & 0 & 0 \\ 0 & 1 & 0 \\ 0 & 0 & \frac{1}{4} \end{bmatrix}$

$\begin{bmatrix} 1 & 0 & 0 \\ 0 & 1 & -\frac{1}{6} \\ 0 & 0 & 1 \end{bmatrix}$ $\quad R_1 + R_3 \rightarrow R_1$ $\quad E_3 = \begin{bmatrix} 1 & 0 & 1 \\ 0 & 1 & 0 \\ 0 & 0 & 1 \end{bmatrix}$

$\begin{bmatrix} 1 & 0 & 0 \\ 0 & 1 & 0 \\ 0 & 0 & 1 \end{bmatrix}$ $\quad R_2 + \left(\frac{1}{6}\right)R_3 \rightarrow R_2$ $\quad E_4 = \begin{bmatrix} 1 & 0 & 0 \\ 0 & 1 & \frac{1}{6} \\ 0 & 0 & 1 \end{bmatrix}$

Use the elementary matrices to find the inverse.

$$A^{-1} = E_4E_3E_2E_1$$
$$= \begin{bmatrix} 1 & 0 & 0 \\ 0 & 1 & \frac{1}{6} \\ 0 & 0 & 1 \end{bmatrix}\begin{bmatrix} 1 & 0 & 1 \\ 0 & 1 & 0 \\ 0 & 0 & 1 \end{bmatrix}\begin{bmatrix} 1 & 0 & 0 \\ 0 & 1 & 0 \\ 0 & 0 & \frac{1}{4} \end{bmatrix}\begin{bmatrix} 1 & 0 & 0 \\ 0 & \frac{1}{6} & 0 \\ 0 & 0 & 1 \end{bmatrix}$$
$$= \begin{bmatrix} 1 & 0 & 1 \\ 0 & 1 & \frac{1}{6} \\ 0 & 0 & 1 \end{bmatrix}\begin{bmatrix} 1 & 0 & 0 \\ 0 & 1 & 0 \\ 0 & 0 & \frac{1}{4} \end{bmatrix}\begin{bmatrix} 1 & 0 & 0 \\ 0 & \frac{1}{6} & 0 \\ 0 & 0 & 1 \end{bmatrix}$$
$$= \begin{bmatrix} 1 & 0 & \frac{1}{4} \\ 0 & 1 & \frac{1}{24} \\ 0 & 0 & \frac{1}{4} \end{bmatrix}\begin{bmatrix} 1 & 0 & 0 \\ 0 & \frac{1}{6} & 0 \\ 0 & 0 & 1 \end{bmatrix} = \begin{bmatrix} 1 & 0 & \frac{1}{4} \\ 0 & \frac{1}{6} & \frac{1}{24} \\ 0 & 0 & \frac{1}{4} \end{bmatrix}$$

© 2013 Cengage Learning. All Rights Reserved. May not be scanned, copied or duplicated, or posted to a publicly accessible website, in whole or in part.

For Exercises 27–33, answers will vary. Sample answers are shown below.

27.

Matrix	Elementary Row Operation	Elementary Matrix
$\begin{bmatrix} 1 & 2 \\ 0 & -2 \end{bmatrix}$	Add -1 times row one to row two.	$E_1 = \begin{bmatrix} 1 & 0 \\ -1 & 1 \end{bmatrix}$
$\begin{bmatrix} 1 & 0 \\ 0 & -2 \end{bmatrix}$	Add row two to row one.	$E_2 = \begin{bmatrix} 1 & 1 \\ 0 & 1 \end{bmatrix}$
$\begin{bmatrix} 1 & 0 \\ 0 & 1 \end{bmatrix}$	Divide row two by -2.	$E_3 = \begin{bmatrix} 1 & 0 \\ 0 & -\frac{1}{2} \end{bmatrix}$

Because $E_3E_2E_1A = I_2$, factor A as follows.

$$A = E_1^{-1}E_2^{-1}E_3^{-1} = \begin{bmatrix} 1 & 0 \\ 1 & 1 \end{bmatrix}\begin{bmatrix} 1 & -1 \\ 0 & 1 \end{bmatrix}\begin{bmatrix} 1 & 0 \\ 0 & -2 \end{bmatrix}$$

Note that this factorization is not unique. For example, another factorization is

$$A = \begin{bmatrix} 0 & 1 \\ 1 & 0 \end{bmatrix}\begin{bmatrix} 1 & 0 \\ 1 & 1 \end{bmatrix}\begin{bmatrix} 1 & 0 \\ 0 & 2 \end{bmatrix}.$$

29.

Matrix	Elementary Row Operation	Elementary Matrix
$\begin{bmatrix} 1 & 0 \\ 3 & -1 \end{bmatrix}$	Add (-1) times row two to row one.	$E_1 = \begin{bmatrix} 1 & -1 \\ 0 & 1 \end{bmatrix}$
$\begin{bmatrix} 1 & 0 \\ 0 & -1 \end{bmatrix}$	Add -3 times row one to row two.	$E_2 = \begin{bmatrix} 1 & 0 \\ -3 & 1 \end{bmatrix}$
$\begin{bmatrix} 1 & 0 \\ 0 & 1 \end{bmatrix}$	Multiply row two by -1.	$E_3 = \begin{bmatrix} 1 & 0 \\ 0 & -1 \end{bmatrix}$

Because $E_3E_2E_1A = I_2$, one way to factor A is as follows.

$$A = E_1^{-1}E_2^{-1}E_3^{-1} = \begin{bmatrix} 1 & 1 \\ 0 & 1 \end{bmatrix}\begin{bmatrix} 1 & 0 \\ 3 & 1 \end{bmatrix}\begin{bmatrix} 1 & 0 \\ 0 & -1 \end{bmatrix}$$

31.

Matrix	Elementary Row Operation	Elementary Matrix
$\begin{bmatrix} 1 & -2 & 0 \\ 0 & 1 & 0 \\ 0 & 0 & 1 \end{bmatrix}$	Add row one to row two.	$E_1 = \begin{bmatrix} 1 & 0 & 0 \\ 1 & 1 & 0 \\ 0 & 0 & 1 \end{bmatrix}$
$\begin{bmatrix} 1 & 0 & 0 \\ 0 & 1 & 0 \\ 0 & 0 & 1 \end{bmatrix}$	Add 2 times row two to row one.	$E_2 = \begin{bmatrix} 1 & 2 & 0 \\ 0 & 1 & 0 \\ 0 & 0 & 1 \end{bmatrix}$

Because $E_2E_1A = I_3$, one way to factor A is

$$A = E_1^{-1}E_2^{-1} = \begin{bmatrix} 1 & 0 & 0 \\ -1 & 1 & 0 \\ 0 & 0 & 1 \end{bmatrix}\begin{bmatrix} 1 & -2 & 0 \\ 0 & 1 & 0 \\ 0 & 0 & 1 \end{bmatrix}.$$

© 2013 Cengage Learning. All Rights Reserved. May not be scanned, copied or duplicated, or posted to a publicly accessible website, in whole or in part.

33. Find a sequence of elementary row operations that can be used to rewrite A in reduced row-echelon form.

$$\begin{bmatrix}1&0&0&1\\0&1&-3&0\\0&0&2&0\\0&0&1&-1\end{bmatrix} \quad -R_2 \to R_2 \qquad E_1 = \begin{bmatrix}1&0&0&0\\0&-1&0&0\\0&0&1&0\\0&0&0&1\end{bmatrix}$$

$$\begin{bmatrix}1&0&0&1\\0&1&-3&0\\0&0&1&0\\0&0&1&-1\end{bmatrix} \quad \left(\tfrac{1}{2}\right)R_3 \to R_3 \qquad E_2 = \begin{bmatrix}1&0&0&0\\0&1&0&0\\0&0&\frac{1}{2}&0\\0&0&0&1\end{bmatrix}$$

$$\begin{bmatrix}1&0&0&1\\0&1&-3&0\\0&0&1&0\\0&0&-1&1\end{bmatrix} \quad -R_4 \to R_4 \qquad E_3 = \begin{bmatrix}1&0&0&0\\0&1&0&0\\0&0&1&0\\0&0&0&-1\end{bmatrix}$$

$$\begin{bmatrix}1&0&0&1\\0&1&-3&0\\0&0&1&0\\0&0&0&1\end{bmatrix} \quad R_4 + R_3 \to R_4 \qquad E_4 = \begin{bmatrix}1&0&0&0\\0&1&0&0\\0&0&1&0\\0&0&1&1\end{bmatrix}$$

$$\begin{bmatrix}1&0&0&1\\0&1&0&0\\0&0&1&0\\0&0&0&1\end{bmatrix} \quad R_2 + 3R_3 \to R_2 \qquad E_5 = \begin{bmatrix}1&0&0&0\\0&1&3&0\\0&0&1&0\\0&0&0&1\end{bmatrix}$$

$$\begin{bmatrix}1&0&0&0\\0&1&0&0\\0&0&1&0\\0&0&0&1\end{bmatrix} \quad R_1 - R_4 \to R_1 \qquad E_6 = \begin{bmatrix}1&0&0&-1\\0&1&0&0\\0&0&1&0\\0&0&0&1\end{bmatrix}$$

So, one way to factor A is

$$A = E_1^{-1}E_2^{-1}E_3^{-1}E_4^{-1}E_5^{-1}E_6^{-1}$$

$$= \begin{bmatrix}1&0&0&0\\0&-1&0&0\\0&0&1&0\\0&0&0&1\end{bmatrix}\begin{bmatrix}1&0&0&0\\0&1&0&0\\0&0&2&0\\0&0&0&1\end{bmatrix}\begin{bmatrix}1&0&0&0\\0&1&0&0\\0&0&1&0\\0&0&0&-1\end{bmatrix}\begin{bmatrix}1&0&0&0\\0&1&0&0\\0&0&1&0\\0&0&-1&1\end{bmatrix}\begin{bmatrix}1&0&0&0\\0&1&-3&0\\0&0&1&0\\0&0&0&1\end{bmatrix}\begin{bmatrix}1&0&0&1\\0&1&0&0\\0&0&1&0\\0&0&0&1\end{bmatrix}.$$

35. (a) True. See "Remark" following the "Definition of an Elementary Matrix" on page 74.

(b) False. Multiplication of a matrix by a scalar is not a single elementary row operation so it cannot be represented by a corresponding elementary matrix.

(c) True. See Theorem 2.13.

37. No. For example, $\begin{bmatrix}1&0\\2&1\end{bmatrix}\begin{bmatrix}1&1\\0&1\end{bmatrix} = \begin{bmatrix}1&1\\2&3\end{bmatrix}$, which is not elementary.

39. $$A^{-1} = \begin{bmatrix}1&0&0\\0&1&0\\0&0&c\end{bmatrix}^{-1}\begin{bmatrix}1&0&0\\b&1&0\\0&0&1\end{bmatrix}^{-1}\begin{bmatrix}1&a&0\\0&1&0\\0&0&1\end{bmatrix}^{-1} = \begin{bmatrix}1&0&0\\0&1&0\\0&0&\frac{1}{c}\end{bmatrix}\begin{bmatrix}1&0&0\\-b&1&0\\0&0&1\end{bmatrix}\begin{bmatrix}1&-a&0\\0&1&0\\0&0&1\end{bmatrix} = \begin{bmatrix}1&-a&0\\-b&1+ab&0\\0&0&\frac{1}{c}\end{bmatrix}.$$

For Exercises 41–45, answers will vary. Sample answers are shown below.

41. Because the matrix is lower triangular, an LU-factorization is

$$\begin{bmatrix}1&0\\-2&1\end{bmatrix}\begin{bmatrix}1&0\\0&1\end{bmatrix}.$$

© 2013 Cengage Learning. All Rights Reserved. May not be scanned, copied or duplicated, or posted to a publicly accessible website, in whole or in part.

43. Matrix | Elementary Matrix

$$\begin{bmatrix} 3 & 0 & 1 \\ 6 & 1 & 1 \\ -3 & 1 & 0 \end{bmatrix} = A$$

$$\begin{bmatrix} 3 & 0 & 1 \\ 0 & 1 & -1 \\ -3 & 1 & 0 \end{bmatrix} \qquad E_1 = \begin{bmatrix} 1 & 0 & 0 \\ -2 & 1 & 0 \\ 0 & 0 & 1 \end{bmatrix}$$

$$\begin{bmatrix} 3 & 0 & 1 \\ 0 & 1 & -1 \\ 0 & 1 & 1 \end{bmatrix} \qquad E_2 = \begin{bmatrix} 1 & 0 & 0 \\ 0 & 1 & 0 \\ 1 & 0 & 1 \end{bmatrix}$$

$$\begin{bmatrix} 3 & 0 & 1 \\ 0 & 1 & -1 \\ 0 & 0 & 2 \end{bmatrix} = U \qquad E_3 = \begin{bmatrix} 1 & 0 & 0 \\ 0 & 1 & 0 \\ 0 & -1 & 1 \end{bmatrix}$$

$E_3E_2E_1A = U$

$$A = E_1^{-1}E_2^{-1}E_3^{-1}U = \begin{bmatrix} 1 & 0 & 0 \\ 2 & 1 & 0 \\ -1 & 1 & 1 \end{bmatrix}\begin{bmatrix} 3 & 0 & 1 \\ 0 & 1 & -1 \\ 0 & 0 & 2 \end{bmatrix} = LU$$

45. (a) Matrix | Elementary Matrix

$$\begin{bmatrix} 2 & 1 & 0 \\ 0 & 1 & -1 \\ -2 & 1 & 1 \end{bmatrix} = A$$

$$\begin{bmatrix} 2 & 1 & 0 \\ 0 & 1 & -1 \\ 0 & 2 & 1 \end{bmatrix} \qquad E_1 = \begin{bmatrix} 1 & 0 & 0 \\ 0 & 1 & 0 \\ 1 & 0 & 1 \end{bmatrix}$$

$$\begin{bmatrix} 2 & 1 & 0 \\ 0 & 1 & -1 \\ 0 & 0 & 3 \end{bmatrix} = U \qquad E_2 = \begin{bmatrix} 1 & 0 & 0 \\ 0 & 1 & 0 \\ 0 & -2 & 1 \end{bmatrix}$$

$E_2E_1A = U$

$$A = E_1^{-1}E_2^{-1}U = \begin{bmatrix} 1 & 0 & 0 \\ 0 & 1 & 0 \\ -1 & 2 & 1 \end{bmatrix}\begin{bmatrix} 2 & 1 & 0 \\ 0 & 1 & -1 \\ 0 & 0 & 3 \end{bmatrix} = LU$$

(b) $L\mathbf{y} = \mathbf{b}$: $\begin{bmatrix} 1 & 0 & 0 \\ 0 & 1 & 0 \\ -1 & 2 & 1 \end{bmatrix}\begin{bmatrix} y_1 \\ y_2 \\ y_3 \end{bmatrix} = \begin{bmatrix} 1 \\ 2 \\ -2 \end{bmatrix}$

$y_1 = 1$, $y_2 = 2$,

and

$-y_1 + 2y_2 + y_3 = -2 \Rightarrow y_3 = -5$

(c) $U\mathbf{x} = \mathbf{y}$: $\begin{bmatrix} 2 & 1 & 0 \\ 0 & 1 & -1 \\ 0 & 0 & 3 \end{bmatrix}\begin{bmatrix} x_1 \\ x_2 \\ x_3 \end{bmatrix} = \begin{bmatrix} 1 \\ 2 \\ -5 \end{bmatrix}$

$x_3 = -\frac{5}{3}$, $x_2 - x_3 = 2 \Rightarrow x_2 = \frac{1}{3}$

and $2x_1 + x_2 = 1 \Rightarrow x_1 = \frac{1}{3}$

So, the solution to the system $A\mathbf{x} = \mathbf{b}$ is

$x_1 = \frac{1}{3}$, $x_2 = \frac{1}{3}$, and $x_3 = -\frac{5}{3}$.

47. You could first factor the matrix $A = LU$. Then, for each right-hand side, $\mathbf{b}_i$, solve $L\mathbf{y} = \mathbf{b}_i$ and $U\mathbf{x} = \mathbf{y}$.

49. $A^2 = \begin{bmatrix} 1 & 0 \\ 0 & 0 \end{bmatrix}^2 = \begin{bmatrix} 1 & 0 \\ 0 & 0 \end{bmatrix} = A$

Because $A^2 = A$, A is idempotent.

51. $A^2 = \begin{bmatrix} 0 & 0 & 1 \\ 0 & 1 & 0 \\ 1 & 0 & 0 \end{bmatrix}\begin{bmatrix} 0 & 0 & 1 \\ 0 & 1 & 0 \\ 1 & 0 & 0 \end{bmatrix} = \begin{bmatrix} 1 & 0 & 0 \\ 0 & 1 & 0 \\ 0 & 0 & 1 \end{bmatrix}$

Because $A^2 \neq A$, A is *not* idempotent.

53. Begin by finding A^2.

$$A^2 = \begin{bmatrix} 1 & 0 \\ a & b \end{bmatrix}\begin{bmatrix} 1 & 0 \\ a & b \end{bmatrix} = \begin{bmatrix} 1 & 0 \\ a(1+b) & b^2 \end{bmatrix}$$

Setting $A^2 = A$ yields the equations $a = a(1 + b)$ and $b = b^2$. The second equation is satisfied when $b = 1$ or $b = 0$. If $b = 1$, then $a = 0$, and if $b = 0$, then a can be any real number.

55. Because A is idempotent and invertible, you have

$$A^2 = A$$
$$A^{-1}A^2 = A^{-1}A$$
$$(A^{-1}A)A = I$$
$$A = I.$$

57. $A = E_k \cdots E_2E_1B$, for $E_1, \ldots, E_k$ elementary

$B = F_t \cdots F_2F_1C$, for $F_1, \ldots, F_t$ elementary

Then

$A = E_k \cdots E_2E_1B = (E_k \cdots E_2E_1)(F_t \cdots F_2F_1C)$,

which shows that A is row-equivalent to C.

59. If B is row-equivalent to A, then

$B = E_k \cdots E_2E_1A$,

where $E_1, \ldots, E_k$ are elementary matrices. Because elementary matrices are nonsingular,

$B^{-1} = (E_k \cdots E_1A)^{-1} = A^{-1}E_1^{-1} \cdots E_k^{-1}$,

which shows that B is also nonsingular.

© 2013 Cengage Learning. All Rights Reserved. May not be scanned, copied or duplicated, or posted to a publicly accessible website, in whole or in part.

Section 2.5 Applications of Matrix Operations

1. The matrix is *not* stochastic because every entry of a stochastic matrix must satisfy the inequality $0 \le a_{ij} \le 1$.

3. This matrix *is* stochastic because each entry is between 0 and 1, and each column adds up to 1.

5. Form the matrix representing the given transition probabilities. Let A represent people who purchased the product and B represent people who did not.

$$P = \overbrace{\begin{matrix} A & B \end{matrix}}^{\text{From}} \quad P = \begin{bmatrix} 0.80 & 0.30 \\ 0.20 & 0.70 \end{bmatrix} \begin{matrix} A \\ B \end{matrix} \Big\} \text{To}$$

The state matrix representing the current population is

$$X = \begin{bmatrix} 100 \\ 900 \end{bmatrix} \begin{matrix} A \\ B \end{matrix}.$$

(a) The state matrix for next month is

$$PX = \begin{bmatrix} 0.80 & 0.30 \\ 0.20 & 0.70 \end{bmatrix} \begin{bmatrix} 100 \\ 900 \end{bmatrix} = \begin{bmatrix} 350 \\ 650 \end{bmatrix}.$$

So, next month 350 people will purchase the product.

(b) The state matrix for the month after next is

$$P(PX) = \begin{bmatrix} 0.80 & 0.30 \\ 0.20 & 0.70 \end{bmatrix} \begin{bmatrix} 350 \\ 650 \end{bmatrix} = \begin{bmatrix} 475 \\ 525 \end{bmatrix}.$$

In 2 months, 475 people will purchase the product.

7. Form the matrix representing the given transition probabilities. Let N represent nonsmokers, S_0 represent those who smoke one pack or less, and S_1 represent those who smoke more than one pack.

$$\overbrace{\begin{matrix} N & S_0 & S_1 \end{matrix}}^{\text{From}} \quad P = \begin{bmatrix} 0.93 & 0.10 & 0.05 \\ 0.05 & 0.80 & 0.10 \\ 0.02 & 0.10 & 0.85 \end{bmatrix} \begin{matrix} N \\ S_0 \\ S_1 \end{matrix} \Bigg\} \text{To}$$

The state matrix representing the current population is

$$X = \begin{bmatrix} 5000 \\ 2500 \\ 2500 \end{bmatrix} \begin{matrix} N \\ S_0 \\ S_1 \end{matrix}.$$

(a) The state matrix for the next month is

$$PX = \begin{bmatrix} 0.93 & 0.10 & 0.05 \\ 0.05 & 0.80 & 0.10 \\ 0.02 & 0.10 & 0.85 \end{bmatrix} \begin{bmatrix} 5000 \\ 2500 \\ 2500 \end{bmatrix} = \begin{bmatrix} 5025 \\ 2500 \\ 2475 \end{bmatrix}.$$

So, next month the population will be grouped as follows: 5025 nonsmokers, 2500 smokers of one pack or less per day, and 2475 smokers of more than one pack per day.

(b) The state matrix for the month after next is

$$P(PX) = \begin{bmatrix} 0.93 & 0.10 & 0.05 \\ 0.05 & 0.80 & 0.10 \\ 0.02 & 0.10 & 0.85 \end{bmatrix} \begin{bmatrix} 5025 \\ 2500 \\ 2475 \end{bmatrix} = \begin{bmatrix} 5047 \\ 2498.75 \\ 2454.25 \end{bmatrix}.$$

In 2 months, the population will be grouped as follows: 5047 nonsmokers, 2499 smokers of one pack or less per day, and 2454 smokers of more than one pack per day.

© 2013 Cengage Learning. All Rights Reserved. May not be scanned, copied or duplicated, or posted to a publicly accessible website, in whole or in part.

9. Form the matrix representing the given transition probabilities. Let A represents users of brand A, B users of brand B, and N users of neither brands.

$$P = \overbrace{\begin{bmatrix} 0.75 & 0.15 & 0.10 \\ 0.20 & 0.75 & 0.15 \\ 0.05 & 0.10 & 0.75 \end{bmatrix}}^{\text{From: } A \quad B \quad N} \left.\begin{matrix} A \\ B \\ N \end{matrix}\right\} \text{To}$$

The state matrix representing the current product usage is

$$X = \begin{bmatrix} 20{,}000 \\ 30{,}000 \\ 50{,}000 \end{bmatrix} \begin{matrix} A \\ B. \\ N \end{matrix}$$

(a) The state matrix for next month is

$$PX = \begin{bmatrix} 0.75 & 0.15 & 0.10 \\ 0.20 & 0.75 & 0.15 \\ 0.05 & 0.10 & 0.75 \end{bmatrix} \begin{bmatrix} 20{,}000 \\ 30{,}000 \\ 50{,}000 \end{bmatrix} = \begin{bmatrix} 24{,}500 \\ 34{,}000 \\ 41{,}500 \end{bmatrix}.$$

So, the next month's users will be grouped as follows: 24,500 for brand A, 34,000 brand B, and 41,500 neither.

Similarly, the state matrices for the following two months are as follows.

(b) $$P(PX) = P\begin{bmatrix} 24{,}500 \\ 34{,}000 \\ 41{,}500 \end{bmatrix} = \begin{bmatrix} 27{,}625 \\ 36{,}625 \\ 35{,}750 \end{bmatrix}$$

In 2 months, the distribution will be 27,625 brand A, 36,625 brand B, and 35,750 neither.

(c) $$P(P(PX)) = \begin{bmatrix} 29{,}788 \\ 38{,}356 \\ 31{,}856 \end{bmatrix}.$$

Finally, in 3 months, the distribution will be 29,788 brand A, 38,356 brand B, and 31,856 neither.

11. Divide the message into groups of three and form the uncoded matrices.

S E L	L _ C	O N S	O L I	D A T	E D _
$[19 \;\; 5 \;\; 12]$	$[12 \;\; 0 \;\; 3]$	$[15 \;\; 14 \;\; 19]$	$[15 \;\; 12 \;\; 9]$	$[4 \;\; 1 \;\; 20]$	$[5 \;\; 4 \;\; 0]$

Multiplying each uncoded row matrix on the right by A yields the following coded row matrices.

$$[19 \;\; 5 \;\; 12]A = [19 \;\; 5 \;\; 12]\begin{bmatrix} 1 & -1 & 0 \\ 1 & 0 & -1 \\ -6 & 2 & 3 \end{bmatrix} = [-48 \;\; 5 \;\; 31]$$

$$[12 \;\; 0 \;\; 3]A = [-6 \;\; -6 \;\; 9]$$

$$[15 \;\; 14 \;\; 19]A = [-85 \;\; 23 \;\; 43]$$

$$[15 \;\; 12 \;\; 9]A = [-27 \;\; 3 \;\; 15]$$

$$[4 \;\; 1 \;\; 20]A = [-115 \;\; 36 \;\; 59]$$

$$[5 \;\; 4 \;\; 0]A = [9 \;\; -5 \;\; -4]$$

So, the coded message is $-48, 5, 31, -6, -6, 9, -85, 23, 43, -27, 3, 15, -115, 36, 59, 9, -5, -4$.

© 2013 Cengage Learning. All Rights Reserved. May not be scanned, copied or duplicated, or posted to a publicly accessible website, in whole or in part.

13. Divide the message into pairs of letters and form the coded matrices.

C O M E _ H O M E _ S O O N

$[3\ \ 15]$ $[13\ \ 5]$ $[0\ \ 8]$ $[15\ \ 13]$ $[5\ \ 0]$ $[19\ \ 15]$ $[15\ \ 14]$

Multiplying each uncoded row matrix on the right by A yields the following coded row matrices.

$$[3\ \ 15]\begin{bmatrix} 1 & 2 \\ 3 & 5 \end{bmatrix} = [48\ \ 81]$$

$$[13\ \ 5]\begin{bmatrix} 1 & 2 \\ 3 & 5 \end{bmatrix} = [28\ \ 51]$$

$$[0\ \ 8]\begin{bmatrix} 1 & 2 \\ 3 & 5 \end{bmatrix} = [24\ \ 40]$$

$$[15\ \ 13]\begin{bmatrix} 1 & 2 \\ 3 & 5 \end{bmatrix} = [54\ \ 95]$$

$$[5\ \ 0]\begin{bmatrix} 1 & 2 \\ 3 & 5 \end{bmatrix} = [5\ \ 10]$$

$$[19\ \ 15]\begin{bmatrix} 1 & 2 \\ 3 & 5 \end{bmatrix} = [64\ \ 113]$$

$$[15\ \ 14]\begin{bmatrix} 1 & 2 \\ 3 & 5 \end{bmatrix} = [57\ \ 100]$$

So, the coded message is 48, 81, 28, 51, 24, 40, 54, 95, 5, 10, 64, 113, 57, 100.

15. Find $A^{-1} = \begin{bmatrix} -5 & 2 \\ 3 & -1 \end{bmatrix}$

and multiply each coded row matrix on the right by A^{-1} to find the associated uncoded row matrix.

$$[11\ \ 21]\begin{bmatrix} -5 & 2 \\ 3 & -1 \end{bmatrix} = [8\ \ 1] \Rightarrow \text{H, A}$$

$$[64\ \ 112]\begin{bmatrix} -5 & 2 \\ 3 & -1 \end{bmatrix} = [16\ \ 16] \Rightarrow \text{P, P}$$

$$[25\ \ 50]\begin{bmatrix} -5 & 2 \\ 3 & -1 \end{bmatrix} = [25\ \ 0] \Rightarrow \text{Y, \_}$$

$$[29\ \ 53]\begin{bmatrix} -5 & 2 \\ 3 & -1 \end{bmatrix} = [14\ \ 5] \Rightarrow \text{N, E}$$

$$[23\ \ 46]\begin{bmatrix} -5 & 2 \\ 3 & -1 \end{bmatrix} = [23\ \ 0] \Rightarrow \text{W, \_}$$

$$[40\ \ 75]\begin{bmatrix} -5 & 2 \\ 3 & -1 \end{bmatrix} = [25\ \ 5] \Rightarrow \text{Y, E}$$

$$[55\ \ 92]\begin{bmatrix} -5 & 2 \\ 3 & -1 \end{bmatrix} = [1\ \ 18] \Rightarrow \text{A, R}$$

So, the message is HAPPY_NEW_YEAR.

17. Find $A^{-1} = \begin{bmatrix} -13 & 6 & 4 \\ 12 & -5 & -3 \\ -5 & 2 & 1 \end{bmatrix}$

and multiply each coded row matrix on the right by A^{-1} to find the associated uncoded row matrix.

$$[13\ \ 19\ \ 10]A^{-1} = [13\ \ 19\ \ 10]\begin{bmatrix} -13 & 6 & 4 \\ 12 & -5 & -3 \\ -5 & 2 & 1 \end{bmatrix}$$

$$= [9\ \ 3\ \ 5] \Rightarrow \text{I, C, E}$$

$$\begin{aligned}
[-1\ \ -33\ \ -77]A^{-1} &= [2\ \ 5\ \ 18] \Rightarrow \text{B, E, R} \\
[3\ \ -2\ \ -14]A^{-1} &= [7\ \ 0\ \ 4] \Rightarrow \text{G, _, D} \\
[4\ \ 1\ \ -9]A^{-1} &= [5\ \ 1\ \ 4] \Rightarrow \text{E, A, D} \\
[-5\ \ -25\ \ -47]A^{-1} &= [0\ \ 1\ \ 8] \Rightarrow \text{_, A, H} \\
[4\ \ 1\ \ -9]A^{-1} &= [5\ \ 1\ \ 4] \Rightarrow \text{E, A, D}
\end{aligned}$$

The message is ICEBERG_DEAD_AHEAD.

© 2013 Cengage Learning. All Rights Reserved. May not be scanned, copied or duplicated, or posted to a publicly accessible website, in whole or in part.

19. Let $A^{-1} = \begin{bmatrix} a & b \\ c & d \end{bmatrix}$ and find that

$$\begin{matrix} & & \_ \quad R \end{matrix}$$

$$\begin{bmatrix} -18 & -18 \end{bmatrix}\begin{bmatrix} a & b \\ c & d \end{bmatrix} = \begin{bmatrix} 0 & 18 \end{bmatrix}$$

$$\begin{bmatrix} -18(a + c) - 18(b + d) \end{bmatrix} = \begin{bmatrix} 0 & 18 \end{bmatrix}.$$

So, $c = -a$ and $d = -1 - b$. Using these values, you find that

$$\begin{matrix} & & O \quad N \end{matrix}$$

$$\begin{bmatrix} 1 & 16 \end{bmatrix}\begin{bmatrix} a & b \\ -a & -(1 + b) \end{bmatrix} = \begin{bmatrix} 15 & 14 \end{bmatrix}$$

$$\begin{bmatrix} -15a & -15b - 16 \end{bmatrix} = \begin{bmatrix} 15 & 14 \end{bmatrix}.$$

So, $a = -1$, $b = -2$, $c = 1$, and $d = 1$. Using the matrix

$$A^{-1} = \begin{bmatrix} -1 & -2 \\ 1 & 1 \end{bmatrix}$$

multiply each coded row matrix to yield the uncoded row matrices

$\begin{bmatrix} 13 & 5 \end{bmatrix}, \begin{bmatrix} 5 & 20 \end{bmatrix}, \begin{bmatrix} 0 & 13 \end{bmatrix}, \begin{bmatrix} 5 & 0 \end{bmatrix}, \begin{bmatrix} 20 & 15 \end{bmatrix},$
$\begin{bmatrix} 14 & 9 \end{bmatrix}, \begin{bmatrix} 7 & 8 \end{bmatrix}, \begin{bmatrix} 20 & 0 \end{bmatrix}, \begin{bmatrix} 0 & 18 \end{bmatrix}, \begin{bmatrix} 15 & 14 \end{bmatrix}.$

This corresponds to the message MEET_ME_TONIGHT_RON.

21. For $A = \begin{bmatrix} 1 & 0 & 2 \\ 2 & -1 & 1 \\ 0 & 1 & 2 \end{bmatrix}$, you have $A^{-1} = \begin{bmatrix} -3 & 2 & 2 \\ -4 & 2 & 3 \\ 2 & -1 & -1 \end{bmatrix}$.

Multiply each coded row matrix on the right by A^{-1} to find the associated uncoded row matrix.

$$\begin{array}{lll}
\begin{bmatrix} 38 & -14 & 29 \end{bmatrix} A^{-1} = \begin{bmatrix} 0 & 19 & 5 \end{bmatrix} & \Rightarrow & _, \ S, \ E \\
\begin{bmatrix} 56 & -15 & 62 \end{bmatrix} A^{-1} = \begin{bmatrix} 16 & 20 & 5 \end{bmatrix} & \Rightarrow & P, \ T, \ E \\
\begin{bmatrix} 17 & 3 & 38 \end{bmatrix} A^{-1} = \begin{bmatrix} 13 & 2 & 5 \end{bmatrix} & \Rightarrow & M, \ B, \ E \\
\begin{bmatrix} 18 & 20 & 76 \end{bmatrix} A^{-1} = \begin{bmatrix} 18 & 0 & 20 \end{bmatrix} & \Rightarrow & R, \ _, \ T \\
\begin{bmatrix} 18 & -5 & 21 \end{bmatrix} A^{-1} = \begin{bmatrix} 8 & 5 & 0 \end{bmatrix} & \Rightarrow & H, \ E, \ _ \\
\begin{bmatrix} 29 & -7 & 32 \end{bmatrix} A^{-1} = \begin{bmatrix} 5 & 12 & 5 \end{bmatrix} & \Rightarrow & E, \ L, \ E \\
\begin{bmatrix} 32 & 9 & 77 \end{bmatrix} A^{-1} = \begin{bmatrix} 22 & 5 & 14 \end{bmatrix} & \Rightarrow & V, \ E, \ N \\
\begin{bmatrix} 36 & -8 & 48 \end{bmatrix} A^{-1} = \begin{bmatrix} 20 & 8 & 0 \end{bmatrix} & \Rightarrow & T, \ H, \ _ \\
\begin{bmatrix} 33 & -5 & 51 \end{bmatrix} A^{-1} = \begin{bmatrix} 23 & 5 & 0 \end{bmatrix} & \Rightarrow & W, \ E, \ _ \\
\begin{bmatrix} 41 & 3 & 79 \end{bmatrix} A^{-1} = \begin{bmatrix} 23 & 9 & 12 \end{bmatrix} & \Rightarrow & W, \ I, \ L \\
\begin{bmatrix} 12 & 1 & 26 \end{bmatrix} A^{-1} = \begin{bmatrix} 12 & 0 & 1 \end{bmatrix} & \Rightarrow & L, \ _, \ A \\
\begin{bmatrix} 58 & -22 & 49 \end{bmatrix} A^{-1} = \begin{bmatrix} 12 & 23 & 1 \end{bmatrix} & \Rightarrow & L, \ W, \ A \\
\begin{bmatrix} 63 & -19 & 69 \end{bmatrix} A^{-1} = \begin{bmatrix} 25 & 19 & 0 \end{bmatrix} & \Rightarrow & Y, \ S, \ _ \\
\begin{bmatrix} 28 & 8 & 67 \end{bmatrix} A^{-1} = \begin{bmatrix} 18 & 5 & 13 \end{bmatrix} & \Rightarrow & R, \ E, \ M \\
\begin{bmatrix} 31 & -11 & 27 \end{bmatrix} A^{-1} = \begin{bmatrix} 5 & 13 & 2 \end{bmatrix} & \Rightarrow & E, \ M, \ B \\
\begin{bmatrix} 41 & -18 & 28 \end{bmatrix} A^{-1} = \begin{bmatrix} 5 & 18 & 0 \end{bmatrix} & \Rightarrow & E, \ R, \ _
\end{array}$$

The message is _SEPTEMBER_THE_ELEVENTH_WE_WILL_ALWAYS_REMEMBER_

23. Use the given information to find D.

$$D = \overset{\text{User}}{\overset{\text{Coal} \quad \text{Steel}}{\begin{bmatrix} 0.10 & 0.20 \\ 0.80 & 0.10 \end{bmatrix}}} \begin{matrix} \text{Coal} \\ \text{Steel} \end{matrix} \Big\} \text{Supplier}$$

The equation $X = DX + E$ may be rewritten in the form $(I - D)X = E$, that is,

$$\begin{bmatrix} 0.90 & -0.20 \\ -0.80 & 0.90 \end{bmatrix} X = \begin{bmatrix} 10{,}000 \\ 20{,}000 \end{bmatrix}.$$

Solve this system by using Gauss-Jordan elimination to obtain $X = \begin{bmatrix} 20{,}000 \\ 40{,}000 \end{bmatrix}$.

25. From the given matrix D, form the linear system $X = DX + E$, which can be written as $(I - D)X = E$

$$\begin{bmatrix} 0.6 & -0.5 & -0.5 \\ -0.3 & 1 & -0.3 \\ -0.2 & -0.2 & 1.0 \end{bmatrix} X = \begin{bmatrix} 1000 \\ 1000 \\ 1000 \end{bmatrix}.$$

Solving this system yields $X = \begin{bmatrix} 8622.0 \\ 4685.0 \\ 3661.4 \end{bmatrix}$.

27. (a) The line that best fits the given points is shown in the graph.

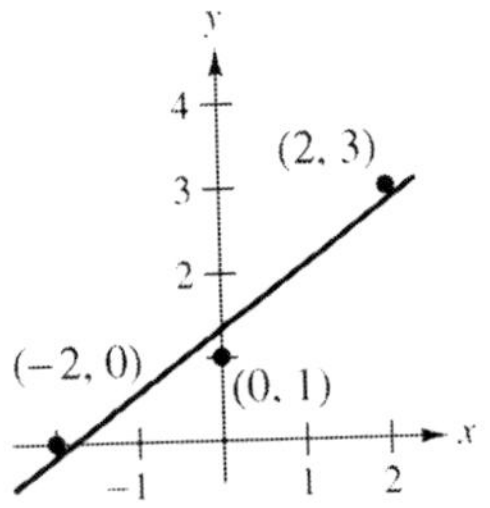

(b) Using the matrices $X = \begin{bmatrix} 1 & -2 \\ 1 & 0 \\ 1 & 2 \end{bmatrix}$ and $Y = \begin{bmatrix} 0 \\ 1 \\ 3 \end{bmatrix}$, you have

$$X^T X = \begin{bmatrix} 1 & 1 & 1 \\ -2 & 0 & 2 \end{bmatrix}\begin{bmatrix} 1 & -2 \\ 1 & 0 \\ 1 & 2 \end{bmatrix} = \begin{bmatrix} 3 & 0 \\ 0 & 8 \end{bmatrix}$$

$$X^T Y = \begin{bmatrix} 1 & 1 & 1 \\ -2 & 0 & 2 \end{bmatrix}\begin{bmatrix} 0 \\ 1 \\ 3 \end{bmatrix} = \begin{bmatrix} 4 \\ 6 \end{bmatrix}$$

$$A = \left(X^T X\right)^{-1} X^T Y = \begin{bmatrix} \frac{1}{3} & 0 \\ 0 & \frac{1}{8} \end{bmatrix}\begin{bmatrix} 4 \\ 6 \end{bmatrix} = \begin{bmatrix} \frac{4}{3} \\ \frac{3}{4} \end{bmatrix}.$$

So, the least squares regression line is $y = \frac{3}{4}x + \frac{4}{3}$.

© 2013 Cengage Learning. All Rights Reserved. May not be scanned, copied or duplicated, or posted to a publicly accessible website, in whole or in part.

(c) Solving $Y = XA + E$ for E,

$$E = Y - XA = \begin{bmatrix} 0 \\ 1 \\ 3 \end{bmatrix} - \begin{bmatrix} 1 & -2 \\ 1 & 0 \\ 1 & 2 \end{bmatrix}\begin{bmatrix} \frac{4}{3} \\ \frac{3}{4} \end{bmatrix} = \begin{bmatrix} \frac{1}{6} \\ -\frac{1}{3} \\ \frac{1}{6} \end{bmatrix}.$$

So, the sum of the squared error is

$$E^T E = \begin{bmatrix} \frac{1}{6} & -\frac{1}{3} & \frac{1}{6} \end{bmatrix}\begin{bmatrix} \frac{1}{6} \\ -\frac{1}{3} \\ \frac{1}{6} \end{bmatrix} = \frac{1}{6}.$$

29. (a) The line that best fits the given points is shown in the graph.

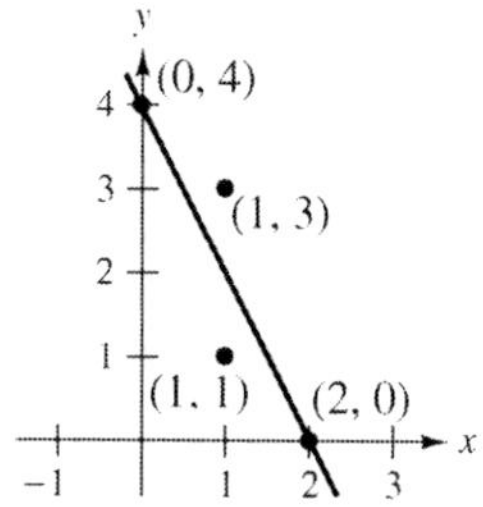

(b) Using the matrices $X = \begin{bmatrix} 1 & 0 \\ 1 & 1 \\ 1 & 1 \\ 1 & 2 \end{bmatrix}$ and $Y = \begin{bmatrix} 4 \\ 3 \\ 1 \\ 0 \end{bmatrix}$, you have

$$X^T X = \begin{bmatrix} 1 & 1 & 1 & 1 \\ 0 & 1 & 1 & 2 \end{bmatrix}\begin{bmatrix} 1 & 0 \\ 1 & 1 \\ 1 & 1 \\ 1 & 2 \end{bmatrix} = \begin{bmatrix} 4 & 4 \\ 4 & 6 \end{bmatrix}$$

$$X^T Y = \begin{bmatrix} 1 & 1 & 1 & 1 \\ 0 & 1 & 1 & 2 \end{bmatrix}\begin{bmatrix} 4 \\ 3 \\ 1 \\ 0 \end{bmatrix} = \begin{bmatrix} 8 \\ 4 \end{bmatrix}$$

$$A = \left(X^T X\right)^{-1} X^T Y = \frac{1}{8}\begin{bmatrix} 6 & -4 \\ -4 & 4 \end{bmatrix}\begin{bmatrix} 8 \\ 4 \end{bmatrix} = \begin{bmatrix} 4 \\ -2 \end{bmatrix}.$$

So, the least squares regression line is $y = 4 - 2x$.

(c) Solving $Y = XA + E$ for E,

$E = Y - XA$

$$= \begin{bmatrix} 4 \\ 3 \\ 1 \\ 0 \end{bmatrix} - \begin{bmatrix} 1 & 0 \\ 1 & 1 \\ 1 & 1 \\ 1 & 2 \end{bmatrix}\begin{bmatrix} 4 \\ -2 \end{bmatrix} = \begin{bmatrix} 4 \\ 3 \\ 1 \\ 0 \end{bmatrix} - \begin{bmatrix} 4 \\ 2 \\ 2 \\ 0 \end{bmatrix} = \begin{bmatrix} 0 \\ 1 \\ -1 \\ 0 \end{bmatrix}.$$

So, the sum of the squared error is

$$E^T E = \begin{bmatrix} 0 & 1 & -1 & 0 \end{bmatrix}\begin{bmatrix} 0 \\ 1 \\ -1 \\ 0 \end{bmatrix} = 2.$$

31. Using the matrices

$$X = \begin{bmatrix} 1 & 0 \\ 1 & 1 \\ 1 & 2 \end{bmatrix} \text{ and } Y = \begin{bmatrix} 0 \\ 1 \\ 4 \end{bmatrix}, \text{ you have}$$

$$X^T X = \begin{bmatrix} 3 & 3 \\ 3 & 5 \end{bmatrix},\ X^T Y = \begin{bmatrix} 5 \\ 9 \end{bmatrix}, \text{ and}$$

$$A = \left(X^T X\right)^{-1}\left(X^T Y\right) = \begin{bmatrix} -\frac{1}{3} \\ 2 \end{bmatrix}.$$

So, the least squares regression line is $y = 2x - \frac{1}{3}$.

33. Using the matrices

$$X = \begin{bmatrix} 1 & -2 \\ 1 & -1 \\ 1 & 0 \\ 1 & 1 \end{bmatrix} \text{ and } Y = \begin{bmatrix} 0 \\ 1 \\ 1 \\ 2 \end{bmatrix}, \text{ you have}$$

$$X^T X = \begin{bmatrix} 4 & -2 \\ -2 & 6 \end{bmatrix},\ X^T Y = \begin{bmatrix} 4 \\ 1 \end{bmatrix}, \text{ and}$$

$$A = \left(X^T X\right)^{-1} X^T Y = \begin{bmatrix} 0.3 & 0.1 \\ 0.1 & 0.2 \end{bmatrix}\begin{bmatrix} 4 \\ 1 \end{bmatrix} = \begin{bmatrix} 1.3 \\ 0.6 \end{bmatrix}.$$

So, the least squares regression line is $y = 0.6x + 1.3$.

35. Using the four given points, the matrices X and Y are

$$X = \begin{bmatrix} 1 & -5 \\ 1 & 1 \\ 1 & 2 \\ 1 & 2 \end{bmatrix} \text{ and } Y = \begin{bmatrix} 1 \\ 3 \\ 3 \\ 5 \end{bmatrix}. \text{ This means that}$$

$$X^T X = \begin{bmatrix} 1 & 1 & 1 & 1 \\ -5 & 1 & 2 & 2 \end{bmatrix}\begin{bmatrix} 1 & -5 \\ 1 & 1 \\ 1 & 2 \\ 1 & 2 \end{bmatrix} = \begin{bmatrix} 4 & 0 \\ 0 & 34 \end{bmatrix} \text{ and}$$

$$X^T Y = \begin{bmatrix} 1 & 1 & 1 & 1 \\ -5 & 1 & 2 & 2 \end{bmatrix}\begin{bmatrix} 1 \\ 3 \\ 3 \\ 5 \end{bmatrix} = \begin{bmatrix} 12 \\ 14 \end{bmatrix}.$$

Now, using $\left(X^T X\right)^{-1}$ to find the coefficient matrix A, you have

$$A = \left(X^T X\right)^{-1} X^T Y$$

$$= \begin{bmatrix} 4 & 0 \\ 0 & 34 \end{bmatrix}^{-1}\begin{bmatrix} 12 \\ 14 \end{bmatrix} = \begin{bmatrix} \frac{1}{4} & 0 \\ 0 & \frac{1}{34} \end{bmatrix}\begin{bmatrix} 12 \\ 14 \end{bmatrix} = \begin{bmatrix} 3 \\ \frac{7}{17} \end{bmatrix}.$$

So, the least squares regression line is

$y = \frac{7}{17}x + 3 = 0.412x + 3.$

© 2013 Cengage Learning. All Rights Reserved. May not be scanned, copied or duplicated, or posted to a publicly accessible website, in whole or in part.

37. Using the five given points, the matrices X and Y are $X = \begin{bmatrix} 1 & -5 \\ 1 & -1 \\ 1 & 3 \\ 1 & 7 \\ 1 & 5 \end{bmatrix}$ and $Y = \begin{bmatrix} 10 \\ 8 \\ 6 \\ 4 \\ 5 \end{bmatrix}$. This means that

$$X^TX = \begin{bmatrix} 1 & 1 & 1 & 1 & 1 \\ -5 & -1 & 3 & 7 & 5 \end{bmatrix}\begin{bmatrix} 1 & -5 \\ 1 & -1 \\ 1 & 3 \\ 1 & 7 \\ 1 & 5 \end{bmatrix} = \begin{bmatrix} 5 & 9 \\ 9 & 109 \end{bmatrix} \text{ and } X^TY = \begin{bmatrix} 1 & 1 & 1 & 1 & 1 \\ -5 & -1 & 3 & 7 & 5 \end{bmatrix}\begin{bmatrix} 10 \\ 8 \\ 6 \\ 4 \\ 5 \end{bmatrix} = \begin{bmatrix} 33 \\ 13 \end{bmatrix}.$$

Now, using $(X^TX)^{-1}$ to find the coefficient matrix A, you have

$$A = (X^TX)^{-1}X^TY = \begin{bmatrix} 5 & 9 \\ 9 & 109 \end{bmatrix}^{-1}\begin{bmatrix} 33 \\ 13 \end{bmatrix} = \begin{bmatrix} \frac{109}{464} & -\frac{9}{464} \\ -\frac{9}{464} & \frac{5}{464} \end{bmatrix}\begin{bmatrix} 33 \\ 13 \end{bmatrix} = \begin{bmatrix} \frac{15}{2} \\ -\frac{1}{2} \end{bmatrix}.$$

So, the least squares regression line is $y = -\frac{1}{2}x + \frac{15}{2} = -0.5x + 7.5$.

39. (a) Using the matrices $X = \begin{bmatrix} 1 & 3.00 \\ 1 & 3.25 \\ 1 & 3.50 \end{bmatrix}$ and $Y = \begin{bmatrix} 4500 \\ 3750 \\ 3300 \end{bmatrix}$, you have

$$X^TX = \begin{bmatrix} 1 & 1 & 1 \\ 3.00 & 3.25 & 3.50 \end{bmatrix}\begin{bmatrix} 1 & 3.00 \\ 1 & 3.25 \\ 1 & 3.50 \end{bmatrix} = \begin{bmatrix} 3 & 9.75 \\ 9.75 & 31.8125 \end{bmatrix} \text{ and } X^TY = \begin{bmatrix} 1 & 1 & 1 \\ 3.00 & 3.25 & 3.50 \end{bmatrix}\begin{bmatrix} 4500 \\ 3750 \\ 3300 \end{bmatrix} = \begin{bmatrix} 11{,}550 \\ 37{,}237.5 \end{bmatrix}.$$

Now, using $(X^TX)^{-1}$ to find the coefficient matrix A, you have

$$A = (X^TX)^{-1}X^TY = \frac{1}{0.375}\begin{bmatrix} 31.8125 & -9.75 \\ -9.75 & 3 \end{bmatrix}\begin{bmatrix} 11{,}550 \\ 37{,}237.5 \end{bmatrix} = \frac{1}{0.375}\begin{bmatrix} 4368.75 \\ -900 \end{bmatrix} = \begin{bmatrix} 11{,}650 \\ -2400 \end{bmatrix}.$$

So, the least squares regression line is $y = 11{,}650 - 2400x$.

(b) When $x = 3.40$, $y = 11{,}650 - 2400(3.40) = 3490$. So, the demand is 3490 gallons.

41. (a) Using the matrices $X = \begin{bmatrix} 1 & 4 \\ 1 & 5 \\ 1 & 6 \\ 1 & 7 \\ 1 & 8 \end{bmatrix}$ and $Y = \begin{bmatrix} 237.2 \\ 241.2 \\ 244.2 \\ 247.3 \\ 248.2 \end{bmatrix}$, you have

$$X^TX = \begin{bmatrix} 1 & 1 & 1 & 1 & 1 \\ 4 & 5 & 6 & 7 & 8 \end{bmatrix}\begin{bmatrix} 1 & 4 \\ 1 & 5 \\ 1 & 6 \\ 1 & 7 \\ 1 & 8 \end{bmatrix} = \begin{bmatrix} 5 & 30 \\ 30 & 190 \end{bmatrix} \text{ and } X^TY = \begin{bmatrix} 1 & 1 & 1 & 1 & 1 \\ 4 & 5 & 6 & 7 & 8 \end{bmatrix}\begin{bmatrix} 237.2 \\ 241.2 \\ 244.2 \\ 247.3 \\ 248.2 \end{bmatrix} = \begin{bmatrix} 1218.1 \\ 7336.7 \end{bmatrix}.$$

Now, using $(X^TX)^{-1}$ to find the coefficient matrix A, you have

$$A = (X^TX)^{T}X^TY = \frac{1}{50}\begin{bmatrix} 190 & -30 \\ -30 & 5 \end{bmatrix}\begin{bmatrix} 1218.1 \\ 7336.7 \end{bmatrix} = \begin{bmatrix} 226.76 \\ 2.81 \end{bmatrix}.$$

So, the least squares regression line is $y = 2.81t + 226.76$.

(b) Using a graphing utility with $L_1 = \{4, 5, 6, 7, 8\}$ and $L_2 = \{237.2, 241.2, 244.2, 247.3, 248.2\}$ gives the same least squares regression line $y = 2.81t + 226.76$.

© 2013 Cengage Learning. All Rights Reserved. May not be scanned, copied or duplicated, or posted to a publicly accessible website, in whole or in part.

43. The sum of squared error is

$$\sum_{i=1}^{n} e_i^2 = \sum_{i=1}^{n} \left(y_i - f(x_i)\right)^2 = \sum_{i=1}^{n} \left(y_i^2 - 2y_i f(x_i) + \left(f(x_i)\right)^2\right) = Y^T Y - 2(XA)^T Y + (XA)^T XA.$$

To find the values for A that minimize the sum, take the derivative with respect to A.

$$\begin{aligned} -2X^T Y + 2X^T XA &= 0 \\ 2X^T XA &= 2X^T Y \\ A &= \left(X^T X\right)^{-1} X^T Y \end{aligned}$$

45. Because the columns in a stochastic matrix add up to 1, you can represent two stochastic matrices as

$$P\begin{bmatrix} a & b \\ 1 - a & 1 - b \end{bmatrix} \text{and } Q = \begin{bmatrix} c & d \\ 1 - c & 1 - d \end{bmatrix}.$$

Then,

$$\begin{aligned} PQ &= \begin{bmatrix} a & b \\ 1 - a & 1 - b \end{bmatrix}\begin{bmatrix} c & d \\ 1 - c & 1 - d \end{bmatrix} \\ &= \begin{bmatrix} ac + b(1 - c) & ad + b(1 - d) \\ c(1 - a) + (1 - b)(1 - c) & d(1 - a) + (1 - b)(1 - d) \end{bmatrix} = \begin{bmatrix} ac + b - bc & ad + b - bd \\ 1 - (ac + b - bc) & 1 - (ad + b - bd) \end{bmatrix}. \end{aligned}$$

The columns of PQ add up to 1, and the entries are nonnegative, because those of P and Q are nonnegative. So, PQ is stochastic.

Review Exercises for Chapter 2

1. $$\begin{bmatrix} 2 & 1 & 0 \\ 0 & 5 & -4 \end{bmatrix} - 3\begin{bmatrix} 5 & 3 & -6 \\ 0 & -2 & 5 \end{bmatrix} = \begin{bmatrix} 2 & 1 & 0 \\ 0 & 5 & -4 \end{bmatrix} - \begin{bmatrix} 15 & 9 & -18 \\ 0 & -6 & 15 \end{bmatrix} = \begin{bmatrix} -13 & -8 & 18 \\ 0 & 11 & -19 \end{bmatrix}$$

3. $$\begin{bmatrix} 1 & 2 \\ 5 & -4 \\ 6 & 0 \end{bmatrix}\begin{bmatrix} 6 & -2 & 8 \\ 4 & 0 & 0 \end{bmatrix} = \begin{bmatrix} 1(6) + 2(4) & 1(-2) + 2(0) & 1(8) + 2(0) \\ 5(6) - 4(4) & 5(-2) - 4(0) & 5(8) - 4(0) \\ 6(6) + 0(4) & 6(-2) + 0(0) & 6(8) + 0(0) \end{bmatrix} = \begin{bmatrix} 14 & -2 & 8 \\ 14 & -10 & 40 \\ 36 & -12 & 48 \end{bmatrix}$$

5. $$\begin{bmatrix} 1 & 3 & 2 \\ 0 & 2 & -4 \\ 0 & 0 & 3 \end{bmatrix}\begin{bmatrix} 4 & -3 & 2 \\ 0 & 3 & -1 \\ 0 & 0 & 2 \end{bmatrix} = \begin{bmatrix} 1(4) & 1(-3) + 3(3) & 1(2) + 3(-1) + 2(2) \\ 0 & 2(3) & 2(-1) + (-4)(2) \\ 0 & 0 & 3(2) \end{bmatrix} = \begin{bmatrix} 4 & 6 & 3 \\ 0 & 6 & -10 \\ 0 & 0 & 6 \end{bmatrix}$$

7. Letting $A = \begin{bmatrix} 2 & 1 \\ 1 & 4 \end{bmatrix}$, $\mathbf{x} = \begin{bmatrix} x_1 \\ x_2 \end{bmatrix}$, and $\mathbf{b} = \begin{bmatrix} -8 \\ -4 \end{bmatrix}$,

the system can be written as

$$A\mathbf{x} = \mathbf{b}$$

$$\begin{bmatrix} 2 & 1 \\ 1 & 4 \end{bmatrix}\begin{bmatrix} x_1 \\ x_2 \end{bmatrix} = \begin{bmatrix} -8 \\ -4 \end{bmatrix}.$$

Using Gaussian elimination, the solution of the system is

$$\mathbf{x} = \begin{bmatrix} -4 \\ 0 \end{bmatrix}.$$

9. Letting $A = \begin{bmatrix} -3 & -1 & 1 \\ 2 & 4 & -5 \\ 1 & -2 & 3 \end{bmatrix}$, $\mathbf{x} = \begin{bmatrix} x_1 \\ x_2 \\ x_3 \end{bmatrix}$, and $\mathbf{b} = \begin{bmatrix} 0 \\ -3 \\ 1 \end{bmatrix}$,

the system can be written as

$$A\mathbf{x} = \mathbf{b}$$

$$\begin{bmatrix} -3 & -1 & 1 \\ 2 & 4 & -5 \\ 1 & -2 & 3 \end{bmatrix}\begin{bmatrix} x_1 \\ x_2 \\ x_3 \end{bmatrix} = \begin{bmatrix} 0 \\ -3 \\ 1 \end{bmatrix}.$$

Using Gaussian elimination, the solution of the system is

$$\mathbf{x} = \begin{bmatrix} \frac{2}{3} \\ -\frac{17}{8} \\ -\frac{11}{3} \end{bmatrix}.$$

© 2013 Cengage Learning. All Rights Reserved. May not be scanned, copied or duplicated, or posted to a publicly accessible website, in whole or in part.

11. $A^T = \begin{bmatrix} 1 & 0 \\ 2 & 1 \\ -3 & 2 \end{bmatrix}$

$$A^T A = \begin{bmatrix} 1 & 0 \\ 2 & 1 \\ -3 & 2 \end{bmatrix} \begin{bmatrix} 1 & 2 & -3 \\ 0 & 1 & 2 \end{bmatrix} = \begin{bmatrix} 1 & 2 & -3 \\ 2 & 5 & -4 \\ -3 & -4 & 13 \end{bmatrix}$$

$$AA^T = \begin{bmatrix} 1 & 2 & -3 \\ 0 & 1 & 2 \end{bmatrix} \begin{bmatrix} 1 & 0 \\ 2 & 1 \\ -3 & 2 \end{bmatrix} = \begin{bmatrix} 14 & -4 \\ -4 & 5 \end{bmatrix}$$

13. $A^T = \begin{bmatrix} 1 & 3 & -1 \end{bmatrix}$

$$A^T A = \begin{bmatrix} 1 & 3 & -1 \end{bmatrix} \begin{bmatrix} 1 \\ 3 \\ -1 \end{bmatrix} = \begin{bmatrix} 11 \end{bmatrix}$$

$$AA^T = \begin{bmatrix} 1 \\ 3 \\ -1 \end{bmatrix} \begin{bmatrix} 1 & 3 & -1 \end{bmatrix} = \begin{bmatrix} 1 & 3 & -1 \\ 3 & 9 & -3 \\ -1 & -3 & 1 \end{bmatrix}$$

15. Use the formula for the inverse of a 2×2 matrix.

$$A^{-1} = \frac{1}{ad - bc} \begin{bmatrix} d & -b \\ -c & a \end{bmatrix} = \frac{1}{3(-1) - (-1)(2)} \begin{bmatrix} -1 & 1 \\ -2 & 3 \end{bmatrix} = \begin{bmatrix} 1 & -1 \\ 2 & -3 \end{bmatrix}$$

17. Begin by adjoining the identity matrix to the given matrix.

$$\begin{bmatrix} A & I \end{bmatrix} = \begin{bmatrix} 2 & 3 & 1 & 1 & 0 & 0 \\ 2 & -3 & -3 & 0 & 1 & 0 \\ 4 & 0 & 3 & 0 & 0 & 1 \end{bmatrix}$$

This matrix reduces to

$$\begin{bmatrix} I & A^{-1} \end{bmatrix} = \begin{bmatrix} 1 & 0 & 0 & \frac{3}{20} & \frac{3}{20} & \frac{1}{10} \\ 0 & 1 & 0 & \frac{3}{10} & -\frac{1}{30} & -\frac{2}{15} \\ 0 & 0 & 1 & -\frac{1}{5} & -\frac{1}{5} & \frac{1}{5} \end{bmatrix}.$$

So, the inverse matrix is

$$A^{-1} = \begin{bmatrix} \frac{3}{20} & \frac{3}{20} & \frac{1}{10} \\ \frac{3}{20} & -\frac{1}{30} & -\frac{2}{15} \\ -\frac{1}{5} & -\frac{1}{5} & \frac{1}{5} \end{bmatrix}.$$

19.

$$\underset{A}{\begin{bmatrix} 5 & 4 \\ -1 & 1 \end{bmatrix}} \underset{\mathbf{x}}{\begin{bmatrix} x \\ y \end{bmatrix}} = \underset{\mathbf{b}}{\begin{bmatrix} -15 \\ -6 \end{bmatrix}}$$

Because $A^{-1} = \frac{1}{9}\begin{bmatrix} 1 & -4 \\ 1 & 5 \end{bmatrix} = \begin{bmatrix} \frac{1}{9} & -\frac{4}{9} \\ \frac{1}{9} & \frac{5}{9} \end{bmatrix}$, solve the equation $A\mathbf{x} = \mathbf{b}$ as follows.

$$\mathbf{x} = A^{-1}\mathbf{b} = \begin{bmatrix} \frac{1}{9} & -\frac{4}{9} \\ \frac{1}{9} & \frac{5}{9} \end{bmatrix} \begin{bmatrix} -15 \\ -6 \end{bmatrix} = \begin{bmatrix} 1 \\ -5 \end{bmatrix}$$

21.

$$\underset{A}{\begin{bmatrix} 0 & 1 & -2 \\ -1 & 3 & 1 \\ 2 & -2 & 4 \end{bmatrix}} \underset{\mathbf{x}}{\begin{bmatrix} x_1 \\ x_2 \\ x_3 \end{bmatrix}} = \underset{\mathbf{b}}{\begin{bmatrix} -1 \\ 0 \\ 2 \end{bmatrix}}$$

Using Gauss-Jordan elimination, you find that

$$A^{-1} = \begin{bmatrix} 1 & 0 & \frac{1}{2} \\ \frac{3}{7} & \frac{2}{7} & \frac{1}{7} \\ -\frac{2}{7} & \frac{1}{7} & \frac{1}{14} \end{bmatrix}.$$

Solve the equation $A\mathbf{x} = \mathbf{b}$ as follows.

$$\mathbf{x} = A^{-1}\mathbf{b} = \begin{bmatrix} 1 & 0 & \frac{1}{2} \\ \frac{3}{7} & \frac{2}{7} & \frac{1}{7} \\ -\frac{2}{7} & \frac{1}{7} & \frac{1}{14} \end{bmatrix} \begin{bmatrix} -1 \\ 0 \\ 2 \end{bmatrix} = \begin{bmatrix} 0 \\ -\frac{1}{7} \\ \frac{3}{7} \end{bmatrix}$$

23. $A\mathbf{x} = \mathbf{b}$

$$\begin{bmatrix} 5 & 4 \\ -1 & 1 \end{bmatrix} \begin{bmatrix} x_1 \\ x_2 \end{bmatrix} = \begin{bmatrix} 2 \\ -22 \end{bmatrix}$$

Because

$$A^{-1} = \frac{1}{5(1) - 4(-1)} \begin{bmatrix} 1 & -4 \\ 1 & 5 \end{bmatrix} = \begin{bmatrix} \frac{1}{9} & -\frac{4}{9} \\ \frac{1}{9} & \frac{5}{9} \end{bmatrix}$$

solve the equation $A\mathbf{x} = \mathbf{b}$ as follows.

$$\mathbf{x} = A^{-1}\mathbf{b} = \begin{bmatrix} \frac{1}{9} & -\frac{4}{9} \\ \frac{1}{9} & \frac{5}{9} \end{bmatrix} \begin{bmatrix} 2 \\ -22 \end{bmatrix} = \begin{bmatrix} 10 \\ -12 \end{bmatrix}$$

25. $A\mathbf{x} = \mathbf{b}$

$$\begin{bmatrix} -1 & 1 & 2 \\ 2 & 3 & 1 \\ 5 & 4 & 2 \end{bmatrix} \begin{bmatrix} x_1 \\ x_2 \\ x_3 \end{bmatrix} = \begin{bmatrix} 1 \\ -2 \\ 4 \end{bmatrix}$$

Using Gauss-Jordan elimination, you find that

$$A^{-1} = \begin{bmatrix} -\frac{2}{15} & -\frac{2}{5} & \frac{1}{3} \\ -\frac{1}{15} & \frac{4}{5} & -\frac{1}{3} \\ \frac{7}{15} & -\frac{3}{5} & \frac{1}{3} \end{bmatrix}.$$

So, solve the equation $A\mathbf{x} = \mathbf{b}$ as follows.

$$\mathbf{x} = A^{-1}\mathbf{b} = \begin{bmatrix} -\frac{2}{15} & -\frac{2}{5} & \frac{1}{3} \\ -\frac{1}{15} & \frac{4}{5} & -\frac{1}{3} \\ \frac{7}{15} & -\frac{3}{5} & \frac{1}{3} \end{bmatrix} \begin{bmatrix} 1 \\ -2 \\ 4 \end{bmatrix} = \begin{bmatrix} 2 \\ -3 \\ 3 \end{bmatrix}$$

© 2013 Cengage Learning. All Rights Reserved. May not be scanned, copied or duplicated, or posted to a publicly accessible website, in whole or in part.

27. Because $(3A)^{-1} = \begin{bmatrix} 4 & -1 \\ 2 & 3 \end{bmatrix}$,

you can use the formula for the inverse of a 2×2 matrix to obtain

$$3A = \begin{bmatrix} 4 & -1 \\ 2 & 3 \end{bmatrix}^{-1} = \frac{1}{4(3) - (-1)(2)} \begin{bmatrix} 3 & 1 \\ -2 & 4 \end{bmatrix} = \frac{1}{14}\begin{bmatrix} 3 & 1 \\ -2 & 4 \end{bmatrix}.$$

So, $A = \frac{1}{42}\begin{bmatrix} 3 & 1 \\ -2 & 4 \end{bmatrix} = \begin{bmatrix} \frac{1}{14} & \frac{1}{42} \\ -\frac{1}{21} & \frac{2}{21} \end{bmatrix}$.

29. A is nonsingular if and only if the second row is not a multiple of the first. That is, A is nonsingular if and only if $x \neq -3$.

Alternatively, you could use the formula for the inverse of a 2×2 matrix to show that A is nonsingular if $ad - bc = 3(-1) - 1(x) \neq 0$. That is, $x \neq -3$.

31. Because the given matrix represents the addition of 4 times the third row to the first row of I_3, reverse the operation and subtract 4 times the third row from the first row.

$$E^{-1} = \begin{bmatrix} 1 & 0 & -4 \\ 0 & 1 & 0 \\ 0 & 0 & 1 \end{bmatrix}$$

For Exercises 33–39, answers will vary. Sample answers are shown below.

33. Begin by finding a sequence of elementary row operations that can be used to write A in reduced row-echelon form.

Matrix	Elementary Row Operation	Elementary Matrix
$\begin{bmatrix} 1 & \frac{3}{2} \\ 0 & 1 \end{bmatrix}$	Divide row one by 2.	$E_1 = \begin{bmatrix} \frac{1}{2} & 0 \\ 0 & 1 \end{bmatrix}$
$\begin{bmatrix} 1 & 0 \\ 0 & 1 \end{bmatrix}$	Subtract $\frac{3}{2}$ times row two from row one.	$E_2 = \begin{bmatrix} 1 & -\frac{3}{2} \\ 0 & 1 \end{bmatrix}$

So, factor A as follows.

$$A = E_1^{-1}E_2^{-1} = \begin{bmatrix} 2 & 0 \\ 0 & 1 \end{bmatrix}\begin{bmatrix} 1 & \frac{3}{2} \\ 0 & 1 \end{bmatrix}$$

35. Begin by finding a sequence of elementary row operations to write A in reduced row-echelon form.

Matrix	Elementary Row Operation	Elementary Matrix
$\begin{bmatrix} 1 & 0 & 1 \\ 0 & 1 & -2 \\ 0 & 0 & 1 \end{bmatrix}$	$\frac{1}{4}$ times row 3.	$E_1 = \begin{bmatrix} 1 & 0 & 0 \\ 0 & 1 & 0 \\ 0 & 0 & \frac{1}{4} \end{bmatrix}$
$\begin{bmatrix} 1 & 0 & 1 \\ 0 & 1 & 0 \\ 0 & 0 & 1 \end{bmatrix}$	Add two times row three to row two.	$E_2 = \begin{bmatrix} 1 & 0 & 0 \\ 0 & 1 & 2 \\ 0 & 0 & 1 \end{bmatrix}$
$\begin{bmatrix} 1 & 0 & 0 \\ 0 & 1 & 0 \\ 0 & 0 & 1 \end{bmatrix}$	Add -1 times row three to row one.	$E_3 = \begin{bmatrix} 1 & 0 & -1 \\ 0 & 1 & 0 \\ 0 & 0 & 1 \end{bmatrix}$

So, factor A as follows.

$$A = E_1^{-1}E_2^{-1}E_3^{-1} = \begin{bmatrix} 1 & 0 & 0 \\ 0 & 1 & 0 \\ 0 & 0 & 4 \end{bmatrix}\begin{bmatrix} 1 & 0 & 0 \\ 0 & 1 & -2 \\ 0 & 0 & 1 \end{bmatrix}\begin{bmatrix} 1 & 0 & 1 \\ 0 & 1 & 0 \\ 0 & 0 & 1 \end{bmatrix}$$

37. Let $A = \begin{bmatrix} a & b \\ c & d \end{bmatrix}$, then $A^2 = \begin{bmatrix} a & b \\ c & d \end{bmatrix}\begin{bmatrix} a & b \\ c & d \end{bmatrix} = \begin{bmatrix} a^2 + bc & ad + bd \\ ca + dc & cb + d^2 \end{bmatrix} = \begin{bmatrix} 1 & 0 \\ 0 & 1 \end{bmatrix}$.

So, many answers are possible: $\begin{bmatrix} 1 & 0 \\ 0 & 1 \end{bmatrix}, \begin{bmatrix} -1 & 0 \\ 0 & -1 \end{bmatrix}, \begin{bmatrix} -1 & 0 \\ 0 & 1 \end{bmatrix}$, etc.

© 2013 Cengage Learning. All Rights Reserved. May not be scanned, copied or duplicated, or posted to a publicly accessible website, in whole or in part.

39. Let $A = \begin{bmatrix} a & b \\ c & d \end{bmatrix}$, then

$$A^2 = \begin{bmatrix} a & b \\ c & d \end{bmatrix}\begin{bmatrix} a & b \\ c & d \end{bmatrix} = \begin{bmatrix} a^2 + bc & b(a + d) \\ c(a + d) & bc + d^2 \end{bmatrix}.$$

Solving $A^2 = A$ gives the system of nonlinear equations

$$a^2 + bc = a$$
$$d^2 + bc = d$$
$$b(a + d) = b$$
$$c(a + d) = c.$$

From this system, conclude that any of the following matrices are solutions to the equation $A^2 = A$.

$$\begin{bmatrix} 0 & 0 \\ 0 & 0 \end{bmatrix}, \begin{bmatrix} 0 & 0 \\ t & 1 \end{bmatrix}, \begin{bmatrix} 0 & t \\ 0 & 1 \end{bmatrix}, \begin{bmatrix} 1 & 0 \\ t & 0 \end{bmatrix}, \begin{bmatrix} 1 & t \\ 0 & 0 \end{bmatrix}, \begin{bmatrix} 1 & 0 \\ 0 & 1 \end{bmatrix}$$

41. (a) Letting $W = aX + bY + cZ$ yields the system of linear equations

$$\begin{aligned} a - b + 3c &= 3 \\ 2a + 4c &= 2 \\ 3b - c &= -4 \\ a + 2b + 2c &= -1 \end{aligned}$$

which has the solution: $a = -1$, $b = -1$, $c = 1$.

(b) Letting $Z = aX + bY$ yields the system of linear equations

$$\begin{aligned} a - b &= 3 \\ 2a &= 4 \\ 3b &= -1 \\ a + 2b &= 2 \end{aligned}$$

which has no solution.

(c) If $aX + bY + cZ = O$, then

$$a\begin{bmatrix} 1 \\ 2 \\ 0 \\ 1 \end{bmatrix} + b\begin{bmatrix} -1 \\ 0 \\ 3 \\ 2 \end{bmatrix} + c\begin{bmatrix} 3 \\ 4 \\ -1 \\ 2 \end{bmatrix} = \begin{bmatrix} 0 \\ 0 \\ 0 \\ 0 \end{bmatrix},$$

which yields the system of equations

$$\begin{aligned} a - b + 3c &= 0 \\ 2a + 4c &= 0 \\ 3b - c &= 0 \\ a + 2b + 2c &= 0. \end{aligned}$$

Solving this homogeneous system, the only solution is $a = b = c = 0$.

43. Answers will vary. *Sample answer*:

Matrix	Elementary Matrix
$\begin{bmatrix} 2 & 5 \\ 6 & 14 \end{bmatrix} = A$	
$\begin{bmatrix} 2 & 5 \\ 0 & -1 \end{bmatrix} = U$	$E = \begin{bmatrix} 1 & 0 \\ -3 & 1 \end{bmatrix}$

$$EA = U$$

$$A = E^{-1}U = \begin{bmatrix} 1 & 0 \\ 3 & 1 \end{bmatrix}\begin{bmatrix} 2 & 5 \\ 0 & -1 \end{bmatrix} = LU$$

45.

Matrix	Elementary Matrix
$\begin{bmatrix} 1 & 0 & 1 \\ 2 & 1 & 2 \\ 3 & 2 & 6 \end{bmatrix} = A$	
$\begin{bmatrix} 1 & 0 & 1 \\ 0 & 1 & 0 \\ 3 & 2 & 6 \end{bmatrix}$	$E_1 = \begin{bmatrix} 1 & 0 & 0 \\ -2 & 1 & 0 \\ 0 & 0 & 1 \end{bmatrix}$
$\begin{bmatrix} 1 & 0 & 1 \\ 0 & 1 & 0 \\ 0 & 2 & 3 \end{bmatrix}$	$E_2 = \begin{bmatrix} 1 & 0 & 0 \\ 0 & 1 & 0 \\ -3 & 0 & 1 \end{bmatrix}$
$\begin{bmatrix} 1 & 0 & 1 \\ 0 & 1 & 0 \\ 0 & 0 & 3 \end{bmatrix} = U$	$E_3 = \begin{bmatrix} 1 & 0 & 0 \\ 0 & 1 & 0 \\ 0 & -2 & 1 \end{bmatrix}$

$$E_3E_2E_1A = U$$

$$A = E_1^{-1}E_2^{-1}E_3^{-1}U = LU = \begin{bmatrix} 1 & 0 & 0 \\ 2 & 1 & 0 \\ 3 & 2 & 1 \end{bmatrix}\begin{bmatrix} 1 & 0 & 1 \\ 0 & 1 & 0 \\ 0 & 0 & 3 \end{bmatrix}.$$

$$L\mathbf{y} = \mathbf{b}: \begin{bmatrix} 1 & 0 & 0 \\ 2 & 1 & 0 \\ 3 & 2 & 1 \end{bmatrix}\begin{bmatrix} y_1 \\ y_2 \\ y_3 \end{bmatrix} = \begin{bmatrix} 3 \\ 7 \\ 8 \end{bmatrix} \Rightarrow \mathbf{y} = \begin{bmatrix} 3 \\ 1 \\ -3 \end{bmatrix}$$

$$U\mathbf{x} = \mathbf{y}: \begin{bmatrix} 1 & 0 & 1 \\ 0 & 1 & 0 \\ 0 & 0 & 3 \end{bmatrix}\begin{bmatrix} x_1 \\ x_2 \\ x_3 \end{bmatrix} = \begin{bmatrix} 3 \\ 1 \\ -3 \end{bmatrix} \Rightarrow \mathbf{x} = \begin{bmatrix} 3 \\ 1 \\ -1 \end{bmatrix}$$

So, $x = 4$, $y = 1$, and $z = -1$.

47. (a) False. See Theorem 2.1, part 1, page 52.

(b) True. See Theorem 2.6, part 2, page 57.

49. (a) False. The matrix $\begin{bmatrix} 1 & 0 \\ 0 & 0 \end{bmatrix}$ is not invertible.

(b) False. See Exercise 65, page 61.

51. $1.1\begin{bmatrix} 100 & 90 & 70 & 30 \\ 40 & 20 & 60 & 60 \end{bmatrix} = \begin{bmatrix} 110 & 99 & 77 & 33 \\ 44 & 22 & 66 & 66 \end{bmatrix}$

© 2013 Cengage Learning. All Rights Reserved. May not be scanned, copied or duplicated, or posted to a publicly accessible website, in whole or in part.

53. (a) $AB = \begin{bmatrix} 580 & 840 & 320 \\ 560 & 420 & 160 \\ 860 & 1020 & 540 \end{bmatrix}\begin{bmatrix} 3.05 & 0.05 \\ 3.15 & 0.08 \\ 3.25 & 0.10 \end{bmatrix} = \begin{bmatrix} 5455 & 128.2 \\ 3551 & 77.6 \\ 7591 & 178.6 \end{bmatrix}$

This matrix shows the total sales of gas each day in the first column and the total profit each day in the second column.

(b) The gasoline sales profit for Friday through Sunday is the sum of the elements in the second column of AB,
$128.2 + 77.6 + 178.6 = \$384.40$.

55. $f(A) = \begin{bmatrix} 5 & 4 \\ 1 & 2 \end{bmatrix}^2 - 7\begin{bmatrix} 5 & 4 \\ 1 & 2 \end{bmatrix} + 6\begin{bmatrix} 1 & 0 \\ 0 & 1 \end{bmatrix}$

$= \begin{bmatrix} 29 & 28 \\ 7 & 8 \end{bmatrix} - \begin{bmatrix} 35 & 28 \\ 7 & 14 \end{bmatrix} + \begin{bmatrix} 6 & 0 \\ 0 & 6 \end{bmatrix}$

$= \begin{bmatrix} 0 & 0 \\ 0 & 0 \end{bmatrix}$

57. The given matrix is not stochastic because the entries in columns two and three do not add up to 1.

59. $PX = \begin{bmatrix} \frac{1}{2} & \frac{1}{4} \\ \frac{1}{2} & \frac{3}{4} \end{bmatrix}\begin{bmatrix} 128 \\ 64 \end{bmatrix} = \begin{bmatrix} 80 \\ 112 \end{bmatrix}$

$P^2X = P\begin{bmatrix} 80 \\ 112 \end{bmatrix} = \begin{bmatrix} 68 \\ 124 \end{bmatrix}$

$P^3X = P\begin{bmatrix} 68 \\ 124 \end{bmatrix} = \begin{bmatrix} 65 \\ 127 \end{bmatrix}$

61. Begin by forming the matrix of transition probabilities.

From Region (columns 1, 2, 3); To Region (rows 1, 2, 3)

$$P = \begin{bmatrix} 0.85 & 0.15 & 0.10 \\ 0.10 & 0.80 & 0.10 \\ 0.05 & 0.05 & 0.80 \end{bmatrix}\begin{matrix} 1 \\ 2 \\ 3 \end{matrix}$$

(a) The population in each region after 1 year is given by

$$PX = \begin{bmatrix} 0.85 & 0.15 & 0.10 \\ 0.10 & 0.80 & 0.10 \\ 0.05 & 0.05 & 0.80 \end{bmatrix}\begin{bmatrix} 100{,}000 \\ 100{,}000 \\ 100{,}000 \end{bmatrix} = \begin{bmatrix} 110{,}000 \\ 100{,}000 \\ 90{,}000 \end{bmatrix}\begin{matrix} \text{Region 1} \\ \text{Region 2.} \\ \text{Region 3} \end{matrix}$$

(b) The population in each region after 3 years is given by

$$P^3X = \begin{bmatrix} 0.665375 & 0.322375 & 0.2435 \\ 0.219 & 0.562 & 0.219 \\ 0.115625 & 0.115625 & 0.5375 \end{bmatrix}\begin{bmatrix} 100{,}000 \\ 100{,}000 \\ 100{,}000 \end{bmatrix} = \begin{bmatrix} 123{,}125 \\ 100{,}000 \\ 76{,}875 \end{bmatrix}\begin{matrix} \text{Region 1} \\ \text{Region 2.} \\ \text{Region 3} \end{matrix}$$

63. The uncoded row matrices are

O N	E _	I F	_ B	Y _	L A	N D
[15 14]	[5 0]	[9 6]	[0 2]	[25 0]	[12 1]	[14 4].

Multiplying each 1×2 matrix on the right by A yields the coded row matrices

[103 44], [25 10], [57 24], [4 2], [125 50], [62 25], [78 32].

So, the coded message is 103, 44, 25, 10, 57, 24, 4, 2, 125, 50, 62, 25, 78, 32.

65. You can find A^{-1} to be $\begin{bmatrix} 3 & 2 \\ 4 & 3 \end{bmatrix}$ and the coded row matrices are

[−45 34], [36 −24], [−43 37], [−23 22], [−37 29], [57 −38], [−39 31].

Multiplying each coded row matrix on the right by A^{-1} yields the uncoded row matrices

A L	L _	S Y	S T	E M	S _	G O
[1 12]	[12 0]	[19 25]	[19 20]	[5 13]	[19 0]	[7 15].

The decoded message is ALL_SYSTEMS_GO.

© 2013 Cengage Learning. All Rights Reserved. May not be scanned, copied or duplicated, or posted to a publicly accessible website, in whole or in part.

67. You can find A^{-1} to be $\begin{bmatrix} -2 & -1 & 0 \\ 0 & 1 & 1 \\ -5 & -3 & -3 \end{bmatrix}$ and the coded row matrices are

$[58 \ \ -3 \ \ -25], [-48 \ \ 28 \ \ 19], [-40 \ \ 13 \ \ 13], [-98 \ \ 39 \ \ 39], [118 \ \ -25 \ \ -48], [28 \ \ -14 \ \ -14].$

Multiplying each coded row matrix on the right by A^{-1} yields the uncoded row matrices

I N V	A S I	O N _	A T _	D A W	N _ _
$[9 \ \ 14 \ \ 22]$	$[1 \ \ 19 \ \ 9]$	$[15 \ \ 14 \ \ 0]$	$[1 \ \ 20 \ \ 0]$	$[4 \ \ 1 \ \ 23]$	$[14 \ \ 0 \ \ 0]$.

So, the message is INVASION_AT_DAWN.

69. Find $A^{-1} = \begin{bmatrix} -1 & -10 & -8 \\ -1 & -6 & -5 \\ 0 & -1 & -1 \end{bmatrix}$,

and multiply each coded row matrix on the right by A^{-1} to find the associated uncoded row matrix.

$$[-2 \ \ 2 \ \ 5]A^{-1} = [-2 \ \ 2 \ \ 5]\begin{bmatrix} -1 & -10 & -8 \\ -1 & -6 & -5 \\ 0 & -1 & -1 \end{bmatrix} = [0 \ \ 3 \ \ 1] \Rightarrow \_, \ C, \ A$$

$$[39 \ \ -53 \ \ -72]\ A^{-1} = [14 \ \ 0 \ \ 25] \Rightarrow N, \ \_, \ Y$$
$$[-6 \ \ -9 \ \ 93]\ A^{-1} = [15 \ \ 21 \ \ 0] \Rightarrow O, \ U, \ \_$$
$$[4 \ \ -12 \ \ 27]\ A^{-1} = [8 \ \ 5 \ \ 1] \Rightarrow H, \ E, \ A$$
$$[31 \ \ -49 \ \ -16]\ A^{-1} = [18 \ \ 0 \ \ 13] \Rightarrow R, \ \_, \ M$$
$$[19 \ \ -24 \ \ -46]\ A^{-1} = [5 \ \ 0 \ \ 14] \Rightarrow E, \ \_, \ N$$
$$[-8 \ \ -7 \ \ 99]\ A^{-1} = [15 \ \ 23 \ \ 0] \Rightarrow O, \ W, \ \_$$

The message is _CAN_YOU_HEAR_ME_NOW_.

71. First, find the input-output matrix D.

$$D = \overbrace{\begin{matrix} A & B \end{matrix}}^{\text{User Industry}} \begin{bmatrix} 0.20 & 0.50 \\ 0.30 & 0.10 \end{bmatrix} \begin{matrix} A \\ B \end{matrix} \Big\} \text{Supplier Industry}$$

Then, solve the equation $X = DX + E$ for X to obtain $(I - D)X = E$, which corresponds to solving the augmented matrix

$\begin{bmatrix} 0.80 & -0.50 & \vdots & 40{,}000 \\ -0.30 & 0.90 & \vdots & 80{,}000 \end{bmatrix}$. The solution to this system gives you $X \approx \begin{bmatrix} 133{,}333 \\ 133{,}333 \end{bmatrix}$.

73. Using the matrices $X = \begin{bmatrix} 1 & 1 \\ 1 & 2 \\ 1 & 3 \end{bmatrix}$ and $Y = \begin{bmatrix} 5 \\ 4 \\ 2 \end{bmatrix}$, you have

$$X^T X = \begin{bmatrix} 3 & 6 \\ 6 & 14 \end{bmatrix}, \ X^T Y = \begin{bmatrix} 11 \\ 19 \end{bmatrix}, \text{ and } A = \left(X^T X\right)^{-1} X^T Y = \begin{bmatrix} \frac{7}{3} & -1 \\ -1 & \frac{1}{2} \end{bmatrix}\begin{bmatrix} 11 \\ 19 \end{bmatrix} = \begin{bmatrix} \frac{20}{3} \\ -\frac{3}{2} \end{bmatrix}.$$

So, the least squares regression line is $y = -\frac{3}{2}x + \frac{20}{3}$.

© 2013 Cengage Learning. All Rights Reserved. May not be scanned, copied or duplicated, or posted to a publicly accessible website, in whole or in part.

75. Using the matrices $X = \begin{bmatrix} 1 & 1 \\ 1 & 1 \\ 1 & 1 \\ 1 & 1 \\ 1 & 2 \end{bmatrix}$ and $Y = \begin{bmatrix} 1 \\ 3 \\ 2 \\ 4 \\ 5 \end{bmatrix}$, you have

$$X^T X = \begin{bmatrix} 5 & 6 \\ 6 & 8 \end{bmatrix},\ X^T Y = \begin{bmatrix} 15 \\ 20 \end{bmatrix}, \text{ and } A = \left(X^T X\right)^{-1} X^T Y = \begin{bmatrix} 2 & -1.5 \\ -1.5 & 1.25 \end{bmatrix}\begin{bmatrix} 15 \\ 20 \end{bmatrix} = \begin{bmatrix} 0 \\ 2.5 \end{bmatrix}.$$

So, the least squares regression line is $y = 2.5x$.

77. (a) Begin by finding the matrices X and Y.

$$X = \begin{bmatrix} 1 & 1.0 \\ 1 & 1.5 \\ 1 & 2.0 \\ 1 & 2.5 \end{bmatrix} \text{ and } Y = \begin{bmatrix} 32 \\ 41 \\ 48 \\ 53 \end{bmatrix}$$

Then $X^T X = \begin{bmatrix} 1 & 1 & 1 & 1 \\ 1.0 & 1.5 & 2.0 & 2.5 \end{bmatrix}\begin{bmatrix} 1 & 1.0 \\ 1 & 1.5 \\ 1 & 2.0 \\ 1 & 2.5 \end{bmatrix} = \begin{bmatrix} 4 & 7 \\ 7 & 13.5 \end{bmatrix}$ and $X^T Y = \begin{bmatrix} 1 & 1 & 1 & 1 \\ 1.0 & 1.5 & 2.0 & 2.5 \end{bmatrix}\begin{bmatrix} 32 \\ 41 \\ 48 \\ 53 \end{bmatrix} = \begin{bmatrix} 174 \\ 322 \end{bmatrix}.$

The matrix of coefficients is $A = \left(X^T X\right)^{-1} X^T Y = \frac{1}{5}\begin{bmatrix} 13.5 & -7 \\ -7 & 4 \end{bmatrix}\begin{bmatrix} 174 \\ 322 \end{bmatrix} = \begin{bmatrix} 19 \\ 14 \end{bmatrix}.$

So, the least squares regression line is $y = 19 + 14x$.

(b) When $x = 1.6$ (160 kilograms per square kilometer), $y = 41.4$ (kilograms per square kilometer).

79. (a) Using the matrices $X = \begin{bmatrix} 1 & 5 \\ 1 & 6 \\ 1 & 7 \\ 1 & 8 \\ 1 & 9 \\ 1 & 10 \end{bmatrix}$ and $Y = \begin{bmatrix} 2.6 \\ 2.9 \\ 2.9 \\ 3.2 \\ 3.2 \\ 3.3 \end{bmatrix}$, you have

$$X^T X = \begin{bmatrix} 1 & 1 & 1 & 1 & 1 & 1 \\ 5 & 6 & 7 & 8 & 9 & 10 \end{bmatrix}\begin{bmatrix} 1 & 5 \\ 1 & 6 \\ 1 & 7 \\ 1 & 8 \\ 1 & 9 \\ 1 & 10 \end{bmatrix} = \begin{bmatrix} 6 & 45 \\ 45 & 355 \end{bmatrix} \text{ and } X^T Y = \begin{bmatrix} 1 & 1 & 1 & 1 & 1 & 1 \\ 5 & 6 & 7 & 8 & 9 & 10 \end{bmatrix}\begin{bmatrix} 2.6 \\ 2.9 \\ 2.9 \\ 3.2 \\ 3.2 \\ 3.3 \end{bmatrix} = \begin{bmatrix} 18.1 \\ 138.1 \end{bmatrix}.$$

Now, using $\left(X^T X\right)^{-1}$ to find the coefficient matrix A, you have

$$A = \left(X^T X\right)^{-1} X^T Y = \frac{1}{105}\begin{bmatrix} 355 & -45 \\ -45 & 6 \end{bmatrix}\begin{bmatrix} 18.1 \\ 138.1 \end{bmatrix} \approx \begin{bmatrix} 2.01 \\ 0.13 \end{bmatrix}.$$

So, the least squares regression line is $y = 0.13x + 2.01$.

(b) Using a graphing utility, the regression line is $y = 0.13x + 2.01$.

(c)

Year	2005	2006	2007	2008	2009	2010
Actual	2.6	2.9	2.9	3.2	3.2	3.3
Estimated	2.7	2.8	2.9	3.1	3.2	3.3

The estimated values are close to the actual values.

© 2013 Cengage Learning. All Rights Reserved. May not be scanned, copied or duplicated, or posted to a publicly accessible website, in whole or in part.

CHAPTER 3
Determinants

© 2013 Cengage Learning. All Rights Reserved. May not be scanned, copied or duplicated, or posted to a publicly accessible website, in whole or in part.

CHAPTER 3
Determinants

Section 3.1 The Determinant of a Matrix

1. The determinant of a matrix of order 1 is the entry in the matrix. So, $\det[1] = 1$.

3. $\begin{vmatrix} 2 & 1 \\ 3 & 4 \end{vmatrix} = 2(4) - 3(1) = 5$

5. $\begin{vmatrix} 5 & 2 \\ -6 & 3 \end{vmatrix} = 5(3) - (-6)(2) = 27$

7. $\begin{vmatrix} -7 & 6 \\ \frac{1}{2} & 3 \end{vmatrix} = -7(3) - \left(\frac{1}{2}\right)(6) = -24$

9. $\begin{vmatrix} 0 & 6 \\ 0 & 3 \end{vmatrix} = 0(3) - 0(6) = 0$

11. $\begin{vmatrix} \lambda - 3 & 2 \\ 4 & \lambda - 1 \end{vmatrix} = (\lambda - 3)(\lambda - 1) - 4(2)$

$= \lambda^2 - 4\lambda - 5$

13. (a) The minors of the matrix are shown below.

$M_{11} = |4| = 4 \qquad M_{12} = |3| = 3$

$M_{21} = |2| = 2 \qquad M_{22} = |1| = 1$

(b) The cofactors of the matrix are shown below.

$C_{11} = (-1)^2 M_{11} = 4 \qquad C_{12} = (-1)^3 M_{12} = -3$

$C_{21} = (-1)^3 M_{21} = -2 \qquad C_{22} = (-1)^4 M_{22} = 1$

15. (a) The minors of the matrix are shown below.

$M_{11} = \begin{vmatrix} 5 & 6 \\ -3 & 1 \end{vmatrix} = 23 \qquad M_{12} = \begin{vmatrix} 4 & 6 \\ 2 & 1 \end{vmatrix} = -8 \qquad M_{13} = \begin{vmatrix} 4 & 5 \\ 2 & -3 \end{vmatrix} = -22$

$M_{21} = \begin{vmatrix} 2 & 1 \\ -3 & 1 \end{vmatrix} = 5 \qquad M_{22} = \begin{vmatrix} -3 & 1 \\ 2 & 1 \end{vmatrix} = -5 \qquad M_{23} = \begin{vmatrix} -3 & 2 \\ 2 & -3 \end{vmatrix} = 5$

$M_{31} = \begin{vmatrix} 2 & 1 \\ 5 & 6 \end{vmatrix} = 7 \qquad M_{32} = \begin{vmatrix} -3 & 1 \\ 4 & 6 \end{vmatrix} = -22 \qquad M_{33} = \begin{vmatrix} -3 & 2 \\ 4 & 5 \end{vmatrix} = -23$

(b) The cofactors of the matrix are shown below.

$C_{11} = (-1)^2 M_{11} = 23 \qquad C_{12} = (-1)^3 M_{12} = 8 \qquad C_{13} = (-1)^4 M_{13} = -22$

$C_{21} = (-1)^3 M_{21} = -5 \qquad C_{22} = (-1)^4 M_{22} = -5 \qquad C_{23} = (-1)^5 M_{23} = -5$

$C_{31} = (-1)^4 M_{31} = 7 \qquad C_{32} = (-1)^5 M_{32} = 22 \qquad C_{33} = (-1)^6 M_{33} = -23$

17. (a) You found the cofactors of the matrix in Exercise 15. Now find the determinant by expanding along the second row.

$\begin{vmatrix} -3 & 2 & 1 \\ 4 & 5 & 6 \\ 2 & -3 & 1 \end{vmatrix} = 4C_{21} + 5C_{22} + 6C_{23}$

$= 4(-5) + 5(-5) + 6(-5)$

$= -75$

(b) Expand along the second column.

$\begin{vmatrix} -3 & 2 & 1 \\ 4 & 5 & 6 \\ 2 & -3 & 1 \end{vmatrix} = 2C_{12} + 5C_{22} - 3C_{32}$

$= 2(8) + 5(-5) - 3(22)$

$= -75$

19. Expand along the second row because it has a zero.

$\begin{vmatrix} 1 & 4 & -2 \\ 3 & 2 & 0 \\ -1 & 4 & 3 \end{vmatrix} = -3\begin{vmatrix} 4 & -2 \\ 4 & 3 \end{vmatrix} + 2\begin{vmatrix} 1 & -2 \\ -1 & 3 \end{vmatrix} - 0\begin{vmatrix} 1 & 4 \\ -1 & 4 \end{vmatrix}$

$= -3(20) + 2(1)$

$= -58$

21. Expand along the first column because it has two zeros.

$\begin{vmatrix} 2 & 4 & 6 \\ 0 & 3 & 1 \\ 0 & 0 & -5 \end{vmatrix} = 2\begin{vmatrix} 3 & 1 \\ 0 & -5 \end{vmatrix} - 0\begin{vmatrix} 4 & 6 \\ 0 & -5 \end{vmatrix} + 0\begin{vmatrix} 4 & 6 \\ 3 & 1 \end{vmatrix}$

$= 2(-15)$

$= -30$

© 2013 Cengage Learning. All Rights Reserved. May not be scanned, copied or duplicated, or posted to a publicly accessible website, in whole or in part.

23. Expand along the first row.

$$\begin{vmatrix} -0.4 & 0.4 & 0.3 \\ 0.2 & 0.2 & 0.2 \\ 0.3 & 0.2 & 0.2 \end{vmatrix} = -0.4\begin{vmatrix} 0.2 & 0.2 \\ 0.2 & 0.2 \end{vmatrix} - 0.4\begin{vmatrix} 0.2 & 0.2 \\ 0.3 & 0.2 \end{vmatrix} + 0.3\begin{vmatrix} 0.2 & 0.2 \\ 0.3 & 0.2 \end{vmatrix}$$

$$= -0.4(0) - 0.4(-0.02) + 0.3(-0.02)$$

$$= 0.002$$

25. Expand along the third row because it has a zero.

$$\begin{vmatrix} x & y & 1 \\ 2 & 3 & 1 \\ 0 & -1 & 1 \end{vmatrix} = -(-1)\begin{vmatrix} x & 1 \\ 2 & 1 \end{vmatrix} + 1\begin{vmatrix} x & y \\ 2 & 3 \end{vmatrix}$$

$$= x - 2 + 3x - 2y$$

$$= 4x - 2y - 2$$

27. Expand along the first column because it has two zeros.

$$\begin{vmatrix} 5 & 3 & 0 & 6 \\ 4 & 6 & 4 & 12 \\ 0 & 2 & -3 & 4 \\ 0 & 1 & -2 & 2 \end{vmatrix} = 5\begin{vmatrix} 6 & 4 & 12 \\ 2 & -3 & 4 \\ 1 & -2 & 2 \end{vmatrix} - 4\begin{vmatrix} 3 & 0 & 6 \\ 2 & -3 & 4 \\ 1 & -2 & 2 \end{vmatrix}$$

The determinants of the two 3×3 matrices are:

$$\begin{vmatrix} 6 & 4 & 12 \\ 2 & -3 & 4 \\ 1 & -2 & 2 \end{vmatrix} = 6\begin{vmatrix} -3 & 4 \\ -2 & 2 \end{vmatrix} - 2\begin{vmatrix} 4 & 12 \\ -2 & 2 \end{vmatrix} + 1\begin{vmatrix} 4 & 12 \\ -3 & 4 \end{vmatrix}$$

$$= 6(2) - 2(32) + 52$$

$$= 0$$

$$\begin{vmatrix} 3 & 0 & 6 \\ 2 & -3 & 4 \\ 1 & -2 & 2 \end{vmatrix} = 3\begin{vmatrix} -3 & 4 \\ -2 & 2 \end{vmatrix} - 0\begin{vmatrix} 2 & 4 \\ 1 & 2 \end{vmatrix} + 6\begin{vmatrix} 2 & -3 \\ 1 & -2 \end{vmatrix}$$

$$= 3(2) + 6(-1)$$

$$= 0$$

So,

$$\begin{vmatrix} 5 & 3 & 0 & 6 \\ 4 & 6 & 4 & 12 \\ 0 & 2 & -3 & 4 \\ 0 & 1 & -2 & 2 \end{vmatrix} = 5(0) - 4(0) = 0.$$

© 2013 Cengage Learning. All Rights Reserved. May not be scanned, copied or duplicated, or posted to a publicly accessible website, in whole or in part.

29. Expand along the first row.

$$\begin{vmatrix} w & x & y & z \\ 21 & -15 & 24 & 30 \\ -10 & 24 & -32 & 18 \\ -40 & 22 & 32 & -35 \end{vmatrix} = w\begin{vmatrix} -15 & 24 & 30 \\ 24 & -32 & 18 \\ 22 & 32 & -35 \end{vmatrix} - x\begin{vmatrix} 21 & 24 & 30 \\ -10 & -32 & 18 \\ -40 & 32 & -35 \end{vmatrix} + y\begin{vmatrix} 21 & -15 & 30 \\ -10 & 24 & 18 \\ -40 & 22 & -35 \end{vmatrix} - z\begin{vmatrix} 21 & -15 & 24 \\ -10 & 24 & -32 \\ -40 & 22 & 32 \end{vmatrix}$$

The determinants of the 3×3 matrices are:

$$\begin{vmatrix} -15 & 24 & 30 \\ 24 & -32 & 18 \\ 22 & 32 & -35 \end{vmatrix} = -15\begin{vmatrix} -32 & 18 \\ 32 & -35 \end{vmatrix} - 24\begin{vmatrix} 24 & 18 \\ 22 & -35 \end{vmatrix} + 30\begin{vmatrix} 24 & -32 \\ 22 & 32 \end{vmatrix}$$

$$= -15(544) - 24(-1236) + 30(1472)$$

$$= 65{,}664$$

$$\begin{vmatrix} 21 & 24 & 30 \\ -10 & -32 & 18 \\ -40 & 32 & -35 \end{vmatrix} = 21\begin{vmatrix} -32 & 18 \\ 32 & -35 \end{vmatrix} - 24\begin{vmatrix} -10 & 18 \\ -40 & -35 \end{vmatrix} + 30\begin{vmatrix} -10 & -32 \\ -40 & 32 \end{vmatrix}$$

$$= 21(544) - 24(1070) + 30(-1600)$$

$$= -62{,}256$$

$$\begin{vmatrix} 21 & -15 & 30 \\ -10 & 24 & 18 \\ -40 & 22 & -35 \end{vmatrix} = 21\begin{vmatrix} 24 & 18 \\ 22 & -35 \end{vmatrix} + 15\begin{vmatrix} -10 & 18 \\ -40 & -35 \end{vmatrix} + 30\begin{vmatrix} -10 & 24 \\ -40 & 22 \end{vmatrix}$$

$$= 21(-1236) + 15(1070) + 30(740)$$

$$= 12{,}294$$

$$\begin{vmatrix} 21 & -15 & 24 \\ -10 & 24 & -32 \\ -40 & 22 & 32 \end{vmatrix} = 21\begin{vmatrix} 24 & -32 \\ 22 & 35 \end{vmatrix} + 15\begin{vmatrix} -10 & -32 \\ -40 & 32 \end{vmatrix} + 24\begin{vmatrix} -10 & 24 \\ -40 & 22 \end{vmatrix}$$

$$= 21(1472) + 15(-1600) + 24(740)$$

$$= 24{,}672$$

So, $\begin{vmatrix} w & x & y & z \\ 21 & -15 & 24 & 30 \\ -10 & 24 & -32 & 18 \\ -40 & 22 & 32 & -35 \end{vmatrix} = 65{,}664w + 62{,}256x + 12{,}294y - 24{,}672z.$

31. Expand along the first column, and then along the first column of the 4×4 matrix.

$$\begin{vmatrix} 5 & 2 & 0 & 0 & -2 \\ 0 & 1 & 4 & 3 & 2 \\ 0 & 0 & 2 & 6 & 3 \\ 0 & 0 & 3 & 4 & 1 \\ 0 & 0 & 0 & 0 & 2 \end{vmatrix} = 5\begin{vmatrix} 1 & 4 & 3 & 2 \\ 0 & 2 & 6 & 3 \\ 0 & 3 & 4 & 1 \\ 0 & 0 & 0 & 2 \end{vmatrix} = 5(1)\begin{vmatrix} 2 & 6 & 3 \\ 3 & 4 & 1 \\ 0 & 0 & 2 \end{vmatrix}$$

Now expand along the third row, and obtain

$$5(1)\begin{vmatrix} 2 & 6 & 3 \\ 3 & 4 & 1 \\ 0 & 0 & 2 \end{vmatrix} = 5(2)\begin{vmatrix} 2 & 6 \\ 3 & 4 \end{vmatrix} = 10(-10) = -100.$$

33. Copy the first two columns and complete the diagonal products as follows.

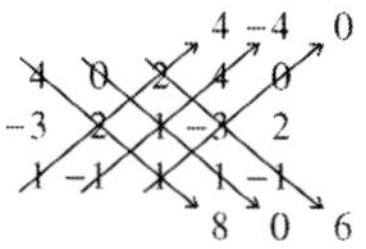

Add the lower three products and subtract the upper three products to find the determinant.

$$\begin{vmatrix} 4 & 0 & 2 \\ -3 & 2 & 1 \\ 1 & -1 & 1 \end{vmatrix} = 8 + 0 + 6 - 4 - (-4) - 0 = 14$$

© 2013 Cengage Learning. All Rights Reserved. May not be scanned, copied or duplicated, or posted to a publicly accessible website, in whole or in part.

35. Using a graphing utility or a software program you have

$$\begin{vmatrix} 0.25 & -1 & 0.6 \\ 0.5 & 0.8 & -0.2 \\ 0.75 & 0.9 & -0.4 \end{vmatrix} = -0.175.$$

37. Using a graphing utility or a software program you have

$$\begin{vmatrix} 1 & 2 & -1 & 4 \\ 0 & 1 & 2 & -2 \\ 0 & 3 & 2 & -1 \\ 1 & 2 & 0 & -2 \end{vmatrix} = 19.$$

39. The determinant of a triangular matrix is the product of the elements on its main diagonal.

$$\begin{vmatrix} -2 & 0 & 0 \\ 4 & 6 & 0 \\ -3 & 7 & 2 \end{vmatrix} = -2(6)(2) = -24$$

41. The determinant of a triangular matrix is the product of the elements on its main diagonal.

$$\begin{vmatrix} 5 & 8 & -4 & 2 \\ 0 & 0 & 6 & 0 \\ 0 & 0 & 2 & 2 \\ 0 & 0 & 0 & -1 \end{vmatrix} = 5(0)(2)(-1) = 0$$

43. (a) False. See "Definition of the Determinant of a 2×2 Matrix," page 104.

(b) True. See the Remark, page 106.

(c) False. See the "Minors and Cofactors of a Matrix" box, page 105.

45.
$$\begin{aligned} (x+3)(x+2) - 1(2) &= 0 \\ x^2 + 5x + 6 - 2 &= 0 \\ x^2 + 5x + 4 &= 0 \\ (x+4)(x+1) &= 0 \\ x &= -4, -1 \end{aligned}$$

47.
$$\begin{aligned} (x-1)(x-2) - 3(2) &= 0 \\ x^2 - 3x + 2 - 6 &= 0 \\ x^2 - 3x - 4 &= 0 \\ (x-4)(x+1) &= 0 \\ x &= 4, -1 \end{aligned}$$

49. $\begin{vmatrix} \lambda + 2 & 2 \\ 1 & \lambda \end{vmatrix} = (\lambda + 2)(\lambda) - 1(2) = \lambda^2 + 2\lambda - 2$

The determinant is zero when $\lambda^2 + 2\lambda - 2 = 0$. Use the Quadratic Formula to find λ.

$$\begin{aligned} \lambda &= \frac{-2 \pm \sqrt{2^2 - 4(1)(-2)}}{2(1)} \\ &= \frac{-2 \pm \sqrt{12}}{2} \\ &= \frac{-2 \pm 2\sqrt{3}}{2} \\ &= -1 \pm \sqrt{3} \end{aligned}$$

51. $\begin{vmatrix} \lambda & 2 & 0 \\ 0 & \lambda + 1 & 2 \\ 0 & 1 & \lambda \end{vmatrix} = \lambda \begin{vmatrix} \lambda + 1 & 2 \\ 1 & \lambda \end{vmatrix}$

$$= \lambda(\lambda^2 + \lambda - 2) = \lambda(\lambda + 2)(\lambda - 1)$$

The determinant is zero when $\lambda(\lambda + 2)(\lambda - 1) = 0$.

So, $\lambda = 0, -2, 1$.

53. $\begin{vmatrix} 4u & -1 \\ -1 & 2v \end{vmatrix} = (4u)(2v) - (-1)(-1) = 8uv - 1$

55. $\begin{vmatrix} e^{2x} & e^{3x} \\ 2e^{2x} & 3e^{3x} \end{vmatrix} = e^{2x}(3e^{3x}) - 2e^{2x}(e^{3x})$

$$= 3e^{5x} - 2e^{5x} = e^{5x}$$

57. $\begin{vmatrix} x & \ln x \\ 1 & 1/x \end{vmatrix} = x(1/x) - 1(\ln x) = 1 - \ln x$

59. Expanding along the first row, the determinant of a 4×4 matrix involves four 3×3 determinants. Each of these 3×3 determinants requires 6 triple products. So, there are $4(6) = 24$ quadruple products.

61. Evaluating the left side yields

$$\begin{vmatrix} w & x \\ y & z \end{vmatrix} = wz - xy.$$

Evaluating the right side yields

$$-\begin{vmatrix} y & z \\ w & x \end{vmatrix} = -(xy - wz) = wz - xy.$$

63. Evaluating the left side yields

$$\begin{vmatrix} w & x \\ y & z \end{vmatrix} = wz - xy.$$

Evaluating the right side yields

$$\begin{vmatrix} w & x + cw \\ y & z + cy \end{vmatrix} = w(z + cy) - y(x + cw) = wz + cwy - xy - cwy = wz - xy.$$

© 2013 Cengage Learning. All Rights Reserved. May not be scanned, copied or duplicated, or posted to a publicly accessible website, in whole or in part.

65. Evaluating the left side yields

$$\begin{vmatrix} 1 & x & x^2 \\ 1 & y & y^2 \\ 1 & z & z^2 \end{vmatrix} = \begin{vmatrix} y & y^2 \\ z & z^2 \end{vmatrix} - \begin{vmatrix} x & x^2 \\ z & z^2 \end{vmatrix} + \begin{vmatrix} x & x^2 \\ y & y^2 \end{vmatrix}$$

$$= yz^2 - y^2z - \left(xz^2 - x^2z\right) + xy^2 - x^2y$$

$$= xy^2 - xz^2 + yz^2 - x^2y + x^2z - y^2z.$$

Expanding the right side yields

$$(y - x)(z - x)(z - y) = \left(yz - xy - xz + x^2\right)(z - y)$$

$$= yz^2 - y^2z - xyz + xy^2 - xz^2 + xyz + x^2z - x^2y$$

$$= xy^2 - xz^2 + yz^2 - x^2y + x^2z - y^2z.$$

67. (a) Expanding along the first row,

$$\begin{vmatrix} x & 0 & c \\ -1 & x & b \\ 0 & -1 & a \end{vmatrix} = x\begin{vmatrix} x & b \\ -1 & a \end{vmatrix} + c\begin{vmatrix} -1 & x \\ 0 & -1 \end{vmatrix} = x(ax + b) + c(1) = ax^2 + bx + c.$$

(b) The right column contains the coefficients a, b, c. So,

$$\begin{vmatrix} x & 0 & 0 & d \\ -1 & x & 0 & c \\ 0 & -1 & x & b \\ 0 & 0 & -1 & a \end{vmatrix} = x\begin{vmatrix} x & 0 & c \\ -1 & x & b \\ 0 & -1 & a \end{vmatrix} + 1\begin{vmatrix} 0 & 0 & d \\ -1 & x & b \\ 0 & -1 & a \end{vmatrix} = x\left(ax^2 + bx + c\right) + d = ax^3 + bx^2 + cx + d.$$

69. To show that

$$x_1 = \frac{b_1a_{22} - b_2a_{12}}{a_{11}a_{22} - a_{21}d_{12}} \text{ and } x_2 = \frac{b_2a_{11} - b_1a_{21}}{a_{11}a_{22} - a_{21}a_{12}}$$

is a solution to the given system, show that both equations are satisfied when the values are substituted.

$$a_{11}x_1 + a_{12}x_1 = a_{11}\left(\frac{b_1a_{22} - b_2a_{12}}{a_{11}a_{22} - a_{21}a_{12}}\right) + a_{12}\left(\frac{b_2a_{11} - b_1a_{21}}{a_{11}a_{22} - a_{21}a_{12}}\right)$$

$$= \frac{a_{11}a_{22}b_1 - a_{11}a_{12}b_2}{a_{11}a_{22} - a_{21}a_{12}} + \frac{a_{11}a_{12}b_2 - a_{21}a_{12}b_1}{a_{11}a_{22} - a_{21}a_{12}}$$

$$= \frac{a_{11}a_{22}b_1 - a_{11}a_{12}b_2 + a_{11}a_{12}b_2 - a_{21}a_{12}b_1}{a_{11}a_{22} - a_{21}a_{12}}$$

$$= \frac{\left(a_{11}a_{22} - a_{21}a_{12}\right)b_1}{a_{11}a_{22} - a_{21}a_{12}}$$

$$= b_1$$

$$a_{21}x_1 + a_{22}x_2 = a_{21}\left(\frac{b_1a_{22} - b_2a_{12}}{a_{11}a_{22} - a_{21}a_{12}}\right) + a_{22}\left(\frac{b_2a_{11} - b_1a_{21}}{a_{11}a_{22} - a_{21}a_{12}}\right)$$

$$= \frac{a_{21}a_{22}b_1 - a_{21}a_{12}b_2}{a_{11}a_{22} - a_{21}a_{12}} + \frac{a_{11}a_{22}b_2 - a_{21}a_{22}b_1}{a_{11}a_{22} - a_{21}a_{12}}$$

$$= \frac{a_{21}a_{22}b_1 - a_{21}a_{12}b_2 + a_{11}a_{22}b_2 - a_{21}a_{22}b_1}{a_{11}a_{22} - a_{21}a_{12}}$$

$$= \frac{\left(a_{11}a_{22} - a_{21}a_{12}\right)b_2}{a_{11}a_{22} - a_{21}a_{12}}$$

$$= b_2$$

© 2013 Cengage Learning. All Rights Reserved. May not be scanned, copied or duplicated, or posted to a publicly accessible website, in whole or in part.

Section 3.2 Determinants and Elementary Operations

1. Because the first row is a multiple of the second row, the determinant is zero.

3. Because the second row is composed of all zeros, the determinant is zero.

5. Because the second and third columns are interchanged, the sign of the determinant is changed.

7. Because 5 has been factored out of the first row, the first determinant is 5 times the second one.

9. Because 4 has been factored out of the second column, and 3 factored out of the third column, the first determinant is 12 times the second one.

11. Because each row in the matrix on the left is divided by 5 to yield the matrix on the right, the determinant of the matrix on the left is 5^3 times the determinant of the matrix on the right.

13. Because a multiple of the first row of the matrix on the left was added to the second row to produce the matrix on the right, the determinants are equal.

15. Because a multiple of the first row of the matrix on the left was added to the second row to produce the matrix on the right, the determinants are equal.

17. Because the second row of the matrix on the left was multiplied by (-1), the sign of the determinant is changed.

19. Because the sixth column is a multiple of the first column, the determinant is zero.

21. Expand by cofactors along the second column.

$$\begin{vmatrix} 1 & 0 & 2 \\ -1 & 1 & 4 \\ 2 & 0 & 3 \end{vmatrix} = 1\begin{vmatrix} 1 & 2 \\ 2 & 3 \end{vmatrix} = 3 - 4 = -1$$

A graphing utility or software program produces the same determinant, -1.

23. Rewrite the matrix in triangular form.

$$\begin{vmatrix} 1 & 2 & 1 & -1 \\ 0 & 1 & 0 & 2 \\ 0 & 3 & -1 & 1 \\ 0 & 0 & 4 & 1 \end{vmatrix} = \begin{vmatrix} 1 & 2 & 1 & -1 \\ 0 & 1 & 0 & 2 \\ 0 & 0 & -1 & -5 \\ 0 & 0 & 4 & 1 \end{vmatrix} = \begin{vmatrix} 1 & 2 & 1 & -1 \\ 0 & 1 & 0 & 2 \\ 0 & 0 & -1 & -5 \\ 0 & 0 & 0 & -19 \end{vmatrix}$$

$$= 1(1)(-1)(-19) = 19$$

A graphing utility or software program produces the same determinant, 19.

25.
$$\begin{vmatrix} 1 & 7 & -3 \\ 1 & 3 & 1 \\ 4 & 8 & 1 \end{vmatrix} = \begin{vmatrix} 1 & 7 & -3 \\ 0 & -4 & 4 \\ 4 & 8 & 1 \end{vmatrix}$$

$$= \begin{vmatrix} 1 & 7 & -3 \\ 0 & -4 & 4 \\ 0 & -20 & 13 \end{vmatrix}$$

$$= \begin{vmatrix} 1 & 7 & -3 \\ 0 & -4 & 4 \\ 0 & 0 & -7 \end{vmatrix} = 1(-4)(-7) = 28$$

27.
$$\begin{vmatrix} 2 & -1 & -1 \\ 1 & 3 & 2 \\ -6 & 3 & 3 \end{vmatrix} = -\begin{vmatrix} 2 & -1 & -1 \\ 1 & 3 & 2 \\ 0 & 0 & 0 \end{vmatrix} = 0$$

29.
$$\begin{vmatrix} 4 & 3 & -2 \\ 5 & 4 & 1 \\ -2 & 3 & 4 \end{vmatrix} = 4\begin{vmatrix} 14 & 11 & 0 \\ 5 & 4 & 1 \\ -22 & -13 & 0 \end{vmatrix}$$

$$= (-1)\begin{vmatrix} 14 & 11 \\ -22 & -13 \end{vmatrix}$$

$$= (-1)\left[(14(-13) - (-22)(11))\right]$$

$$= (-1)(60) = -60$$

31.
$$\begin{vmatrix} 4 & -7 & 9 & 1 \\ 6 & 2 & 7 & 0 \\ 3 & 6 & -3 & 3 \\ 0 & 7 & 4 & -1 \end{vmatrix} = \begin{vmatrix} 4 & -7 & 9 & 1 \\ 6 & 2 & 7 & 0 \\ -9 & 27 & -30 & 0 \\ 4 & 0 & 13 & 0 \end{vmatrix}$$

$$= -\begin{vmatrix} 6 & 2 & 7 \\ -9 & 27 & -30 \\ 4 & 0 & 13 \end{vmatrix}$$

$$= 3\begin{vmatrix} 6 & 2 & 7 \\ 3 & -9 & 10 \\ 4 & 0 & 13 \end{vmatrix}$$

$$= 3\begin{vmatrix} 0 & 20 & -13 \\ 3 & -9 & 10 \\ 4 & 0 & 13 \end{vmatrix}$$

$$= 3\left[(-3)260 + 4(200 - 117)\right]$$

$$= -1344$$

© 2013 Cengage Learning. All Rights Reserved. May not be scanned, copied or duplicated, or posted to a publicly accessible website, in whole or in part.

33. $\begin{vmatrix} 1 & -2 & 7 & 9 \\ 3 & -4 & 5 & 5 \\ 3 & 6 & 1 & -1 \\ 4 & 5 & 3 & 2 \end{vmatrix} = \begin{vmatrix} 1 & -2 & 7 & 9 \\ 0 & 2 & -16 & -22 \\ 0 & 12 & -20 & -28 \\ 0 & 13 & -25 & -34 \end{vmatrix}$

$= 2\begin{vmatrix} 1 & -2 & 7 & 9 \\ 0 & 1 & -8 & -11 \\ 0 & 12 & -20 & -28 \\ 0 & 13 & -25 & -34 \end{vmatrix}$

$= 2\begin{vmatrix} 1 & -2 & 7 & 9 \\ 0 & 1 & -8 & -11 \\ 0 & 0 & 76 & 104 \\ 0 & 0 & 79 & 109 \end{vmatrix}$

$= 2(1)(1)\begin{vmatrix} 76 & 104 \\ 79 & 109 \end{vmatrix}$

$= 2[76(109) - 79(104)]$

$= 136$

35. $\begin{vmatrix} 1 & -1 & 8 & 4 & 2 \\ 2 & 6 & 0 & -4 & 3 \\ 2 & 0 & 2 & 6 & 2 \\ 0 & 2 & 8 & 0 & 0 \\ 0 & 1 & 1 & 2 & 2 \end{vmatrix} = \begin{vmatrix} 1 & -1 & 8 & 4 & 2 \\ 0 & 8 & -16 & -12 & -1 \\ 0 & 2 & -14 & -2 & -2 \\ 0 & 2 & 8 & 0 & 0 \\ 0 & 1 & 1 & 2 & 2 \end{vmatrix}$

$= \begin{vmatrix} 1 & -1 & 8 & 4 & 2 \\ 0 & 0 & -24 & -28 & -17 \\ 0 & 0 & -16 & -6 & -6 \\ 0 & 0 & 6 & -4 & -4 \\ 0 & 1 & 1 & 2 & 2 \end{vmatrix}$

$= (-1)\begin{vmatrix} -24 & -28 & -17 \\ -16 & -6 & -6 \\ 6 & -4 & -4 \end{vmatrix}$

$= 2\begin{vmatrix} -24 & -28 & -17 \\ -16 & -6 & -6 \\ -3 & 2 & 2 \end{vmatrix}$

$= 2\begin{vmatrix} -24 & -28 & -17 \\ -25 & 0 & 0 \\ -3 & 2 & 2 \end{vmatrix}$

$= 50(-56 + 34)$

$= -1100$

37. (a) True. See Theorem 3.3, part 1, page 113.

(b) True. See Theorem 3.3, part 3, page 113.

(c) True. See Theorem 3.4, part 2, page 115.

39. $\begin{vmatrix} 1 & 0 & 0 \\ 0 & k & 0 \\ 0 & 0 & 1 \end{vmatrix} = k\begin{vmatrix} 1 & 0 & 0 \\ 0 & 1 & 0 \\ 0 & 0 & 1 \end{vmatrix} = k$

41. $\begin{vmatrix} 1 & 0 & 0 \\ k & 1 & 0 \\ 0 & 0 & 1 \end{vmatrix} = \begin{vmatrix} 1 & 0 & 0 \\ 0 & 1 & 0 \\ 0 & 0 & 1 \end{vmatrix} = 1$

43. Expand the two determinants on the left.

$$\begin{vmatrix} a_{11} & a_{12} & a_{13} \\ a_{21} & a_{22} & a_{23} \\ a_{31} & a_{32} & a_{33} \end{vmatrix} + \begin{vmatrix} b_{11} & a_{12} & a_{13} \\ b_{21} & a_{22} & a_{23} \\ b_{31} & a_{32} & a_{33} \end{vmatrix}$$

$$= a_{11}\begin{vmatrix} a_{22} & a_{23} \\ a_{32} & a_{33} \end{vmatrix} - a_{21}\begin{vmatrix} a_{12} & a_{13} \\ a_{32} & a_{33} \end{vmatrix} + a_{31}\begin{vmatrix} a_{12} & a_{13} \\ a_{22} & a_{23} \end{vmatrix} + b_{11}\begin{vmatrix} a_{22} & a_{23} \\ a_{32} & a_{33} \end{vmatrix} - b_{21}\begin{vmatrix} a_{12} & a_{13} \\ a_{32} & a_{33} \end{vmatrix} + b_{31}\begin{vmatrix} a_{12} & a_{13} \\ a_{22} & a_{23} \end{vmatrix}$$

$$= (a_{11} + b_{11})\begin{vmatrix} a_{22} & a_{23} \\ a_{32} & a_{33} \end{vmatrix} - (a_{21} + b_{21})\begin{vmatrix} a_{12} & a_{13} \\ a_{32} & a_{33} \end{vmatrix} + (a_{31} + b_{31})\begin{vmatrix} a_{12} & a_{13} \\ a_{22} & a_{23} \end{vmatrix}$$

$$= \begin{vmatrix} (a_{11} + b_{11}) & a_{12} & a_{13} \\ (a_{21} + b_{21}) & a_{22} & a_{23} \\ (a_{31} + b_{31}) & a_{32} & a_{33} \end{vmatrix}$$

45. (a) $\begin{vmatrix} \cos\theta & \sin\theta \\ -\sin\theta & \cos\theta \end{vmatrix} = \cos\theta(\cos\theta) - (-\sin\theta)(\sin\theta) = \cos^2\theta + \sin^2\theta = 1$

(b) $\begin{vmatrix} \sin\theta & 1 \\ 1 & \sin\theta \end{vmatrix} = (\sin\theta)(\sin\theta) - 1(1) = \sin^2\theta - 1 = -\cos^2\theta$

© 2013 Cengage Learning. All Rights Reserved. May not be scanned, copied or duplicated, or posted to a publicly accessible website, in whole or in part.

47. Suppose B is obtained from A by adding a multiple of a row of A to another row of A. More specifically, suppose c times the jth row of A is added to the ith row of A.

$$B = \begin{bmatrix} a_{11} & \cdots & a_{1n} \\ \vdots & & \\ (a_{i1} + ca_{j1}) & \cdots & (a_{in} + ca_{jn}) \\ \vdots & & \\ a_{n1} & \cdots & a_{nn} \end{bmatrix}$$

Expand along this row.

$$\det B = (a_{i1} + ca_{j1})C_{i1} + \cdots + (a_{in} + ca_{jn})C_{in} = [a_{i1}C_{i1} + \cdots + a_{in}C_{in}] + [ca_{j1}C_{i1} + \cdots + ca_{jn}C_{in}]$$

The first bracketed expression is det A, so prove that the second bracketed expression is zero. Use mathematical induction. For $n = 2$ (assuming $i = 2$ and $j = 1$),

$$ca_{11}C_{21} + ca_{12}C_{22} = \det\begin{bmatrix} a_{11} & a_{12} \\ ca_{11} & ca_{12} \end{bmatrix} = 0 \text{ (because row 2 is a multiple of row 1).}$$

Assuming the expression is true for $n - 1$, then

$$ca_{j1}C_{i1} + \cdots + ca_{jn}C_{in} = 0$$

by expanding along any row different from c and j and applying the induction hypothesis.

Section 3.3 Properties of Determinants

1. (a) $|A| = \begin{vmatrix} -2 & 1 \\ 4 & -2 \end{vmatrix} = 0$

(b) $|B| = \begin{vmatrix} 1 & 1 \\ 0 & -1 \end{vmatrix} = -1$

(c) $AB = \begin{bmatrix} -2 & 1 \\ 4 & -2 \end{bmatrix}\begin{bmatrix} 1 & 1 \\ 0 & -1 \end{bmatrix} = \begin{bmatrix} -2 & -3 \\ 4 & 6 \end{bmatrix}$

(d) $|AB| = \begin{vmatrix} -2 & -3 \\ 4 & 6 \end{vmatrix} = 0$

Notice that $|A||B| = 0(-1) = 0 = |AB|$.

3. (a) $|A| = \begin{vmatrix} -1 & 2 & 1 \\ 1 & 0 & 1 \\ 0 & 1 & 0 \end{vmatrix} = 2$

(b) $|B| = \begin{vmatrix} -1 & 0 & 0 \\ 0 & 2 & 0 \\ 0 & 0 & 3 \end{vmatrix} = -6$

(c) $AB = \begin{bmatrix} -1 & 2 & 1 \\ 1 & 0 & 1 \\ 0 & 1 & 0 \end{bmatrix}\begin{bmatrix} -1 & 0 & 0 \\ 0 & 2 & 0 \\ 0 & 0 & 3 \end{bmatrix} = \begin{bmatrix} 1 & 4 & 3 \\ -1 & 0 & 3 \\ 0 & 2 & 0 \end{bmatrix}$

(d) $|AB| = \begin{vmatrix} 1 & 4 & 3 \\ -1 & 0 & 3 \\ 0 & 2 & 0 \end{vmatrix} = -12$

Notice that $|A||B| = 2(-6) = -12 = |AB|$.

© 2013 Cengage Learning. All Rights Reserved. May not be scanned, copied or duplicated, or posted to a publicly accessible website, in whole or in part.

5. (a) $|A| = \begin{vmatrix} 2 & 0 & 1 & 1 \\ 1 & -1 & 0 & 1 \\ 2 & 3 & 1 & 0 \\ 1 & 2 & 3 & 0 \end{vmatrix} = \begin{vmatrix} 1 & 1 & 1 & 0 \\ 1 & -1 & 0 & 1 \\ 2 & 3 & 1 & 0 \\ 1 & 2 & 3 & 0 \end{vmatrix} = \begin{vmatrix} 1 & 1 & 1 \\ 2 & 3 & 1 \\ 1 & 2 & 3 \end{vmatrix} = \begin{vmatrix} 1 & 1 & 1 \\ 0 & 1 & -1 \\ 1 & 2 & 3 \end{vmatrix} = \begin{vmatrix} 1 & 1 & 1 \\ 0 & 1 & -1 \\ 0 & 1 & 2 \end{vmatrix} = 3$

(b) $|B| = \begin{vmatrix} 1 & 0 & -1 & 1 \\ 2 & 1 & 0 & 2 \\ 1 & 1 & -1 & 0 \\ 3 & 2 & 1 & 0 \end{vmatrix} = \begin{vmatrix} 1 & 0 & -1 & 1 \\ 0 & 1 & 2 & 0 \\ 1 & 1 & -1 & 0 \\ 3 & 2 & 1 & 0 \end{vmatrix} = -\begin{vmatrix} 0 & 1 & 2 \\ 1 & 1 & -1 \\ 3 & 2 & 1 \end{vmatrix} = -\begin{vmatrix} 0 & 1 & 2 \\ 1 & 1 & -1 \\ 0 & -1 & 4 \end{vmatrix} = 6$

(c) $AB = \begin{bmatrix} 2 & 0 & 1 & 1 \\ 1 & -1 & 0 & 1 \\ 2 & 3 & 1 & 0 \\ 1 & 2 & 3 & 0 \end{bmatrix}\begin{bmatrix} 1 & 0 & -1 & 1 \\ 2 & 1 & 0 & 2 \\ 1 & 1 & -1 & 0 \\ 3 & 2 & 1 & 0 \end{bmatrix} = \begin{bmatrix} 6 & 3 & -2 & 2 \\ 2 & 1 & 0 & -1 \\ 9 & 4 & -3 & 8 \\ 8 & 5 & -4 & 5 \end{bmatrix}$

(d) $|AB| = \begin{vmatrix} 6 & 3 & -2 & 2 \\ 2 & 1 & 0 & -1 \\ 9 & 4 & -3 & 8 \\ 8 & 5 & -4 & 5 \end{vmatrix} = \begin{vmatrix} 0 & 0 & -2 & 5 \\ 2 & 1 & 0 & -1 \\ 9 & 4 & -3 & 8 \\ 8 & 5 & -4 & 5 \end{vmatrix} = -2\begin{vmatrix} 2 & 1 & -1 \\ 9 & 4 & 8 \\ 8 & 5 & 5 \end{vmatrix} - 5\begin{vmatrix} 2 & 1 & 0 \\ 9 & 4 & -3 \\ 8 & 5 & -4 \end{vmatrix} = -2\begin{vmatrix} 2 & 1 & -1 \\ 1 & 0 & 12 \\ 8 & 5 & 5 \end{vmatrix} - 5\begin{vmatrix} 2 & 1 & 0 \\ 3 & \frac{1}{4} & 0 \\ 8 & 5 & -4 \end{vmatrix}$

$= -2\begin{vmatrix} 2 & 1 & -1 \\ 1 & 0 & 12 \\ -2 & 0 & 10 \end{vmatrix} - 5(-4)\left(\frac{1}{2} - 3\right)$

$= -2(-34) - 50 = 18$

Notice that $|A||B| = 3 \cdot 6 = 18 = |AB|$.

7. $|A| = \begin{vmatrix} 4 & 2 \\ 6 & -8 \end{vmatrix} = 2^2\begin{vmatrix} 2 & 1 \\ 3 & -4 \end{vmatrix} = 4(-11) = -44$

9. $|A| = \begin{vmatrix} -3 & 6 & 9 \\ 6 & 9 & 12 \\ 9 & 12 & 15 \end{vmatrix} = 3^3\begin{vmatrix} -1 & 2 & 3 \\ 2 & 3 & 4 \\ 3 & 4 & 5 \end{vmatrix} = 3^3\begin{vmatrix} -1 & 2 & 3 \\ 0 & 7 & 10 \\ 0 & 10 & 14 \end{vmatrix}$

$= (-27)(-2) = 54$

11. $|A| = \begin{vmatrix} 2 & -4 & 6 \\ -4 & 6 & -8 \\ 6 & -8 & 10 \end{vmatrix}$

$= 2^3\begin{vmatrix} 1 & -2 & 3 \\ -2 & 3 & -4 \\ 3 & -4 & 5 \end{vmatrix}$

$= 2^3\begin{vmatrix} 1 & -2 & 3 \\ 0 & -1 & 2 \\ 0 & 2 & -4 \end{vmatrix}$

$= 8(0)$

$= 0$

13. (a) $|A| = \begin{vmatrix} -1 & 1 \\ 2 & 0 \end{vmatrix} = -2$

(b) $|B| = \begin{vmatrix} 1 & -1 \\ -2 & 0 \end{vmatrix} = -2$

(c) $|A + B| = \left|\begin{bmatrix} -1 & 1 \\ 2 & 0 \end{bmatrix} + \begin{bmatrix} 1 & -1 \\ -2 & 0 \end{bmatrix}\right| = \begin{vmatrix} 0 & 0 \\ 0 & 0 \end{vmatrix} = 0$

Notice that

$|A| + |B| = -2 + (-2) = -4 \neq |A + B|$.

15. (a) $|A| = \begin{vmatrix} 1 & 0 & 1 \\ -1 & 2 & 1 \\ 0 & 1 & 1 \end{vmatrix} = \begin{vmatrix} 1 & 0 & 1 \\ 0 & 2 & 2 \\ 0 & 1 & 1 \end{vmatrix} = \begin{vmatrix} 1 & 0 & 1 \\ 0 & 2 & 2 \\ 0 & 0 & 0 \end{vmatrix} = 0$

(b) $|B| = \begin{vmatrix} -1 & 0 & 2 \\ 0 & 1 & 2 \\ 1 & 1 & 1 \end{vmatrix} = \begin{vmatrix} 0 & 1 & 3 \\ 0 & 1 & 2 \\ 1 & 1 & 1 \end{vmatrix} = -1$

(c) $|A + B| = \left|\begin{bmatrix} 1 & 0 & 1 \\ -1 & 2 & 1 \\ 0 & 1 & 1 \end{bmatrix} + \begin{bmatrix} -1 & 0 & 2 \\ 0 & 1 & 2 \\ 1 & 1 & 1 \end{bmatrix}\right| = \begin{vmatrix} 0 & 0 & 3 \\ -1 & 3 & 3 \\ 1 & 2 & 2 \end{vmatrix}$

$= -15$

Notice that $|A| + |B| = 0 + (-1) = -1 \neq |A + B|$.

© 2013 Cengage Learning. All Rights Reserved. May not be scanned, copied or duplicated, or posted to a publicly accessible website, in whole or in part.

17. Because

$$\begin{vmatrix} 5 & 4 \\ 10 & 8 \end{vmatrix} = 0,$$

the matrix is singular.

19. Because

$$\begin{vmatrix} 14 & 5 & 7 \\ -2 & 0 & 3 \\ 1 & -5 & -10 \end{vmatrix} = 195 \neq 0,$$

the matrix is nonsingular.

21. Because

$$\begin{vmatrix} \frac{1}{2} & \frac{3}{2} & 2 \\ \frac{2}{3} & -\frac{1}{3} & 0 \\ 1 & 1 & 1 \end{vmatrix} = \frac{5}{6} \neq 0,$$

the matrix is nonsingular.

23. Because

$$\begin{vmatrix} 1 & 0 & -8 & 2 \\ 0 & 8 & -1 & 10 \\ 0 & 0 & 0 & 1 \\ 0 & 0 & 0 & 2 \end{vmatrix} = 0,$$

the matrix is singular.

25. $A^{-1} = \frac{1}{5}\begin{bmatrix} 4 & -3 \\ -1 & 2 \end{bmatrix} = \begin{bmatrix} \frac{4}{5} & -\frac{3}{5} \\ -\frac{1}{5} & \frac{2}{5} \end{bmatrix}$

$$\left|A^{-1}\right| = \frac{4}{5}\left(\frac{2}{5}\right) - \left(-\frac{1}{5}\right)\left(-\frac{3}{5}\right) = \frac{8}{25} - \frac{3}{25} = \frac{1}{5}$$

Notice that $|A| = 5$, so $\left|A^{-1}\right| = \frac{1}{|A|} = \frac{1}{5}$.

27. $A^{-1} = \begin{bmatrix} -2 & 2 & -1 \\ \frac{1}{2} & 0 & -\frac{1}{2} \\ \frac{3}{2} & -1 & \frac{1}{2} \end{bmatrix}$

$$\left|A^{-1}\right| = \begin{vmatrix} -2 & 2 & -1 \\ \frac{1}{2} & 0 & -\frac{1}{2} \\ \frac{3}{2} & -1 & \frac{1}{2} \end{vmatrix} = \begin{vmatrix} 1 & 0 & 0 \\ \frac{1}{2} & 0 & -\frac{1}{2} \\ \frac{3}{2} & -1 & \frac{1}{2} \end{vmatrix} = -\frac{1}{2}$$

Notice that $|A| = \begin{vmatrix} 1 & 0 & 2 \\ 2 & -1 & 3 \\ 1 & -2 & 2 \end{vmatrix} = \begin{vmatrix} 1 & 0 & 2 \\ 2 & -1 & 3 \\ -3 & 0 & -4 \end{vmatrix} = -2$, so

$$\left|A^{-1}\right| = \frac{1}{|A|} = -\frac{1}{2}.$$

29. $A^{-1} = \begin{bmatrix} -\frac{1}{8} & -\frac{5}{8} & \frac{7}{8} & 0 \\ \frac{5}{12} & \frac{5}{12} & -\frac{1}{4} & -\frac{1}{3} \\ \frac{3}{8} & \frac{7}{8} & -\frac{5}{8} & 0 \\ \frac{1}{2} & \frac{1}{2} & -\frac{1}{2} & 0 \end{bmatrix}$

$$\left|A^{-1}\right| = \begin{vmatrix} -\frac{1}{8} & -\frac{5}{8} & \frac{7}{8} & 0 \\ \frac{5}{12} & \frac{5}{12} & -\frac{1}{4} & -\frac{1}{3} \\ \frac{3}{8} & \frac{7}{8} & -\frac{5}{8} & 0 \\ \frac{1}{2} & \frac{1}{2} & -\frac{1}{2} & 0 \end{vmatrix} = -\frac{1}{3}\begin{vmatrix} -\frac{1}{8} & -\frac{5}{8} & \frac{7}{8} \\ \frac{3}{8} & \frac{7}{8} & -\frac{5}{8} \\ \frac{1}{2} & \frac{1}{2} & -\frac{1}{2} \end{vmatrix} = -\frac{1}{3}\begin{vmatrix} -\frac{1}{8} & -\frac{5}{8} & \frac{7}{8} \\ 0 & -1 & 2 \\ \frac{1}{2} & \frac{1}{2} & -\frac{1}{2} \end{vmatrix} = -\frac{1}{3}\begin{vmatrix} -\frac{1}{8} & -\frac{5}{8} & \frac{7}{8} \\ 0 & -1 & 2 \\ 0 & -2 & 3 \end{vmatrix} = \frac{1}{24}$$

Notice that $|A| = \begin{vmatrix} 1 & 0 & -1 & 3 \\ 1 & 0 & 3 & -2 \\ 2 & 0 & 2 & -1 \\ 1 & -3 & 1 & 2 \end{vmatrix} = -3\begin{vmatrix} 1 & -1 & 3 \\ 1 & 3 & -2 \\ 2 & 2 & -1 \end{vmatrix} = -3\begin{vmatrix} 1 & -1 & 3 \\ 0 & 4 & -5 \\ 2 & 2 & -1 \end{vmatrix} = -3\begin{vmatrix} 1 & -1 & 3 \\ 0 & 4 & -5 \\ 0 & 4 & -7 \end{vmatrix} = 24.$

So, $\left|A^{-1}\right| = \frac{1}{|A|} = \frac{1}{24}$.

© 2013 Cengage Learning. All Rights Reserved. May not be scanned, copied or duplicated, or posted to a publicly accessible website, in whole or in part.

31. The coefficient matrix of the system is

$$\begin{bmatrix} 1 & -3 \\ 2 & 1 \end{bmatrix}.$$

Because the determinant of this matrix is 7, not zero, the system has a unique solution.

33. The coefficient matrix of the system is

$$\begin{bmatrix} 1 & -1 & 1 \\ 2 & -1 & 1 \\ 3 & -2 & 2 \end{bmatrix}$$

which has a determinant of

$$\begin{vmatrix} 1 & -1 & 1 \\ 2 & -1 & 1 \\ 3 & -2 & 2 \end{vmatrix} = 0.$$

Because the determinant is zero, the system does not have a unique solution.

35. The coefficient matrix of the system is

$$\begin{vmatrix} 2 & 1 & 5 & 1 \\ 1 & 1 & -3 & -4 \\ 2 & 2 & 2 & -3 \\ 1 & 5 & -6 & 0 \end{vmatrix}.$$

Because the determinant of this matrix is 115, and not zero, the system has a unique solution.

37. First observe that $|A| = \begin{vmatrix} 6 & -11 \\ 4 & -5 \end{vmatrix} = 14.$

(a) $|A^T| = |A| = 14$

(b) $|A^2| = |A||A| = |A|^2 = 196$

(c) $|AA^T| = |A||A^T| = 14(14) = 196$

(d) $|2A| = 4|A| = 56$

(e) $|A^{-1}| = \dfrac{1}{|A|} = \dfrac{1}{14}$

39. First observe that $|A| = \begin{vmatrix} 5 & 0 & 0 \\ 1 & -3 & 0 \\ 0 & -1 & 2 \end{vmatrix} = -30.$

(a) $|A^T| = |A| = -30$

(b) $|A^2| = |A||A| = 900$

(c) $|AA^T| = |A||A^T| = 900$

(d) $|2A| = 2^3|A| = -240$

(e) $|A^{-1}| = \dfrac{1}{|A|} = \dfrac{1}{30}$

41. First observe that $|A| = \begin{vmatrix} 2 & 0 & 5 \\ 4 & -1 & 6 \\ 3 & 2 & 1 \end{vmatrix} = 29.$

(a) $|A^T| = |A| = 29$

(b) $|A^2| = |A||A| = 29^2 = 841$

(c) $|AA^T| = |A||A^T| = 29(29) = 841$

(d) $|2A| = 2^3|A| = 8(29) = 232$

(e) $|A^{-1}| = \dfrac{1}{|A|} = \dfrac{1}{29}$

43. First observe that $|A| = \begin{vmatrix} 2 & 0 & 0 & 0 \\ 0 & -3 & 0 & 0 \\ 0 & 0 & 4 & 0 \\ 0 & 0 & 0 & 1 \end{vmatrix} = -24.$

(a) $|A^T| = |A| = -24$

(b) $|A^2| = |A||A| = 576$

(c) $|AA^T| = |A||A^T| = 576$

(d) $|2A| = 2^4|A| = -384$

(e) $|A^{-1}| = \dfrac{1}{|A|} = -\dfrac{1}{24}$

45. (a) $|A| = \begin{vmatrix} 4 & 2 \\ -1 & 5 \end{vmatrix} = 22$

(b) $|A^T| = \begin{vmatrix} 4 & -1 \\ 2 & 5 \end{vmatrix} = 22$

(c) $|A^2| = \begin{vmatrix} 14 & 18 \\ -9 & 23 \end{vmatrix} = 484$

(d) $|2A| = \begin{vmatrix} 8 & 4 \\ -2 & 10 \end{vmatrix} = 88$

(e) $|A^{-1}| = \begin{vmatrix} \frac{5}{22} & -\frac{1}{11} \\ \frac{1}{22} & \frac{2}{11} \end{vmatrix} = \frac{1}{22}$

47. (a) $|A| = \begin{vmatrix} 3 & 1 & -2 \\ 2 & -1 & 3 \\ -3 & 1 & 2 \end{vmatrix} = -26$

(b) $|A^T| = |A| = -26$

(c) $|A^2| = |A||A| = 676$

(d) $|2A| = 2^3|A| = -208$

(e) $|A^{-1}| = \dfrac{1}{|A|} = \dfrac{1}{-26}$

© 2013 Cengage Learning. All Rights Reserved. May not be scanned, copied or duplicated, or posted to a publicly accessible website, in whole or in part.

49. (a) $|A| = \begin{vmatrix} 4 & -2 & 1 & 5 \\ 3 & 8 & 2 & -1 \\ 6 & 8 & 9 & 2 \\ 2 & 3 & -1 & 0 \end{vmatrix} = -115$

(b) $|A^T| = \begin{vmatrix} 4 & 3 & 6 & 2 \\ -2 & 8 & 8 & 3 \\ 1 & 2 & 9 & -1 \\ 5 & -1 & 2 & 0 \end{vmatrix} = -115$

(c) $|A^2| = \begin{vmatrix} 26 & -1 & 4 & 24 \\ 46 & 71 & 38 & 11 \\ 106 & 130 & 101 & 40 \\ 11 & 12 & -1 & 5 \end{vmatrix} = 13{,}225$

(d) $|2A| = \begin{vmatrix} 8 & -4 & 2 & 10 \\ 6 & 16 & 4 & -2 \\ 12 & 16 & 18 & 4 \\ 4 & 6 & -2 & 0 \end{vmatrix} = -1840$

(e) $|A^{-1}| = \begin{vmatrix} -\frac{63}{115} & -\frac{173}{115} & \frac{71}{115} & 2 \\ \frac{38}{115} & \frac{108}{115} & -\frac{41}{115} & -1 \\ -\frac{12}{115} & -\frac{22}{115} & \frac{19}{115} & 0 \\ \frac{91}{115} & \frac{186}{115} & -\frac{77}{115} & -2 \end{vmatrix} = -\frac{1}{115}$

51. (a) $|A^2| = |A|^2 = (-5)^2 = 25$

(b) $|B^2| = |B|^2 = (3)^2 = 9$

(c) $|A^3| = |A|^3 = (-5)^3 = -125$

(d) $|B^4| = |B|^4 = (3)^4 = 81$

53. Find the values of k necessary to make A singular by setting $|A| = 0$.

$$\begin{aligned} |A| &= \begin{vmatrix} k-1 & 3 \\ 2 & k-2 \end{vmatrix} \\ &= (k-1)(k-2) - 6 \\ &= k^2 - 3k - 4 \\ &= (k-4)(k+1) = 0 \end{aligned}$$

So, $|A| = 0$ when $k = -1, 4$.

55. Find the value of k necessary to make A singular by setting $|A| = 0$.

$$|A| = \begin{vmatrix} 1 & 0 & 3 \\ 2 & -1 & 0 \\ 4 & 2 & k \end{vmatrix} = 1(-k) + 3(8) = 0$$

So, $k = 24$.

57. $AB = I$, which implies that $|AB| = |A||B| = |I| = 1$. So, both $|A|$ and $|B|$ must be nonzero, because their product is 1.

59. Let

$$A = \begin{bmatrix} 1 & 0 \\ 0 & 0 \end{bmatrix} \text{ and } B = \begin{bmatrix} 0 & 1 \\ 0 & 0 \end{bmatrix}.$$

Then

$|A| + |B| = 0 + 0 = 0$, and $|A + B| = \begin{vmatrix} 1 & 1 \\ 0 & 0 \end{vmatrix} = 0.$

(The answer is not unique.)

61. For each i, $i = 1, 2, \ldots, n$, the ith row of A can be written as

$$a_{i1},\ a_{i2},\ \ldots,\ a_{in-1},\ -\sum_{j=1}^{n-1} a_{ij}.$$

Therefore, the last column can be reduced to all zeros by adding the other columns of A to it. Because A can be reduced to a matrix with a column of zeros, $|A| = 0$.

63. Let $\det(A) = x$ and $\det(A^{-1}) = y$. First note that

$$\begin{aligned} xy &= \det(A) \cdot \det(A^{-1}) \\ &= \det(AA^{-1}) \\ &= \det(I) \\ &= 1. \end{aligned}$$

Assume that all of the entries of A and A^{-1} are integers. Because a determinant is a product of the entries of a matrix, $x = \det(A)$ and $y = \det(A^{-1})$ are integers. Therefore it must be that x and y are each ± 1 because these are the only integer solutions to $xy = 1$.

65. (a) False. See Theorem 3.6, page 121.
(b) True. See Theorem 3.8, page 122.
(c) True. See "Equivalent Conditions for a Nonsingular Matrix," parts 1 and 2, page 123.

67. $P^{-1}AP \neq A$ in general. For example,

$$P = \begin{bmatrix} 1 & 2 \\ 3 & 5 \end{bmatrix},\ P^{-1} = \begin{bmatrix} -5 & 2 \\ 3 & -1 \end{bmatrix},\ A = \begin{bmatrix} 2 & 1 \\ -1 & 0 \end{bmatrix},$$

$$P^{-1}AP = \begin{bmatrix} -27 & -49 \\ 16 & 29 \end{bmatrix} \neq A.$$

However, the determinants $|A|$ and $|P^{-1}AP|$ are equal.

$$|P^{-1}AP| = |P^{-1}||A||P| = |P^{-1}||P||A| = \frac{1}{|P|}|P||A| = |A|$$

69. Let A be an $n \times n$ matrix satisfying $A^T = -A$. Then,

$$|A| = |A^T| = |-A| = (-1)^n|A|.$$

71. The inverse of this matrix is

$$\begin{bmatrix} 0 & 1 \\ 1 & 0 \end{bmatrix}^{-1} = \begin{bmatrix} 0 & 1 \\ 1 & 0 \end{bmatrix}.$$

Because $A^T = A^{-1}$, $\begin{bmatrix} 0 & 1 \\ 1 & 0 \end{bmatrix}$ *is* orthogonal.

© 2013 Cengage Learning. All Rights Reserved. May not be scanned, copied or duplicated, or posted to a publicly accessible website, in whole or in part.

73. Because the matrix does not have an inverse (its determinant is 0), it is *not* orthogonal.

75. The inverse of this elementary matrix is

$$A^{-1} = \begin{bmatrix} 1 & 0 & 0 \\ 0 & 0 & 1 \\ 0 & 1 & 0 \end{bmatrix}.$$

Because $A^{-1} = A^T$, the matrix *is* orthogonal.

77. If $A^T = A^{-1}$, then $|A^T| = |A^{-1}|$ and so

$$|I| = |AA^{-1}| = |A||A^{-1}| = |A||A^T| = |A|^2 = 1 \Rightarrow |A| = \pm 1.$$

79. $A = \begin{bmatrix} \frac{2}{3} & -\frac{2}{3} & \frac{1}{3} \\ \frac{2}{3} & \frac{1}{3} & -\frac{2}{3} \\ \frac{1}{3} & \frac{2}{3} & \frac{2}{3} \end{bmatrix}$

Using a graphing utility you have

(a), (b) $A^{-1} = \begin{bmatrix} \frac{2}{3} & \frac{2}{3} & \frac{1}{3} \\ -\frac{2}{3} & \frac{1}{3} & \frac{2}{3} \\ \frac{1}{3} & -\frac{2}{3} & \frac{2}{3} \end{bmatrix} = A^T$

(c) As shown in Exercise 77, if A is an orthogonal matrix, then $|A| = \pm 1$. For this given A you have $|A| = 1$. Because $A^{-1} = A^T$, A is an orthogonal matrix.

81. $|SB| = |S||B| = 0|B| = 0 \Rightarrow SB$ is singular.

Section 3.4 Applications of Determinants

1. The matrix of cofactors is

$$\begin{bmatrix} 4 & -3 \\ -2 & 1 \end{bmatrix} = \begin{bmatrix} 4 & -3 \\ -2 & 1 \end{bmatrix}.$$

So, the adjoint of A is

$$\text{adj}(A) = \begin{bmatrix} 4 & -3 \\ -2 & 1 \end{bmatrix}^T = \begin{bmatrix} 4 & -2 \\ -3 & 1 \end{bmatrix}.$$

Because $|A| = -2$, the inverse of A is

$$A^{-1} = \frac{1}{|A|}\text{adj}(A) = -\frac{1}{2}\begin{bmatrix} 4 & -2 \\ -3 & 1 \end{bmatrix} = \begin{bmatrix} -2 & 1 \\ \frac{3}{2} & -\frac{1}{2} \end{bmatrix}.$$

3. The matrix of cofactors is

$$\begin{bmatrix} \begin{vmatrix} 2 & 6 \\ -4 & -12 \end{vmatrix} & -\begin{vmatrix} 0 & 6 \\ 0 & -12 \end{vmatrix} & \begin{vmatrix} 0 & 2 \\ 0 & -4 \end{vmatrix} \\ -\begin{vmatrix} 0 & 0 \\ -4 & -12 \end{vmatrix} & \begin{vmatrix} 1 & 0 \\ 0 & -12 \end{vmatrix} & -\begin{vmatrix} 1 & 0 \\ 0 & -4 \end{vmatrix} \\ \begin{vmatrix} 0 & 0 \\ 2 & 6 \end{vmatrix} & -\begin{vmatrix} 1 & 0 \\ 0 & 6 \end{vmatrix} & \begin{vmatrix} 1 & 0 \\ 0 & 2 \end{vmatrix} \end{bmatrix} = \begin{bmatrix} 0 & 0 & 0 \\ 0 & -12 & 4 \\ 0 & -6 & 2 \end{bmatrix}.$$

So, the adjoint of A is $\text{adj}(A) = \begin{bmatrix} 0 & 0 & 0 \\ 0 & -12 & -6 \\ 0 & 4 & 2 \end{bmatrix}$.

Because row 3 of A is a multiple of row 2, the determinant is zero, and A has no inverse.

5. The matrix of cofactors is

$$\begin{bmatrix} \begin{vmatrix} 4 & 3 \\ 1 & -1 \end{vmatrix} & -\begin{vmatrix} 2 & 3 \\ 0 & -1 \end{vmatrix} & \begin{vmatrix} 2 & 4 \\ 0 & 1 \end{vmatrix} \\ -\begin{vmatrix} -5 & -7 \\ 1 & -1 \end{vmatrix} & \begin{vmatrix} -3 & -7 \\ 0 & -1 \end{vmatrix} & -\begin{vmatrix} -3 & -5 \\ 0 & 1 \end{vmatrix} \\ \begin{vmatrix} -5 & -7 \\ 4 & 3 \end{vmatrix} & -\begin{vmatrix} -3 & -7 \\ 2 & 3 \end{vmatrix} & \begin{vmatrix} -3 & -5 \\ 2 & 4 \end{vmatrix} \end{bmatrix} = \begin{bmatrix} -7 & 2 & 2 \\ -12 & 3 & 3 \\ 13 & -5 & -2 \end{bmatrix}.$$

So, the adjoint is $\text{adj}(A) = \begin{bmatrix} -7 & -12 & 13 \\ 2 & 3 & -5 \\ 2 & 3 & -2 \end{bmatrix}$. Because $|A| = -3$, the inverse of A is

$$A^{-1} = \frac{1}{|A|}\text{adj}(A) = -\frac{1}{3}\begin{bmatrix} -7 & -12 & 13 \\ 2 & 3 & -5 \\ 2 & 3 & -2 \end{bmatrix} = \begin{bmatrix} \frac{7}{3} & 4 & -\frac{13}{3} \\ -\frac{2}{3} & -1 & \frac{5}{3} \\ -\frac{2}{3} & -1 & \frac{2}{3} \end{bmatrix}.$$

© 2013 Cengage Learning. All Rights Reserved. May not be scanned, copied or duplicated, or posted to a publicly accessible website, in whole or in part.

7. The matrix of cofactors is

$$\begin{bmatrix} \begin{vmatrix} -1 & 4 & 1 \\ 0 & 1 & 2 \\ 1 & 1 & 2 \end{vmatrix} & -\begin{vmatrix} 3 & 4 & 1 \\ 0 & 1 & 2 \\ -1 & 1 & 2 \end{vmatrix} & \begin{vmatrix} 3 & -1 & 1 \\ 0 & 0 & 2 \\ -1 & 1 & 2 \end{vmatrix} & -\begin{vmatrix} 3 & -1 & 4 \\ 0 & 0 & 1 \\ -1 & 1 & 1 \end{vmatrix} \\ -\begin{vmatrix} 2 & 0 & 1 \\ 0 & 1 & 2 \\ 1 & 1 & 2 \end{vmatrix} & \begin{vmatrix} -1 & 0 & 1 \\ 0 & 1 & 2 \\ -1 & 1 & 2 \end{vmatrix} & -\begin{vmatrix} -1 & 2 & 1 \\ 0 & 0 & 2 \\ -1 & 1 & 2 \end{vmatrix} & \begin{vmatrix} -1 & 2 & 0 \\ 0 & 0 & 1 \\ -1 & 1 & 1 \end{vmatrix} \\ \begin{vmatrix} 2 & 0 & 1 \\ -1 & 4 & 1 \\ 1 & 1 & 2 \end{vmatrix} & -\begin{vmatrix} -1 & 0 & 1 \\ 3 & 4 & 1 \\ -1 & 1 & 2 \end{vmatrix} & \begin{vmatrix} -1 & 2 & 1 \\ 3 & -1 & 1 \\ -1 & 1 & 2 \end{vmatrix} & -\begin{vmatrix} -1 & 2 & 0 \\ 3 & -1 & 4 \\ -1 & 1 & 1 \end{vmatrix} \\ -\begin{vmatrix} 2 & 0 & 1 \\ -1 & 4 & 1 \\ 0 & 1 & 2 \end{vmatrix} & \begin{vmatrix} -1 & 0 & 1 \\ 3 & 4 & 1 \\ 0 & 1 & 2 \end{vmatrix} & -\begin{vmatrix} -1 & 2 & 1 \\ 3 & -1 & 1 \\ 0 & 0 & 2 \end{vmatrix} & \begin{vmatrix} -1 & 2 & 0 \\ 3 & -1 & 4 \\ 0 & 0 & 1 \end{vmatrix} \end{bmatrix} = \begin{bmatrix} 7 & 7 & -4 & 2 \\ 1 & 1 & 2 & -1 \\ 9 & 0 & -9 & 9 \\ -13 & -4 & 10 & -5 \end{bmatrix}.$$

So, the adjoint of A is $\operatorname{adj}(A) = \begin{bmatrix} 7 & 1 & 9 & -13 \\ 7 & 1 & 0 & -4 \\ -4 & 2 & -9 & 10 \\ 2 & -1 & 9 & -5 \end{bmatrix}$. Because $\det(A) = 9$, the inverse of A is

$$A^{-1} = \frac{1}{|A|}\operatorname{adj}(A) = \begin{bmatrix} \frac{7}{9} & \frac{1}{9} & 1 & -\frac{13}{9} \\ \frac{7}{9} & \frac{1}{9} & 0 & -\frac{4}{9} \\ -\frac{4}{9} & \frac{2}{9} & -1 & \frac{10}{9} \\ \frac{2}{9} & -\frac{1}{9} & 1 & -\frac{5}{9} \end{bmatrix}.$$

9. If all the entries of A are integers, then so are those of the adjoint of A.

Because $A^{-1} = \frac{1}{|A|}\operatorname{adj}(A)$, and $|A| = 1$, the entries of A^{-1} must be integers.

11. Because $\operatorname{adj}(A) = |A|A^{-1}$,

$$|\operatorname{adj}(A)| = \left||A|A^{-1}\right| = |A|^n\left|A^{-1}\right| = |A|^n\frac{1}{|A|} = |A|^{n-1}.$$

13. $|\operatorname{adj}(A)| = \begin{vmatrix} -2 & 0 \\ -1 & 1 \end{vmatrix} = -2$

$$|A| = \begin{vmatrix} 1 & 0 \\ 1 & -2 \end{vmatrix} = -2$$

So, $|\operatorname{adj}(A)| = |A|$.

15. Because $\operatorname{adj}(A^{-1}) = |A^{-1}|A$ and

$$(\operatorname{adj}(A))^{-1} = \left(|A|A^{-1}\right)^{-1} = \frac{1}{|A|}A,$$

you have $\operatorname{adj}(A^{-1}) = (\operatorname{adj}(A))^{-1}$.

17. The coefficient matrix is

$$A = \begin{bmatrix} 1 & 2 \\ -1 & 1 \end{bmatrix}, \text{ and } |A| = 3.$$

Because $|A| \neq 0$, you can use Cramer's Rule.

$$A_1 = \begin{bmatrix} 5 & 2 \\ 1 & 1 \end{bmatrix}, \qquad |A_1| = 3$$

$$A_2 = \begin{bmatrix} 1 & 5 \\ -1 & 1 \end{bmatrix}, \qquad |A_2| = 6$$

The solution is

$$x_1 = \frac{|A_1|}{|A|} = \frac{3}{3} = 1$$

$$x_2 = \frac{|A_2|}{|A|} = \frac{6}{3} = 2.$$

© 2013 Cengage Learning. All Rights Reserved. May not be scanned, copied or duplicated, or posted to a publicly accessible website, in whole or in part.

19. The coefficient matrix is

$$A = \begin{bmatrix} 3 & 4 \\ 5 & 3 \end{bmatrix}, \quad \text{and } |A| = -11.$$

Because $|A| \neq 0$, you can use Cramer's Rule.

$$A_1 = \begin{bmatrix} -2 & 4 \\ 4 & 3 \end{bmatrix}, \quad |A_1| = -22$$

$$A_2 = \begin{bmatrix} 3 & -2 \\ 5 & 4 \end{bmatrix}, \quad |A_2| = 22$$

The solution is

$$x_1 = \frac{|A_1|}{|A|} = \frac{-22}{-11} = 2$$

$$x_2 = \frac{|A_2|}{|A|} = \frac{22}{-11} = -2.$$

21. The coefficient matrix is

$$A = \begin{bmatrix} 20 & 8 \\ 12 & -24 \end{bmatrix}, \quad \text{and } |A| = -576.$$

Because $|A| \neq 0$, you can use Cramer's Rule.

$$A_1 = \begin{bmatrix} 11 & 8 \\ 21 & -24 \end{bmatrix}, \quad |A_1| = -432$$

$$A_2 = \begin{bmatrix} 20 & 11 \\ 12 & 21 \end{bmatrix}, \quad |A_2| = 288$$

The solution is

$$x_1 = \frac{|A_1|}{|A|} = \frac{-432}{-576} = \frac{3}{4}$$

$$x_2 = \frac{|A_2|}{|A|} = \frac{288}{-576} = -\frac{1}{2}.$$

23. The coefficient matrix is

$$A = \begin{bmatrix} -0.4 & 0.8 \\ 2 & -4 \end{bmatrix}, \quad \text{and } |A| = 0.$$

Because $|A| = 0$, Cramer's Rule cannot be applied. (The system does not have a solution.)

25. The coefficient matrix is

$$A = \begin{bmatrix} 4 & -1 & -1 \\ 2 & 2 & 3 \\ 5 & -2 & -2 \end{bmatrix}, \quad \text{and } |A| = 3.$$

Because $|A| \neq 0$, you can use Cramer's Rule.

$$A_1 = \begin{bmatrix} 1 & -1 & -1 \\ 10 & 2 & 3 \\ -1 & -2 & -2 \end{bmatrix}, \quad |A_1| = 3$$

$$A_2 = \begin{bmatrix} 4 & 1 & -1 \\ 2 & 10 & 3 \\ 5 & -1 & -2 \end{bmatrix}, \quad |A_2| = 3$$

$$A_3 = \begin{bmatrix} 4 & -1 & 1 \\ 2 & 2 & 10 \\ 5 & -2 & -1 \end{bmatrix}, \quad |A_3| = 6$$

The solution is

$$x_1 = \frac{|A_1|}{|A|} = \frac{3}{3} = 1$$

$$x_2 = \frac{|A_2|}{|A|} = \frac{3}{3} = 1$$

$$x_3 = \frac{|A_3|}{|A|} = \frac{6}{3} = 2.$$

27. The coefficient matrix is

$$A = \begin{bmatrix} 3 & 4 & 4 \\ 4 & -4 & 6 \\ 6 & -6 & 0 \end{bmatrix}, \quad \text{and } |A| = 252.$$

Because $|A| \neq 0$, you can use Cramer's Rule.

$$A_1 = \begin{bmatrix} 11 & 4 & 4 \\ 11 & -4 & 6 \\ 3 & -6 & 0 \end{bmatrix}, \quad |A_1| = 252$$

$$A_2 = \begin{bmatrix} 3 & 11 & 4 \\ 4 & 11 & 6 \\ 6 & 3 & 0 \end{bmatrix}, \quad |A_2| = 126$$

$$A_3 = \begin{bmatrix} 3 & 4 & 11 \\ 4 & -4 & 11 \\ 6 & -6 & 3 \end{bmatrix}, \quad |A_3| = 378$$

The solution is

$$x_1 = \frac{|A_1|}{|A|} = \frac{252}{252} = 1$$

$$x_2 = \frac{|A_2|}{|A|} = \frac{126}{252} = \frac{1}{2}$$

$$x_3 = \frac{|A_3|}{|A|} = \frac{378}{252} = \frac{3}{2}.$$

© 2013 Cengage Learning. All Rights Reserved. May not be scanned, copied or duplicated, or posted to a publicly accessible website, in whole or in part.

29. The coefficient matrix is

$$A = \begin{bmatrix} 4 & -1 & 1 \\ 2 & 2 & 3 \\ 5 & -2 & 6 \end{bmatrix}, \text{ and } |A| = 55.$$

Because $|A| \neq 0$, you can use Cramer's Rule.

$$A_1 = \begin{bmatrix} -5 & -1 & 1 \\ 10 & 2 & 3 \\ 1 & -2 & 6 \end{bmatrix}, \quad |A_1| = -55$$

$$A_2 = \begin{bmatrix} 4 & -5 & 1 \\ 2 & 10 & 3 \\ 5 & 1 & 6 \end{bmatrix}, \quad |A_2| = 165$$

$$A_3 = \begin{bmatrix} 4 & -1 & -5 \\ 2 & 2 & 10 \\ 5 & -2 & 1 \end{bmatrix}, \quad |A_3| = 110$$

The solution is

$$x_1 = \frac{|A_1|}{|A|} = \frac{-55}{55} = -1$$

$$x_2 = \frac{|A_2|}{|A|} = \frac{165}{55} = 3$$

$$x_3 = \frac{|A_3|}{|A|} = \frac{110}{55} = 2.$$

31. The coefficient matrix is $A = \begin{bmatrix} \frac{5}{6} & -1 \\ \frac{4}{3} & -\frac{7}{2} \end{bmatrix}$.

$$A_1 = \begin{bmatrix} -20 & -1 \\ -51 & -\frac{7}{2} \end{bmatrix}, \quad A_2 = \begin{bmatrix} \frac{5}{6} & -20 \\ \frac{4}{3} & -51 \end{bmatrix}$$

Using a graphing utility, $|A| = -\frac{19}{12}$, $|A_1| = 19$, and $|A_2| = -\frac{95}{6}$.

So, $x_1 = \dfrac{|A_1|}{|A|} = \dfrac{19}{\left(-\frac{19}{12}\right)} = -12$ and

$$x_2 = \frac{|A_2|}{|A|} = \frac{\left(-\frac{95}{6}\right)}{\left(-\frac{19}{12}\right)} = 10.$$

33. The coefficient matrix is $A = \begin{bmatrix} 3 & -2 & 9 & 4 \\ -1 & 0 & -9 & -6 \\ 0 & 0 & 3 & 1 \\ 2 & 2 & 0 & 8 \end{bmatrix}$.

$$A_1 = \begin{bmatrix} 35 & -2 & 9 & 4 \\ -17 & 0 & -9 & -6 \\ 5 & 0 & 3 & 1 \\ -4 & 2 & 0 & 8 \end{bmatrix}, \quad A_2 = \begin{bmatrix} 3 & 35 & 9 & 4 \\ -1 & -17 & -9 & -6 \\ 0 & 5 & 3 & 1 \\ 2 & -4 & 0 & 8 \end{bmatrix},$$

$$A_3 = \begin{bmatrix} 3 & -2 & 35 & 4 \\ -1 & 0 & -17 & -6 \\ 0 & 0 & 5 & 1 \\ 2 & 2 & -4 & 8 \end{bmatrix}, \quad A_4 = \begin{bmatrix} 3 & -2 & 9 & 35 \\ -1 & 0 & -9 & -17 \\ 0 & 0 & 3 & 5 \\ 2 & 2 & 0 & -4 \end{bmatrix}$$

Using a graphing utility, $|A| = 36$, $|A_1| = 180$, $|A_2| = -108$, $|A_3| = 72$, and $|A_4| = -36$.

So,

$$x_1 = \frac{|A_1|}{|A|} = \frac{180}{36} = 5, \quad x_2 = \frac{|A_2|}{|A|} = \frac{-108}{36} = -3,$$

$$x_3 = \frac{|A_3|}{|A|} = \frac{72}{36} = 2, \text{ and } x_4 = \frac{|A_4|}{|A|} = \frac{-36}{36} = -1.$$

35. The coefficient matrix is

$$A = \begin{bmatrix} k & 1-k \\ 1-k & k \end{bmatrix}, \text{ and } |A| = k^2 - (1-k)^2 = 2k - 1.$$

Replacing the ith column of A with the column of constants yields A_i.

$$A_1 = \begin{bmatrix} 1 & 1-k \\ 3 & k \end{bmatrix}, \quad |A_1| = 4k - 3$$

$$A_2 = \begin{bmatrix} k & 1 \\ 1-k & 3 \end{bmatrix}, \quad |A_2| = 4k - 1$$

The solution is

$$x = \frac{|A_1|}{|A|} = \frac{4k-3}{2k-1}$$

$$y = \frac{|A_2|}{|A|} = \frac{4k-1}{2k-1}.$$

Notice that when $k = \frac{1}{2}$, $|A| = 2k - 1 = 0$ and the system will be inconsistent.

37. Use the formula for area as follows.

$$\text{Area} = \pm\frac{1}{2}\begin{vmatrix} x_1 & y_1 & 1 \\ x_2 & y_2 & 1 \\ x_3 & y_3 & 1 \end{vmatrix} = \pm\frac{1}{2}\begin{vmatrix} 0 & 0 & 1 \\ 2 & 0 & 1 \\ 0 & 3 & 1 \end{vmatrix} = \pm\frac{1}{2}(6) = 3$$

© 2013 Cengage Learning. All Rights Reserved. May not be scanned, copied or duplicated, or posted to a publicly accessible website, in whole or in part.

39. Use the formula for area as follows.

$$\text{Area} = \pm\tfrac{1}{2}\begin{vmatrix} x_1 & y_1 & 1 \\ x_2 & y_2 & 1 \\ x_3 & y_3 & 1 \end{vmatrix} = \pm\tfrac{1}{2}\begin{vmatrix} -1 & 2 & 1 \\ 2 & 2 & 1 \\ -2 & 4 & 1 \end{vmatrix} = \pm\tfrac{1}{2}(6) = 3$$

41. Use the fact that

$$\begin{vmatrix} x_1 & y_1 & 1 \\ x_2 & y_2 & 1 \\ x_3 & y_3 & 1 \end{vmatrix} = \begin{vmatrix} 1 & 2 & 1 \\ 3 & 4 & 1 \\ 5 & 6 & 1 \end{vmatrix} = 0$$

to determine that the three points are collinear.

43. Use the fact that

$$\begin{vmatrix} x_1 & y_1 & 1 \\ x_2 & y_2 & 1 \\ x_3 & y_3 & 1 \end{vmatrix} = \begin{vmatrix} -2 & 5 & 1 \\ 0 & -1 & 1 \\ 3 & -9 & 1 \end{vmatrix} = 2$$

to determine that the three points are not collinear.

45. Find the equation as follows.

$$0 = \begin{vmatrix} x & y & 1 \\ x_1 & y_1 & 1 \\ x_2 & y_2 & 1 \end{vmatrix} = \begin{vmatrix} x & y & 1 \\ 0 & 0 & 1 \\ 3 & 4 & 1 \end{vmatrix} = 3y - 4x = 0$$

So, an equation of the line is $-4x + 3y = 0$.

47. Find the equation as follows.

$$0 = \begin{vmatrix} x & y & 1 \\ x_1 & y_1 & 1 \\ x_2 & y_2 & 1 \end{vmatrix} = \begin{vmatrix} x & y & 1 \\ -2 & 3 & 1 \\ -2 & -4 & 1 \end{vmatrix} = 7x + 14$$

So, an equation of the line is $x = -2$.

49. Use the formula for volume as follows.

$$\text{Volume} = \pm\tfrac{1}{6}\begin{vmatrix} x_1 & y_1 & z_1 & 1 \\ x_2 & y_2 & z_2 & 1 \\ x_3 & y_3 & z_3 & 1 \\ x_4 & y_4 & z_4 & 1 \end{vmatrix} = \pm\tfrac{1}{6}\begin{vmatrix} 1 & 0 & 0 & 1 \\ 0 & 1 & 0 & 1 \\ 0 & 0 & 1 & 1 \\ 1 & 1 & 1 & 1 \end{vmatrix} = \pm\tfrac{1}{6}(-2) = \tfrac{1}{3}$$

51. Use the formula for volume as follows.

$$\text{Volume} = \pm\tfrac{1}{6}\begin{vmatrix} x_1 & y_1 & z_1 & 1 \\ x_2 & y_2 & z_2 & 1 \\ x_3 & y_3 & z_3 & 1 \\ x_4 & y_4 & z_4 & 1 \end{vmatrix} = \pm\tfrac{1}{6}\begin{vmatrix} 3 & -1 & 1 & 1 \\ 4 & -4 & 4 & 1 \\ 1 & 1 & 1 & 1 \\ 0 & 0 & 1 & 1 \end{vmatrix} = \pm\tfrac{1}{6}(-12) = 2$$

53. Use the fact that

$$\begin{vmatrix} x_1 & y_1 & z_1 & 1 \\ x_2 & y_2 & z_2 & 1 \\ x_3 & y_3 & z_3 & 1 \\ x_4 & y_4 & z_4 & 1 \end{vmatrix} = \begin{vmatrix} -4 & 1 & 0 & 1 \\ 0 & 1 & 2 & 1 \\ 4 & 3 & -1 & 1 \\ 0 & 0 & 1 & 1 \end{vmatrix} = 28$$

to determine that the four points are not coplanar.

55. Use the fact that

$$\begin{vmatrix} x_1 & y_1 & z_1 & 1 \\ x_2 & y_2 & z_2 & 1 \\ x_3 & y_3 & z_3 & 1 \\ x_4 & y_4 & z_4 & 1 \end{vmatrix} = \begin{vmatrix} 0 & 0 & -1 & 1 \\ 0 & -1 & 0 & 1 \\ 1 & 1 & 0 & 1 \\ 2 & 1 & 2 & 1 \end{vmatrix} = 0$$

to determine that the four points are coplanar.

57. Find the equation as follows.

$$0 = \begin{vmatrix} x & y & z & 1 \\ x_1 & y_1 & z_1 & 1 \\ x_2 & y_2 & z_2 & 1 \\ x_3 & y_3 & z_3 & 1 \end{vmatrix} = \begin{vmatrix} x & y & z & 1 \\ 1 & -2 & 1 & 1 \\ -1 & -1 & 7 & 1 \\ 2 & -1 & 3 & 1 \end{vmatrix} = x\begin{vmatrix} -2 & 1 & 1 \\ -1 & 7 & 1 \\ -1 & 3 & 1 \end{vmatrix} - y\begin{vmatrix} 1 & 1 & 1 \\ -1 & 7 & 1 \\ 2 & 3 & 1 \end{vmatrix} + z\begin{vmatrix} 1 & -2 & 1 \\ -1 & -1 & 1 \\ 2 & -1 & 1 \end{vmatrix} - \begin{vmatrix} 1 & -2 & 1 \\ -1 & -1 & 7 \\ 2 & -1 & 3 \end{vmatrix} = 0, \text{ or } 4x - 10y + 3z = 27$$

59. Find the equation as follows.

$$0 = \begin{vmatrix} x & y & z & 1 \\ x_1 & y_1 & z_1 & 1 \\ x_2 & y_2 & z_2 & 1 \\ x_3 & y_3 & z_3 & 1 \end{vmatrix} = \begin{vmatrix} x & y & z & 1 \\ 0 & 0 & 0 & 1 \\ 1 & -1 & 0 & 1 \\ 0 & 1 & -1 & 1 \end{vmatrix} = x\begin{vmatrix} 0 & 0 & 1 \\ -1 & 0 & 1 \\ 1 & -1 & 1 \end{vmatrix} - y\begin{vmatrix} 0 & 0 & 1 \\ 1 & 0 & 1 \\ 0 & -1 & 1 \end{vmatrix} + z\begin{vmatrix} 0 & 0 & 1 \\ 1 & -1 & 1 \\ 0 & 1 & 1 \end{vmatrix} - \begin{vmatrix} 0 & 0 & 0 \\ 1 & -1 & 0 \\ 0 & 1 & -1 \end{vmatrix} = x + y + z = 0$$

61. The given use of Cramer's Rule to solve for y is not correct. The numerator and denominator have been reversed. The determinant of the coefficient matrix should be in the denominator.

© 2013 Cengage Learning. All Rights Reserved. May not be scanned, copied or duplicated, or posted to a publicly accessible website, in whole or in part.

63. (a) $49a + 7b + c = 10{,}697$

$64a + 8b + c = 11{,}162$

$81a + 9b + c = 9891$

(b) The coefficient matrix is

$$A = \begin{bmatrix} 49 & 7 & 1 \\ 64 & 8 & 1 \\ 81 & 9 & 1 \end{bmatrix} \text{ and } |A| = -2.$$

$$\text{Also, } A_1 = \begin{bmatrix} 10{,}697 & 7 & 1 \\ 11{,}162 & 8 & 1 \\ 9891 & 9 & 1 \end{bmatrix} \text{ and } |A_1| = 1736,$$

$$A_2 = \begin{bmatrix} 49 & 10{,}697 & 1 \\ 64 & 11{,}162 & 1 \\ 81 & 9891 & 1 \end{bmatrix} \text{ and } |A_2| = -26{,}970,$$

$$A_3 = \begin{bmatrix} 49 & 7 & 10{,}697 \\ 64 & 8 & 11{,}162 \\ 81 & 9 & 9891 \end{bmatrix} \text{ and } |A_3| = 82{,}332.$$

$$\text{So, } a = \frac{1736}{-2} = -868,\ b = \frac{-26{,}970}{-2} = 13{,}485, \text{ and } c = \frac{82{,}332}{-2} = -41{,}166.$$

(c)

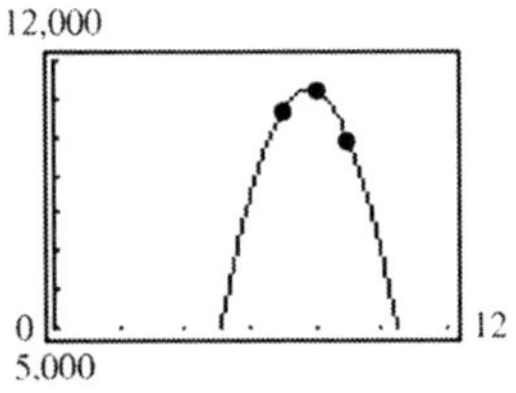

(d) The function fits the data exactly.

Review Exercises for Chapter 3

1. Using the formula for the determinant of a 2×2 matrix,

$$\begin{vmatrix} 4 & -1 \\ 2 & 2 \end{vmatrix} = 4(2) - 2(-1) = 10.$$

3. Using the formula for the determinant of a 2×2 matrix,

$$\begin{vmatrix} -3 & 1 \\ 6 & -2 \end{vmatrix} = (-3)(-2) - 6(1) = 0.$$

5. Expansion by cofactors along the first column produces

$$\begin{vmatrix} 1 & 4 & -2 \\ 0 & -3 & 1 \\ 1 & 1 & -1 \end{vmatrix} = 1\begin{vmatrix} -3 & 1 \\ 1 & -1 \end{vmatrix} - 0\begin{vmatrix} 4 & -2 \\ 1 & -1 \end{vmatrix} + 1\begin{vmatrix} 4 & -2 \\ -3 & 1 \end{vmatrix}$$

$$= 1(2) + (-2) = 0.$$

7. The determinant of a diagonal matrix is the product of the entries along the main diagonal.

$$\begin{vmatrix} -2 & 0 & 0 \\ 0 & -3 & 0 \\ 0 & 0 & -1 \end{vmatrix} = (-2)(-3)(-1) = -6$$

9. Expansion by cofactors along the first column produces

$$\begin{vmatrix} -3 & 6 & 9 \\ 9 & 12 & -3 \\ 0 & 15 & -6 \end{vmatrix} = -3\begin{vmatrix} 12 & -3 \\ 15 & -6 \end{vmatrix} - 9\begin{vmatrix} 6 & 9 \\ 15 & -6 \end{vmatrix} + 0\begin{vmatrix} 6 & 9 \\ 12 & -3 \end{vmatrix}$$

$$= 81 + 1539 + 0 = 1620.$$

11. Expansion by cofactors along the second column produces

$$\begin{vmatrix} 2 & 0 & -1 & 4 \\ -1 & 2 & 0 & 3 \\ 3 & 0 & 1 & 2 \\ -2 & 0 & 3 & 1 \end{vmatrix} = 2\begin{vmatrix} 2 & -1 & 4 \\ 3 & 1 & 2 \\ -2 & 3 & 1 \end{vmatrix} = 2\begin{vmatrix} 5 & 0 & 6 \\ 3 & 1 & 2 \\ -11 & 0 & -5 \end{vmatrix} = 2\begin{vmatrix} 5 & 6 \\ -11 & -5 \end{vmatrix} = 2(-25 + 66) = 82.$$

© 2013 Cengage Learning. All Rights Reserved. May not be scanned, copied or duplicated, or posted to a publicly accessible website, in whole or in part.

13.

$$\begin{vmatrix} -4 & 1 & 2 & 3 \\ 1 & -2 & 1 & 2 \\ 2 & -1 & 3 & 4 \\ 1 & 2 & 2 & -1 \end{vmatrix} = -\begin{vmatrix} 1 & 2 & 2 & -1 \\ 1 & -2 & 1 & 2 \\ 2 & -1 & 3 & 4 \\ -4 & 1 & 2 & 3 \end{vmatrix}$$

$$= -\begin{vmatrix} 1 & 2 & 2 & -1 \\ 0 & -4 & -1 & 3 \\ 0 & -5 & -1 & 6 \\ 0 & 9 & 10 & -1 \end{vmatrix}$$

$$= -\begin{vmatrix} 1 & 2 & 2 & -1 \\ 0 & 1 & 0 & -3 \\ 0 & -5 & -1 & 6 \\ 0 & 9 & 10 & -1 \end{vmatrix}$$

$$= -\begin{vmatrix} 1 & 2 & 2 & -1 \\ 0 & 1 & 0 & -3 \\ 0 & 0 & -1 & -9 \\ 0 & 0 & 10 & 26 \end{vmatrix}$$

$$= -\begin{vmatrix} 1 & 2 & 2 & -1 \\ 0 & 1 & 0 & -3 \\ 0 & 0 & -1 & -9 \\ 0 & 0 & 0 & -64 \end{vmatrix}$$

$$= -64$$

15.

$$\begin{vmatrix} -1 & 1 & -1 & 0 & 0 \\ 0 & 1 & -1 & 0 & 1 \\ 1 & 0 & 1 & -1 & 0 \\ 0 & -1 & 0 & 1 & -1 \\ 0 & 1 & 1 & -1 & 1 \end{vmatrix} = \begin{vmatrix} -1 & 1 & -1 & 0 & 0 \\ 0 & 1 & -1 & 0 & 1 \\ 0 & 1 & 0 & -1 & 0 \\ 0 & -1 & 0 & 1 & -1 \\ 0 & 1 & 1 & -1 & 1 \end{vmatrix}$$

$$= (-1)\begin{vmatrix} 1 & -1 & 0 & 1 \\ 1 & 0 & -1 & 0 \\ -1 & 0 & 1 & -1 \\ 1 & 1 & -1 & 1 \end{vmatrix}$$

$$= (-1)\begin{vmatrix} 1 & -1 & 0 & 1 \\ 1 & 0 & -1 & 0 \\ -1 & 0 & 1 & -1 \\ 2 & 0 & -1 & 2 \end{vmatrix}$$

$$= (-1)\begin{vmatrix} 1 & -1 & 0 \\ -1 & 1 & -1 \\ 2 & -1 & 2 \end{vmatrix}$$

$$= (-1)(1) = -1$$

17. The determinant of a diagonal matrix is the product of its main diagonal entries. So,

$$\begin{vmatrix} -1 & 0 & 0 & 0 & 0 \\ 0 & -1 & 0 & 0 & 0 \\ 0 & 0 & -1 & 0 & 0 \\ 0 & 0 & 0 & -1 & 0 \\ 0 & 0 & 0 & 0 & -1 \end{vmatrix} = (-1)^5 = -1.$$

19. Because the second row is a multiple of the first row, the determinant is zero.

21. Because -4 has been factored out of the second column, and 3 factored out of the third column, the first determinant is -12 times the second one.

23. (a) $|A| = \begin{vmatrix} -1 & 2 \\ 0 & 1 \end{vmatrix} = -1$

(b) $|B| = \begin{vmatrix} 3 & 4 \\ 2 & 1 \end{vmatrix} = -5$

(c) $AB = \begin{bmatrix} -1 & 2 \\ 0 & 1 \end{bmatrix}\begin{bmatrix} 3 & 4 \\ 2 & 1 \end{bmatrix} = \begin{bmatrix} 1 & -2 \\ 2 & 1 \end{bmatrix}$

(d) $|AB| = \begin{vmatrix} 1 & -2 \\ 2 & 1 \end{vmatrix} = 5$

Notice that $|A||B| = |AB| = 5$.

25. First find

$$|A| = \begin{vmatrix} -2 & 6 \\ 1 & 3 \end{vmatrix} = -12.$$

(a) $|A^T| = |A| = -12$

(b) $|A^3| = |A|^3 = (-12)^3 = -1728$

(c) $|A^TA| = |A^T||A| = -12(-12) = 144$

(d) $|5A| = 5^2|A| = 25(-12) = -300$

27. (a) $|A| = \begin{vmatrix} 1 & 0 & -4 \\ 0 & 3 & 2 \\ -2 & 7 & 6 \end{vmatrix} = \begin{vmatrix} 1 & 0 & -4 \\ 0 & 3 & 2 \\ 0 & 7 & -2 \end{vmatrix} = \begin{vmatrix} 3 & 2 \\ 7 & -2 \end{vmatrix} = -20$

(b) $|A^{-1}| = \dfrac{1}{|A|} = -\dfrac{1}{20}$

29. $A^{-1} = \dfrac{1}{6}\begin{bmatrix} 4 & 1 \\ -2 & 1 \end{bmatrix} = \begin{bmatrix} \frac{2}{3} & \frac{1}{6} \\ -\frac{1}{3} & \frac{1}{6} \end{bmatrix}$

$$|A^{-1}| = \frac{2}{3}\left(\frac{1}{6}\right) - \left(-\frac{1}{3}\right)\left(\frac{1}{6}\right) = \frac{1}{9} + \frac{1}{18} = \frac{1}{6}$$

Notice that $|A| = 6$, so $|A^{-1}| = \dfrac{1}{|A|} = \dfrac{1}{6}$.

© 2013 Cengage Learning. All Rights Reserved. May not be scanned, copied or duplicated, or posted to a publicly accessible website, in whole or in part.

31. $A^{-1} = \begin{bmatrix} \frac{12}{5} & -\frac{3}{5} & -\frac{1}{10} \\ -\frac{4}{5} & \frac{1}{5} & \frac{1}{5} \\ -\frac{7}{5} & \frac{3}{5} & \frac{1}{10} \end{bmatrix}$

$$|A^{-1}| = \begin{vmatrix} \frac{12}{5} & -\frac{3}{5} & -\frac{1}{10} \\ -\frac{4}{5} & \frac{1}{5} & \frac{1}{5} \\ -\frac{7}{5} & \frac{3}{5} & \frac{1}{10} \end{vmatrix} = \begin{vmatrix} 1 & 0 & 0 \\ -\frac{4}{5} & \frac{1}{5} & \frac{1}{5} \\ -\frac{7}{5} & \frac{3}{5} & \frac{1}{10} \end{vmatrix} = -\frac{1}{10}$$

Notice that $|A| = \begin{vmatrix} 1 & 0 & 1 \\ 2 & -1 & 4 \\ 2 & 6 & 0 \end{vmatrix} = 1(0 - 24) + 1(12 + 2) = -10$, so $|A^{-1}| = \dfrac{1}{|A|} = -\dfrac{1}{10}$.

33. (a) $\begin{bmatrix} 3 & 3 & 5 & 1 \\ 3 & 5 & 9 & 2 \\ 5 & 9 & 17 & 4 \end{bmatrix} \Rightarrow \begin{bmatrix} 1 & 1 & \frac{5}{3} & \frac{1}{3} \\ 3 & 5 & 9 & 2 \\ 5 & 9 & 17 & 4 \end{bmatrix} \Rightarrow \begin{bmatrix} 1 & 1 & \frac{5}{3} & \frac{1}{3} \\ 0 & 2 & 4 & 1 \\ 0 & 4 & \frac{26}{3} & \frac{7}{3} \end{bmatrix} \Rightarrow \begin{bmatrix} 1 & 1 & \frac{5}{3} & \frac{1}{3} \\ 0 & 1 & 2 & \frac{1}{2} \\ 0 & 0 & 1 & \frac{1}{2} \end{bmatrix}$

So, $x_3 = \frac{1}{2}$, $x_2 = \frac{1}{2} - 2\left(\frac{1}{2}\right) = -\frac{1}{2}$, and $x_1 = \frac{1}{3} - \frac{5}{3}\left(\frac{1}{2}\right) - 1\left(-\frac{1}{2}\right) = 0$.

(b) $\begin{bmatrix} 1 & 1 & \frac{5}{3} & \frac{1}{3} \\ 0 & 1 & 2 & \frac{1}{2} \\ 0 & 0 & 1 & \frac{1}{2} \end{bmatrix} \Rightarrow \begin{bmatrix} 1 & 0 & -\frac{1}{3} & -\frac{1}{6} \\ 0 & 1 & 2 & \frac{1}{2} \\ 0 & 0 & 1 & \frac{1}{2} \end{bmatrix} \Rightarrow \begin{bmatrix} 1 & 0 & 0 & 0 \\ 0 & 1 & 2 & \frac{1}{2} \\ 0 & 0 & 1 & \frac{1}{2} \end{bmatrix} \Rightarrow \begin{bmatrix} 1 & 0 & 0 & 0 \\ 0 & 1 & 0 & -\frac{1}{2} \\ 0 & 0 & 1 & \frac{1}{2} \end{bmatrix}$

So, $x_1 = 0$, $x_2 = -\frac{1}{2}$, and $x_3 = \frac{1}{2}$.

(c) The coefficient matrix is

$$A = \begin{bmatrix} 3 & 3 & 5 \\ 3 & 5 & 9 \\ 5 & 9 & 17 \end{bmatrix} \text{ and } |A| = 4.$$

Also, $A_1 = \begin{bmatrix} 1 & 3 & 5 \\ 2 & 5 & 9 \\ 4 & 9 & 17 \end{bmatrix}$ and $|A_1| = 0$,

$A_2 = \begin{bmatrix} 3 & 1 & 5 \\ 3 & 2 & 9 \\ 5 & 4 & 17 \end{bmatrix}$ and $|A_2| = -2$,

$A_3 = \begin{bmatrix} 3 & 3 & 1 \\ 3 & 5 & 2 \\ 5 & 9 & 4 \end{bmatrix}$ and $|A_3| = 2$.

So, $x_1 = \dfrac{0}{4} = 0$, $x_2 = \dfrac{-2}{4} = -\dfrac{1}{2}$, and $x_3 = \dfrac{2}{4} = \dfrac{1}{2}$.

© 2013 Cengage Learning. All Rights Reserved. May not be scanned, copied or duplicated, or posted to a publicly accessible website, in whole or in part.

35. (a) $\begin{bmatrix} 1 & 2 & -1 & -7 \\ 2 & -2 & -2 & -8 \\ -1 & 3 & 4 & 8 \end{bmatrix} \Rightarrow \begin{bmatrix} 1 & 2 & -1 & -7 \\ 0 & -6 & 0 & 6 \\ 0 & 5 & 3 & 1 \end{bmatrix} \Rightarrow \begin{bmatrix} 1 & 2 & -1 & -7 \\ 0 & 1 & 0 & -1 \\ 0 & 5 & 3 & 1 \end{bmatrix} \Rightarrow \begin{bmatrix} 1 & 2 & -1 & -7 \\ 0 & 1 & 0 & -1 \\ 0 & 0 & 1 & 2 \end{bmatrix}$

So, $x_3 = 2$, $x_2 = -1$, and $x_1 = -7 + 1(2) - 2(-1) = -3$.

(b) $\begin{bmatrix} 1 & 2 & -1 & -7 \\ 0 & 1 & 0 & -1 \\ 0 & 0 & 1 & 2 \end{bmatrix} \Rightarrow \begin{bmatrix} 1 & 0 & -1 & -5 \\ 0 & 1 & 0 & -1 \\ 0 & 0 & 1 & 2 \end{bmatrix} \Rightarrow \begin{bmatrix} 1 & 0 & 0 & -3 \\ 0 & 1 & 0 & -1 \\ 0 & 0 & 1 & 2 \end{bmatrix}$

So, $x_1 = -3$, $x_2 = -1$, and $x_3 = 2$.

(c) The coefficient matrix is

$$A = \begin{bmatrix} 1 & 2 & -1 \\ 2 & -2 & -2 \\ -1 & 3 & 4 \end{bmatrix} \text{ and } |A| = -18.$$

Also, $A_1 = \begin{bmatrix} -7 & 2 & -1 \\ -8 & -2 & -2 \\ 8 & 3 & 4 \end{bmatrix}$ and $|A_1| = 54$,

$$A_2 = \begin{bmatrix} 1 & -7 & -1 \\ 2 & -8 & -2 \\ -1 & 8 & 4 \end{bmatrix} \text{ and } |A_2| = 18,$$

$$A_3 = \begin{bmatrix} 1 & 2 & -7 \\ 2 & -2 & -8 \\ -1 & 3 & 8 \end{bmatrix} \text{ and } |A_3| = -36.$$

So, $x_1 = \dfrac{54}{-18} = -3$, $x_2 = \dfrac{18}{-18} = -1$, and $x_3 = \dfrac{-36}{-18} = 2$.

37. Because the determinant of the coefficient matrix is

$$\begin{vmatrix} 5 & 4 \\ -1 & 1 \end{vmatrix} = 9 \neq 0,$$

the system has a unique solution.

39. Because the determinant of the coefficient matrix is

$$\begin{vmatrix} -1 & 1 & 2 \\ 2 & 3 & 1 \\ 5 & 4 & 2 \end{vmatrix} = -15 \neq 0,$$

the system has a unique solution.

41. Because the determinant of the coefficient matrix is

$$\begin{vmatrix} 1 & 2 & 6 \\ 2 & 5 & 15 \\ 3 & 1 & 3 \end{vmatrix} = 0,$$

the system does not have a unique solution.

43. (a) $|BA| = |B||A| = 2 \cdot 4 = 8$

(b) $|B^2| = |B|^2 = 2^2 = 4$

(c) $|2A| = 2^4|A| = 16 \cdot 4 = 64$

(d) $|(AB)^T| = |B^T A^T| = |B^T||A^T| = |B||A| = 2 \cdot 4 = 8$

Equivalently,

$|(AB)^T| = |AB| = |A||B| = 4 \cdot 2 = 8$

(e) $|B^{-1}| = \dfrac{1}{|B|} = \dfrac{1}{2}$

© 2013 Cengage Learning. All Rights Reserved. May not be scanned, copied or duplicated, or posted to a publicly accessible website, in whole or in part.

45. Expand the determinant on the left along the third row.

$$\begin{vmatrix} a_{11} & a_{12} & a_{13} \\ a_{21} & a_{22} & a_{23} \\ (a_{31} + c_{31}) & (a_{32} + c_{32}) & (a_{33} + c_{33}) \end{vmatrix}$$

$$= (a_{31} + c_{31})\begin{vmatrix} a_{12} & a_{13} \\ a_{22} & a_{23} \end{vmatrix} - (a_{32} + c_{32})\begin{vmatrix} a_{11} & a_{13} \\ a_{21} & a_{23} \end{vmatrix} + (a_{33} + c_{33})\begin{vmatrix} a_{11} & a_{12} \\ a_{21} & a_{22} \end{vmatrix}$$

$$= a_{31}\begin{vmatrix} a_{12} & a_{13} \\ a_{22} & a_{23} \end{vmatrix} + c_{31}\begin{vmatrix} a_{12} & a_{13} \\ a_{22} & a_{23} \end{vmatrix} - a_{32}\begin{vmatrix} a_{11} & a_{13} \\ a_{21} & a_{23} \end{vmatrix} - c_{32}\begin{vmatrix} a_{11} & a_{13} \\ a_{21} & a_{23} \end{vmatrix} + a_{33}\begin{vmatrix} a_{11} & a_{12} \\ a_{21} & a_{22} \end{vmatrix} + c_{33}\begin{vmatrix} a_{11} & a_{12} \\ a_{21} & a_{22} \end{vmatrix}$$

The first, third, and fifth terms in this sum correspond to the determinant

$$\begin{vmatrix} a_{11} & a_{12} & a_{13} \\ a_{21} & a_{22} & a_{23} \\ a_{31} & a_{32} & a_{33} \end{vmatrix}$$

expanded along the third row. Similarly, the second, fourth, and sixth terms of the sum correspond to the determinant

$$\begin{vmatrix} a_{11} & a_{12} & a_{13} \\ a_{21} & a_{22} & a_{23} \\ c_{31} & c_{32} & c_{33} \end{vmatrix}$$

expanded along the third row.

47. Each row consists of $n - 1$ ones and one element equal to $1 - n$. The sum of these elements is then

$$(n - 1)1 + (1 - n) = 0.$$

In Section 3.3, Exercise 61, you showed that a matrix whose rows each add up to zero has a determinant of zero. So, the determinant of this matrix is zero.

49. By definition of the Jacobian,

$$J(u, v) = \begin{vmatrix} \frac{\partial x}{\partial u} & \frac{\partial x}{\partial v} \\ \frac{\partial y}{\partial u} & \frac{\partial y}{\partial v} \end{vmatrix} = \begin{vmatrix} -\frac{1}{2} & \frac{1}{2} \\ \frac{1}{2} & \frac{1}{2} \end{vmatrix} = -\frac{1}{4} - \frac{1}{4} = -\frac{1}{2}.$$

51. $$J(u, v, w) = \begin{vmatrix} \frac{1}{2} & \frac{1}{2} & 0 \\ \frac{1}{2} & -\frac{1}{2} & 0 \\ 2vw & 2uw & 2uv \end{vmatrix} = 2uv\left(-\tfrac{1}{4} - \tfrac{1}{4}\right) = -uv$$

53. Row reduction is generally preferred for matrices with few zeros. For a matrix with many zeros, it is often easier to expand along a row or column having many zeros.

55. $$\begin{vmatrix} \cos x & 0 & \sin x \\ \sin x & 0 & \cos x \\ \sin x - \cos x & 1 & \sin x - \cos x \end{vmatrix} = -1\begin{vmatrix} \cos x & \sin x \\ \sin x & \cos x \end{vmatrix}$$

$$= -1(\cos^2 x - \sin^2 x)$$

$$= -\cos 2x$$

$$-\cos 2x = 0 \Rightarrow 2x = \frac{\pi}{2} + n\pi \text{ , } n \text{ is an integer.}$$

$$x = \frac{\pi}{4} + \frac{n\pi}{2}$$

57. The matrix of cofactors is given by $\begin{bmatrix} 1 & 2 \\ -1 & 0 \end{bmatrix}$.

So, the adjoint is $\text{adj}\begin{bmatrix} 0 & 1 \\ -2 & 1 \end{bmatrix} = \begin{bmatrix} 1 & -1 \\ 2 & 0 \end{bmatrix}$.

59. The determinant of the coefficient matrix is

$$\begin{vmatrix} 0.2 & -0.1 \\ 0.4 & -0.5 \end{vmatrix} = -0.06 \neq 0.$$

So, the system has a unique solution. Using Cramer's Rule

$$A_1 = \begin{bmatrix} 0.07 & -0.1 \\ -0.01 & -0.5 \end{bmatrix}, \quad |A_1| = -0.036$$

$$A_2 = \begin{bmatrix} 0.2 & 0.07 \\ 0.4 & -0.01 \end{bmatrix}, \quad |A_2| = -0.03.$$

So,

$$x = \frac{|A_1|}{|A|} = \frac{-0.036}{-0.06} = 0.6$$

$$y = \frac{|A_2|}{|A|} = \frac{-0.03}{-0.06} = 0.5.$$

© 2013 Cengage Learning. All Rights Reserved. May not be scanned, copied or duplicated, or posted to a publicly accessible website, in whole or in part.

61. The determinant of the coefficient matrix is

$$\begin{vmatrix} 2 & 3 & 3 \\ 6 & 6 & 12 \\ 12 & 9 & -1 \end{vmatrix} = 168 \neq 0.$$

So, the system has a unique solution. Using Cramer's Rule

$$A_1 = \begin{bmatrix} 3 & 3 & 3 \\ 13 & 6 & 12 \\ 2 & 9 & -1 \end{bmatrix}, \quad |A_1| = 84$$

$$A_2 = \begin{bmatrix} 2 & 3 & 3 \\ 6 & 13 & 12 \\ 12 & 2 & -1 \end{bmatrix}, \quad |A_2| = -56$$

$$A_3 = \begin{bmatrix} 2 & 3 & 3 \\ 6 & 6 & 13 \\ 12 & 9 & 2 \end{bmatrix}, \quad |A_3| = 168.$$

So,

$$x_1 = \frac{|A_1|}{|A|} = \frac{84}{168} = \frac{1}{2}$$

$$x_2 = \frac{|A_2|}{|A|} = \frac{-56}{168} = -\frac{1}{3}$$

$$x_3 = \frac{|A_3|}{|A|} = \frac{168}{168} = 1.$$

63. The coefficient matrix is $A = \begin{bmatrix} 0.2 & -0.6 \\ -1 & 1.4 \end{bmatrix}$.

$$A_1 = \begin{bmatrix} 2.4 & -0.6 \\ -8.8 & 1.4 \end{bmatrix}, \quad A_2 = \begin{bmatrix} 0.2 & 2.4 \\ -1 & -8.8 \end{bmatrix}$$

Using a graphing utility, $|A| = -0.32$, $|A_1| = -1.92$, and $|A_2| = 0.64$.

So, $x_1 = \dfrac{|A_1|}{|A|} = \dfrac{-1.92}{(-0.32)} = 6$ and

$$x_1 = \frac{|A_2|}{|A|} = \frac{0.64}{(-0.32)} = -2.$$

65. The formula for area yields

$$\text{Area} = \pm\frac{1}{2}\begin{vmatrix} x_1 & y_1 & 1 \\ x_2 & y_2 & 1 \\ x_3 & y_3 & 1 \end{vmatrix} = \pm\frac{1}{2}\begin{vmatrix} 1 & 0 & 1 \\ 5 & 0 & 1 \\ 5 & 8 & 1 \end{vmatrix}$$

$$= \pm\tfrac{1}{2}(-8)(1 - 5) = 16.$$

67. Find the equation as follows.

$$0 = \begin{vmatrix} x & y & 1 \\ x_1 & y_1 & 1 \\ x_2 & y_2 & 1 \end{vmatrix} = \begin{vmatrix} x & y & 1 \\ -4 & 0 & 1 \\ 4 & 4 & 1 \end{vmatrix} = -4x + 8y - 16 = 0$$

So, an equation of the line is $x - 2y = -4$.

69. Find the equation as follows.

$$0 = \begin{vmatrix} x & y & z & 1 \\ 0 & 0 & 0 & 1 \\ 1 & 0 & 3 & 1 \\ 0 & 3 & 4 & 1 \end{vmatrix} = \begin{vmatrix} x & y & z \\ 1 & 0 & 3 \\ 0 & 3 & 4 \end{vmatrix} = 0, \text{ or } 9x + 4y - 3z = 0$$

71. Cramer's Rule was not used correctly to solve for z. The given setup solves for x not z.

73. (a) False. See the "Minors and Cofactors of a Matrix" box, page 105.
(b) False. See Theorem 3.3, part 1, page 113.
(c) True. See Theorem 3.4, part 3, page 115.
(d) False. See Theorem 3.9, page 124.

75. (a) False. See Theorem 3.11, page 131.
(b) False. See "Test for Collinear Points in the xy-Plane," page 133.

Cumulative Test for Chapters 1–3

1. Because the equation cannot be written in the form $a_1x + a_2y = b$, it is *not* linear in the variables x and y.

2. Because the equation is in the form $a_1x + a_2x = b$, it *is* linear in the variables x and y.

3. Add -3 times the first equation to the second equation.

$$\begin{aligned} x - 2y &= 5 \\ 3x + y &= 1 \end{aligned} \quad \Rightarrow \quad \begin{aligned} x - 2y &= 5 \\ 7y &= -14 \end{aligned}$$

So, $y = -2$ and $x - 2(-2) = 5 \Rightarrow x = 1$.

© 2013 Cengage Learning. All Rights Reserved. May not be scanned, copied or duplicated, or posted to a publicly accessible website, in whole or in part.

4. Interchange the first equation and the third equation.

$$\begin{aligned} x_1 + x_2 + x_3 &= -3 \\ 2x_1 - 3x_2 + 2x_3 &= 9 \\ 4x_1 + x_2 - 3x_3 &= 11 \end{aligned}$$

Adding -2 times the first equation to the second equation produces a new second equation.

$$\begin{aligned} x_1 + x_2 + x_3 &= -3 \\ -5x_2 &= 15 \\ 4x_1 + x_2 - 3x_3 &= 11 \end{aligned}$$

Adding -4 times the first equation to the third equation produces a new third equation.

$$\begin{aligned} x_1 + x_2 + x_3 &= -3 \\ -5x_2 &= 15 \\ -3x_2 - 7x_3 &= 23 \end{aligned}$$

Dividing the second equation by -5 produces a new second equation.

$$\begin{aligned} x_1 + x_2 + x_3 &= -3 \\ x_2 &= -3 \\ -3x_2 - 7x_3 &= 23 \end{aligned}$$

Adding 3 times the second equation to the third equation produces a new third equation.

$$\begin{aligned} x_1 + x_2 + x_3 &= -3 \\ x_2 &= -3 \\ -7x_3 &= 14 \end{aligned}$$

Dividing the third equation by -7 produces a new third equation.

$$\begin{aligned} x_1 + x_2 + x_3 &= -3 \\ x_2 &= -3 \\ x_3 &= -2 \end{aligned}$$

Using back-substitution, the answers are found to be $x_1 = 2$, $x_2 = -3$, and $x_3 = -2$.

5. Using a software program or graphing utility, you obtain $x = 10$, $y = -20$, $z = 40$, $w = -12$.

6. $\begin{bmatrix} 0 & 1 & -1 & 0 & 2 \\ 1 & 0 & 2 & -1 & 0 \\ 1 & 2 & 0 & -1 & 4 \end{bmatrix} \Rightarrow \begin{bmatrix} 1 & 0 & 2 & -1 & 0 \\ 0 & 1 & -1 & 0 & 2 \\ 0 & 0 & 0 & 0 & 0 \end{bmatrix}$

$$\begin{aligned} x_1 &= s - 2t \\ x_2 &= 2 + t \\ x_3 &= t \\ x_4 &= s \end{aligned}$$

7. $\begin{bmatrix} 1 & 2 & 1 & -2 \\ 0 & 0 & 2 & -4 \\ -2 & -4 & 1 & -2 \end{bmatrix} \Rightarrow \begin{bmatrix} 1 & 2 & 0 & 0 \\ 0 & 0 & 1 & -2 \\ 0 & 0 & 0 & 0 \end{bmatrix}$

$$\begin{aligned} x_1 &= -2s \\ x_2 &= s \\ x_3 &= 2t \\ x_4 &= t \end{aligned}$$

8. $\begin{bmatrix} 1 & 2 & -1 & 3 \\ -1 & -1 & 1 & 2 \\ -1 & 1 & 1 & k \end{bmatrix} \Rightarrow \begin{bmatrix} 1 & 2 & -1 & 3 \\ 0 & 1 & 0 & 5 \\ 0 & 3 & 0 & 3 + k \end{bmatrix} \Rightarrow \begin{bmatrix} 1 & 2 & -1 & 3 \\ 0 & 1 & 0 & 5 \\ 0 & 0 & 0 & -12 + k \end{bmatrix}$

$k = 12$ (for consistent system)

9. $2A - B = \begin{bmatrix} -2 & 2 \\ 4 & 6 \end{bmatrix} - \begin{bmatrix} x & 2 \\ y & 5 \end{bmatrix} = \begin{bmatrix} 1 & 0 \\ 0 & 1 \end{bmatrix} \Rightarrow$

$$\begin{aligned} -2 - x &= 1 \\ 4 - y &= 0 \\ x &= -3 \\ y &= 4 \end{aligned}$$

© 2013 Cengage Learning. All Rights Reserved. May not be scanned, copied or duplicated, or posted to a publicly accessible website, in whole or in part.

10. $A^TA = \begin{bmatrix} 17 & 22 & 27 \\ 22 & 29 & 36 \\ 27 & 36 & 45 \end{bmatrix}$

11. $\begin{bmatrix} -2 & 3 \\ 4 & 6 \end{bmatrix}^{-1} = -\frac{1}{24}\begin{bmatrix} 6 & -3 \\ -4 & -2 \end{bmatrix}\begin{bmatrix} -\frac{1}{4} & \frac{1}{8} \\ \frac{1}{6} & \frac{1}{12} \end{bmatrix}$

12. $\begin{bmatrix} -2 & 3 \\ 3 & 6 \end{bmatrix}^{-1} = -\frac{1}{21}\begin{bmatrix} 6 & -3 \\ -3 & -2 \end{bmatrix} = \begin{bmatrix} -\frac{2}{7} & \frac{1}{7} \\ \frac{1}{7} & \frac{2}{21} \end{bmatrix}$

13. $\begin{bmatrix} -1 & 0 & 0 \\ 0 & \frac{1}{2} & 0 \\ 0 & 0 & 3 \end{bmatrix}^{-1} = \begin{bmatrix} -1 & 0 & 0 \\ 0 & 2 & 0 \\ 0 & 0 & \frac{1}{3} \end{bmatrix}$

14. $\begin{bmatrix} 1 & 1 & 0 \\ -3 & 6 & 5 \\ 0 & 1 & 0 \end{bmatrix}^{-1} = \begin{bmatrix} 1 & 0 & -1 \\ 0 & 0 & 1 \\ \frac{3}{5} & \frac{1}{5} & -\frac{9}{5} \end{bmatrix}$

15. The coefficient matrix for the system is $A = \begin{bmatrix} 1 & 2 \\ 1 & -2 \end{bmatrix}$ and the formula for the inverse of a 2×2 matrix produces

$$A^{-1} = \frac{1}{-2-2}\begin{bmatrix} -2 & -2 \\ -1 & 1 \end{bmatrix} = \begin{bmatrix} \frac{1}{2} & \frac{1}{2} \\ \frac{1}{4} & -\frac{1}{4} \end{bmatrix}.$$

$$\mathbf{x} = A^{-1}\mathbf{b} = \begin{bmatrix} \frac{1}{2} & \frac{1}{2} \\ \frac{1}{4} & -\frac{1}{4} \end{bmatrix}\begin{bmatrix} -3 \\ 0 \end{bmatrix} = \begin{bmatrix} -\frac{3}{2} \\ -\frac{3}{4} \end{bmatrix}$$

The solution is: $x = -\frac{3}{2}$ and $y = -\frac{3}{4}$.

16. The coefficient matrix for the system is $A = \begin{bmatrix} 2 & -1 \\ 2 & 1 \end{bmatrix}$ and the formula for the inverse of a 2×2 matrix produces

$$A^{-1} = \frac{1}{2+2}\begin{bmatrix} 1 & 1 \\ -2 & 2 \end{bmatrix} = \begin{bmatrix} \frac{1}{4} & \frac{1}{4} \\ -\frac{1}{2} & \frac{1}{2} \end{bmatrix}.$$

$$\mathbf{x} = A^{-1}\mathbf{b} = \begin{bmatrix} \frac{1}{4} & \frac{1}{4} \\ -\frac{1}{2} & \frac{1}{2} \end{bmatrix}\begin{bmatrix} 6 \\ 10 \end{bmatrix} = \begin{bmatrix} 4 \\ 2 \end{bmatrix}$$

The solution is: $x = 4$ and $y = 2$.

17. $\begin{bmatrix} 2 & -4 \\ 1 & 0 \end{bmatrix} \Rightarrow \begin{bmatrix} 1 & 0 \\ 2 & -4 \end{bmatrix} \Rightarrow \begin{bmatrix} 1 & 0 \\ 0 & -4 \end{bmatrix} \Rightarrow \begin{bmatrix} 1 & 0 \\ 0 & 1 \end{bmatrix}$

$$\begin{bmatrix} 1 & 0 \\ 0 & -\frac{1}{4} \end{bmatrix}\begin{bmatrix} 1 & 0 \\ -2 & 1 \end{bmatrix}\begin{bmatrix} 0 & 1 \\ 1 & 0 \end{bmatrix}\begin{bmatrix} 2 & -4 \\ 1 & 0 \end{bmatrix} = \begin{bmatrix} 1 & 0 \\ 0 & 1 \end{bmatrix}$$

$$A = \begin{bmatrix} 0 & 1 \\ 1 & 0 \end{bmatrix}\begin{bmatrix} 1 & 0 \\ 2 & 1 \end{bmatrix}\begin{bmatrix} 1 & 0 \\ 0 & -4 \end{bmatrix}$$

(The answer is not unique.)

18. Because the fourth row already has two zeros, choose it for cofactor expansion. An additional zero can be created by adding 4 times the first column to the fourth column.

$$\begin{vmatrix} 5 & 1 & 2 & 24 \\ 1 & 0 & -2 & 1 \\ 1 & 1 & 6 & 5 \\ 1 & 0 & 0 & 0 \end{vmatrix} = 1(-1)^5\begin{vmatrix} 1 & 2 & 24 \\ 0 & -2 & 1 \\ 1 & 6 & 5 \end{vmatrix} = -\begin{vmatrix} 1 & 2 & 24 \\ 0 & -2 & 1 \\ 1 & 6 & 5 \end{vmatrix}$$

Because the first column already has a zero, choose it for the next cofactor expansion. An additional zero can be created by adding -1 times the first row to the third row.

$$-\begin{vmatrix} 1 & 2 & 24 \\ 0 & -2 & 1 \\ 0 & 4 & -19 \end{vmatrix} = -(1)(-1)^2\begin{vmatrix} -2 & 1 \\ 4 & -19 \end{vmatrix} = -(38-4) = -34$$

19. (a) $|A| = 14$

(b) $|B| = -10$

(c) $|AB| = \begin{vmatrix}\begin{bmatrix} 1 & -3 \\ 4 & 2 \end{bmatrix}\begin{bmatrix} -2 & 1 \\ 0 & 5 \end{bmatrix}\end{vmatrix} = \begin{vmatrix} -2 & -14 \\ -8 & 14 \end{vmatrix}$

(d) $|AB| = -140$

20. (a) $|A| = 84$

(b) $|A^{-1}| = \frac{1}{|A|} = \frac{1}{84}$

21. (a) $|3A| = 3^4 \cdot 7 = 567$

(b) $|A^T| = |A| = 7$

(c) $|A^{-1}| = \frac{1}{7}$

(d) $|A^3| = 7^3 = 343$

22. The matrix of cofactors is

$$\begin{bmatrix} \begin{vmatrix} -2 & 1 \\ 0 & 2 \end{vmatrix} & -\begin{vmatrix} 0 & 1 \\ 1 & 2 \end{vmatrix} & \begin{vmatrix} 0 & -2 \\ 1 & 0 \end{vmatrix} \\ -\begin{vmatrix} -5 & -1 \\ 0 & 2 \end{vmatrix} & \begin{vmatrix} 1 & -1 \\ 1 & 2 \end{vmatrix} & -\begin{vmatrix} 1 & -5 \\ 1 & 0 \end{vmatrix} \\ \begin{vmatrix} -5 & -1 \\ -2 & 1 \end{vmatrix} & \begin{vmatrix} 1 & -1 \\ 0 & 1 \end{vmatrix} & \begin{vmatrix} 1 & -5 \\ 0 & -2 \end{vmatrix} \end{bmatrix} = \begin{bmatrix} -4 & 1 & 2 \\ 10 & 3 & -5 \\ -7 & -1 & -2 \end{bmatrix}.$$

So, the adjoint of A is $\begin{bmatrix} -4 & 10 & -7 \\ 1 & 3 & -1 \\ 2 & -5 & -2 \end{bmatrix}$.

Because $|A| = -11$, the inverse of A is

$$A^{-1} = \frac{1}{|A|}\operatorname{adj}(A) = -\frac{1}{11}\begin{bmatrix} -4 & 10 & -7 \\ 1 & 3 & -1 \\ 2 & -5 & -2 \end{bmatrix} = \begin{bmatrix} \frac{4}{11} & -\frac{10}{11} & \frac{7}{11} \\ -\frac{1}{11} & -\frac{3}{11} & \frac{1}{11} \\ -\frac{2}{11} & \frac{5}{11} & \frac{2}{11} \end{bmatrix}.$$

© 2013 Cengage Learning. All Rights Reserved. May not be scanned, copied or duplicated, or posted to a publicly accessible website, in whole or in part.

23. $a\begin{bmatrix}1\\0\\1\end{bmatrix} + b\begin{bmatrix}1\\1\\0\end{bmatrix} + c\begin{bmatrix}0\\1\\1\end{bmatrix} = \begin{bmatrix}1\\2\\3\end{bmatrix}$

The solution of this system is $a = 1$, $b = 0$, and $c = 2$.

(The answer is not unique.)

24.
$$\begin{aligned} a - b + c &= 2 \\ c &= 1 \\ 4a + 2b + c &= 6 \end{aligned}$$

The solution of this system is $a = \frac{7}{6}$, $b = \frac{1}{6}$, and $c = 1$, so $y = \frac{7}{6}x^2 + \frac{1}{6}x + 1$.

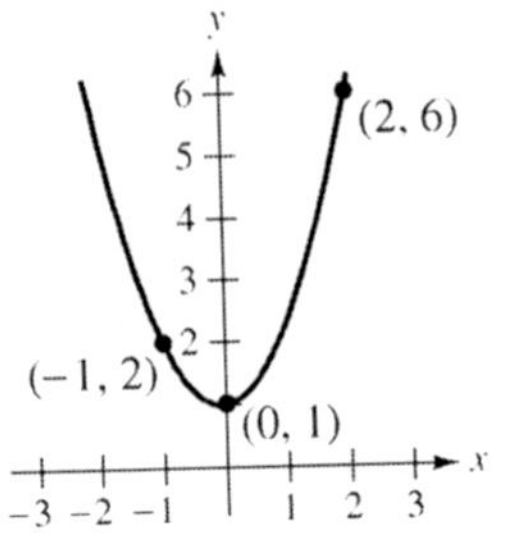

25. Find the equation as follows.

$$0 = \begin{vmatrix} x & y & 1 \\ x_1 & y_1 & 1 \\ x_2 & y_2 & 1 \end{vmatrix} = \begin{vmatrix} x & y & 1 \\ 1 & 4 & 1 \\ 5 & -2 & 1 \end{vmatrix} = x(6) - y(-4) + 1(-22) = 6x + 4y - 22$$

So, an equation of the line is $6x + 4y - 22 = 0$, or $3x + 2y = 11$.

26. Use the formula for area.

$$\text{Area} = \pm\frac{1}{2}\begin{vmatrix} x_1 & y_1 & 1 \\ x_2 & y_2 & 1 \\ x_3 & y_3 & 1 \end{vmatrix} = \pm\frac{1}{2}\begin{vmatrix} 3 & 1 & 1 \\ 7 & 1 & 1 \\ 7 & 9 & 1 \end{vmatrix} = \pm\frac{1}{2}\begin{vmatrix} 3 & 1 & 1 \\ 4 & 0 & 0 \\ 7 & 9 & 1 \end{vmatrix} = \frac{1}{2}(-4)(-8) = 16$$

27. Applying Kirchoff's first law to either junction produces

$I_1 + I_3 = I_2$

and applying the second law to the two paths produces

$R_1I_1 + R_2I_2 = 4I_1 + I_2 = 16$

$R_2I_2 + R_3I_3 = I_2 + 4I_3 = 8.$

Rearrange these equations, form the augmented matrix, and use Gauss-Jordan elimination.

$$\begin{bmatrix} 1 & -1 & 1 & 0 \\ 4 & 1 & 0 & 16 \\ 0 & 1 & 4 & 8 \end{bmatrix} \Rightarrow \begin{bmatrix} 1 & 0 & 0 & 3 \\ 0 & 1 & 0 & 4 \\ 0 & 0 & 1 & 1 \end{bmatrix}$$

So, the solution is: $I_1 = 3$, $I_2 = 4$, and $I_3 = 1$.

28. $BA = \begin{bmatrix}12.50 & 9.00 & 21.50\end{bmatrix}\begin{bmatrix} 200 & 300 \\ 600 & 350 \\ 250 & 400 \end{bmatrix} = \begin{bmatrix}13{,}275.00 & 15{,}500.00\end{bmatrix}$

This product represents the total value of the three products sent to the two warehouses.

29. No. C could be singular. $\underset{A}{\begin{bmatrix} 0 & 0 \\ 0 & 1 \end{bmatrix}}\underset{C}{\begin{bmatrix} 0 & 1 \\ 0 & 0 \end{bmatrix}} = \underset{B}{\begin{bmatrix} 0 & 1 \\ 0 & 0 \end{bmatrix}}\underset{C}{\begin{bmatrix} 0 & 1 \\ 0 & 0 \end{bmatrix}}$

© 2013 Cengage Learning. All Rights Reserved. May not be scanned, copied or duplicated, or posted to a publicly accessible website, in whole or in part.

CHAPTER 4
Vector Spaces

© 2013 Cengage Learning. All Rights Reserved. May not be scanned, copied or duplicated, or posted to a publicly accessible website, in whole or in part.

CHAPTER 4
Vector Spaces

Section 4.1 Vectors in R^n

1. $\mathbf{v} = (4, 5)$

3.

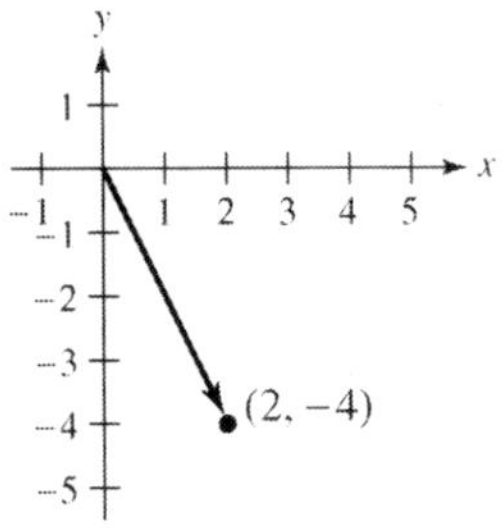

5.

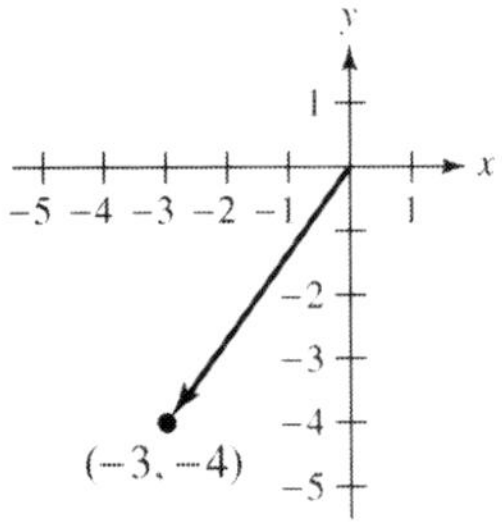

7. $\mathbf{u} + \mathbf{v} = (1, 3) + (2, -2)$
$= (1 + 2, 3 - 2)$
$= (3, 1)$

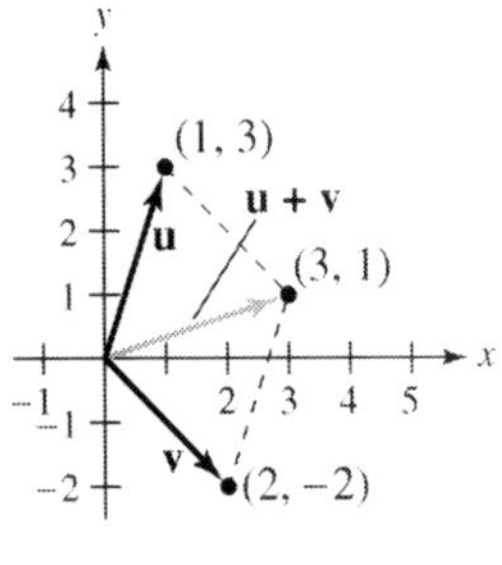

9. $\mathbf{u} + \mathbf{v} = (2, -3) + (-3, -1)$
$= (2 - 3, -3 - 1)$
$= (-1, -4)$

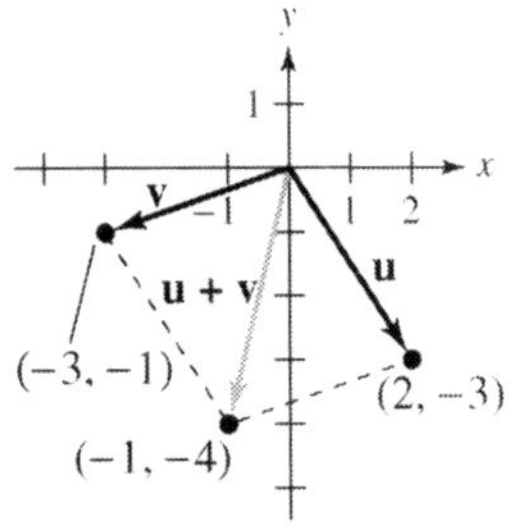

11. $\mathbf{v} = \frac{3}{2}\mathbf{u} = \frac{3}{2}(-2, 3) = \left(-3, \frac{9}{2}\right)$

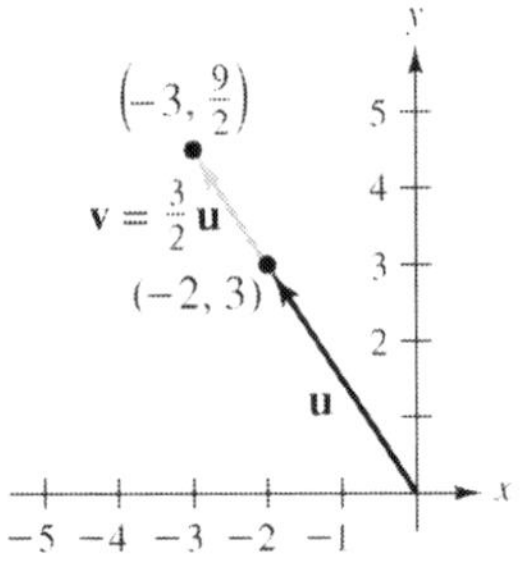

13. $\mathbf{v} = \mathbf{u} + 2\mathbf{w}$
$= (-2, 3) + 2(-3, -2)$
$= (-2, 3) + (-6, -4)$
$= (-2 - 6, 3 - 4)$
$= (-8, -1)$

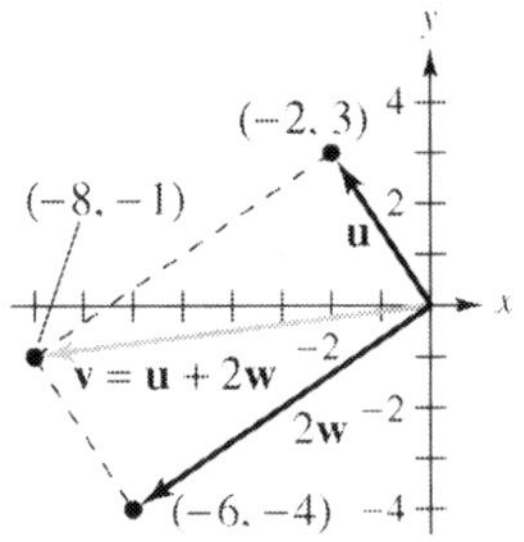

15. $\mathbf{v} = \frac{1}{2}(3\mathbf{u} + \mathbf{w})$
$= \frac{1}{2}(3(-2, 3) + (-3, -2))$
$= \frac{1}{2}((-6, 9) + (-3, -2))$
$= \frac{1}{2}(-9, 7) = \left(-\frac{9}{2}, \frac{7}{2}\right)$

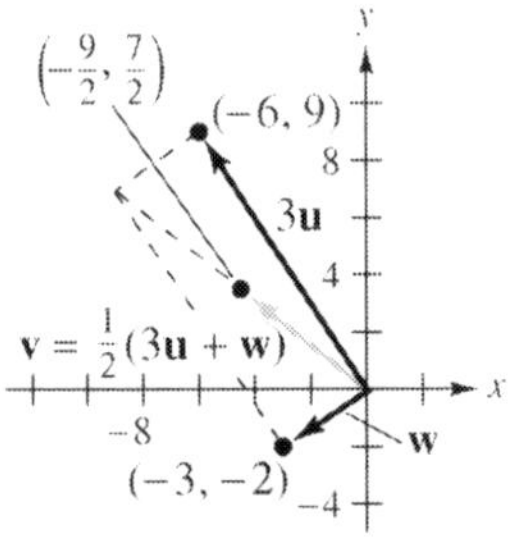

© 2013 Cengage Learning. All Rights Reserved. May not be scanned, copied or duplicated, or posted to a publicly accessible website, in whole or in part.

17. (a) $2\mathbf{v} = 2(2, 1) = (2(2), 2(1)) = (4, 2)$

(b) $-3\mathbf{v} = -3(2, 1) = (-3(2), -3(1)) = (-6, -3)$

(c) $\frac{1}{2}\mathbf{v} = \frac{1}{2}(2, 1) = \left(\frac{1}{2}(2), \frac{1}{2}(1)\right) = \left(1, \frac{1}{2}\right)$

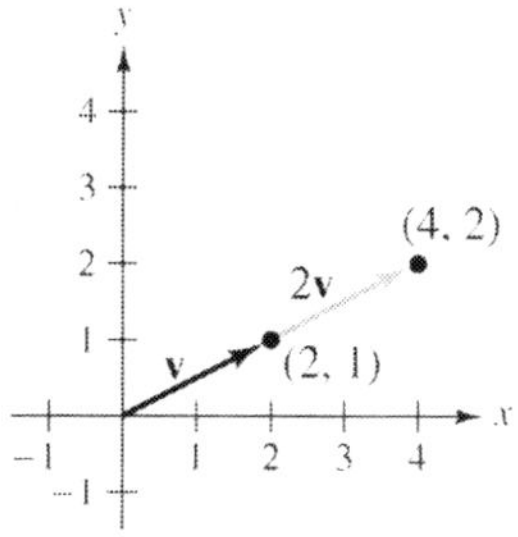

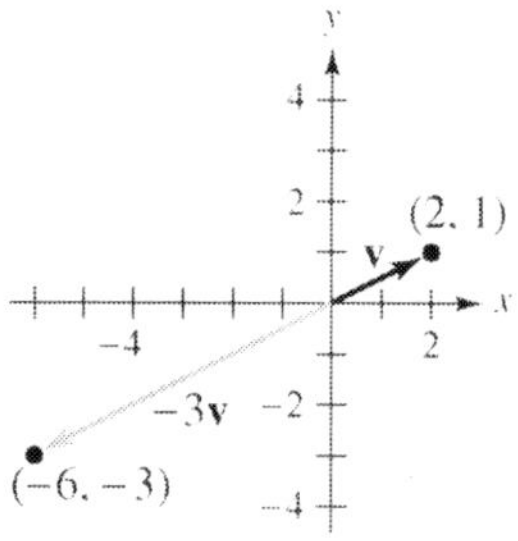

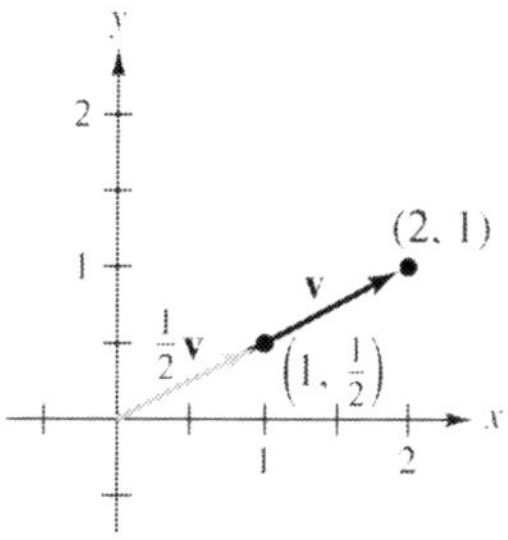

19. $\mathbf{u} - \mathbf{v} = (1, 2, 3) - (2, 2, -1) = (-1, 0, 4)$

$\mathbf{v} - \mathbf{u} = (2, 2, -1) - (1, 2, 3) = (1, 0, -4)$

21. $2\mathbf{u} + 4\mathbf{v} - \mathbf{w} = 2(1, 2, 3) + 4(2, 2, -1) - (4, 0, -4)$

$= (2, 4, 6) + (8, 8, -4) - (4, 0, -4)$

$= (2 + 8 - 4, 4 + 8 - 0, 6 + (-4) - (-4))$

$= (6, 12, 6)$

23. $2\mathbf{z} - 3\mathbf{u} = \mathbf{w}$ implies that $2\mathbf{z} = 3\mathbf{u} + \mathbf{w}$, or $\mathbf{z} = \frac{3}{2}\mathbf{u} + \frac{1}{2}\mathbf{w}$.

So, $\mathbf{z} = \frac{3}{2}(1, 2, 3) + \frac{1}{2}(4, 0, -4) = \left(\frac{3}{2}, 3, \frac{9}{2}\right) + (2, 0, -2) = \left(\frac{7}{2}, 3, \frac{5}{2}\right)$.

25. (a) $2\mathbf{v} = 2(1, 2, 2) = (2, 4, 4)$

(b) $-\mathbf{v} = -(1, 2, 2) = (-1, -2, -2)$

(c) $\frac{1}{2}\mathbf{v} = \frac{1}{2}(1, 2, 2) = \left(\frac{1}{2}, 1, 1\right)$

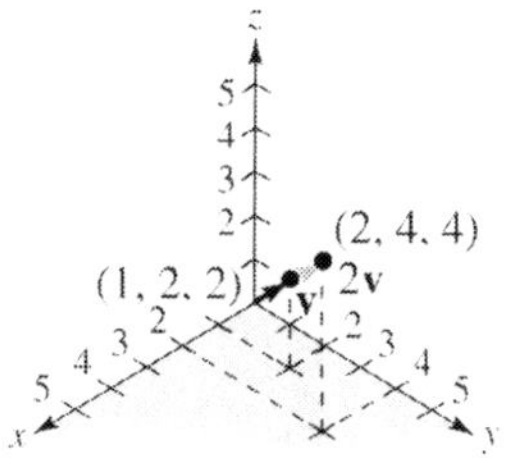

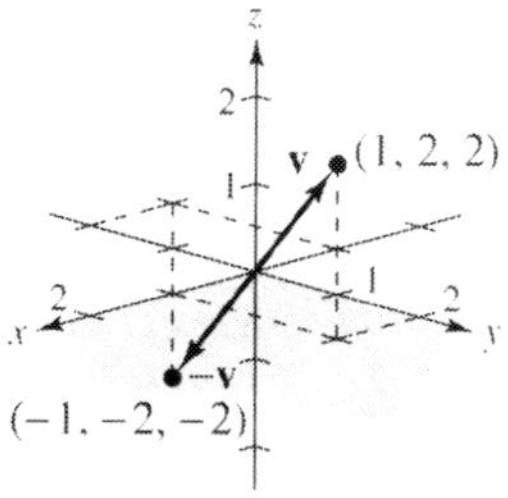

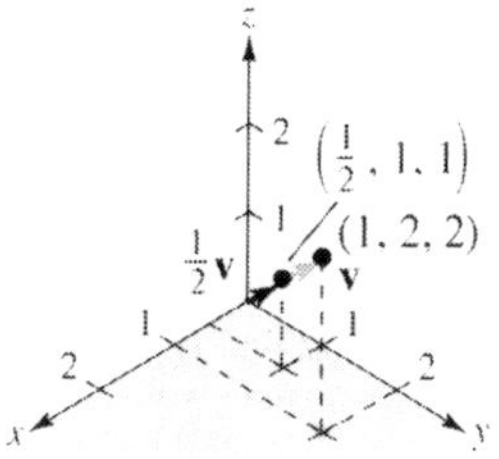

27. (a) Because $\left(2, \frac{4}{3}, -\frac{10}{3}\right) = \frac{2}{3}(3, 2, -5)$, **v** is a scalar multiple of **z**.

(b) Because $(6, 4, 10) \neq c(3, 2, -5)$ for any c, **w** is *not* a scalar multiple of **z**.

29. (a) $\mathbf{u} - \mathbf{v} = (4, 0, -3, 5) - (0, 2, 5, 4)$

$= (4 - 0, 0 - 2, -3 - 5, 5 - 4)$

$= (4, -2, -8, 1)$

(b) $2(\mathbf{u} + 3\mathbf{v}) = 2[(4, 0, -3, 5) + 3(0, 2, 5, 4)]$

$= 2[(4, 0, -3, 5) + (0, 6, 15, 12)]$

$= 2(4 + 0, 0 + 6, -3 + 15, 5 + 12)$

$= 2(4, 6, 12, 17)$

$= (8, 12, 24, 34)$

(c) $2\mathbf{v} - \mathbf{u} = 2(0, 2, 5, 4) - (4, 0, -3, 5)$

$= (0, 4, 10, 8) - (4, 0, -3, 5)$

$= (-4, 4, 13, 3)$

31. Using a graphing utility with $\mathbf{u} = (1, 2, -3, 1)$, $\mathbf{v} = (0, 2, -1, -2)$, and $\mathbf{w} = (2, -2, 1, 3)$, you have

(a) $\mathbf{u} + 2\mathbf{v} = (1, 6, -5, -3)$

(b) $\mathbf{w} - 3\mathbf{u} = (-1, -8, 10, 0)$

(c) $4\mathbf{v} + \frac{1}{2}\mathbf{u} - \mathbf{w} = (-1.5, 11, -6.5, -10.5)$

33. $2\mathbf{w} = \mathbf{u} - 3\mathbf{v}$

$\mathbf{w} = \frac{1}{2}\mathbf{u} - \frac{3}{2}\mathbf{v}$

$= \frac{1}{2}(1, -1, 0, 1) - \frac{3}{2}(0, 2, 3, -1)$

$= \left(\frac{1}{2}, -\frac{1}{2}, 0, \frac{1}{2}\right) - \left(0, 3, \frac{9}{2}, -\frac{3}{2}\right)$

$= \left(\frac{1}{2} - 0, -\frac{1}{2} - 3, 0 - \frac{9}{2}, \frac{1}{2} - \left(-\frac{3}{2}\right)\right)$

$= \left(\frac{1}{2}, -\frac{7}{2}, -\frac{9}{2}, 2\right)$

© 2013 Cengage Learning. All Rights Reserved. May not be scanned, copied or duplicated, or posted to a publicly accessible website, in whole or in part.

35. $\frac{1}{2}\mathbf{w} = 2\mathbf{u} + 3\mathbf{v}$

$$\begin{aligned}\mathbf{w} &= 4\mathbf{u} + 6\mathbf{v}\\ &= 4(1, -1, 0, 1) + 6(0, 2, 3, -1)\\ &= (4, -4, 0, 4) + (0, 12, 18, -6)\\ &= (4, 8, 18, -2)\end{aligned}$$

37. $2\mathbf{u} + \mathbf{v} - 3\mathbf{w} = \mathbf{0}$

$$\begin{aligned}\mathbf{w} &= \tfrac{2}{3}\mathbf{u} + \tfrac{1}{3}\mathbf{v}\\ &= \tfrac{2}{3}(0, 2, 7, 5) + \tfrac{1}{3}(-3, 1, 4, -8)\\ &= \left(0, \tfrac{4}{3}, \tfrac{14}{3}, \tfrac{10}{3}\right) + \left(-1, \tfrac{1}{3}, \tfrac{4}{3}, -\tfrac{8}{3}\right)\\ &= \left(0 + (-1), \tfrac{4}{3} + \tfrac{1}{3}, \tfrac{14}{3} + \tfrac{4}{3}, \tfrac{10}{3} + \left(-\tfrac{8}{3}\right)\right)\\ &= \left(-1, \tfrac{5}{3}, 6, \tfrac{2}{3}\right)\end{aligned}$$

39. The equation

$$a\mathbf{u} + b\mathbf{w} = \mathbf{v}$$
$$a(1, 2) + b(1, -1) = (2, 1)$$

yields the system

$$a + b = 2$$
$$2a - b = 1.$$

Solving this system produces $a = 1$ and $b = 1$. So, $\mathbf{v} = \mathbf{u} + \mathbf{w}$.

41. The equation

$$a\mathbf{u} + b\mathbf{w} = \mathbf{v}$$
$$a(1, 2) + b(1, -1) = (3, 0)$$

yields the system

$$a + b = 3$$
$$2a - b = 0.$$

Solving this system produces $a = 1$ and $b = 2$. So, $\mathbf{v} = \mathbf{u} + 2\mathbf{w}$.

43. The equation

$$a\mathbf{u} + b\mathbf{w} = \mathbf{v}$$
$$a(1, 2) + b(1, -1) = (-1, -2)$$

yields the system

$$a + b = -1$$
$$2a - b = -2.$$

Solving this system produces $a = -1$ and $b = 0$. So, $\mathbf{v} = -\mathbf{u}$.

45. The equation

$$a\mathbf{u}_1 + b\mathbf{u}_2 + c\mathbf{u}_3 = \mathbf{v}$$
$$a(2, 3, 5) + b(1, 2, 4) + c(-2, 2, 3) = (10, 1, 4)$$

yields the system

$$\begin{aligned}2a + b - 2c &= 10\\ 3a + 2b + 2c &= 1\\ 5a + 4b + 3c &= 4.\end{aligned}$$

Solving this system produces $a = 1$, $b = 2$, and $c = -3$. So, $\mathbf{v} = \mathbf{u}_1 + 2\mathbf{u}_2 - 3\mathbf{u}_3$.

47. The equation

$$a\mathbf{u}_1 + b\mathbf{u}_2 + c\mathbf{u}_3 = \mathbf{v}$$
$$a(1, 1, 2, 2) + b(2, 3, 5, 6) + c(-3, 1, -4, 2) = (0, 5, 3, 0)$$

yields the system

$$\begin{aligned}a + 2b - 3c &= 0\\ a + 3b + c &= 5\\ 2a + 5b - 4c &= 3\\ 2a + 6b + 2c &= 0.\end{aligned}$$

The second and fourth equations cannot both be true. So, the system has no solution. It is not possible to write $\mathbf{v}$ as a linear combination of $\mathbf{u}_1$, $\mathbf{u}_2$, and $\mathbf{u}_3$.

49. Write a matrix using the given $\mathbf{u}_1, \mathbf{u}_2, \ldots, \mathbf{u}_5$ as columns and augment this matrix with $\mathbf{v}$ as a column.

$$A = \begin{bmatrix} 1 & 1 & 0 & 2 & 0 & 5\\ 2 & 2 & 1 & 1 & 2 & 3\\ -3 & 0 & 1 & -1 & 2 & -11\\ 4 & 2 & 1 & 2 & -1 & 11\\ -1 & 1 & -4 & 1 & -1 & 9\end{bmatrix}$$

The reduced row-echelon form for A is

$$A = \begin{bmatrix} 1 & 0 & 0 & 0 & 0 & 2\\ 0 & 1 & 0 & 0 & 0 & 1\\ 0 & 0 & 1 & 0 & 0 & -2\\ 0 & 0 & 0 & 1 & 0 & 1\\ 0 & 0 & 0 & 0 & 1 & -1\end{bmatrix}.$$

So, $\mathbf{v} = 2\mathbf{u}_1 + \mathbf{u}_2 - 2\mathbf{u}_3 + \mathbf{u}_4 - \mathbf{u}_5$. Verify the solution by showing that

$2(1, 2, -3, 4, -1) + (1, 2, 0, 2, 1) - 2(0, 1, 1, 1, -4) + (2, 1, -1, 2, 1) - (0, 2, 2, -1, -1)$ equals $(5, 3, -11, 11, 9)$.

© 2013 Cengage Learning. All Rights Reserved. May not be scanned, copied or duplicated, or posted to a publicly accessible website, in whole or in part.

51. (a) True. See the discussion before "Definition of Vector Addition and Scalar Multiplication in R^n," page 149.

(b) False. The vector $-\mathbf{v}$ is called the additive inverse of the vector $\mathbf{v}$.

53. The equation

$$a\mathbf{v}_1 + b\mathbf{v}_2 + c\mathbf{v}_3 = \mathbf{0}$$
$$a(1, 0, 1) + b(-1, 1, 2) + c(0, 1, 4) = (0, 0, 0)$$

yields the homogeneous system

$$\begin{aligned} a - b \quad\quad &= 0 \\ b + c &= 0 \\ a + 2b + 4c &= 0. \end{aligned}$$

This system has only the trivial solution $a = b = c = 0$. So, you cannot find a nontrivial way of writing $\mathbf{0}$ as a combination of $\mathbf{v}_1$, $\mathbf{v}_2$, and $\mathbf{v}_3$.

55. (1) $\mathbf{u} + \mathbf{v} = (2, -1, 3, 6) + (1, 4, 0, 1) = (3, 3, 3, 7)$ is a vector in R^4.

(2) $\mathbf{u} + \mathbf{v} = (2, -1, 3, 6) + (1, 4, 0, 1) = (3, 3, 3, 7)$

$\mathbf{v} + \mathbf{u} = (1, 4, 0, 1) + (2, -1, 3, 6) = (3, 3, 3, 7)$

So, $\mathbf{u} + \mathbf{v} = \mathbf{v} + \mathbf{u}$.

(3) $(\mathbf{u} + \mathbf{v}) + \mathbf{w} = [(2, -1, 3, 6) + (1, 4, 0, 1)] + (3, 0, 2, 0) = (3, 3, 3, 7) + (3, 0, 2, 0) = (6, 3, 5, 7)$

$\mathbf{u} + (\mathbf{v} + \mathbf{w}) = (2, -1, 3, 6) + [(1, 4, 0, 1) + (3, 0, 2, 0)] = (2, -1, 3, 6) + (4, 4, 2, 1) = (6, 3, 5, 7)$

So, $(\mathbf{u} + \mathbf{v}) + \mathbf{w} = \mathbf{u} + (\mathbf{v} + \mathbf{w})$.

(4) $\mathbf{u} + \mathbf{0} = (2, -1, 3, 6) + (0, 0, 0, 0) = (2, -1, 3, 6) = \mathbf{u}$

(5) $\mathbf{u} + (-\mathbf{u}) = (2, -1, 3, 6) + (-2, 1, -3, -6) = (0, 0, 0, 0) = \mathbf{0}$

(6) $c\mathbf{u} = 5(2, -1, 3, 6) = (10, -5, 15, 30)$ is a vector in R^4.

(7) $c(\mathbf{u} + \mathbf{v}) = 5[(2, -1, 3, 6) + (1, 4, 0, 1)] = 5(3, 3, 3, 7) = (15, 15, 15, 35)$

$c\mathbf{u} + c\mathbf{v} = 5(2, -1, 3, 6) + 5(1, 4, 0, 1) = (10, -5, 15, 30) + (5, 20, 0, 5) = (15, 15, 15, 35)$

So, $c(\mathbf{u} + \mathbf{v}) = c\mathbf{u} + c\mathbf{v}$.

(8) $(c + d)\mathbf{u} = (5 + (-2))(2, -1, 3, 6) = 3(2, -1, 3, 6) = (6, -3, 9, 18)$

$c\mathbf{u} + d\mathbf{u} = 5(2, -1, 3, 6) + (-2)(2, -1, 3, 6) = (10, -5, 15, 30) + (-4, 2, -6, -12) = (6, -3, 9, 18)$

So, $(c + d)\mathbf{u} = c\mathbf{u} + d\mathbf{u}$.

(9) $c(d\mathbf{u}) = 5((-2)(2, -1, 3, 6)) = 5(-4, 2, -6, -12) = (-20, 10, -30, -60)$

$(cd)\mathbf{u} = (5(-2))(2, -1, 3, 6) = -10(2, -1, 3, 6) = (-20, 10, -30, -60)$

So, $c(d\mathbf{u}) = (cd)\mathbf{u}$.

(10) $1(\mathbf{u}) = 1(2, -1, 3, 6) = (2, -1, 3, 6) = \mathbf{u}$

57. Prove the remaining eight properties.

(1) $\mathbf{u} + \mathbf{v} = (u_1, u_2) + (v_1, v_2) = (u_1 + v_1, u_2 + v_2)$ is a vector in the plane.

(2) $\mathbf{u} + \mathbf{v} = (u_1, u_2) + (v_1, v_2) = (u_1 + v_1, u_2 + v_2) = (v_1 + u_1, v_2 + u_2) = (v_1, v_2) + (u_1, u_2) = \mathbf{v} + \mathbf{u}$

(4) $\mathbf{u} + \mathbf{0} = (u_1, u_2) + (0, 0) = (u_1 + 0, u_2 + 0) = (u_1, u_2) = \mathbf{u}$

(5) $\mathbf{u} + (-\mathbf{u}) = (u_1, u_2) + (-u_1, -u_2) = (u_1 - u_1, u_2 - u_2) = (0, 0) = \mathbf{0}$

(6) $c\mathbf{u} = c(u_1, u_2) = (cu_1, cu_2)$ is a vector in the plane.

© 2013 Cengage Learning. All Rights Reserved. May not be scanned, copied or duplicated, or posted to a publicly accessible website, in whole or in part.

(7) $c(\mathbf{u} + \mathbf{v}) = c[(u_1, u_2) + (v_1, v_2)] = c(u_1 + v_1, u_2 + v_2)$

$= (c(u_1 + v_1), c(u_2 + v_2)) = (cu_1 + cv_1, cu_2 + cv_2)$

$= (cu_1, cu_2) + (cv_1, cv_2)$

$= c(u_1, u_2) + c(v_1, v_2) = c\mathbf{u} + c\mathbf{v}$

(9) $c(d\mathbf{u}) = c(d(u_1, u_2)) = c(du_1, du_2) = (cdu_1, cdu_2) = (cd)(u_1, u_2) = (cd)\mathbf{u}$

(10) $1(\mathbf{u}) = 1(u_1, u_2) = (u_1, u_2) = \mathbf{u}$

59. If $\mathbf{b} = x_1\mathbf{a}_1 + \cdots + x_n\mathbf{a}_n$ is a linear combination of the columns of A, then a solution of $A\mathbf{x} = \mathbf{b}$ is

$$\mathbf{x} = \begin{bmatrix} x_1 \\ \vdots \\ x_n \end{bmatrix}.$$

The system $A\mathbf{x} = \mathbf{b}$ is inconsistent if $\mathbf{b}$ is not a linear combination of the columns of A.

61. *Justification for each step:*

(a) Use the fact that $c + 0 = 0$ for any real number c, so, in particular, $0 = 0 + 0$.

(b) Use property 8 of Theorem 4.2.

(c) An equality remains an equality if you add the same vector to both sides; you also used property 6 of Theorem 4.2 to conclude that $0\mathbf{v}$ is a vector in R^n, so, $-0\mathbf{v}$ is a vector in R^n.

(d) Property 5 of Theorem 4.2 is applied to the left-hand side. Property 3 of Theorem 4.2 is applied to the right-hand side.

(e) Use properties 5 and 6 of Theorem 4.2.

(f) Use properties 4 of Theorem 4.2.

63. *Justification for each step:*

(a) Equality remains if both sides are multiplied by a nonzero constant; property 6 of Theorem 4.2 assures you that the results are still in R^n.

(b) Use property 9 of Theorem 4.2 on the left side and property 4 of Theorem 4.3 (proved in Exercise 62) on the right side.

(c) Use the property of reals that states that the product of multiplicative inverses is 1.

(d) Use property 10 of Theorem 4.2.

65. You can describe vector subtraction $\mathbf{u} - \mathbf{v}$ as follows.

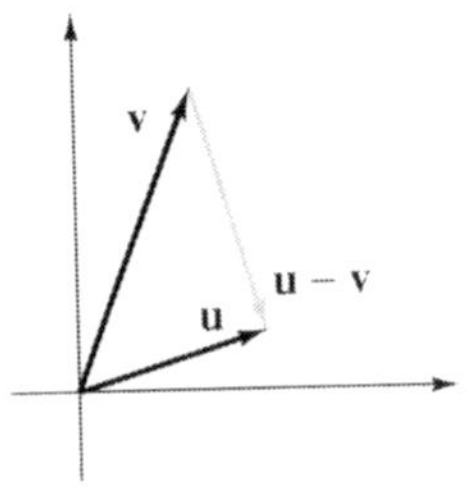

Or, write subtraction in terms of addition, $\mathbf{u} - \mathbf{v} = \mathbf{u} + (-1)\mathbf{v}$.

Section 4.2 Vector Spaces

1. The additive identity of R^4 is the vector (0, 0, 0, 0).

3. The additive identity of $M_{2,3}$ is the 2×3 zero matrix

$$\begin{bmatrix} 0 & 0 & 0 \\ 0 & 0 & 0 \end{bmatrix}.$$

5. P_3 is the set of all polynomials of degree less than or equal to 3. Its additive identity is

$0x^3 + 0x^2 + 0x + 0 = 0.$

7. In R^4, the additive inverse of (v_1, v_2, v_3, v_4) is $(-v_1, -v_2, -v_3, -v_4)$.

9. $M_{2,3}$ is the set of all 2×3 matrices. The additive inverse of

$$\begin{bmatrix} a_{11} & a_{12} & a_{13} \\ a_{21} & a_{22} & a_{23} \end{bmatrix}$$

is $-\begin{bmatrix} a_{11} & a_{12} & a_{13} \\ a_{21} & a_{22} & a_{23} \end{bmatrix} = \begin{bmatrix} -a_{11} & -a_{12} & -a_{13} \\ -a_{21} & -a_{22} & -a_{23} \end{bmatrix}$.

11. P_3 is the set of all polynomials of degree less than or equal to 3. The additive inverse of $a_3x^3 + a_2x^2 + a_1x + a_0$ is

$-(a_3x^3 + a_2x^2 + a_1x + a_0) = -a_3x^3 - a_2x^2 - a_1x - a_0.$

13. $M_{4,6}$ with the standard operations is a vector space. All ten vector space axioms hold.

15. This set is *not* a vector space. Axiom 1 fails. The set is not closed under addition. For example, $(-x^3 + 4x^2) + (x^3 + 2x) = 4x^2 + 2x$ is not a third-degree polynomial.

17. This set is *not* a vector space. Axiom 1 fails. For example, given $f(x) = x$ and $g(x) = -x$, $f(x) + g(x) = 0$ is not of the form ax, where $a \neq 0$.

19. This set is a vector space. All ten vector space axioms hold.

© 2013 Cengage Learning. All Rights Reserved. May not be scanned, copied or duplicated, or posted to a publicly accessible website, in whole or in part.

21. This set is *not* a vector space. The set is not closed under scalar multiplication. For example, $(-1)(3, 2) = (-3, -2)$ is not in the set.

23. This set is a vector space. All ten vector space axioms hold.

25. This set is a vector space. All ten vector space axioms hold.

27. This set is a vector space. All ten vector space axioms hold.

29. This set is *not* a vector space because it is not closed under addition. A counterexample is

$$\begin{bmatrix} 1 & 0 \\ 0 & 0 \end{bmatrix} + \begin{bmatrix} 0 & 0 \\ 0 & 1 \end{bmatrix} = \begin{bmatrix} 1 & 0 \\ 0 & 1 \end{bmatrix}.$$

Each matrix on the left is singular, while the sum is nonsingular.

31. This set is a vector space. All ten vector space axioms hold.

33. $C[0, 1]$ is a vector space. All ten vector space axioms hold.

35. (a) Axiom 8 fails. For example,

$$(1 + 2)(1, 1) = 3(1, 1) = (3, 1) \qquad (\text{Because } c(x, y) = (cx, y))$$
$$1(1, 1) + 2(1, 1) = (1, 1) + (2, 1) = (3, 2).$$

So, R^2 is *not* a vector space with these operations.

(b) Axiom 2 fails. For example,

$$(1, 2) + (2, 1) = (1, 0)$$
$$(2, 1) + (1, 2) = (2, 0).$$

So, R^2 is *not* a vector space with these operations.

(c) Axiom 6 fails. For example, $(-1)(1, 1) = (\sqrt{-1}, \sqrt{-1})$, which is not in R^2.

So, R^2 is *not* a vector space with these operations.

37. Verify the ten axioms in the definition of vector space.

(1) $\mathbf{u} + \mathbf{v} = \begin{bmatrix} u_1 & u_2 \\ u_3 & u_4 \end{bmatrix} + \begin{bmatrix} v_1 & v_2 \\ v_3 & v_4 \end{bmatrix} = \begin{bmatrix} u_1 + v_1 & u_2 + v_2 \\ u_3 + v_3 & u_4 + v_4 \end{bmatrix}$ is in $M_{2,2}$.

(2)
$$\begin{aligned}
\mathbf{u} + \mathbf{v} &= \begin{bmatrix} u_1 & u_2 \\ u_3 & u_4 \end{bmatrix} + \begin{bmatrix} v_1 & v_2 \\ v_3 & v_4 \end{bmatrix} = \begin{bmatrix} u_1 + v_1 & u_2 + v_2 \\ u_3 + v_3 & u_4 + v_4 \end{bmatrix} \\
&= \begin{bmatrix} v_1 + u_1 & v_2 + u_2 \\ v_3 + u_3 & v_4 + u_4 \end{bmatrix} = \begin{bmatrix} v_1 & v_2 \\ v_3 & v_4 \end{bmatrix} + \begin{bmatrix} u_1 & u_2 \\ u_3 & u_4 \end{bmatrix} = \mathbf{v} + \mathbf{u}
\end{aligned}$$

(3)
$$\begin{aligned}
\mathbf{u} + (\mathbf{v} + \mathbf{w}) &= \begin{bmatrix} u_1 & u_2 \\ u_3 & u_4 \end{bmatrix} + \left(\begin{bmatrix} v_1 & v_2 \\ v_3 & v_4 \end{bmatrix} + \begin{bmatrix} w_1 & w_2 \\ w_3 & w_4 \end{bmatrix}\right) \\
&= \begin{bmatrix} u_1 & u_2 \\ u_3 & u_4 \end{bmatrix} + \begin{bmatrix} v_1 + w_1 & v_2 + w_2 \\ v_3 + w_3 & v_4 + w_4 \end{bmatrix} \\
&= \begin{bmatrix} u_1 + (v_1 + w_1) & u_2 + (v_2 + w_2) \\ u_3 + (v_3 + w_3) & u_4 + (v_4 + w_4) \end{bmatrix} \\
&= \begin{bmatrix} (u_1 + v_1) + w_1 & (u_2 + v_2) + w_2 \\ (u_3 + v_3) + w_3 & (u_4 + v_4) + w_4 \end{bmatrix} \\
&= \begin{bmatrix} u_1 + v_1 & u_2 + v_2 \\ u_3 + v_3 & u_4 + v_4 \end{bmatrix} + \begin{bmatrix} w_1 & w_2 \\ w_3 & w_4 \end{bmatrix} \\
&= \left(\begin{bmatrix} u_1 & u_2 \\ u_3 & u_4 \end{bmatrix} + \begin{bmatrix} v_1 & v_2 \\ v_3 & v_4 \end{bmatrix}\right) + \begin{bmatrix} w_1 & w_2 \\ w_3 & w_4 \end{bmatrix} = (\mathbf{u} + \mathbf{v}) + \mathbf{w}
\end{aligned}$$

© 2013 Cengage Learning. All Rights Reserved. May not be scanned, copied or duplicated, or posted to a publicly accessible website, in whole or in part.

(4) The zero vector is

$$\mathbf{0} = \begin{bmatrix} 0 & 0 \\ 0 & 0 \end{bmatrix}. \text{ So,}$$

$$\mathbf{u} + \mathbf{0} = \begin{bmatrix} u_1 & u_2 \\ u_3 & u_4 \end{bmatrix} + \begin{bmatrix} 0 & 0 \\ 0 & 0 \end{bmatrix} = \begin{bmatrix} u_1 & u_2 \\ u_3 & u_4 \end{bmatrix} = \mathbf{u}.$$

(5) For every

$$\mathbf{u} = \begin{bmatrix} u_1 & u_2 \\ u_3 & u_4 \end{bmatrix}, \text{ you have } -\mathbf{u} = \begin{bmatrix} -u_1 & -u_2 \\ -u_3 & -u_4 \end{bmatrix}.$$

$$\begin{aligned} \mathbf{u} + (-\mathbf{u}) &= \begin{bmatrix} u_1 & u_2 \\ u_3 & u_4 \end{bmatrix} + \begin{bmatrix} -u_1 & -u_2 \\ -u_3 & -u_4 \end{bmatrix} \\ &= \begin{bmatrix} 0 & 0 \\ 0 & 0 \end{bmatrix} \\ &= \mathbf{0} \end{aligned}$$

(6) $c\mathbf{u} = c\begin{bmatrix} u_1 & u_2 \\ u_3 & u_4 \end{bmatrix} = \begin{bmatrix} cu_1 & cu_2 \\ cu_3 & cu_4 \end{bmatrix}$ is in $M_{2,2}$.

(7)
$$\begin{aligned} c(\mathbf{u} + \mathbf{v}) &= c\left(\begin{bmatrix} u_1 & u_2 \\ u_3 & u_4 \end{bmatrix} + \begin{bmatrix} v_1 & v_2 \\ v_3 & v_4 \end{bmatrix}\right) = c\begin{bmatrix} u_1 + v_1 & u_2 + v_2 \\ u_3 + v_3 & u_4 + v_4 \end{bmatrix} \\ &= \begin{bmatrix} c(u_1 + v_1) & c(u_2 + v_2) \\ c(u_3 + v_3) & c(u_4 + v_4) \end{bmatrix} = \begin{bmatrix} cu_1 + cv_1 & cu_2 + cv_2 \\ cu_3 + cv_3 & cu_4 + cv_4 \end{bmatrix} \\ &= \begin{bmatrix} cu_1 & cu_2 \\ cu_3 & cu_4 \end{bmatrix} + \begin{bmatrix} cv_1 & cv_2 \\ cv_3 & cv_4 \end{bmatrix} = c\begin{bmatrix} u_1 & u_2 \\ u_3 & u_4 \end{bmatrix} + c\begin{bmatrix} v_1 & v_2 \\ v_3 & v_4 \end{bmatrix} \\ &= c\mathbf{u} + c\mathbf{v} \end{aligned}$$

(8)
$$\begin{aligned} (c + d)\mathbf{u} &= (c + d)\begin{bmatrix} u_1 & u_2 \\ u_3 & u_4 \end{bmatrix} = \begin{bmatrix} (c + d)u_1 & (c + d)u_2 \\ (c + d)u_3 & (c + d)u_4 \end{bmatrix} \\ &= \begin{bmatrix} cu_1 + du_1 & cu_2 + du_2 \\ cu_3 + du_3 & cu_4 + du_4 \end{bmatrix} = \begin{bmatrix} cu_1 & cu_2 \\ cu_3 & cu_4 \end{bmatrix} + \begin{bmatrix} du_1 & du_2 \\ du_3 & du_4 \end{bmatrix} \\ &= c\begin{bmatrix} u_1 & u_2 \\ u_3 & u_4 \end{bmatrix} + d\begin{bmatrix} u_1 & u_2 \\ u_3 & u_4 \end{bmatrix} = c\mathbf{u} + d\mathbf{u} \end{aligned}$$

(9)
$$\begin{aligned} c(d\mathbf{u}) &= c\left(d\begin{bmatrix} u_1 & u_2 \\ u_3 & u_4 \end{bmatrix}\right) = c\begin{bmatrix} du_1 & du_2 \\ du_3 & du_4 \end{bmatrix} = \begin{bmatrix} c(du_1) & c(du_2) \\ c(du_3) & c(du_4) \end{bmatrix} \\ &= \begin{bmatrix} (cd)u_1 & (cd)u_2 \\ (cd)u_3 & (cd)u_4 \end{bmatrix} = (cd)\begin{bmatrix} u_1 & u_2 \\ u_3 & u_4 \end{bmatrix} = (cd)\mathbf{u} \end{aligned}$$

(10) $1(\mathbf{u}) = 1\begin{bmatrix} u_1 & u_2 \\ u_3 & u_4 \end{bmatrix} = \begin{bmatrix} 1u_1 & 1u_2 \\ 1u_3 & 1u_4 \end{bmatrix} = \mathbf{u}$

39. This set is *not* a vector space because Axiom 5 fails. The additive identity is (1, 1) and so (0, 0) has no additive inverse. Axioms 7 and 8 also fail.

41. Verify the ten axioms in the definition of vector space. Let $x(t) = a_1 \sin \omega t + a_2 \cos \omega t$, $y(t) = b_1 \sin \omega t + b_2 \cos \omega t$, and $z(t) = c_1 \sin \omega t + c_2 \cos \omega t$, and let d_1 and d_2 be scalars.

(1)
$$\begin{aligned} x(t) + y(t) &= (a_1 \sin \omega t + a_2 \cos \omega t) + (b_1 \sin \omega t + b_2 \cos \omega t) \\ &= a_1 \sin \omega t + b_1 \sin \omega t + a_2 \cos \omega t + b_2 \cos \omega t \\ &= (a_1 + b_1)\sin \omega t + (a_2 + b_2)\cos \omega t \text{ is in the set.} \end{aligned}$$

© 2013 Cengage Learning. All Rights Reserved. May not be scanned, copied or duplicated, or posted to a publicly accessible website, in whole or in part.

(2) $x(t) + y(t) = (a_1 + b_1)\sin \omega t + (a_2 + b_2)\cos \omega t$

$= (b_1 + a_1)\sin \omega t + (b_2 + a_2)\cos \omega t$

$= y(t) + x(t)$

(3) $x(t) + (y(t) + z(t)) = (a_1 \sin \omega t + a_2 \cos \omega t) + ((b_1 + c_1) \sin \omega t + (b_2 + c_2) \cos \omega t)$

$= (a_1 + (b_1 + c_1)) \sin \omega t + (a_2 + (b_2 + c_2)) \cos \omega t$

$= ((a_1 + b_1) + c_1) \sin \omega t + ((a_2 + b_2) + c_2) \cos \omega t$

$= (a_1 + b_1) \sin \omega t + (a_2 + b_2) \cos \omega t + c_1 \sin \omega t + c_2 \cos \omega t$

$= (x(t) + y(t)) + z(t)$

(4) The zero vector is

$\mathbf{0} = 0 \sin \omega t + 0 \cos \omega t$

$x(t) + \mathbf{0} = a_1 \sin \omega t + a_2 \sin \omega t + 0 \sin \omega t + 0 \cos \omega t = a_1 \sin \omega t + a_2 \cos \omega t = x(t).$

(5) The additive inverse of $x(t)$ is

$-x(t) = -a_1 \sin \omega t - a_2 \cos \omega t$

$x(t) + (-x(t)) = (a_1 + (-a_1))\sin \omega t + (a_2 + (-a_2))\cos \omega t = 0.$

(6) $d_1(a_1 \sin \omega t + a_2 \cos \omega t) = d_1 a_1 \sin \omega t + d_1 a_2 \cos \omega t$ is in the set.

(7) $d_1(x(t) + y(t)) = d_1((a_1 + b_1) \sin \omega t + (a_2 + b_2) \cos \omega t)$

$= d_1(a_1 + b_1) \sin \omega t + d_1(a_2 + b_2) \cos \omega t$

$= d_1 a_1 \sin \omega t + d_1 b_1 \sin \omega t + d_1 a_2 \cos \omega t + d_1 b_2 \cos \omega t$

$= d_1(a_1 \sin \omega t + a_2 \cos \omega t) + d_1(b_1 \sin \omega t + b_2 \cos \omega t)$

$= d_1 x(t) + d_1 y(t)$

(8) $(d_1 + d_2)x(t) = (d_1 + d_2)(a_1 \sin \omega t + a_2 \cos \omega t)$

$= (d_1 + d_2)a_1 \sin \omega t + (d_1 + d_2)a_2 \cos \omega t$

$= d_1 a_1 \sin \omega t + d_2 a_1 \sin \omega t + d_1 a_2 \cos \omega t + d_2 a_2 \cos \omega t$

$= d_1(a_1 \sin \omega t + a_2 \cos \omega t) + d_2(a_1 \sin \omega t + a_2 \cos \omega t)$

$= d_1\, x(t) + d_2\, x(t)$

(9) $d_1(d_2 x(t)) = d_1(d_2(a_1 \sin \omega t + a_2 \cos \omega t))$

$= d_1(d_2 a_1 \sin \omega t + d_2 a_2 \cos \omega t)$

$= d_1 d_2 a_1 \sin \omega t + d_1 d_2 a_2 \cos \omega t$

$= (d_1 d_2)(a_1 \sin \omega t + a_2 \cos \omega t)$

$= (d_1 d_2)x(t)$

(10) $1(x(t)) = 1(a_1 \sin \omega t + a_2 \cos \omega t)$

$= 1a_1 \sin \omega t + 1a_2 \cos \omega t$

$= a_1 \sin \omega t + a_2 \cos \omega t$

$= x(t)$

© 2013 Cengage Learning. All Rights Reserved. May not be scanned, copied or duplicated, or posted to a publicly accessible website, in whole or in part.

43. Yes, V is a vector space. Verify the ten axioms.

(1) $x, y \in V \Rightarrow x + y = xy \in V$ (V = positive real numbers)

(2) $x + y = xy = yx = y + x$

(3) $x + (y + z) = x + (yz) = x(yz) = (xy)z = (x + y)z = (x + y) + z$

(4) $x + 1 = x1 = x = 1x = 1 + x$ (Zero vector is 1.)

(5) $x + \frac{1}{x} = x\left(\frac{1}{x}\right) = 1$ $\left(\text{Additive inverse of } x \text{ is } \frac{1}{x}.\right)$

(6) For $c \in R, x \in V, cx = x^c \in V$.

(7) $c(x + y) = (x + y)^c = (xy)^c = x^c y^c = x^c + y^c = cx + cy$

(8) $(c + d)x = x^{c+d} = x^c x^d = x^c + x^d = cx + dx$

(9) $c(dx) = (dx)^c = (x^d)^c = x^{(dc)} = (dc)x = (cd)x$

(10) $1x = x^1 = x$

45. (a) True. See the paragraph below the "Definition of a Vector Space" box, page 155.

(b) False. See Example 6, page 159.

(c) False. With standard operations on R^3, the additive inverse axiom is not satisfied.

47. Prove that $0\mathbf{v} = \mathbf{0}$ for any element $\mathbf{v}$ of a vector space V. First note that $0\mathbf{v}$ is a vector in V by property 6 of the definition of a vector space. Because $0 = 0 + 0$, you have $0\mathbf{v} = (0 + 0)\mathbf{v} = 0\mathbf{v} + 0\mathbf{v}$. The last equality holds by property 8 of the definition of a vector space. Add $(-0\mathbf{v})$ to both sides of the last equality to obtain

$0\mathbf{v} + (-0\mathbf{v}) = (0\mathbf{v} + 0\mathbf{v}) + (-0\mathbf{v})$.

Apply property 3 of the definition of a vector space to the right hand side to obtain

$0\mathbf{v} + (-0\mathbf{v}) = 0\mathbf{v} + (0\mathbf{v} + (-0\mathbf{v}))$.

Apply property 5 of the definition of a vector space to both sides to obtain

$\mathbf{0} = 0\mathbf{v} + \mathbf{0}$.

But the right-hand side is equal to $0\mathbf{v}$ by property 4 of the definition of a vector space, and so you see that $\mathbf{0} = 0\mathbf{v}$, as required.

Section 4.3 Subspaces of Vector Spaces

1. Because W is nonempty and $W \subset R^4$, you need only check that W is closed under addition and scalar multiplication. Given

$(x_1, x_2, x_3, 0) \in W$ and $(y_1, y_2, y_3, 0) \in W$,

it follows that

$(x_1, x_2, x_3, 0) + (y_1, y_2, y_3, 0) = (x_1 + y_1, x_2 + y_2, x_3 + y_3, 0) \in W$.

Furthermore, for any real number c and $(x_1, x_2, x_3, 0) \in W$, it follows that

$c(x_1, x_2, x_3, 0) = (cx_1, cx_2, cx_3, 0) \in W$.

3. Because W is nonempty and $W \subset M_{2,2}$, you need only check that W is closed under addition and scalar multiplication. Given

$\begin{bmatrix} 0 & a_1 \\ b_1 & 0 \end{bmatrix} \in W$ and $\begin{bmatrix} 0 & a_2 \\ b_2 & 0 \end{bmatrix} \in W$, it follows that $\begin{bmatrix} 0 & a_1 \\ b_1 & 0 \end{bmatrix} + \begin{bmatrix} 0 & a_2 \\ b_2 & 0 \end{bmatrix} = \begin{bmatrix} 0 & a_1 + a_2 \\ b_1 + b_2 & 0 \end{bmatrix} \in W$.

Furthermore, for any real number c and $\begin{bmatrix} 0 & a \\ b & o \end{bmatrix} \in W$, it follows that $c\begin{bmatrix} 0 & a \\ b & 0 \end{bmatrix} = \begin{bmatrix} 0 & ca \\ cb & 0 \end{bmatrix} \in W$.

© 2013 Cengage Learning. All Rights Reserved. May not be scanned, copied or duplicated, or posted to a publicly accessible website, in whole or in part.

5. Recall from calculus that continuity implies integrability, so $W \subset V$. Furthermore, because W is nonempty, you need only check that W is closed under addition and scalar multiplication. Given continuous functions $f, g \in W$, it follows that $f + g$ is continuous, so $f + g \in W$. Also, for any real number c and for a continuous function $f \in W$, cf is continuous. So, $cf \in W$.

7. The vectors in W are of the form $(a, b, -1)$. This set is not closed under addition or scalar multiplication. For example,

$(0, 0, -1) + (0, 0, -1) = (0, 0, -2) \notin W$

and

$2(0, 0, -1) = (0, 0, -2) \notin W$.

9. This set is not closed under scalar multiplication. For example,

$\sqrt{2}, (1, 1) = (\sqrt{2}, \sqrt{2}) \notin W$.

11. Consider $f(x) = e^x$, which is continuous and nonnegative. So, $f \in W$. The function $(-1)f = -f$ is negative. So, $-f \notin W$, and W is not closed under scalar multiplication.

13. This set is not closed under scalar multiplication. For example,

$(-2)(1, 1, 1) = (-2, -2, -2) \notin W$.

15. The set is not closed under scalar multiplication. For instance,

$$3\begin{bmatrix} 0 & 2 & 3 \\ 2 & 0 & 3 \\ 1 & 4 & 1 \end{bmatrix} = \begin{bmatrix} 0 & 6 & 9 \\ 6 & 0 & 9 \\ 3 & 12 & 3 \end{bmatrix} \notin W.$$

17. The set is not closed under addition. For instance, when $n = 2$,

$$\begin{bmatrix} 1 & 0 \\ 0 & 1 \end{bmatrix} + \begin{bmatrix} -1 & 0 \\ 0 & -1 \end{bmatrix} = \begin{bmatrix} 0 & 0 \\ 0 & 0 \end{bmatrix} \notin W.$$

19. The vectors in W are of the form (a, a^3). This set is not closed under addition or scalar multiplication. For example,

$(1, 1) + (2, 8) = (3, 9) \notin W$

and

$3(2, 8) = (6, 24) \notin W$.

21. This set is *not* a subspace because it is not closed under scalar multiplication.

23. This set is a subspace of $C(-\infty, \infty)$ because it is closed under addition and scalar multiplication.

25. This set is a subspace of $C(-\infty, \infty)$ because it is closed under addition and scalar multiplication.

27. This set is a subspace of $C(-\infty, \infty)$ because it is closed under addition and scalar multiplication.

29. This set is a subspace of $M_{m,n}$ because it is closed under addition and scalar multiplication.

31. This set is *not* a subspace because it is not closed under scalar multiplication.

33. This set is *not* a subspace because it is not closed under addition.

35. This set is a subspace of $M_{m,n}$ because it is closed under addition and scalar multiplication.

37. W is a subspace of R^3 because it is nonempty and closed under addition and scalar multiplication.

39. Note that $W \subset R^3$ and W is nonempty. If $(a_1, b_1, a_1 + 2b_1)$ and $(a_2, b_2, a_2 + 2b_2)$ are vectors in W, then their sum

$$(a_1, b_1, a_1 + 2b_1) + (a_2, b_2, a_2 + 2b_2) = (a_1 + a_2, b_1 + b_2, (a_1 + a_2) + 2(b_1 + b_2))$$

is also in W. Furthermore, for any real number c and $(a, b, a + 2b)$ in W,

$$c(a, b, a + 2b) = (ca, cb, ca + 2cb)$$

is in W. Because W is closed under addition and scalar multiplication, W is a subspace of R^3.

41. W is not a subspace of R^3 because it is not closed under addition or scalar multiplication. For example, $(1, 1, 1) \in W$, but

$(1, 1, 1) + (1, 1, 1) = (2, 2, 2) \notin W$.

Or,

$2(1, 1, 1) = (2, 2, 2) \notin W$.

43. (a) True. See "Remark," page 163.
(b) True. See Theorem 4.6, page 164.
(c) False. There may be elements of W that are not elements of U or vice-versa.

45. If $j \leq k$, $P_j C P_k$. Since P_j is a vector space, it is closed under addition and scalar multiplication. So, P_j is a subspace of P_k.

© 2013 Cengage Learning. All Rights Reserved. May not be scanned, copied or duplicated, or posted to a publicly accessible website, in whole or in part.

47. (a) Since every continuous function is a function, $C(-\infty, \infty) \subset F(-\infty, \infty)$. Because $C(-\infty, \infty)$ is a vector space (see page 157), it is a subspace.

(b) The sum of two differentiable functions is differentiable, and a differentiable function times a constant is differentiable. The set is closed under addition and scalar multiplication, so it is a subspace.

(c) The set is nonempty because when $f = 0$, $f' - 3f = 0 - 3(0) = 0$. If f and g are such that

$f' - 3f = 0$ and $g' - 3g = 0$, then $(f + g)' - 3(f + g) = f' + g' - 3f - 3g = f' - 3f + g' - 3g = 0 - 0 = 0.$

So, the set is closed under addition. If c is a scalar, then

$(cf)' - 3(cf) = cf' - 3cf = c(f' - 3f) = c(0) = 0.$

So, the set is closed under scalar multiplication as well, and the set is a subspace.

49. Let $\ell_1, \ell_2 \notin W$ be represented by $\ell_1 = (a_1t, b_1t, c_2t)$ and $\ell_2 = (a_2t, b_2t, c_2t)$. W is nonempty because $0 = (0t, 0t, 0t) \notin W$.

$\ell_1 + \ell_2 = (a_1t + a_2t, b_1t + b_2t, c_1t + c_2t) = ((a_1 + a_2)t, (b_1 + b_2)t, (c_1 + c_2)t) \notin W$

If $d \in R$ is a scalar, $d\ell_1 = (da_1t, db_1t, dc_1t) \in W$.

So, W is a subspace.

51. Let W be a nonempty subset of a vector space V. On the one hand, if W is a subspace of V, then for any scalars a, b, and any vectors $\mathbf{x}, \mathbf{y} \in W$, $a\mathbf{x} \in W$ and $b\mathbf{y} \in W$, and so, $a\mathbf{x} + b\mathbf{y} \in W$.

On the other hand, assume that $a\mathbf{x} + b\mathbf{y}$ is an element of W where a, b are scalars, and $\mathbf{x}, \mathbf{y} \in W$. To show that W is a subspace, you must verify the closure axioms. If $\mathbf{x}, \mathbf{y} \in W$, then $\mathbf{x} + \mathbf{y} \in W$ (by taking $a = b = 1$). Finally, if a is a scalar, $a\mathbf{x} \in W$ (by taking $b = 0$).

53. Assume A is a fixed 2×3 matrix. Assuming W is nonempty, let $\mathbf{x} \in W$. Then $A\mathbf{x} = \begin{bmatrix} 1 \\ 2 \end{bmatrix}$. Now, let c be a nonzero scalar such that $c \neq 1$. Then $c\mathbf{x} \in R^3$ and

$$A(c\mathbf{x}) = cA\mathbf{x} = c\begin{bmatrix} 1 \\ 2 \end{bmatrix} = \begin{bmatrix} c \\ 2c \end{bmatrix}.$$

So, $c\mathbf{x} \notin \mathbf{W}$. Therefore, W is not a subspace of R^3.

55. Let W be a subspace of the vector space V and let $\mathbf{0}_v$ be the zero vector of V and $\mathbf{0}_w$ be the zero vector of W. Because $\mathbf{0}_w \in W \subset V$,

$\mathbf{0}_w = \mathbf{0}_w + (-\mathbf{0}_w) = \mathbf{0}_v.$

So, the zero vector in V is also the zero vector in W.

57. The set W is a nonempty subset of $M_{2,2}$. (For instance, $O, B \in W$.) To show closure, let $X, Y \in W \Rightarrow BAX = XAB$ and $BAY = YAB$. Then

$(X + Y)AB = XAB + YAB = BAX + BAY = BA(X + Y) \Rightarrow X + Y \in W.$

Similarly, if c is a scalar, then $(cX)AB = c(XAB) = c(BAX) = BA(cX) \Rightarrow cX \in W$.

59. Let c be scalar and $\mathbf{u} \in V \cap W$. Then $\mathbf{u} \in V$ and $\mathbf{u} \in W$, which are both subspaces. So, $c\mathbf{u} \in V$ and $c\mathbf{u} \in W$, which implies that $c\mathbf{u} \in V \cap W$.

© 2013 Cengage Learning. All Rights Reserved. May not be scanned, copied or duplicated, or posted to a publicly accessible website, in whole or in part.

Section 4.4 Spanning Sets and Linear Independence

1. (a) Solving the equation

$$c_1(2, -1, 3) + c_2(5, 0, 4) = (-1, -2, 2)$$

for c_1 and c_2 yields the system

$$\begin{aligned} 2c_1 + 5c_2 &= -1 \\ -c_1 \quad &= -2 \\ 3c_1 + 4c_2 &= 2. \end{aligned}$$

The solution of this system is $c_1 = 2$ and $c_2 = -1$. So, **z** can be written as a linear combination of vectors in S.

(b) Proceed as in (a), substituting $\left(8, -\frac{1}{4}, \frac{27}{4}\right)$ for $(-1, -2, 2)$, which yields the system

$$\begin{aligned} 2c_1 + 5c_2 &= 8 \\ -c_1 \quad &= -\tfrac{1}{4} \\ 3c_1 + 4c_2 &= \tfrac{27}{4}. \end{aligned}$$

The solution of this system is $c_1 = \frac{1}{4}$ and $c_2 = \frac{3}{2}$. So, **v** can be written as a linear combination of vectors in S.

(c) Proceed as in (a), substituting $(1, -8, 12)$ for $(-1, -2, 2)$, which yields the system

$$\begin{aligned} 2c_1 + 5c_2 &= 1 \\ -c_1 \quad &= -8 \\ 3c_1 + 4c_2 &= 12. \end{aligned}$$

The solution of this system is $c_1 = 8$ and $c_2 = -3$. So, **w** can be written as a linear combination of the vectors in S.

(d) Proceed as in (a), substituting $(1, 1, -1)$ for $(-1, -2, 2)$, which yields the system

$$\begin{aligned} 2c_1 + 5c_2 &= 1 \\ -c_1 \quad &= 1 \\ 3c_1 + 4c_2 &= -1. \end{aligned}$$

This system has no solution. So, **u** cannot be written as a linear combination of vectors in S.

3. (a) Solving the equation

$$c_1(2, 0, 7) + c_2(2, 4, 5) + c_3(2, -12, 13) = (-1, 5, -6)$$

for c_1, c_2, and c_3, yields the system

$$\begin{aligned} 2c_1 + 2c_2 + 2c_3 &= -1 \\ 4c_2 - 12c_3 &= 5 \\ 7c_1 + 5c_2 + 13c_3 &= -6. \end{aligned}$$

One solution is $c_1 = -\frac{7}{4}$, $c_2 = \frac{5}{4}$, and $c_3 = 0$. So, **u** can be written as a linear combination of vectors in S.

(b) Proceed as in (a), substituting $(-3, 15, 18)$ for $(-1, 5, -6)$, which yields the system

$$\begin{aligned} 2c_1 + 2c_2 + 2c_3 &= -3 \\ 4c_2 - 12c_3 &= 15 \\ 7c_1 + 5c_2 + 13c_3 &= 18. \end{aligned}$$

This system has no solution. So, **v** cannot be written as a linear combination of vectors in S.

(c) Proceed as in (a), substituting $\left(\frac{1}{3}, \frac{4}{3}, \frac{1}{2}\right)$ for $(-1, 5, -6)$, which yields the system

$$\begin{aligned} 2c_1 + 2c_2 + 2c_3 &= \tfrac{1}{3} \\ 4c_2 - 12c_3 &= \tfrac{4}{3} \\ 7c_1 + 5c_2 + 13c_3 &= \tfrac{1}{2}. \end{aligned}$$

One solution is $c_1 = -\frac{1}{6}$, $c_2 = \frac{1}{3}$, and $c_3 = 0$. So, **w** can be written as a linear combination of vectors in S.

(d) Proceed as in (a), substituting $(2, 20, -3)$ for $(-1, 5, -6)$, which yields the system

$$\begin{aligned} 2c_1 + 2c_2 + 2c_3 &= 2 \\ 4c_2 - 12c_3 &= 20 \\ 7c_1 + 5c_2 + 13c_3 &= -3 \end{aligned}$$

One solution is $c_1 = -4$, $c_2 = 5$, and $c_3 = 0$. So, **z** can be written as a linear combination of vectors in S.

© 2013 Cengage Learning. All Rights Reserved. May not be scanned, copied or duplicated, or posted to a publicly accessible website, in whole or in part.

5. From the vector equation

$$c_1\begin{bmatrix}2 & -3\\4 & 1\end{bmatrix} + c_2\begin{bmatrix}0 & 5\\1 & -2\end{bmatrix} = \begin{bmatrix}6 & -19\\10 & 7\end{bmatrix}$$

you obtain the linear system

$$\begin{aligned}2c_1 \qquad &= 6\\ -3c_1 + 5c_2 &= -19\\ 4c_1 + c_2 &= 10\\ c_1 - 2c_2 &= 7.\end{aligned}$$

The solution of this system is $c_1 = 3$ and $c_2 = -2$.

$$\begin{bmatrix}6 & -19\\10 & 7\end{bmatrix} = 3\begin{bmatrix}2 & -3\\4 & 1\end{bmatrix} - 2\begin{bmatrix}0 & 5\\1 & -2\end{bmatrix} = 3A - 2B$$

and so the matrix is a linear combination of A and B.

7. From the vector equation

$$c_1\begin{bmatrix}2 & -3\\4 & 1\end{bmatrix} + c_2\begin{bmatrix}0 & 5\\1 & -2\end{bmatrix} = \begin{bmatrix}-2 & 28\\1 & -11\end{bmatrix}$$

you obtain the linear system

$$\begin{aligned}2c_1 \qquad &= -2\\ -3c_1 + 5c_2 &= 28\\ 4c_1 + c_2 &= 1\\ c_1 - 2c_2 &= -11.\end{aligned}$$

The solution of this system is $c_1 = -1$ and $c_2 = 5$. So,

$$\begin{bmatrix}-2 & 28\\1 & -11\end{bmatrix} = -\begin{bmatrix}2 & -3\\4 & 1\end{bmatrix} + 5\begin{bmatrix}0 & 5\\1 & -2\end{bmatrix} = -A + 5B$$

and so the matrix is a linear combination of A and B.

9. Let $\mathbf{u} = (u_1, u_2)$ be any vector in R^2. Solving the equation

$$c_1(2, 1) + c_2(-1, 2) = (u_1, u_2)$$

for c_1 and c_2 yields the system

$$\begin{aligned}2c_1 - c_2 &= u_1\\ c_1 + 2c_2 &= u_2.\end{aligned}$$

This system has a unique solution because the determinant of the coefficient matrix is nonzero. So, S spans R^2.

11. Let $\mathbf{u} = (u_1, u_2)$ be any vector in R^2. Solving the equation

$$c_1(5, 0) + c_2(5, -4) = (u_1, u_2)$$

for c_1 and c_2 yields the system

$$\begin{aligned}5c_1 + 5c_2 &= u_1\\ -4c_2 &= u_2.\end{aligned}$$

This system has a unique solution because the determinant of the coefficient matrix is nonzero. So, S spans R^2.

13. S does not span R^2 because only vectors of the form $t(-3, 5)$ are in span(S). For example, $(0, 1)$ is not in span(S). S spans a line in R^2.

15. S does not span R^2 because only vectors of the form $t(1, 3)$ are in span(S). For example, $(0, 1)$ is not in span(S). S spans a line in R^2.

17. S does not span R^2 because only vectors of the form $t(1, -2)$ are in span(S). For example, $(0, 1)$ is not in span(S). S spans a line in R^2.

19. S spans R^2. Let $\mathbf{u} = (u_1, u_2)$ be any vector in R^2. Solving the equation

$$c_1(-1, 4) + c_2(4, -1) + c_3(1, 1) = (u_1, u_2)$$

for c_1, c_2, and c_3 yields the system

$$\begin{aligned}-c_1 + 4c_2 + c_3 &= u_1\\ 4c_1 - c_2 + c_3 &= u_2.\end{aligned}$$

This system is equivalent to

$$\begin{aligned}c_1 - 4c_2 - c_3 &= -u_1\\ 15c_2 - 5c_3 &= 4u_1 + u_2.\end{aligned}$$

So, for any $\mathbf{u} = (u_1, u_2)$ in R^2, you can take

$c_3 = 0, c_2 = (4u_1 + u_2)/15$ and

$c_1 = 4c_2 - u_1 = (u_1 + 4u_2)/15$.

21. Let $\mathbf{u} = (u_1, u_2, u_3)$ be any vector in R^3. Solving the equation

$$c_1(4, 7, 3) + c_2(-1, 2, 6) + c_3(2, -3, 5) = (u_1, u_2, u_3)$$

for c_1, c_2, and c_3 yields the system

$$\begin{aligned}4c_1 - c_2 + 2c_3 &= u_1\\ 7c_1 + 2c_2 - 3c_3 &= u_2\\ 3c_1 + 6c_2 + 5c_3 &= u_3.\end{aligned}$$

This system has a unique solution because the determinant of the coefficient matrix is nonzero. So, S spans R^3.

23. This set does not span R^3. S spans a plane in R^3.

© 2013 Cengage Learning. All Rights Reserved. May not be scanned, copied or duplicated, or posted to a publicly accessible website, in whole or in part.

25. Let $\mathbf{u} = (u_1, u_2, u_3)$ be any vector in R^3. Solving the equation

$$c_1(1, -2, 0) + c_2(0, 0, 1) + c_3(-1, 2, 0) = (u_1, u_2, u_3)$$

for c_1, c_2, and c_3 yields the system

$$\begin{aligned} c_1 \quad\quad - c_3 &= u_1 \\ -2c_1 \quad\quad + 2c_3 &= u_2 \\ c_2 \quad\quad &= u_3. \end{aligned}$$

This system has an infinite number of solutions if $u_2 = -2u_1$, otherwise it has no solution.

For instance $(1, 1, 1)$ is not in the span of S. So, S does not span R^3. The subspace spanned by S is

$\text{span}(S) = \{(a, -2a, b)\text{: } a \text{ and } b \text{ are any real numbers}\}$,

which is a plane in R^3.

27. S does not span P_2 because only vectors of the form $s(x^2) + t(1)$ are in span(S). For example, $1 + x + x^2$ is not in span(S).

29. Because $(-2, 2)$ is not a scalar multiple of $(3, 5)$, the set S is linearly independent.

31. The set S is linearly dependent because

$$1(0, 0) + 0(1, -1) = (0, 0).$$

33. Because $(1, -4, 1)$ is not a scalar multiple of $(6, 3, 2)$, the set S is linearly independent.

35. Because these vectors are multiples of each other, the set S is linearly dependent.

37. From the vector equation

$$c_1(-4, -3, 4) + c_2(1, -2, 3) + c_3(6, 0, 0) = \mathbf{0}$$

you obtain the homogenous system

$$\begin{aligned} -4c_1 + c_2 + 6c_3 &= 0 \\ -3c_1 - 2c_2 \quad\quad &= 0 \\ 4c_1 + 3c_2 \quad\quad &= 0. \end{aligned}$$

This system has only the trivial solution $c_1 = c_2 = c_3 = 0$. So, the set S is linearly independent.

39. From the vector equation

$$c_1(4, -3, 6, 2) + c_2(1, 8, 3, 1) + c_3(3, -2, -1, 0) = (0, 0, 0, 0)$$

you obtain the homogeneous system

$$\begin{aligned} 4c_1 + c_2 + 3c_3 &= 0 \\ -3c_1 + 8c_2 - 2c_3 &= 0 \\ 6c_1 + 3c_2 - c_3 &= 0 \\ 2c_1 + c_2 \quad\quad &= 0. \end{aligned}$$

This system has only the trivial solution $c_1 = c_2 = c_3 = 0$. So, the set S is linearly independent.

41. From the vector equation $c_1(2 - x) + c_2(2x - x^2) + c_3(6 - 5x + x^2) = 0 + 0x + 0x^2$, you obtain the homogeneous system

$$\begin{aligned} 2c_1 \quad\quad + 6c_3 &= 0 \\ -c_1 + 2c_2 - 5c_3 &= 0 \\ -c_2 + c_3 &= 0. \end{aligned}$$

This system has infinitely many solutions. For instance, $c_1 = -3$, $c_2 = 1$, and $c_3 = 1$. So, S is linearly dependent.

43. From the vector equation $c_1(x^2 + 3x + 1) + c_2(2x^2 + x - 1) + c_3(4x) = 0 + 0x + 0x^2$, you obtain the homogeneous system

$$\begin{aligned} c_1 - c_2 \quad\quad &= 0 \\ 3c_1 + c_2 + 4c_3 &= 0 \\ c_1 + 2c_2 \quad\quad &= 0. \end{aligned}$$

This system has only the trivial solution. So, S is linearly independent.

45. The set is linearly dependent because

$$-2A + B = -2\begin{bmatrix} 1 & 0 \\ 0 & -2 \end{bmatrix} + \begin{bmatrix} 0 & 1 \\ 1 & 0 \end{bmatrix} = \begin{bmatrix} -2 & 1 \\ 1 & 4 \end{bmatrix} = C.$$

© 2013 Cengage Learning. All Rights Reserved. May not be scanned, copied or duplicated, or posted to a publicly accessible website, in whole or in part.

47. From the vector equation

$$c_1\begin{bmatrix}1 & -1\\4 & 5\end{bmatrix} + c_2\begin{bmatrix}4 & 3\\-2 & 3\end{bmatrix} + c_3\begin{bmatrix}1 & -8\\22 & 23\end{bmatrix} = \begin{bmatrix}0 & 0\\0 & 0\end{bmatrix}$$

you obtain the homogeneous system

$$\begin{aligned} c_1 + 4c_2 + c_3 &= 0\\ -c_1 + 3c_2 - 8c_3 &= 0\\ 4c_1 - 2c_2 + 22c_3 &= 0\\ 5c_1 + 3c_2 + 23c_3 &= 0. \end{aligned}$$

Because this system has only the trivial solution $c_1 = c_2 = c_3 = 0$, the set of vectors is linearly independent.

49. One example of a nontrivial linear combination of vectors in S whose sum is the zero vector is

$2(3, 4) - 8(-1, 1) - 7(2, 0) = (0, 0)$. Solving this equation for $(2, 0)$ yields $(2, 0) = \frac{2}{7}(3, 4) - \frac{8}{7}(-1, 1)$.

51. One example of a nontrivial linear combination of vectors in S whose sum is the zero vector is

$$(1, 1, 1) - (1, 1, 0) - 0(0, 1, 1) - (0, 0, 1) = (0, 0, 0).$$

Solving this equation for $(1, 1, 1)$ yields

$$(1, 1, 1) = (1, 1, 0) + (0, 0, 1) + 0(0, 1, 1).$$

53. (a) From the vector equation

$$c_1(t, 1, 1) + c_2(1, t, 1) + c_3(1, 1, t) = (0, 0, 0)$$

you obtain the homogeneous system

$$\begin{aligned} tc_1 + c_2 + c_3 &= 0\\ c_1 + tc_2 + c_3 &= 0\\ c_1 + c_2 + tc_3 &= 0. \end{aligned}$$

The coefficient matrix of this system will have a nonzero determinant if $t^3 - 3t + 2 \neq 0$. So, the vectors will be linearly independent for all values of t other than $t = -2$ or $t = 1$.

(b) Proceeding as in (a), you obtain the homogeneous system

$$\begin{aligned} tc_1 + c_2 + c_3 &= 0\\ c_1 \quad\quad + c_3 &= 0\\ c_1 + c_2 + 3tc_3 &= 0. \end{aligned}$$

The coefficient matrix of this system will have a nonzero determinant if $2 - 4t \neq 0$.

So, the vectors will be linearly independent for all values of t other than $t = \frac{1}{2}$.

55. Let U be another subspace of V that contains S. To show that $\text{span}(S) \subset U$, let $\mathbf{u} \in \text{span}(S)$. Then

$\mathbf{u} = \sum_{i=1}^{k} c_i\mathbf{v}_i$, where $\mathbf{v}_i \in S$. So, $\mathbf{v}_i \in U$,

because U contains S. Because U is a subspace, $\mathbf{u} \in U$.

57. Because the matrix $\begin{bmatrix}1 & 2 & -1\\0 & 1 & 1\\2 & 5 & -1\end{bmatrix}$ row reduces to $\begin{bmatrix}1 & 0 & -3\\0 & 1 & 1\\0 & 0 & 0\end{bmatrix}$ and $\begin{bmatrix}-2 & -6 & 0\\1 & 1 & -2\end{bmatrix}$ row reduces to $\begin{bmatrix}1 & 0 & -3\\0 & 1 & 1\end{bmatrix}$, you see that S_1 and S_2 span the same subspace. You could also verify this by showing that each vector in S_1 is in the span of S_2, and conversely, each vector in S_2 is in the span of S_1. For example, $(1, 2, -1) = -\frac{1}{4}(-2, -6, 0) + \frac{1}{2}(1, 1, -2)$.

59. (a) False. See "Definition of Linear Dependence and Linear Independence," page 173.

(b) True. Any vector $\mathbf{u} = (u_1, u_2, u_3, u_4)$ in R^4 can be written as

$$\mathbf{u} = u_1(1, 0, 0, 0) - u_2(0, -1, 0, 0) + u_3(0, 0, 1, 0) + u_4(0, 0, 0, 1).$$

61. The matrix $\begin{bmatrix}1 & 1 & 1\\1 & 1 & 0\\1 & 0 & 0\end{bmatrix}$ row reduces to $\begin{bmatrix}1 & 0 & 0\\0 & 1 & 0\\0 & 0 & 1\end{bmatrix}$, which shows that the equation

$$c_1(1, 1, 1) + c_2(1, 1, 0) + c_3(1, 0, 0) = (0, 0, 0)$$

only has the trivial solution. So, the three vectors are linearly independent. Furthermore, the vectors span R^3 because the coefficient matrix of the linear system

$$\begin{bmatrix}1 & 1 & 1\\1 & 1 & 0\\1 & 0 & 0\end{bmatrix}\begin{bmatrix}c_1\\c_2\\c_3\end{bmatrix} = \begin{bmatrix}u_1\\u_2\\u_3\end{bmatrix}$$

is nonsingular.

63. Let S be a set of linearly independent vectors and $T \subset S$. If $T = \{\mathbf{v}_1, \cdots, \mathbf{v}_k\}$ and T were linearly dependent, then there would exist constants $c_1, \cdots, c_k$, not all zero, satisfying $c_1\mathbf{v}_1 + \cdots + c_k\mathbf{v}_k = \mathbf{0}$. But, $\mathbf{v}_i \in S$, and S is linearly independent, which is impossible. So, T is linearly independent.

65. If a set of vectors $\{\mathbf{v}_1, \mathbf{v}_2, \cdots\}$ contains the zero vector, then $\mathbf{0} = 0\mathbf{v}_1 + \ldots + 0\mathbf{v}_k + 1 \cdot \mathbf{0}$, which implies that the set is linearly dependent.

© 2013 Cengage Learning. All Rights Reserved. May not be scanned, copied or duplicated, or posted to a publicly accessible website, in whole or in part.

67. If the set $\{\mathbf{v}_1, \cdots, \mathbf{v}_{k-1}\}$ spanned $\mathbf{v}$, then

$\mathbf{v}_k = c_1\mathbf{v}_1 + \cdots + c_{k-1}\mathbf{v}_{k-1}$

for some scalars $c_1, \cdots, c_{k-1}$. So,

$c\mathbf{v}_1 + \cdots + c_{k-1}\mathbf{v}_{k-1} - \mathbf{v}_k = \mathbf{0},$

which is impossible because $\{\mathbf{v}_1, \cdots, \mathbf{v}_k\}$ is linearly independent.

69. Consider the vector equation

$c_1(\mathbf{u} + \mathbf{v}) + c_2(\mathbf{u} - \mathbf{v}) = \mathbf{0}.$

Regrouping, you have

$(c_1 + c_2)\mathbf{u} + (c_1 - c_2)\mathbf{v} = \mathbf{0}.$

Because **u** and **v** are linearly independent, $c_1 + c_2 = c_1 - c_2 = 0$. So, $c_1 = c_2 = 0$, and the vectors $\mathbf{u} + \mathbf{v}$ and $\mathbf{u} - \mathbf{v}$ are linearly independent.

71. Consider

$$\begin{aligned} c_1A\mathbf{v}_1 + c_2A\mathbf{v}_2 + c_3A\mathbf{v}_3 &= \mathbf{0} \\ A(c_1\mathbf{v}_1 + c_2\mathbf{v}_2 + c_3\mathbf{v}_3) &= \mathbf{0} \\ A^{-1}A(c_1\mathbf{v}_1 + c_2\mathbf{v}_2 + c_3\mathbf{v}_3) &= A^{-1}\mathbf{0} \\ c_1\mathbf{v}_1 + c_2\mathbf{v}_2 + c_3\mathbf{v}_3 &= \mathbf{0}. \end{aligned}$$

Because $\{\mathbf{v}_1, \mathbf{v}_2, \mathbf{v}_3\}$ are linearly independent, $c_1 = c_2 = c_3 = 0$, proving that $\{A\mathbf{v}_1, A\mathbf{v}_2, A\mathbf{v}_3\}$ are linearly independent.

If $A = \mathbf{0}$, then $\{A\mathbf{v}_1, A\mathbf{v}_2, A\mathbf{v}_3\} = \{\mathbf{0}\}$ is linearly dependent.

73. On the one hand, if **u** and **v** are linearly dependent, then there exist constants c_1 and c_2, not both zero, such that $c_1\mathbf{u} + c_2\mathbf{v} = \mathbf{0}$. Without loss of generality, you can assume $c_1 \neq 0$, and obtain $\mathbf{u} = -\frac{c_2}{c_1}\mathbf{v}.$

On the other hand, if one vector is a scalar multiple of another, $\mathbf{u} = c\mathbf{v}$, then $\mathbf{u} - c\mathbf{v} = \mathbf{0}$, which implies that **u** and **v** are linearly dependent.

Section 4.5 Basis and Dimension

1. There are six vectors in the standard basis for R^6.

$\{(1, 0, 0, 0, 0, 0), (0, 1, 0, 0, 0, 0), (0, 0, 1, 0, 0, 0),$
$(0, 0, 0, 1, 0, 0), (0, 0, 0, 0, 1, 0), (0, 0, 0, 0, 0, 1)\}$

3. There are eight vectors in the standard basis for $M_{2,4}$.

$$\left\{\begin{bmatrix}1&0&0&0\\0&0&0&0\end{bmatrix}, \begin{bmatrix}0&1&0&0\\0&0&0&0\end{bmatrix}, \begin{bmatrix}0&0&1&0\\0&0&0&0\end{bmatrix}, \begin{bmatrix}0&0&0&1\\0&0&0&0\end{bmatrix},\right.$$
$$\left.\begin{bmatrix}0&0&0&0\\1&0&0&0\end{bmatrix}, \begin{bmatrix}0&0&0&0\\0&1&0&0\end{bmatrix}, \begin{bmatrix}0&0&0&0\\0&0&1&0\end{bmatrix}, \begin{bmatrix}0&0&0&0\\0&0&0&1\end{bmatrix}\right\}$$

5. There are five vectors in the standard basis for P_4.

$\{1, x, x^2, x^3, x^4\}$

7. A basis for R^2 can only have two vectors. Because S has three vectors, it is not a basis for R^2.

9. S is linearly dependent $((0, 0) \in S)$ and does not span R^2. For instance, $(1, 1) \notin \operatorname{span}(S)$.

11. S is linearly dependent and does not span R^2. For instance, $(1, 1) \notin \operatorname{span}(S)$.

13. S does not span R^2, although it is linearly independent. For instance, $(1, 1) \notin \operatorname{span}(S)$.

15. A basis for R^3 contains three linearly independent vectors. Because

$-2(1, 3, 0) + (4, 1, 2) + (-2, 5, -2) = (0, 0, 0)$

S is linearly dependent and is, therefore, not a basis for R^3.

17. S does not span R^3, although it is linearly independent. For instance, $(0, 1, 0) \notin \operatorname{span}(S)$.

19. S is linearly dependent and does not span R^3. For instance, $(0, 0, 1) \notin \operatorname{span}(S)$.

21. A basis for P_2 can have only three vectors. Because S has four vectors, it is not a basis for P_2.

© 2013 Cengage Learning. All Rights Reserved. May not be scanned, copied or duplicated, or posted to a publicly accessible website, in whole or in part.

23. S is not a basis because the vectors are linearly dependent.

$$-2(1 - x) + 3(1 - x^2) + (3x^2 - 2x - 1) = 0$$

25. A basis for $M_{2,2}$ must have four vectors. Because S has only two vectors, it is not a basis for $M_{2,2}$.

27. S is not a basis because the vectors are linearly dependent.

$$5\begin{bmatrix} 1 & 0 \\ 0 & 0 \end{bmatrix} - 4\begin{bmatrix} 0 & 1 \\ 1 & 0 \end{bmatrix} + 3\begin{bmatrix} 1 & 0 \\ 0 & 1 \end{bmatrix} - \begin{bmatrix} 8 & -4 \\ -4 & 3 \end{bmatrix} = \begin{bmatrix} 0 & 0 \\ 0 & 0 \end{bmatrix}$$

Also, S does not span $M_{2,2}$.

29. Because $\{\mathbf{v}_1, \mathbf{v}_2\}$ consists of exactly two linearly independent vectors, it is a basis for R^2.

31. Because $\mathbf{v}_1$ and $\mathbf{v}_2$ are multiples of each other, they do not form a basis for R^2.

33. Because the vectors in S are not scalar multiples of one another, they are linearly independent. Because S consists of exactly two linearly independent vectors, it is a basis for R^2.

35. To determine if the vectors in S are linearly independent, find the solution of

$$c_1(1, 5, 3) + c_2(0, 1, 2) + c_3(0, 0, 6) = (0, 0, 0)$$

which corresponds to the solution of

$$\begin{aligned} c_1 \phantom{{}+ 2c_2 + 6c_3} &= 0 \\ 5c_1 + c_2 \phantom{{}+ 6c_3} &= 0 \\ 3c_1 + 2c_2 + 6c_3 &= 0. \end{aligned}$$

This system has only the trivial solution. So, S consists of exactly three linearly independent vectors, and is therefore a basis for R^3.

37. To determine if the vectors in S are linearly independent, find the solution of

$$c_1(0, 3, -2) + c_2(4, 0, 3) + c_3(-8, 15, -16) = (0, 0, 0)$$

which corresponds to the solution of

$$\begin{aligned} 4c_2 - 8c_3 &= 0 \\ 3c_1 + 15c_3 &= 0 \\ -2c_1 + 3c_2 - 16c_3 &= 0. \end{aligned}$$

Because this system has nontrivial solutions (for instance, $c_1 = -5, c_2 = 2,$ and $c_3 = 1$), the vectors are linearly dependent, and S is not a basis for R^3.

39. To determine if the vectors of S are linearly independent, find the solution of

$$c_1(-1, 2, 0, 0) + c_2(2, 0, -1, 0) + c_3(3, 0, 0, 4) + c_4(0, 0, 5, 0) = (0, 0, 0, 0)$$

which corresponds to the solution of

$$\begin{aligned} -c_1 + 2c_2 + 3c_3 &= 0 \\ 2c_1 &= 0 \\ -c_2 + 5c_4 &= 0 \\ 4c_3 &= 0. \end{aligned}$$

This system has only the trivial solution. So, S consists of exactly four linearly independent vectors, and is therefore a basis for R^4.

41. Form the equation

$$c_1(t^3 - 2t^2 + 1) + c_2(t^2 - 4) + c_3(t^3 + 2t) + c_4(5t) = 0 + 0t + 0t^2 + 0t^3$$

which yields the homogeneous system

$$\begin{aligned} c_1 + c_3 &= 0 \\ -2c_1 + c_2 &= 0 \\ 2c_3 + 5c_4 &= 0 \\ c_1 - 4c_2 &= 0. \end{aligned}$$

This system has only the trivial solution. So, S consists of exactly four linearly independent vectors, and is therefore a basis for P_3.

43. Because a basis for P_3 can contain only four basis vectors and set S contains five vectors, S is not a basis for P_3.

© 2013 Cengage Learning. All Rights Reserved. May not be scanned, copied or duplicated, or posted to a publicly accessible website, in whole or in part.

45. Form the equation

$$c_1\begin{bmatrix}2 & 0\\0 & 3\end{bmatrix} + c_2\begin{bmatrix}1 & 4\\0 & 1\end{bmatrix} + c_3\begin{bmatrix}0 & 1\\3 & 2\end{bmatrix} + c_4\begin{bmatrix}0 & 1\\2 & 0\end{bmatrix} = \begin{bmatrix}0 & 0\\0 & 0\end{bmatrix}$$

which yields the homogeneous system

$$\begin{aligned} 2c_1 + c_2 &= 0\\ 4c_2 + c_3 + c_4 &= 0\\ 3c_3 + 2c_4 &= 0\\ 3c_1 + c_2 + 2c_3 &= 0. \end{aligned}$$

This system has only the trivial solution. So, S consists of exactly four linearly independent vectors, and is therefore a basis for $M_{2,2}$.

47. Form the equation

$$c_1(4, 3, 2) + c_2(0, 3, 2) + c_3(0, 0, 2) = (0, 0, 0)$$

which yields the homogeneous system

$$\begin{aligned} 4c_1 &= 0\\ 3c_1 + 3c_2 &= 0\\ 2c_1 + 2c_2 + 2c_3 &= 0. \end{aligned}$$

This system has only the trivial solution, so S is a basis for R^3. Solving the system

$$\begin{aligned} 4c_1 &= 8\\ 3c_1 + 3c_2 &= 3\\ 2c_1 + 2c_2 + 2c_3 &= 8 \end{aligned}$$

yields $c_1 = 2$, $c_2 = -1$, and $c_3 = 3$. So,

$$\mathbf{u} = 2(4, 3, 2) - (0, 3, 2) + 3(0, 0, 2) = (8, 3, 8).$$

49. The set S contains the zero vector, and is therefore linearly dependent.

$$1(0, 0, 0) + 0(1, 3, 4) + 0(6, 1, -2) = (0, 0, 0)$$

So, S is not a basis for R^3.

51. Because a basis for R^6 has six linearly independent vectors, the dimension of R^6 is 6.

53. Because a basis for P_7 has eight linearly independent vectors, the dimension of P_7 is 8.

55. Because a basis for $M_{2,3}$ has six linearly independent vectors, the dimension of $M_{2,3}$ is 6.

57. One basis for $D_{3,3}$ is

$$\left\{\begin{bmatrix}1 & 0 & 0\\0 & 0 & 0\\0 & 0 & 0\end{bmatrix}, \begin{bmatrix}0 & 0 & 0\\0 & 1 & 0\\0 & 0 & 0\end{bmatrix}, \begin{bmatrix}0 & 0 & 0\\0 & 0 & 0\\0 & 0 & 1\end{bmatrix}\right\}.$$

Because a basis for $D_{3,3}$ has three vectors, $\dim(D_{3,3}) = 3$.

59. The following subsets of two vectors form a basis for R^2.

$$\{(1, 0), (0, 1)\}, \{(1, 0), (1, 1)\}, \{(0, 1), (1, 1)\}$$

61. Any vector that is not a multiple of $(2, 2)$ will form a basis with $(2, 2)$. For instance, $\{(2, 2), (1, 0)\}$ is a basis for R^2.

63. (a) W is a line through the origin.

(b) A basis for W is $\{(2, 1)\}$.

(c) The dimension of W is 1.

65. (a) W is a line through the origin.

(b) A basis for W is $\{(2, 1, -1)\}$.

(c) The dimension of W is 1.

67. (a) A basis for W is $\{(2, 1, 0, 1), (-1, 0, 1, 0)\}$.

(b) The dimension of W is 2.

69. (a) A basis for W is $\{(0, 6, 1, -1)\}$.

(b) The dimension of W is 1.

71. (a) False. See paragraph before "Definition of Dimension of a Vector Space," page 185.

(b) True. Find a set of n basis vectors in V that will span V and add any other vector.

73. Because the set $S_1 = \{c\mathbf{v}_1, \cdots, c\mathbf{v}_n\}$ has n vectors, you only need to show that they are linearly independent. Consider the equation

$$\begin{aligned} a_1(c\mathbf{v}_1) + a_2(c\mathbf{v}_2) + \cdots + a_n(c\mathbf{v}_n) &= \mathbf{0}\\ c(a_1\mathbf{v}_1 + a_2\mathbf{v}_2 + \cdots + a_n\mathbf{v}_n) &= \mathbf{0}\\ a_1\mathbf{v}_1 + a_2\mathbf{v}_2 + \cdots + a_n\mathbf{v}_n &= \mathbf{0}. \end{aligned}$$

Because $\{\mathbf{v}_1, \cdots, \mathbf{v}_n\}$ are linearly independent, the coefficients $a_1, \cdots, a_n$ must all be zero. So, S_1 is linearly independent.

75. Let $W \subset V$ and $\dim(V) = n$. Let $\mathbf{w}_1, \cdots, \mathbf{w}_k$ be a basis for W. Because $W \subset V$, the vectors $\mathbf{w}_1, \cdots, \mathbf{w}_k$ are linearly independent in V. If $\operatorname{span}(\mathbf{w}_1, \cdots, \mathbf{w}_k) = V$, then $\dim(W) = \dim(V)$. If not, let $\mathbf{v} \in V$, $\mathbf{v} \notin W$. Then $\dim(W) < \dim(V)$.

77. If S spans V, you are done. If not, let $\mathbf{v}_1 \notin \operatorname{span}(S)$, and consider the linearly independent set $S_1 = S \cup \{\mathbf{v}_1\}$. If S_1 spans V you are done. If not, let $\mathbf{v}_2 \notin \operatorname{span}(S_1)$ and continue as before. Because the vector space is finite-dimensional, this process will ultimately produce a basis of V containing S.

© 2013 Cengage Learning. All Rights Reserved. May not be scanned, copied or duplicated, or posted to a publicly accessible website, in whole or in part.

Section 4.6 Rank of a Matrix and Systems of Linear Equations

1. (a) $(0, -2), (1, -3)$

(b) $\begin{bmatrix} 0 \\ 1 \end{bmatrix}, \begin{bmatrix} -2 \\ -3 \end{bmatrix}$

3. (a) $(4, 3, 1), (1, -4, 0)$

(b) $\begin{bmatrix} 4 \\ 1 \end{bmatrix}, \begin{bmatrix} 3 \\ -4 \end{bmatrix}, \begin{bmatrix} 1 \\ 0 \end{bmatrix}$

5. (a) A basis for the row space is $\{(1, 0), (0, 1)\}$.

(b) Because this matrix row reduces to

$$\begin{bmatrix} 1 & 0 \\ 0 & 1 \end{bmatrix}$$

the rank of the matrix is 2.

7. (a) A basis for the row space is $\left\{\left(1, 0, \frac{1}{2}\right), \left(0, 1, -\frac{1}{2}\right)\right\}$.

(b) Because this matrix row reduces to

$$\begin{bmatrix} 1 & 0 & \frac{1}{2} \\ 0 & 1 & -\frac{1}{2} \end{bmatrix}$$

the rank of the matrix is 2.

9. (a) A basis for the row space of the matrix is $\{(1, 2, -2, 0), (0, 0, 0, 1)\}$.

(b) Because this matrix row reduces to

$$\begin{bmatrix} 1 & 2 & -2 & 0 \\ 0 & 0 & 0 & 1 \\ 0 & 0 & 0 & 0 \end{bmatrix}$$

the rank of the matrix is 2.

11. Use $\mathbf{v}_1$, $\mathbf{v}_2$, and $\mathbf{v}_3$ to form the rows of matrix A. Then write A in row-echelon form.

$$A = \begin{bmatrix} 1 & 2 & 4 \\ -1 & 3 & 4 \\ 2 & 3 & 1 \end{bmatrix} \begin{matrix} \mathbf{v}_1 \\ \mathbf{v}_2 \\ \mathbf{v}_3 \end{matrix} \rightarrow B = \begin{bmatrix} 1 & 0 & 0 \\ 0 & 1 & 0 \\ 0 & 0 & 1 \end{bmatrix} \begin{matrix} \mathbf{w}_1 \\ \mathbf{w}_2 \\ \mathbf{w}_3 \end{matrix}$$

So, the nonzero row vectors of B, $\mathbf{w}_1 = (1, 0, 0)$, $\mathbf{w}_2 = (0, 1, 0)$, and $\mathbf{w}_3 = (0, 0, 1)$, form a basis for the row space of A. That is, they form a basis for the subspace spanned by S.

13. Use $\mathbf{v}_1$, $\mathbf{v}_2$, and $\mathbf{v}_3$ to form the rows of matrix A. Then write A in row-echelon form.

$$A = \begin{bmatrix} 4 & 4 & 8 \\ 1 & 1 & 2 \\ 1 & 1 & 1 \end{bmatrix} \begin{matrix} \mathbf{v}_1 \\ \mathbf{v}_2 \\ \mathbf{v}_3 \end{matrix} \rightarrow B = \begin{bmatrix} 1 & 1 & 0 \\ 0 & 0 & 1 \\ 0 & 0 & 0 \end{bmatrix} \begin{matrix} \mathbf{w}_1 \\ \mathbf{w}_2 \\ \end{matrix}$$

So, the nonzero row vectors of B, $\mathbf{w}_1 = (1, 1, 0)$ and $\mathbf{w}_2 = (0, 0, 1)$, form a basis for the row space of A. That is, they form a basis for the subspace spanned by S.

15. Begin by forming the matrix whose rows are vectors in S.

$$\begin{bmatrix} 2 & 9 & -2 & 53 \\ -3 & 2 & 3 & -2 \\ 8 & -3 & -8 & 17 \\ 0 & -3 & 0 & 15 \end{bmatrix}$$

This matrix reduces to

$$\begin{bmatrix} 1 & 0 & -1 & 0 \\ 0 & 1 & 0 & 0 \\ 0 & 0 & 0 & 1 \\ 0 & 0 & 0 & 0 \end{bmatrix}.$$

So, a basis for span(S) is $\{(1, 0, -1, 0), (0, 1, 0, 0), (0, 0, 0, 1)\}$.

17. Begin by forming the matrix whose rows are the vectors in S.

$$\begin{bmatrix} -3 & 2 & 5 & 28 \\ -6 & 1 & -8 & -1 \\ 14 & -10 & 12 & -10 \\ 0 & 5 & 12 & 50 \end{bmatrix}$$

This matrix reduces to

$$\begin{bmatrix} 1 & 0 & 0 & 0 \\ 0 & 1 & 0 & 0 \\ 0 & 0 & 1 & 0 \\ 0 & 0 & 0 & 1 \end{bmatrix}.$$

So, a basis for span(S) is

$\{(1, 0, 0, 0), (0, 1, 0, 0), (0, 0, 1, 0), (0, 0, 0, 1)\}$.

19. (a) Row-reducing the transpose of the original matrix produces

$$\begin{bmatrix} 1 & \frac{1}{2} \\ 0 & 1 \end{bmatrix} \text{ or } \begin{bmatrix} 1 & 0 \\ 0 & 1 \end{bmatrix}.$$

So, a basis for the column space of the matrix is $\left\{\left(1, \frac{1}{2}\right), (0, 1)\right\}$ $\left(\text{or } \{(1, 0), (0, 1)\}\right)$.

(b) Because this matrix row reduces to

$$\begin{bmatrix} 1 & 2 \\ 0 & 1 \end{bmatrix} \text{ or } \begin{bmatrix} 1 & 0 \\ 0 & 1 \end{bmatrix}$$

the rank of the matrix is 2.

© 2013 Cengage Learning. All Rights Reserved. May not be scanned, copied or duplicated, or posted to a publicly accessible website, in whole or in part.

21. (a) Row-reducing the transpose of the original matrix produces

$$\begin{bmatrix} 1 & 0 \\ 0 & 1 \\ 0 & 0 \end{bmatrix}.$$

So, a basis for the column space of the matrix is $\{(1, 0), (0, 1)\}$.

(b) Because this matrix row reduces to

$$\begin{bmatrix} 1 & 0 & \frac{3}{2} \\ 0 & 1 & \frac{5}{4} \end{bmatrix}$$

the rank of the matrix is 2.

23. (a) Row-reducing the transpose of the original matrix produces

$$\begin{bmatrix} 1 & 0 & \frac{5}{9} & \frac{2}{9} \\ 0 & 1 & -\frac{4}{9} & \frac{2}{9} \\ 0 & 0 & 0 & 0 \\ 0 & 0 & 0 & 0 \end{bmatrix}.$$

So, a basis for the column space is $\left\{\left(1, 0, \frac{5}{9}, \frac{2}{9}\right), \left(0, 1, -\frac{4}{9}, \frac{2}{9}\right)\right\}$, or, the first and third columns of the original matrix.

(b) Because this matrix row reduces to

$$\begin{bmatrix} 1 & 2 & 0 & 3 \\ 0 & 0 & 1 & 4 \\ 0 & 0 & 0 & 0 \\ 0 & 0 & 0 & 0 \end{bmatrix}$$

the rank of the matrix is 2.

25. Solving the system $A\mathbf{x} = \mathbf{0}$ yields solutions of the form $(t, 2t)$, where t is any real number. The dimension of the solution space is 1, and a basis is $\{(1, 2)\}$.

27. Solving the system $A\mathbf{x} = \mathbf{0}$ yields solutions of the form $(-2s - 3t, s, t)$, where s and t are any real numbers. The dimension of the solution space is 2, and a basis is $\{(-2, 1, 0), (-3, 0, 1)\}$.

29. Solving the system $A\mathbf{x} = \mathbf{0}$ yields solutions of the form $(-3t, 0, t)$, where t is any real number. The dimension of the solution space is 1, and a basis is $\{(-3, 0, 1)\}$.

31. Solving the system $A\mathbf{x} = \mathbf{0}$ yields solutions of the form $(-t, 2t, t)$, where t is any real number. The dimension of the solution space is 1, and a basis for the solution space is $\{(-1, 2, 1)\}$.

33. Solving the system $A\mathbf{x} = \mathbf{0}$ yields solutions of the form $(-s + 2t, s - 2t, s, t)$, where s and t are any real numbers. The dimension of the solution space is 2, and a basis is $\{(2, -2, 0, 1), (-1, 1, 1, 0)\}$.

35. The only solution to the system $A\mathbf{x} = \mathbf{0}$ is the trivial solution. So, the solution space is $\{(0, 0, 0, 0)\}$ whose dimension is 0.

37. (a) This system yields solutions of the form $(4t, t)$, where t is any real number and a basis for the solution space is $\{(4, 1)\}$.

(b) The dimension of the solution space is 1.

39. (a) This system yields solutions of the form $(-t, -3t, 2t)$, where t is any real number and a basis is $\{(-1, -3, 2)\}$.

(b) The dimension of the solution space is 1.

41. (a) This system yields solutions of the form $(2s - 3t, s, t)$, where s and t are any real numbers and a basis for the solution space is $\{(2, 1, 0), (-3, 0, 1)\}$.

(b) The dimension of the solution space is 2.

43. (a) This system yields solutions of the form $\left(-4s - 3t, -s - \frac{2}{3}t, s, t\right)$, where s and t are any real numbers and a basis for the solution space is $\left\{(-4, -1, 1, 0), \left(-3, -\frac{2}{3}, 0, 1\right)\right\}$.

(b) The dimension of the solution space is 2.

45. (a) This system yields solutions of the form $\left(\frac{4}{3}t, -\frac{3}{2}t, -t, t\right)$, where t is any real number and a basis for the solution space is $\left\{\left(\frac{4}{3}, -\frac{3}{2}, -1, 1\right)\right\}$ or $\{(8, -9, -6, 6)\}$.

(b) The dimension of the solution space is 1.

© 2013 Cengage Learning. All Rights Reserved. May not be scanned, copied or duplicated, or posted to a publicly accessible website, in whole or in part.

47. (a) $\text{rank}(A) = \text{rank}(B) = 3$

$\text{nullity}(A) = n - \text{rank}(A) = 5 - 3 = 2$

(b) Matrix B represents the system of linear equations

$$\begin{aligned} x_1 \quad + 3x_3 \quad - 4x_5 &= 0 \\ x_2 - x_3 \quad + 2x_5 &= 0 \\ x_4 - 2x_5 &= 0. \end{aligned}$$

Choosing the nonessential variables x_3 and x_5 as the free variables s and t produces the solution

$$\begin{aligned} x_1 &= -3s + 4t \\ x_2 &= s - 2t \\ x_3 &= s \\ x_4 &= 2t \\ x_5 &= t. \end{aligned}$$

So, a basis for the null space of A is

$$\left\{ \begin{bmatrix} -3 \\ 1 \\ 1 \\ 0 \\ 0 \end{bmatrix}, \begin{bmatrix} 4 \\ -2 \\ 0 \\ 2 \\ 1 \end{bmatrix} \right\}.$$

(c) By Theorem 4.14, the nonzero row vectors of B form a basis for the row space of A: $\{(1, 0, 3, 0, -4), (0, 1, -1, 0, 2), (0, 0, 0, 1, -2)\}$.

(d) The columns of B with leading ones are the first, second, and fourth. The corresponding columns of A form a basis for the column space of A: $\{(1, 2, 3, 4), (2, 5, 7, 9), (0, 1, 2, -1)\}$.

(e) Because the row-equivalent form B contains a row of zeros, the rows of A are linearly dependent.

(f) Because $a_3 = 3a_1 - a_2$, $\{a_1, a_2, a_3\}$ is linearly dependent. So, (i) and (iii) are linearly independent.

49. (a) The system $A\mathbf{x} = \mathbf{b}$ is consistent because its augmented matrix reduces to

$$\begin{bmatrix} 1 & 0 & -2 & 3 \\ 0 & 1 & 4 & 5 \\ 0 & 0 & 0 & 0 \\ 0 & 0 & 0 & 0 \end{bmatrix}.$$

(b) The solutions of $A\mathbf{x} = \mathbf{b}$ are of the form $(3 + 2t, 5 - 4t, t)$, where t is any real number. That is,

$$\mathbf{x} = \begin{bmatrix} 3 \\ 5 \\ 0 \end{bmatrix} + t\begin{bmatrix} 2 \\ -4 \\ 1 \end{bmatrix},$$

where

$$\mathbf{x}_p = \begin{bmatrix} 3 \\ 5 \\ 0 \end{bmatrix} \text{ and } \mathbf{x}_h = t\begin{bmatrix} 2 \\ -4 \\ 1 \end{bmatrix}.$$

51. The system $A\mathbf{x} = \mathbf{b}$ is inconsistent because its augmented matrix reduces to

$$\begin{bmatrix} 1 & 0 & 0 & -2 & 0 \\ 0 & 1 & 0 & -\frac{1}{2} & 0 \\ 0 & 0 & 1 & \frac{1}{2} & 0 \\ 0 & 0 & 0 & 0 & 1 \end{bmatrix}.$$

53. (a) The system $A\mathbf{x} = \mathbf{b}$ is consistent because its augmented matrix reduces to

$$\begin{bmatrix} 1 & 2 & 0 & 0 & -5 & 1 \\ 0 & 0 & 1 & 0 & 6 & 2 \\ 0 & 0 & 0 & 1 & 4 & -3 \\ 0 & 0 & 0 & 0 & 0 & 0 \end{bmatrix}.$$

(b) The solutions of the system are of the form

$(1 - 2s + 5t, s, 2 - 6t, -3 - 4t, t)$

where s and t are any real numbers. That is,

$$\mathbf{x} = \begin{bmatrix} 1 \\ 0 \\ 2 \\ -3 \\ 0 \end{bmatrix} + s\begin{bmatrix} -2 \\ 1 \\ 0 \\ 0 \\ 0 \end{bmatrix} + t\begin{bmatrix} 5 \\ 0 \\ -6 \\ -4 \\ 1 \end{bmatrix},$$

where

$$\mathbf{x}_p = \begin{bmatrix} 1 \\ 0 \\ 2 \\ -3 \\ 0 \end{bmatrix} \text{ and } \mathbf{x}_h = s\begin{bmatrix} -2 \\ 1 \\ 0 \\ 0 \\ 0 \end{bmatrix} + t\begin{bmatrix} 5 \\ 0 \\ -6 \\ -4 \\ 1 \end{bmatrix}.$$

55. The vector $\mathbf{b}$ is in the column space of A if the equation $A\mathbf{x} = \mathbf{b}$ is consistent. Because $A\mathbf{x} = \mathbf{b}$ has the solution

$$\mathbf{x} = \begin{bmatrix} 1 \\ 2 \end{bmatrix}$$

$\mathbf{b}$ is in the column space of A. Furthermore,

$$\mathbf{b} = 1\begin{bmatrix} -1 \\ 4 \end{bmatrix} + 2\begin{bmatrix} 2 \\ 0 \end{bmatrix} = \begin{bmatrix} 3 \\ 4 \end{bmatrix}.$$

57. The vector $\mathbf{b}$ is not in the column space of A because the linear system $A\mathbf{x} = \mathbf{b}$ is inconsistent.

59. Assume that A is an $m \times n$ matrix where $n > m$. Then the set of n column vectors of A are vectors in R^m and must be linearly dependent. Similarly, if $m > n$, then the set of m row vectors of A are vectors in R^n, and must be linearly dependent.

© 2013 Cengage Learning. All Rights Reserved. May not be scanned, copied or duplicated, or posted to a publicly accessible website, in whole or in part.

61. (a) Let

$$A = \begin{bmatrix} 1 & 0 \\ 0 & 1 \end{bmatrix} \quad \text{and} \quad B = \begin{bmatrix} 0 & 1 \\ 1 & 0 \end{bmatrix}. \quad \text{Then } A + B = \begin{bmatrix} 1 & 1 \\ 1 & 1 \end{bmatrix}.$$

Note that $\text{rank}(A) = \text{rank}(B) = 2$, and $\text{rank}(A + B) = 1$.

(b) Let

$$A = \begin{bmatrix} 1 & 0 \\ 0 & 0 \end{bmatrix} \quad \text{and} \quad B = \begin{bmatrix} 0 & 1 \\ 0 & 0 \end{bmatrix}. \quad \text{Then } A + B = \begin{bmatrix} 1 & 1 \\ 0 & 0 \end{bmatrix}.$$

Note that $\text{rank}(A) = \text{rank}(B) = 1$, and $\text{rank}(A + B) = 1$.

(c) Let

$$A = \begin{bmatrix} 1 & 0 \\ 0 & 0 \end{bmatrix} \quad \text{and} \quad B = \begin{bmatrix} 0 & 0 \\ 0 & 1 \end{bmatrix}. \quad \text{Then } A + B = \begin{bmatrix} 1 & 0 \\ 0 & 1 \end{bmatrix}.$$

Note that $\text{rank}(A) = \text{rank}(B) = 1$, and $\text{rank}(A + B) = 2$.

63. (a) Because the row (or column) space has dimension no larger than the smaller of m and n, $r \le m$ (because $m < n$)

(b) There are r vectors in a basis for the row space of A.

(c) There are r vectors in a basis for the column space of A.

(d) The row space of A is a subspace of R^n.

(e) The column space of A is a subspace of R^m.

65. Consider the first row of the product AB.

$$\underset{A}{\begin{bmatrix} a_{11} & \cdots & a_{1n} \\ \vdots & & \vdots \\ a_{m1} & \cdots & a_{mn} \end{bmatrix}} \underset{B}{\begin{bmatrix} b_{11} & \cdots & b_{1k} \\ \vdots & & \vdots \\ b_{n1} & \cdots & b_{nk} \end{bmatrix}} = \underset{AB}{\begin{bmatrix} (a_{11}b_{11} + \cdots + a_{1n}b_{n1}) & \cdots & (a_{11}b_{1k} + \cdots + a_{1n}b_{nk}) \\ \vdots & & \\ (a_{m1}b_{11} + \cdots + a_{mn}b_{n1}) & \cdots & (a_{m1}b_{1k} + \cdots + a_{mn}b_{nk}) \end{bmatrix}}.$$

The first row of AB is $\left[(a_{11}b_{11} + \cdots + a_{1n}b_{n1}), \cdots, (a_{11}b_{1k} + \cdots + a_{1n}b_{nk})\right]$.

You can express this first row as $a_{11}(b_{11}, b_{12}, \cdots b_{1k}) + a_{12}(b_{21}, \cdots, b_{2k}) + \cdots + a_{1n}(b_{n1}, b_{n2}, \cdots, b_{nk})$,

which means that the first row of AB is in the row space of B. The same argument applies to the other rows of A. A similar argument can be used to show that the column vectors of AB are in the column space of A. The first column of AB is

$$\begin{bmatrix} a_{11}b_{11} + \cdots + a_{1n}b_{n1} \\ \vdots \\ a_{m1}b_{11} + \cdots + a_{mn}b_{n1} \end{bmatrix} = b_{11}\begin{bmatrix} a_{11} \\ \vdots \\ a_{m1} \end{bmatrix} + \cdots + b_{n1}\begin{bmatrix} a_{1n} \\ \vdots \\ a_{mn} \end{bmatrix}.$$

67. Let $A\mathbf{x} = \mathbf{b}$ be a system of linear equations in n variables.

(a) If $\text{rank}(A) = \text{rank}([A \vdots \mathbf{b}]) = n$, then $\mathbf{b}$ is in the column space of A, and so $A\mathbf{x} = \mathbf{b}$ has a unique solution.

(b) If $\text{rank}(A) = \text{rank}([A \vdots \mathbf{b}]) < n$, then $\mathbf{b}$ is in the column space of A and $\text{rank}(A) < n$, which implies that $A\mathbf{x} = \mathbf{b}$ has an infinite number of solutions.

(c) If $\text{rank}(A) < \text{rank}([A \vdots \mathbf{b}])$, then $\mathbf{b}$ is *not* in the column space of A, and the system is inconsistent.

69. (a) True. The null space of A is also called the solution space of the system $A\mathbf{x} = \mathbf{0}$.

(b) True. The null space of A is the solution space of the homogeneous system $A\mathbf{x} = \mathbf{0}$.

71. (a) False. See "Remark," page 190.

(b) False. See Theorem 4.19, page 198.

(c) True. The columns of A become the rows of the transpose, A^T. So, the columns of A span the same space as the rows of A^T.

73. (a) $A\mathbf{x} = B\mathbf{x} \Rightarrow (A - B)(\mathbf{x}) = \mathbf{0}$ for all $\mathbf{x} \in R^n \Rightarrow \text{nullity}(A - B) = n$ and $\text{rank}(A - B) = 0$

(b) So, $A - B = O \Rightarrow A = B$.

© 2013 Cengage Learning. All Rights Reserved. May not be scanned, copied or duplicated, or posted to a publicly accessible website, in whole or in part.

71. Let A and B be $2m \times n$ row equivalent matrices. The dependency relationships among the columns of A can be expressed in the form $A\mathbf{x} = \mathbf{0}$, while those of B in the form $B\mathbf{x} = \mathbf{0}$. Because A and B are row-equivalent, $A\mathbf{x} = \mathbf{0}$ and $B\mathbf{x} = \mathbf{0}$ have the same solution sets, and therefore the same dependency relationships.

Section 4.7 Coordinates and Change of Basis

1. $\begin{bmatrix} 5 \\ -2 \end{bmatrix}$

3. $\begin{bmatrix} 7 \\ -4 \\ -1 \\ 2 \end{bmatrix}$

5. Because $[\mathbf{x}]_B = \begin{bmatrix} 4 \\ 1 \end{bmatrix}$, you can write

$\mathbf{x} = 4(2, -1) + 1(0, 1) = (8, -3).$

Moreover, because $(8, -3) = 8(1, 0) - 3(0, 1)$, it follows that the coordinates of $\mathbf{x}$ relative to S are

$[\mathbf{x}]_S = \begin{bmatrix} 8 \\ -3 \end{bmatrix}.$

7. Because $[\mathbf{x}]_B = \begin{bmatrix} 2 \\ 3 \\ 1 \end{bmatrix}$, you can write

$\mathbf{x} = 2(1, 0, 1) + 3(1, 1, 0) + 1(0, 1, 1) = (5, 4, 3).$

Moreover, because

$(5, 4, 3) = 5(1, 0, 0) + 4(0, 1, 0) + 3(0, 0, 1)$, it follows that the coordinates of $\mathbf{x}$ relative to S are

$[\mathbf{x}]_S = \begin{bmatrix} 5 \\ 4 \\ 3 \end{bmatrix}.$

9. Because $[\mathbf{x}]_B = \begin{bmatrix} 1 \\ -2 \\ 3 \\ -1 \end{bmatrix}$, you can write

$$\begin{aligned} \mathbf{x} &= 1(0, 0, 0, 1) - 2(0, 0, 1, 1) + 3(0, 1, 1, 1) - 1(1, 1, 1, 1) \\ &= (-1, 2, 0, 1) \end{aligned}$$

which implies that the coordinates of $\mathbf{x}$ relative to the standard basis S are

$[\mathbf{x}]_S = \begin{bmatrix} -1 \\ 2 \\ 0 \\ 1 \end{bmatrix}.$

11. Begin by writing $\mathbf{x}$ as a linear combination of the vectors in B.

$\mathbf{x} = (12, 6) = c_1(4, 0) + c_2(0, 3)$

Equating corresponding components yields the following system of linear equations.

$$\begin{aligned} 4c_1 \quad\quad &= 12 \\ 3c_2 &= 6 \end{aligned}$$

The solution of this system is $c_1 = 3$ and $c_2 = 2$. So, $\mathbf{x} = 3(4, 0) + 2(0, 3)$, and the coordinate vector of $\mathbf{x}$ relative to B is $[\mathbf{x}]_B = \begin{bmatrix} 3 \\ 2 \end{bmatrix}.$

13. Begin by writing $\mathbf{x}$ as a linear combination of the vector in B.

$\mathbf{x} = (3, 19, 2) = c_1(8, 11, 0) + c_2(7, 0, 10) + c_3(1, 4, 6)$

Equating corresponding components yields the following system of linear equations.

$$\begin{aligned} 8c_1 + 7c_2 + c_3 &= 3 \\ 11c_1 \quad\quad + 4c_3 &= 19 \\ 10c_2 + 6c_3 &= 2 \end{aligned}$$

The solution of this system is $c_1 = 1$, $c_2 = -1$, and $c_3 = 2$. So, $\mathbf{x} = 1(8, 11, 0) + (-1)(7, 0, 10) + 2(1, 4, 6)$, and the coordinate vector of $\mathbf{x}$ relative to B is

$[\mathbf{x}]_B = \begin{bmatrix} 1 \\ -1 \\ 2 \end{bmatrix}.$

15. Begin by writing $\mathbf{x}$ as a linear combination of the vector in B.

$\mathbf{x} = (11, 18, -7) = c_1(4, 3, 3) + c_2(-11, 0, 11) + c_3(0, 9, 2)$

Equating corresponding components yields the following system of linear equations.

$$\begin{aligned} 4c_1 - 11c_2 \quad\quad &= 11 \\ 3c_1 \quad\quad + 9c_3 &= 18 \\ 3c_1 + 11c_2 + 2c_3 &= -7 \end{aligned}$$

The solution of this system is $c_1 = 0$, $c_2 = -1$, and $c_3 = 2$. So,

$\mathbf{x} = (11, 18, -7) = 0(4, 3, 3) - 1(-11, 0, 11) + 2(0, 9, 2)$

and the coordinate vector of x relative to B is

$[\mathbf{x}]_B = \begin{bmatrix} 0 \\ -1 \\ 2 \end{bmatrix}.$

© 2013 Cengage Learning. All Rights Reserved. May not be scanned, copied or duplicated, or posted to a publicly accessible website, in whole or in part.

17. Begin by forming the matrix

$$[B' \ B] = \begin{bmatrix} 2 & 1 & 1 & 0 \\ 4 & 3 & 0 & 1 \end{bmatrix}$$

and then use Gauss-Jordan elimination to produce

$$[I_2 \ P^{-1}] = \begin{bmatrix} 1 & 0 & \frac{3}{2} & -\frac{1}{2} \\ 0 & 1 & -2 & 1 \end{bmatrix}.$$

So, the transition matrix from B to B' is

$$P^{-1} = \begin{bmatrix} \frac{3}{2} & -\frac{1}{2} \\ -2 & 1 \end{bmatrix}.$$

19. Begin by forming the matrix

$$[B' \ B] = \begin{bmatrix} 1 & 0 & 2 & -1 \\ 0 & 1 & 4 & 3 \end{bmatrix}.$$

Because this matrix is already in the form $[I_2 \ P^{-1}]$, you see that the transition matrix from B to B' is

$$P^{-1} = \begin{bmatrix} 2 & -1 \\ 4 & 3 \end{bmatrix}.$$

21. Begin by forming the matrix

$$[B' \ B] = \begin{bmatrix} 1 & 0 & 6 & 1 & 0 & 0 \\ 0 & 2 & 0 & 0 & 1 & 0 \\ 0 & 8 & 12 & 0 & 0 & 1 \end{bmatrix}$$

and then use Gauss-Jordan elimination to produce

$$[I_3 \ P^{-1}] = \begin{bmatrix} 1 & 0 & 0 & 1 & 2 & -\frac{1}{2} \\ 0 & 1 & 0 & 0 & \frac{1}{2} & 0 \\ 0 & 0 & 1 & 0 & -\frac{1}{3} & \frac{1}{12} \end{bmatrix}.$$

So, the transition matrix from B to B' is

$$P^{-1} = \begin{bmatrix} 1 & 2 & -\frac{1}{2} \\ 0 & \frac{1}{2} & 0 \\ 0 & -\frac{1}{3} & \frac{1}{12} \end{bmatrix}.$$

23. Begin by forming the matrix

$$[B' \ B] = \begin{bmatrix} 1 & 0 & 0 & 3 & -2 & 1 \\ 0 & 1 & 0 & 4 & -1 & 0 \\ 0 & 0 & 1 & 0 & 1 & -3 \end{bmatrix}.$$

Because this matrix is already in the form $[I_3 \ P^{-1}]$, the transition matrix from B to B' is

$$P^{-1} = \begin{bmatrix} 3 & -2 & 1 \\ 4 & -1 & 0 \\ 0 & 1 & -3 \end{bmatrix}.$$

25. Begin by forming the matrix

$$[B' \ B] = \begin{bmatrix} 2 & -1 & 2 & 1 \\ 1 & 2 & 5 & 2 \end{bmatrix}$$

and then use Gauss-Jordan elimination to produce

$$[I_2 \ P^{-1}] = \begin{bmatrix} 1 & 0 & \frac{9}{5} & \frac{4}{5} \\ 0 & 1 & \frac{8}{5} & \frac{3}{5} \end{bmatrix}.$$

So, the transition matrix from B to B' is

$$P^{-1} = \begin{bmatrix} \frac{9}{5} & \frac{4}{5} \\ \frac{8}{5} & \frac{3}{5} \end{bmatrix}.$$

27. Begin by forming the matrix

$$[B' \ B] = \begin{bmatrix} 1 & 1 & 1 & 1 & 0 & 0 \\ 3 & 5 & 4 & 0 & 1 & 0 \\ 3 & 6 & 5 & 0 & 0 & 1 \end{bmatrix}$$

and then use Gauss-Jordan elimination to produce

$$[I_3 \ P^{-1}] = \begin{bmatrix} 1 & 0 & 0 & 1 & 1 & -1 \\ 0 & 1 & 0 & -3 & 2 & -1 \\ 0 & 0 & 1 & 3 & -3 & 2 \end{bmatrix}.$$

So, the transition matrix from B to B' is

$$P^{-1} = \begin{bmatrix} 1 & 1 & -1 \\ -3 & 2 & -1 \\ 3 & -3 & 2 \end{bmatrix}.$$

29. Begin by forming the matrix

$$[B' \ B] = \begin{bmatrix} 0 & -2 & 1 & 1 & -1 & 2 \\ 2 & 1 & 1 & 2 & 2 & 4 \\ 1 & 0 & 1 & 4 & 0 & 0 \end{bmatrix}$$

and then use Gauss-Jordan elimination to produce

$$[I_3 \ P^{-1}] = \begin{bmatrix} 1 & 0 & 0 & -7 & 3 & 10 \\ 0 & 1 & 0 & 5 & -1 & -6 \\ 0 & 0 & 1 & 11 & -3 & -10 \end{bmatrix}.$$

So, the transition matrix from B to B' is

$$P^{-1} = \begin{bmatrix} -7 & 3 & 10 \\ 5 & -1 & -6 \\ 11 & -3 & -10 \end{bmatrix}.$$

© 2013 Cengage Learning. All Rights Reserved. May not be scanned, copied or duplicated, or posted to a publicly accessible website, in whole or in part.

31. Begin by forming the matrix

$$[B' \ B] = \begin{bmatrix} 1 & -2 & -1 & -2 & 1 & 0 & 0 & 0 \\ 3 & -5 & -2 & -3 & 0 & 1 & 0 & 0 \\ 2 & -5 & -2 & -5 & 0 & 0 & 1 & 0 \\ -1 & 4 & 4 & 11 & 0 & 0 & 0 & 1 \end{bmatrix}$$

and then use Gauss-Jordan elimination to produce

$$[I_4 \ P^{-1}] = \begin{bmatrix} 1 & 0 & 0 & 0 & -24 & 7 & 1 & -2 \\ 0 & 1 & 0 & 0 & -10 & 3 & 0 & -1 \\ 0 & 0 & 1 & 0 & -29 & 7 & 3 & -2 \\ 0 & 0 & 0 & 1 & 12 & -3 & -1 & 1 \end{bmatrix}.$$

So, the transition matrix from B to B' is

$$P^{-1} = \begin{bmatrix} -24 & 7 & 1 & -2 \\ -10 & 3 & 0 & -1 \\ -29 & 7 & 3 & -2 \\ 12 & -3 & -1 & 1 \end{bmatrix}.$$

33. Begin by forming the matrix

$$[B' \ B] = \begin{bmatrix} 1 & -2 & 0 & 0 & 1 & 1 & 0 & 0 & 0 & 0 \\ 2 & -3 & 1 & 1 & -1 & 0 & 1 & 0 & 0 & 0 \\ 4 & 4 & 2 & 2 & 0 & 0 & 0 & 1 & 0 & 0 \\ -1 & 2 & -2 & 2 & 1 & 0 & 0 & 0 & 1 & 0 \\ 2 & 1 & 1 & 1 & 2 & 0 & 0 & 0 & 0 & 1 \end{bmatrix}$$

and then use Gauss-Jordan elimination to produce

$$[I_5 \ P^{-1}] = \begin{bmatrix} 1 & 0 & 0 & 0 & 0 & 1 & -\frac{3}{11} & \frac{5}{11} & 0 & -\frac{7}{11} \\ 0 & 1 & 0 & 0 & 0 & 0 & -\frac{2}{11} & \frac{3}{22} & 0 & -\frac{1}{11} \\ 0 & 0 & 1 & 0 & 0 & -\frac{5}{4} & \frac{9}{22} & -\frac{19}{44} & -\frac{1}{4} & \frac{21}{22} \\ 0 & 0 & 0 & 1 & 0 & -\frac{3}{4} & \frac{1}{2} & -\frac{1}{4} & \frac{1}{4} & \frac{1}{2} \\ 0 & 0 & 0 & 0 & 1 & 0 & -\frac{1}{11} & -\frac{2}{11} & 0 & \frac{5}{11} \end{bmatrix}.$$

So, the transition matrix from B to B' is

$$P^{-1} = \begin{bmatrix} 1 & -\frac{3}{11} & \frac{5}{11} & 0 & -\frac{7}{11} \\ 0 & -\frac{2}{11} & \frac{3}{22} & 0 & -\frac{1}{11} \\ -\frac{5}{4} & \frac{9}{22} & -\frac{19}{44} & -\frac{1}{4} & \frac{21}{22} \\ -\frac{3}{4} & \frac{1}{2} & -\frac{1}{4} & \frac{1}{4} & \frac{1}{2} \\ 0 & -\frac{1}{11} & -\frac{2}{11} & 0 & \frac{5}{11} \end{bmatrix}.$$

35. (a) $[B' \ B] = \begin{bmatrix} -12 & -4 & 1 & -2 \\ 0 & 4 & 3 & -2 \end{bmatrix} \Rightarrow \begin{bmatrix} 1 & 0 & -\frac{1}{3} & \frac{1}{3} \\ 0 & 1 & \frac{3}{4} & -\frac{1}{2} \end{bmatrix} = [I \ P^{-1}]$

(b) $[B \ B'] = \begin{bmatrix} 1 & -2 & -12 & -4 \\ 3 & -2 & 0 & 4 \end{bmatrix} \Rightarrow \begin{bmatrix} 1 & 0 & 6 & 4 \\ 0 & 1 & 9 & 4 \end{bmatrix} = [I \ P]$

(c) $P^{-1}P = \begin{bmatrix} -\frac{1}{3} & \frac{1}{3} \\ \frac{3}{4} & -\frac{1}{2} \end{bmatrix}\begin{bmatrix} 6 & 4 \\ 9 & 4 \end{bmatrix} = \begin{bmatrix} 1 & 0 \\ 0 & 1 \end{bmatrix}$

(d) $[\mathbf{x}]_B = P[\mathbf{x}]_{B'} = \begin{bmatrix} 6 & 4 \\ 9 & 4 \end{bmatrix}\begin{bmatrix} -1 \\ 3 \end{bmatrix} = \begin{bmatrix} 6 \\ 3 \end{bmatrix}$

© 2013 Cengage Learning. All Rights Reserved. May not be scanned, copied or duplicated, or posted to a publicly accessible website, in whole or in part.

37. (a) $[B' \ B] = \begin{bmatrix} 2 & 1 & 0 & 1 & 0 & 1 \\ 1 & 0 & 2 & 0 & 1 & 1 \\ 1 & 0 & 1 & 2 & 3 & 1 \end{bmatrix} \Rightarrow \begin{bmatrix} 1 & 0 & 0 & 4 & 5 & 1 \\ 0 & 1 & 0 & -7 & -10 & -1 \\ 0 & 0 & 1 & -2 & -2 & 0 \end{bmatrix} = [I \ P^{-1}]$

(b) $[B \ B'] = \begin{bmatrix} 1 & 0 & 1 & 2 & 1 & 0 \\ 0 & 1 & 1 & 1 & 0 & 2 \\ 2 & 3 & 1 & 1 & 0 & 1 \end{bmatrix} \Rightarrow \begin{bmatrix} 1 & 0 & 0 & \frac{1}{2} & \frac{1}{2} & -\frac{5}{4} \\ 0 & 1 & 0 & -\frac{1}{2} & -\frac{1}{2} & \frac{3}{4} \\ 0 & 0 & 1 & \frac{3}{2} & \frac{1}{2} & \frac{5}{4} \end{bmatrix} = [I \ P]$

(c) $P^{-1}P = \begin{bmatrix} 4 & 5 & 1 \\ -7 & -10 & -1 \\ -2 & -2 & 0 \end{bmatrix} \begin{bmatrix} \frac{1}{2} & \frac{1}{2} & -\frac{5}{4} \\ -\frac{1}{2} & -\frac{1}{2} & \frac{3}{4} \\ \frac{3}{2} & \frac{1}{2} & \frac{5}{4} \end{bmatrix} = \begin{bmatrix} 1 & 0 & 0 \\ 0 & 1 & 0 \\ 0 & 0 & 1 \end{bmatrix}$

(d) $[\mathbf{x}]_B = P[\mathbf{x}]_{B'} = \begin{bmatrix} \frac{1}{2} & \frac{1}{2} & -\frac{5}{4} \\ -\frac{1}{2} & -\frac{1}{2} & \frac{3}{4} \\ \frac{3}{2} & \frac{1}{2} & \frac{5}{4} \end{bmatrix} \begin{bmatrix} 1 \\ 2 \\ -1 \end{bmatrix} = \begin{bmatrix} \frac{11}{4} \\ -\frac{9}{4} \\ \frac{5}{4} \end{bmatrix}$

39. (a) $[B' \ B] = \begin{bmatrix} 1 & 4 & 2 & 4 & 6 & 2 \\ 0 & 2 & 5 & 2 & -5 & -1 \\ 4 & 8 & -2 & -4 & -6 & 8 \end{bmatrix}$

$[I \ P^{-1}] = \begin{bmatrix} 1 & 0 & 0 & -\frac{48}{5} & -24 & \frac{4}{5} \\ 0 & 1 & 0 & 4 & 10 & \frac{1}{2} \\ 0 & 0 & 1 & -\frac{6}{5} & -5 & -\frac{2}{5} \end{bmatrix}$

So, the transition matrix from B to B' is

$P^{-1} = \begin{bmatrix} -\frac{48}{5} & -24 & \frac{4}{5} \\ 4 & 10 & \frac{1}{2} \\ -\frac{6}{5} & -5 & -\frac{2}{5} \end{bmatrix}.$

(b) $[B \ B'] = \begin{bmatrix} 4 & 6 & 2 & 1 & 4 & 2 \\ 2 & -5 & -1 & 0 & 2 & 5 \\ -4 & -6 & 8 & 4 & 8 & -2 \end{bmatrix}$

$[I \ P] = \begin{bmatrix} 1 & 0 & 0 & \frac{3}{32} & \frac{17}{20} & \frac{5}{4} \\ 0 & 1 & 0 & \frac{1}{16} & -\frac{3}{10} & -\frac{1}{2} \\ 0 & 0 & 1 & \frac{1}{2} & \frac{6}{5} & 0 \end{bmatrix}$

So, the transition matrix from B' to B is

$P = \begin{bmatrix} \frac{3}{32} & \frac{17}{20} & \frac{5}{4} \\ \frac{1}{16} & -\frac{3}{10} & -\frac{1}{2} \\ \frac{1}{2} & \frac{6}{5} & 0 \end{bmatrix}.$

(c) Using a graphing utility, you have $PP^{-1} = I$.

(d) $[\mathbf{x}]_B = P[\mathbf{x}]_{B'} = P\begin{bmatrix} 1 \\ -1 \\ 2 \end{bmatrix} = \begin{bmatrix} \frac{279}{160} \\ -\frac{61}{80} \\ -\frac{7}{10} \end{bmatrix}$

41. (a) $[B' \ B] = \begin{bmatrix} 0 & -1 & -3 & 2 & 0 & 1 \\ -1 & 3 & -2 & 0 & -1 & -3 \\ -3 & -2 & 0 & -1 & 3 & -2 \end{bmatrix}$

$[I \ P^{-1}] = \begin{bmatrix} 1 & 0 & 0 & \frac{19}{39} & -\frac{9}{13} & \frac{44}{39} \\ 0 & 1 & 0 & -\frac{3}{13} & -\frac{6}{13} & -\frac{9}{13} \\ 0 & 0 & 1 & -\frac{23}{39} & \frac{2}{13} & -\frac{4}{39} \end{bmatrix}$

So, the transition matrix from B to B' is

$P^{-1} = \begin{bmatrix} \frac{19}{39} & -\frac{9}{13} & \frac{44}{39} \\ -\frac{3}{13} & -\frac{6}{13} & -\frac{9}{13} \\ -\frac{23}{39} & \frac{2}{13} & -\frac{4}{39} \end{bmatrix}.$

(b) $[B \ B'] = \begin{bmatrix} 2 & 0 & 1 & 0 & -1 & -3 \\ 0 & -1 & -3 & -1 & 3 & -2 \\ -1 & 3 & -2 & -3 & -2 & 0 \end{bmatrix}$

$[I \ P] = \begin{bmatrix} 1 & 0 & 0 & -\frac{2}{7} & -\frac{4}{21} & -\frac{13}{7} \\ 0 & 1 & 0 & -\frac{5}{7} & -\frac{8}{7} & -\frac{1}{7} \\ 0 & 0 & 1 & \frac{4}{7} & -\frac{13}{21} & \frac{5}{7} \end{bmatrix}$

So, the transition matrix from B' to B is

$P = \begin{bmatrix} -\frac{2}{7} & -\frac{4}{21} & -\frac{13}{7} \\ -\frac{5}{7} & -\frac{8}{7} & -\frac{1}{7} \\ \frac{4}{7} & -\frac{13}{21} & \frac{5}{7} \end{bmatrix}.$

(c) Using a graphing utility, you have $PP^{-1} = I$.

(d) $[\mathbf{x}]_B = P[\mathbf{x}]_{B'} = P\begin{bmatrix} 4 \\ -3 \\ -2 \end{bmatrix} = \begin{bmatrix} \frac{22}{7} \\ \frac{6}{7} \\ \frac{19}{7} \end{bmatrix}$

© 2013 Cengage Learning. All Rights Reserved. May not be scanned, copied or duplicated, or posted to a publicly accessible website, in whole or in part.

43. The standard basis for P_3 is $S = \{1, x, x^2, x^3\}$ and because

$$p = 4(1) + 11(x) + 1(x^2) + 2(x^3)$$

it follows that

$$[p]_S = \begin{bmatrix} 4 \\ 11 \\ 1 \\ 2 \end{bmatrix}.$$

45. The standard basis for P_3 is $S = \{1, x, x^2, x^3\}$ and because

$$p = 1(1) + 5(x) - 2(x^2) + 1(x^3)$$

it follows that

$$[p]_S = \begin{bmatrix} 1 \\ 5 \\ -2 \\ 1 \end{bmatrix}.$$

47. The standard basis in $M_{3,1}$ is

$$S = \left\{\begin{bmatrix} 1 \\ 0 \\ 0 \end{bmatrix}, \begin{bmatrix} 0 \\ 1 \\ 0 \end{bmatrix}, \begin{bmatrix} 0 \\ 0 \\ 1 \end{bmatrix}\right\}$$

and because

$$X = 0\begin{bmatrix} 1 \\ 0 \\ 0 \end{bmatrix} + 3\begin{bmatrix} 0 \\ 1 \\ 0 \end{bmatrix} + 2\begin{bmatrix} 0 \\ 0 \\ 1 \end{bmatrix}$$

it follows that

$$[X]_S = \begin{bmatrix} 0 \\ 3 \\ 2 \end{bmatrix}.$$

49. The standard basis in $M_{3,1}$ is

$$S = \left\{\begin{bmatrix} 1 \\ 0 \\ 0 \end{bmatrix}, \begin{bmatrix} 0 \\ 1 \\ 0 \end{bmatrix}, \begin{bmatrix} 0 \\ 0 \\ 1 \end{bmatrix}\right\}$$

and because

$$X = 1\begin{bmatrix} 1 \\ 0 \\ 0 \end{bmatrix} + 2\begin{bmatrix} 0 \\ 1 \\ 0 \end{bmatrix} - 1\begin{bmatrix} 0 \\ 0 \\ 1 \end{bmatrix}$$

it follows that

$$[X]_S = \begin{bmatrix} 1 \\ 2 \\ -1 \end{bmatrix}.$$

51. (a) False. See Theorem 4.20, page 204.

(b) True. See paragraph before Example 1, page 202.

53. If P is the transition matrix from B'' to B', then $P[\mathbf{x}]_{B''} = [\mathbf{x}]_{B'}$. If Q is the transition matrix from B' to B, then $Q[\mathbf{x}]_{B'} = [\mathbf{x}]_B$. So,

$$[\mathbf{x}]_B = Q[\mathbf{x}]_{B'} = QP[\mathbf{x}]_{B''}$$

which means that QP is the transition matrix from B'' to B.

Section 4.8 Applications of Vector Spaces

1. (a) If $y = e^x$, then $y'' = e^x$ and $y'' + y = 2e^x \neq 0$. So, e^x *is not* a solution of the equation.

(b) If $y = \sin x$, then $y'' = -\sin x$ and $y'' + y = 0$. So, $\sin x$ *is* a solution of the equation.

(c) If $y = \cos x$, then $y'' = -\cos x$ and $y'' + y = 0$. So, $\cos x$ *is* a solution of the equation.

(d) If $y = \sin x - \cos x$, then $y'' = -\sin x + \cos x$ and $y'' + y = 0$. So, $\sin x - \cos x$ *is* a solution of the equation.

3. (a) If $y = x$, then $y' = 1$, $y'' = y''' = 0$ and $y''' + y'' + y' + y = x + 1 \neq 0$. So, x *is not* a solution of the equation.

(b) If $y = e^x$, then $y' = y'' = y''' = e^x$ and $y''' + y'' + y' + y = 4e^x \neq 0$. So, e^x *is not* a solution of the equation.

(c) If $y = e^{-x}$, then $y' = y''' = -e^x$, $y'' = e^x$ and $y''' + y'' + y' + y = 0$. So, e^{-x} *is* a solution of the equation.

(d) If $y = xe^{-x}$, $y' = (1 - x)e^{-x}$, $y'' = (x - 2)e^{-x}$, $y''' = (3 - x)e^{-x}$, and $y''' + y'' + y' + y = 2e^{-x} \neq 0$. So, xe^{-x} *is not* a solution of the equation.

© 2013 Cengage Learning. All Rights Reserved. May not be scanned, copied or duplicated, or posted to a publicly accessible website, in whole or in part.

5. (a) If $y = 1$, then $y'' = y''' = y^{(4)} = 0$ and $y^{(4)} + y''' - 2y'' = 0$. So, 1 *is* a solution of the equation.

(b) If $y = x$, then $y'' = y''' = y^{(4)} = 0$ and $y^{(4)} + y''' - 2y'' = 0$. So, x *is* a solution of the equation.

(c) If $y = x^2$, then $y'' = 2$, $y''' = y^{(4)} = 0$ and $y^{(4)} + y''' - 2y'' = -4 \neq 0$. So, x^2 *is not* a solution of the equation.

(d) If $y = e^x$, then $y'' = y''' = y^{(4)} = e^x$ and $y^{(4)} + y''' - 2y'' = 0$. So, e^x *is* a solution of the equation.

7. (a) If $y = \frac{1}{x^2}$, then $y'' = \frac{6}{x^4}$. So, $x^2y'' - 2y = x^2\left(\frac{6}{x^4}\right) - 2\left(\frac{1}{x^2}\right) \neq 0$, and $y = \frac{1}{x^2}$ *is not* a solution of the equation.

(b) If $y = x^2$, then $y'' = 2$. So, $x^2y'' - 2y = x^2(2) - 2x^2 = 0$, and $y = x^2$ *is* a solution of the equation.

(c) If $y = e^{x^2}$, then $y'' = 4x^2e^{x^2} + 2e^{x^2}$. So, $x^2y'' - 2y = x^2\left(4x^2e^{x^2} + 2e^{x^2}\right) - 2\left(e^{x^2}\right) \neq 0$, and $y = e^{x^2}$ *is not* a solution of the equation.

(d) If $y = e^{-x^2}$, then $y'' = 4x^2e^{-x^2} - 2e^{-x^2}$. So, $x^2y'' - 2y = x^2\left(4x^2e^{-x^2} - 2e^{-x^2}\right) - 2\left(e^{-x^2}\right) \neq 0$, and $y = e^{-x^2}$ *is not* a solution of the equation.

9. (a) If $y = \sqrt{x}$, then $y' = \frac{1}{2}x^{-1/2}$ and $xy' - 2y = -\frac{3}{2}\sqrt{x} \neq 0$. So, $\sqrt{x}$ *is not* a solution of the equation.

(b) If $y = x$, then $y' = 1$ and $xy' - 2y = -x \neq 0$. So, x *is not* a solution of the equation.

(c) If $y = x^2$, then $y' = 2x$ and $xy' - 2y = 0$. So, x^2 *is* a solution of the equation.

(d) If $y = x^3$, then $y' = 3x^2$ and $xy' - 2y = x^3 \neq 0$. So, x^3 *is not* a solution of the equation.

11. (a) If $y = xe^{2x}$, then $y' = 2xe^{2x} + e^{2x}$ and $y'' = 4xe^{2x} + 4e^{2x}$. So,

$y'' - y' - 2y = 4xe^{2x} + 4e^{2x} - \left(2xe^{2x} + e^{2x}\right) - 2\left(xe^{2x}\right) \neq 0$, and $y = xe^{2x}$ *is not* a solution of the equation.

(b) If $y = 2e^{2x}$, then $y' = 4e^{2x}$ and $y'' = 8e^{2x}$. So, $y'' - y' - 2y = 8e^{2x} - 4e^{2x} - 2\left(2e^{2x}\right) = 0$, and $y = 2e^{2x}$ *is* a solution of the equation.

(c) If $y = 2e^{-2x}$, then $y' = -4e^{-2x}$ and $y'' = 8e^{-2x}$. So, $y'' - y' - 2y = 8e^{-2x} - \left(-4e^{-2x}\right) - 2\left(2e^{-2x}\right) \neq 0$, and $y = 2e^{-2x}$ *is not* a solution of the equation.

(d) If $y = xe^{-x}$, then $y' = e^{-x} - xe^{-x}$ and $y'' = xe^{-x} - 2e^{-x}$. So,

$y'' - y' - 2y = xe^{-x} - 2e^{-x} - \left(e^{-x} - xe^{-x}\right) - 2\left(xe^{-x}\right) \neq 0$, and $y = xe^{-x}$ *is not* a solution of the equation.

13. $$W(x, \cos x) = \begin{vmatrix} x & \cos x \\ 1 & -\sin x \end{vmatrix} = -x\sin x - \cos x$$

15. $$W\left(e^x, e^{-x}\right) = \begin{vmatrix} e^x & e^{-x} \\ \frac{d}{dx}\left(e^x\right) & \frac{d}{dx}\left(e^{-x}\right) \end{vmatrix} = \begin{vmatrix} e^x & e^{-x} \\ e^x & -e^{-x} \end{vmatrix} = -2$$

17. $$W(x, \sin x, \cos x) = \begin{vmatrix} x & \sin x & \cos x \\ 1 & \cos x & -\sin x \\ 0 & -\sin x & -\cos x \end{vmatrix} = x\left(-\cos^2 x - \sin^2 x\right) - 1(0) = -x$$

19. $$W\left(e^{-x}, xe^{-x}, (x+3)e^{-x}\right) = \begin{vmatrix} e^{-x} & xe^{-x} & (x+3)e^{-x} \\ -e^{-x} & (1-x)e^{-x} & (-x-2)e^{-x} \\ e^{-x} & (x-2)e^{-x} & (x+1)e^{-x} \end{vmatrix} = e^{-3x}\begin{vmatrix} 1 & x & x+3 \\ -1 & 1-x & -x-2 \\ 1 & x-2 & x+1 \end{vmatrix} = e^{-3x}\begin{vmatrix} 1 & x & x+3 \\ 0 & 1 & 1 \\ 0 & -2 & -2 \end{vmatrix} = 0$$

21. $$W\left(1, e^x, e^{2x}\right) = \begin{vmatrix} 1 & e^x & e^{2x} \\ 0 & e^x & 2e^{2x} \\ 0 & e^x & 4e^{2x} \end{vmatrix} = 4e^{3x} - 2e^{3x} = 2e^{3x}$$

© 2013 Cengage Learning. All Rights Reserved. May not be scanned, copied or duplicated, or posted to a publicly accessible website, in whole or in part.

23. $W(1, x, x^2, x^3) = \begin{vmatrix} 1 & x & x^2 & x^3 \\ 0 & 1 & 2x & 3x^2 \\ 0 & 0 & 2 & 6x \\ 0 & 0 & 0 & 6 \end{vmatrix} = 12$

25. $W(1, x, \cos x, e^{-x}) = \begin{vmatrix} 1 & x & \cos x & e^{-x} \\ 0 & 1 & -\sin x & -e^{-x} \\ 0 & 0 & -\cos x & e^{-x} \\ 0 & 0 & \sin x & -e^{-x} \end{vmatrix} = e^{-x}\cos x - e^{-x}\sin x$

27. (a) $y = \sin 4x \Rightarrow y'' = -16\sin 4x \Rightarrow y'' + 16y = 0$

$y = \cos 4x \Rightarrow y'' = -16\sin 4x \Rightarrow y'' + 16y = 0$

(b) Because $W(\sin 4x, \cos 4x) = \begin{vmatrix} \sin 4x & \cos 4x \\ 4\cos 4x & -4\sin 4x \end{vmatrix} = -4\sin^2 4x - 4\cos^2 4x = -4 \neq 0,$

the set is linearly independent.

(c) $y = C_1 \sin 4x + C_2 \cos 4x$

29. (a) $y = e^{-2x} \Rightarrow y' = -2e^{-2x}, y'' = 4e^{-2x}, y''' = -8e^{-2x} \Rightarrow y''' + 4y'' + 4y' = 0$

$y = xe^{-2x} \Rightarrow y' = (1 - 2x)e^{-2x}, y'' = (4x - 4)e^{-2x}, y''' = (12 - 8x)e^{-2x} \Rightarrow y''' + 4y'' + 4y' = 0$

$y = (2x + 1)e^{-2x} \Rightarrow y' = -4xe^{-2x}, y'' = (8x - 4)e^{-2x}, y''' = (16 - 16x)e^{-2x} \Rightarrow y''' + 4y'' + 4y' = 0$

(b) Because $W(e^{-2x}, xe^{-2x}, (2x + 1)e^{-2x}) = \begin{vmatrix} e^{-2x} & xe^{-2x} & (2x + 3)e^{-2x} \\ -2e^{-2x} & (1 - 2x)e^{-2x} & -4xe^{-2x} \\ 4e^{-2x} & (4x - 4)e^{-2x} & (8x - 4)e^{-2x} \end{vmatrix}$

$$= e^{-6x}\begin{vmatrix} 1 & x & 2x + 1 \\ -2 & 1 - 2x & -4x \\ 4 & 4x - 4 & 8x - 4 \end{vmatrix}$$

$$= e^{-6x}\begin{vmatrix} 1 & x & 2x + 1 \\ 0 & 1 & 2 \\ 0 & -4 & -8 \end{vmatrix}$$

$$= 0,$$

the set is linearly dependent.

(c) Not applicable

31. (a) $y = 1 \Rightarrow y' = y''' = 0 \Rightarrow y''' + 4y' = 0$

$y = \sin 2x \Rightarrow y' = 2\cos 2x, y''' = -8\cos 2x \Rightarrow y''' + 4y' = 0$

$y = \cos 2x \Rightarrow y' = -2\sin 2x, y''' = 8\cos 2x \Rightarrow y''' + 4y' = 0$

(b) Because $W(1, \sin 2x, \cos 2x) = \begin{vmatrix} 1 & \sin 2x & \cos 2x \\ 0 & 2\cos 2x & -2\sin 2x \\ 0 & -4\sin 2x & -4\cos 2x \end{vmatrix}$

$$= -8\cos^2 2x - 8\sin^2 2x = -8 \neq 0,$$

the set is linearly independent.

(c) $y = C_1 + C_2 \sin 2x + C_3 \cos 2x$

© 2013 Cengage Learning. All Rights Reserved. May not be scanned, copied or duplicated, or posted to a publicly accessible website, in whole or in part.

33. (a) $y = e^{-x} \Rightarrow y' = -e^{-x}, y'' = e^{-x}, y''' = -e^{-x} \Rightarrow y''' + 3y'' + 3y' + y = 0$

$y = xe^{-x} \Rightarrow y' = (1 - x)e^{-x}, y'' = (x - 2)e^{-x}, y''' = (3 - x)e^{-x} \Rightarrow y''' + 3y'' + 3y' + y = 0$

$y = e^{-x} + xe^{-x} \Rightarrow y' = -xe^{-x}, y'' = (x - 1)e^{-x}, y''' = (2 - x)e^{-x} \Rightarrow y''' + 3y'' + 3y' + y = 0$

(b) Note that $e^{-x} + xe^{-x}$ is the sum of the first two expressions in the set. So, the set is linearly dependent.

(c) Not applicable

35. (a) $\theta(t) = \sin\sqrt{\frac{g}{L}}t \Rightarrow \frac{d^2\theta}{dt^2} = -\frac{g}{L}\sin\sqrt{\frac{g}{L}}t \Rightarrow \frac{d^2\theta}{dt^2} + \frac{g}{L}\theta = 0$

$\theta(t) = \cos\sqrt{\frac{g}{L}}t \Rightarrow \frac{d^2\theta}{dt^2} = -\frac{g}{L}\cos\sqrt{\frac{g}{L}}t \Rightarrow \frac{d^2\theta}{dt^2} + \frac{g}{L}\theta = 0$

Because $W\left(\sin\sqrt{\frac{g}{L}}t, \cos\sqrt{\frac{g}{L}}t\right) = \begin{vmatrix} \sin\sqrt{\frac{g}{L}}t & \cos\sqrt{\frac{g}{L}}t \\ \sqrt{\frac{g}{L}}\cos\sqrt{\frac{g}{L}}t & -\sqrt{\frac{g}{L}}\sin\sqrt{\frac{g}{L}}t \end{vmatrix}$

$$= -\sqrt{\frac{g}{L}}\sin^2\sqrt{\frac{g}{L}}t - \sqrt{\frac{g}{L}}\cos^2\sqrt{\frac{g}{L}}t$$

$$= -\sqrt{\frac{g}{L}} \neq 0,$$

the solutions are linearly independent.

(b) $\frac{d^2\theta}{dt^2} + \frac{g}{L}\theta = 0, \frac{g}{L} > 0$

$$\theta(t) = C_1\sin\left(\sqrt{\frac{g}{L}}t\right) + C_2\cos\left(\sqrt{\frac{g}{L}}t\right)$$

Let ϕ be given by $\tan\left(\sqrt{\frac{g}{L}}\phi\right) = -\frac{C_1}{C_2}, -\frac{\pi}{2} < \phi < \frac{\pi}{2}$.

Then $C_2\sin\left(\sqrt{\frac{g}{L}}\phi\right) = -C_1\cos\left(\sqrt{\frac{g}{L}}\phi\right)$.

Let $A = \dfrac{C_2}{\cos\left(\sqrt{\frac{g}{L}}\phi\right)} = -\dfrac{C_1}{\sin\left(\sqrt{\frac{g}{L}}\phi\right)}$

$$\theta(t) = C_1\sin\left(\sqrt{\frac{g}{L}}t\right) + C_2\cos\left(\sqrt{\frac{g}{L}}t\right) = -A\sin\left(\sqrt{\frac{g}{L}}\phi\right)\sin\left(\sqrt{\frac{g}{L}}t\right) + A\cos\left(\sqrt{\frac{g}{L}}\phi\right)\cos\left(\sqrt{\frac{g}{L}}t\right) = A\cos\left[\sqrt{\frac{g}{L}}(t + \phi)\right]$$

37. First, calculate the Wronskian of the two functions.

$$W(e^{ax}, e^{bx}) = \begin{vmatrix} e^{ax} & e^{bx} \\ ae^{ax} & be^{bx} \end{vmatrix} = (b - a)e^{(a+b)x}$$

If $a \neq b$, then $W(e^{ax}, e^{bx}) \neq 0$. Because e^{ax} and e^{bx} are solutions to $y'' - (a + b)y' + aby = 0$, the functions are linearly independent. On the other hand, if $a = b$, then $e^{ax} = e^{bx}$, and the functions are linearly dependent.

39. First, calculate the Wronskian.

$$W(e^{ax}\cos bx, e^{ax}\sin bx) = \begin{vmatrix} e^{ax}\cos bx & e^{ax}\sin bx \\ e^{ax}(a\cos bx - b\sin bx) & e^{ax}(a\sin bx + b\cos bx) \end{vmatrix}$$

$$= be^{2ax} \neq 0, \qquad \text{because } b \neq 0$$

Because these functions satisfy the differential equation $y'' - 2ay' + (a^2 + b^2)y = 0$, they are linearly independent.

© 2013 Cengage Learning. All Rights Reserved. May not be scanned, copied or duplicated, or posted to a publicly accessible website, in whole or in part.

41. No. For instance, consider the nonhomogeneous differential equation $y'' = 1$. Two solutions are

$y_1 = \dfrac{x^2}{2}$ and $y_2 = \dfrac{x^2}{2} + 1$, but $y_1 + y_2$ is not a solution.

43. The graph of this equation is a parabola $x = -y^2$ with the vertex at the origin. The parabola opens to the left.

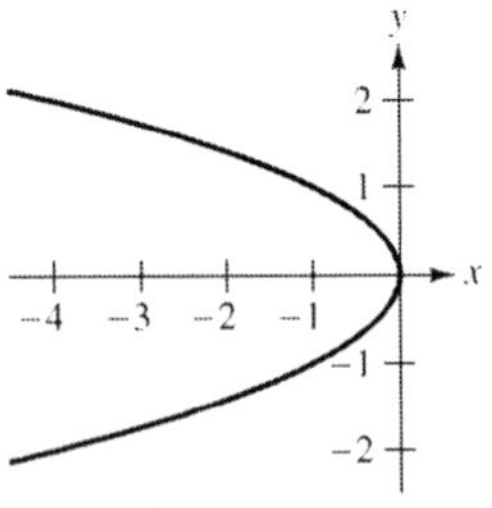

$y^2 + x = 0$

45. First, rewrite the equation.

$$\frac{x^2}{16} + \frac{y^2}{4} = 1$$

You see that this is the equation of an ellipse centered at the origin, with major axis falling along the x-axis.

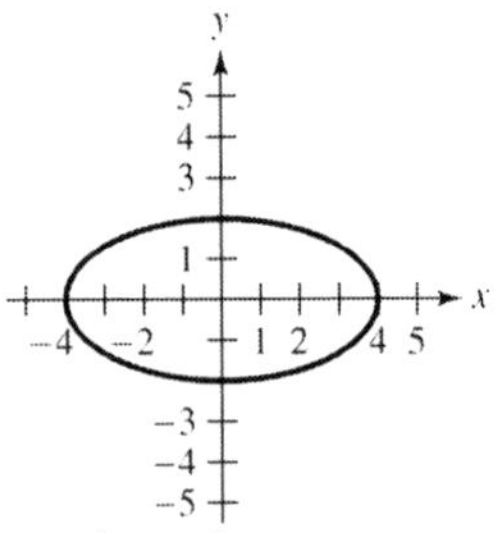

$x^2 + 4y^2 - 16 = 0$

47. First, rewrite the equation.

$$\frac{x^2}{9} - \frac{y^2}{16} = 1$$

The graph of this equation is a hyperbola centered at the origin, with transverse axis along the x-axis.

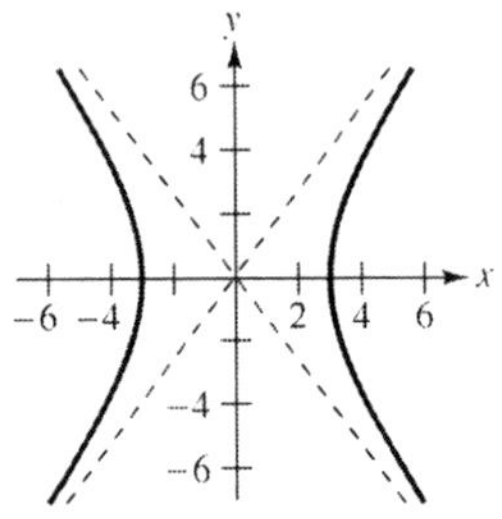

$\frac{x^2}{9} - \frac{y^2}{16} - 1 = 0$

49. First, complete the square to find the standard form.

$$(x - 1)^2 = 4(-2)(y + 2)$$

You see that this is the equation of a parabola, with vertex at $(1, -2)$ and opening downward.

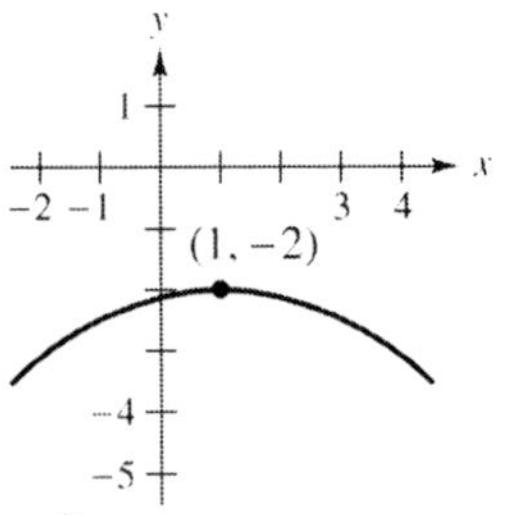

$x^2 - 2x + 8y + 17 = 0$

51. First, complete the square to find the standard form.

$$(3x - 6)^2 + (5y - 5)^2 = 0$$

The graph of this equation is the single point $(2, 1)$.

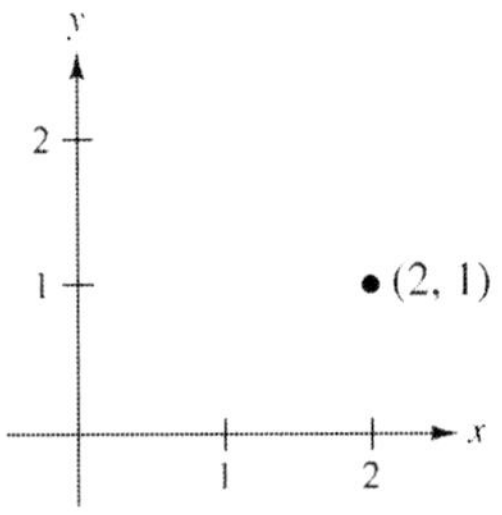

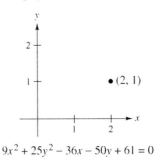

$9x^2 + 25y^2 - 36x - 50y + 61 = 0$

53. First, complete the square to find the standard form.

$$\frac{(x + 3)^2}{\left(\frac{1}{3}\right)^2} - \frac{(y - 5)^2}{1} = 1$$

You see that this is the equation of a hyperbola centered at $(-3, 5)$, with transverse axis parallel to the x-axis.

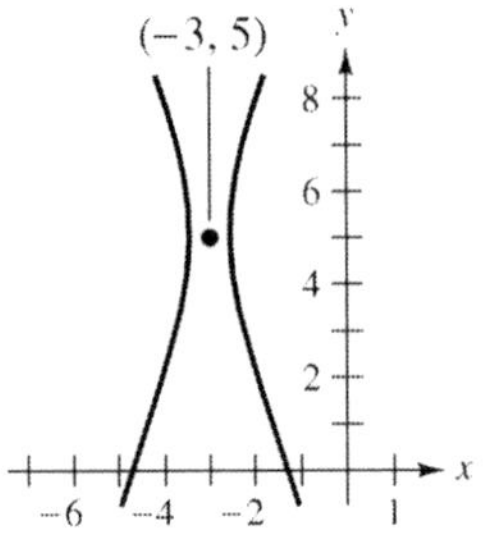

$9x^2 - y^2 + 54x + 10y + 55 = 0$

© 2013 Cengage Learning. All Rights Reserved. May not be scanned, copied or duplicated, or posted to a publicly accessible website, in whole or in part.

55. First, complete the square to find the standard form.

$$\frac{(x+2)^2}{2^2} + \frac{(y+4)^2}{1^2} = 1$$

You see that this is the equation of an ellipse centered at $(-2, -4)$, with major axis parallel to the x-axis.

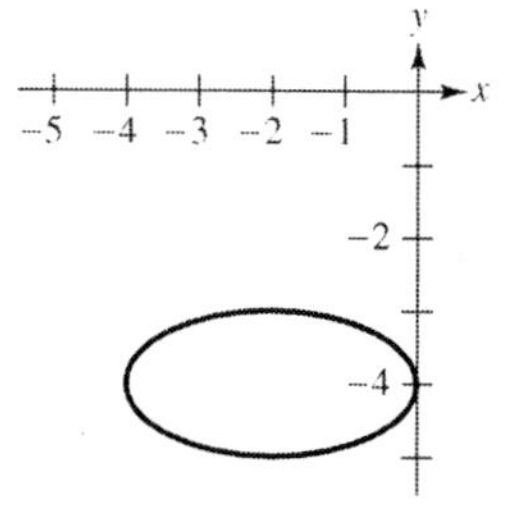

$x^2 + 4y^2 + 4x + 32y + 64 = 0$

57. First, complete the square to find the standard form.

$$\frac{(y-5)^2}{1} - \frac{(x+1)^2}{\frac{1}{2}} = 1$$

You see that this is the equation of a hyperbola centered at $(-1, 5)$, with transverse axis parallel to the y-axis.

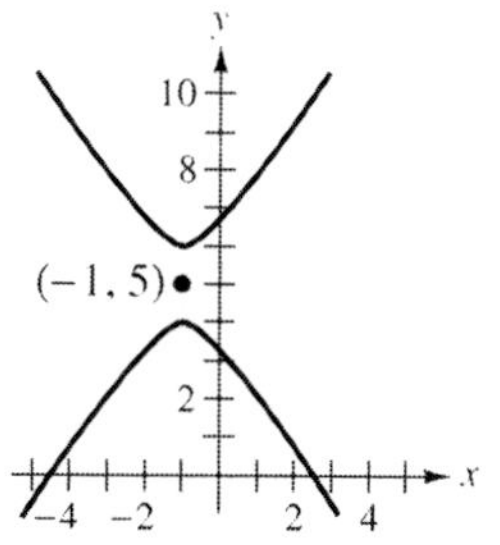

$2x^2 - y^2 + 4x + 10y - 22 = 0$

59. First, complete the square to find the standard form.

$$(x+2)^2 = -6(y-1)$$

This is the equation of a parabola with vertex at $(-2, 1)$ and opening downward.

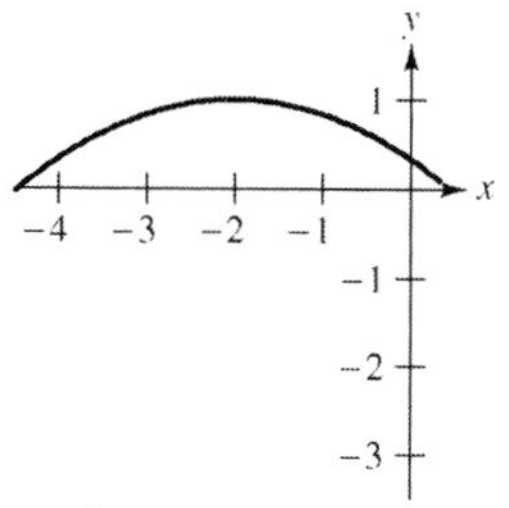

$x^2 + 4x + 6y - 2 = 0$

61. $xy + 2 = 0$

$B^2 - 4AC = 1 \Rightarrow$ The graph is a hyperbola.

$$\cot 2\theta = \frac{A - C}{B} = 0 \Rightarrow \theta = 45°$$

Matches graph (c).

63. $x^2 - xy + 3y^2 - 5 = 0$

$A = 1, B = -1, C = 3$

$B^2 - 4AC = (-1)^2 - 4(1)(3) = -11$

The graph is an ellipse.

$$\cot 2\theta = \frac{A - C}{B} = \frac{1 - 3}{-1} = 2 \Rightarrow \theta \approx 13.28°$$

Matches graph (a).

65. Begin by finding the rotation angle θ, where

$$\cot 2\theta = \frac{a - c}{b} = \frac{0 - 0}{1} = 0, \text{ implying that } \theta = \frac{\pi}{4}.$$

So, $\sin\theta = \frac{1}{\sqrt{2}}$ and $\cos\theta = \frac{1}{\sqrt{2}}$. By substituting

$$x = x'\cos\theta - y'\sin\theta = \frac{1}{\sqrt{2}}(x' - y')$$

and

$$y = x'\sin\theta + y'\cos\theta = \frac{1}{\sqrt{2}}(x' + y')$$

into

$xy + 1 = 0$

and simplifying, you obtain

$(x')^2 - (y')^2 + 2 = 0.$

In standard form,

$$\frac{(y')^2}{2} - \frac{(x')^2}{2} = 1.$$

This is the equation of a hyperbola with a transverse axis along the y'-axis.

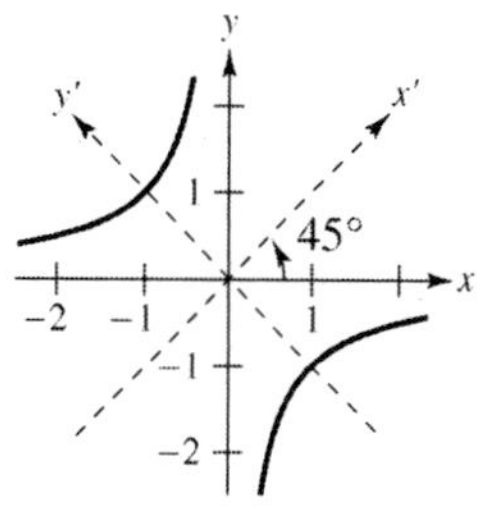

© 2013 Cengage Learning. All Rights Reserved. May not be scanned, copied or duplicated, or posted to a publicly accessible website, in whole or in part.

67. Begin by finding the rotation angle θ, where

$$\cot 2\theta = \frac{a-c}{b} = \frac{4-4}{2} = 0 \Rightarrow \theta = \frac{\pi}{4}.$$

So, $\sin\theta = \frac{1}{\sqrt{2}}$ and $\cos\theta = \frac{1}{\sqrt{2}}$. By substituting

$$x = x'\cos\theta - y'\sin\theta = \frac{1}{\sqrt{2}}(x' - y')$$

and

$$y = x'\sin\theta + y'\cos\theta = \frac{1}{\sqrt{2}}(x' + y')$$

into $4x^2 + 2xy + 4y^2 - 15 = 0$ and simplifying, you obtain $\frac{(x')^2}{3} + \frac{(y')^2}{5} = 1$, which is an ellipse with major axis along the y'-axis.

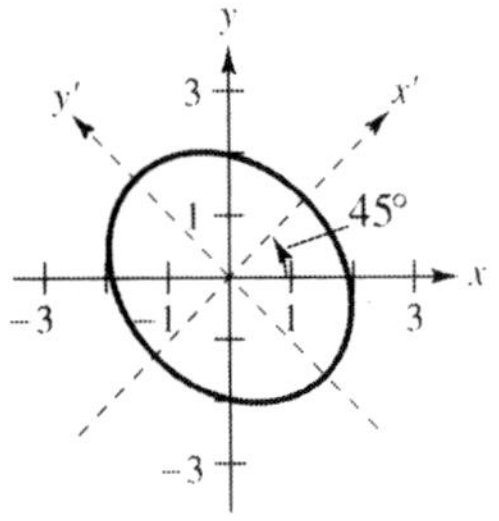

69. Begin by finding the rotation angle θ, where

$$\cot 2\theta = \frac{a-c}{b} = -\frac{4}{3}, \text{ implying that } \theta \approx 71.57°.$$

So, $\sin\theta = \frac{3}{\sqrt{10}}$ and $\cos\theta = \frac{1}{\sqrt{10}}$. By substituting

$$x = \frac{x' - 3y'}{\sqrt{10}} \text{ and } y = \frac{3x' + y'}{\sqrt{10}}$$

into $2x^2 - 3xy - 2y^2 + 10 = 0$ and simplifying, you obtain $\frac{(x')^2}{4} - \frac{(y')^2}{4} = 1$, which is a hyperbola with a transverse axis along the x'-axis.

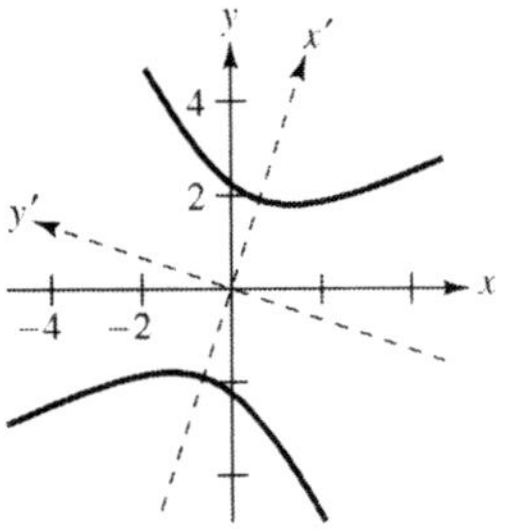

71. Begin by finding the rotation angle θ, where

$$\cot 2\theta = \frac{a-c}{b} = -\frac{7}{24}, \text{ implying that } \theta \approx 53.13°.$$

So, $\sin\theta = \frac{4}{5}$ and $\cos\theta = \frac{3}{5}$. By substituting

$$x = \frac{3x' - 4y'}{5} \text{ and } y = \frac{4x' + 3y}{5}$$

into $9x^2 + 24xy + 16y^2 + 80x - 60y = 0$ and simplifying, you obtain $\frac{1}{4}(x')^2 = y'$, which is a parabola.

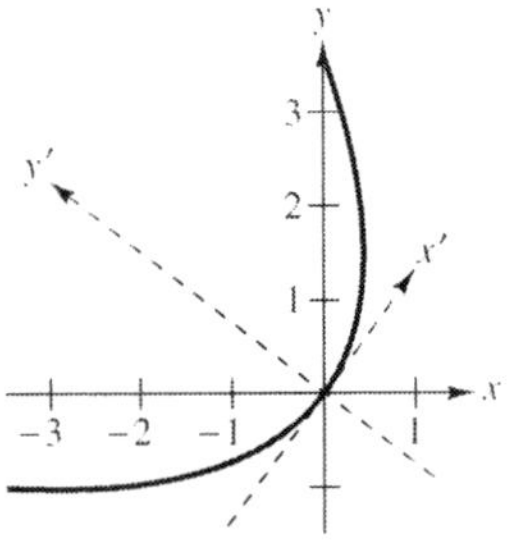

73. Begin by finding the rotation angle θ, where

$$\cot 2\theta = \frac{a-c}{b} = \frac{13-7}{6\sqrt{3}} = \frac{1}{\sqrt{3}} \Rightarrow 2\theta = \frac{\pi}{3} \Rightarrow \theta = \frac{\pi}{6}.$$

So, $\sin\theta = \frac{1}{2}$ and $\cos\theta = \frac{\sqrt{3}}{2}$. By substituting

$$x = x'\cos\theta - y'\sin\theta = \frac{\sqrt{3}}{2}x' - \frac{1}{2}y'$$

and

$$y = x'\sin\theta + y'\cos\theta = \frac{1}{2}x' + \frac{\sqrt{3}}{2}y'$$

into $13x^2 + 6\sqrt{3}xy + 7y^2 - 16 = 0$ and simplifying, you obtain $(x')^2 + \frac{(y')^2}{4} = 1$, which is an ellipse with major axis along the y'-axis.

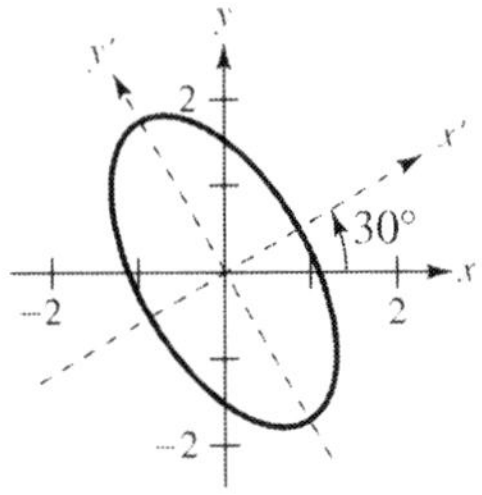

© 2013 Cengage Learning. All Rights Reserved. May not be scanned, copied or duplicated, or posted to a publicly accessible website, in whole or in part.

75. Begin by finding the rotation angle θ, where

$$\cot 2\theta = \frac{a-c}{b} = \frac{3-1}{-2\sqrt{3}} = \frac{-1}{\sqrt{3}} \Rightarrow 2\theta = \frac{2\pi}{3} \Rightarrow \theta = \frac{\pi}{3}.$$

So, $\sin\theta = \dfrac{\sqrt{3}}{2}$ and $\cos\theta = \dfrac{1}{2}$. By substituting

$$x = x'\cos\theta - y'\sin\theta = \frac{1}{2}x' - \frac{\sqrt{3}}{2}y'$$

and

$$y = x'\sin\theta + y'\cos\theta = \frac{\sqrt{3}}{2}x' + \frac{1}{2}y'$$

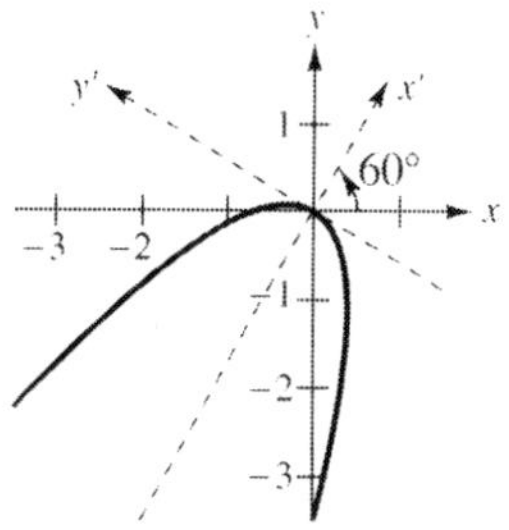

into $3x^2 - 2\sqrt{3}xy + y^2 + 2x + 2\sqrt{3}y = 0$ and simplifying, you obtain $x' = -(y')^2$, which is a parabola.

77. Begin by finding the rotation angle θ, where

$$\cot 2\theta = \frac{1-1}{-2} = 0, \text{ implying that } \theta = \frac{\pi}{4}.$$

So, $\sin\theta = \dfrac{1}{\sqrt{2}}$ and $\cos\theta = \dfrac{1}{\sqrt{2}}$. By substituting

$$x = x'\cos\theta - y'\sin\theta = \frac{1}{\sqrt{2}}(x' - y')$$

and

$$y = x'\sin\theta + y'\cos\theta = \frac{1}{\sqrt{2}}(x' + y')$$

into $x^2 - 2xy + y^2 = 0$ and simplifying, you obtain $2(y')^2 = 0$.

The graph of this equation is the line $y' = 0$.

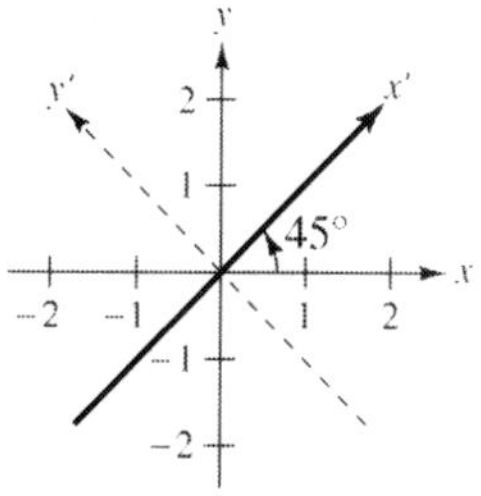

79. Begin by finding the rotation angle θ, where

$$\cot 2\theta = \frac{1-1}{2} = 0, \text{ implying that } \theta = \frac{\pi}{4}.$$

So, $\sin\theta = \dfrac{1}{\sqrt{2}}$ and $\cos\theta = \dfrac{1}{\sqrt{2}}$. By substituting

$$x = x'\cos\theta - y'\sin\theta = \frac{1}{\sqrt{2}}(x' - y')$$

and

$$y = x'\sin\theta - y'\cos\theta = \frac{1}{\sqrt{2}}(x' + y')$$

into

$x^2 + 2xy + y^2 - 1 = 0$ and simplifying, you obtain $2(x')^2 - 1 = 0$.

The graph of this equation is two lines $x' = \pm\dfrac{\sqrt{2}}{2}$.

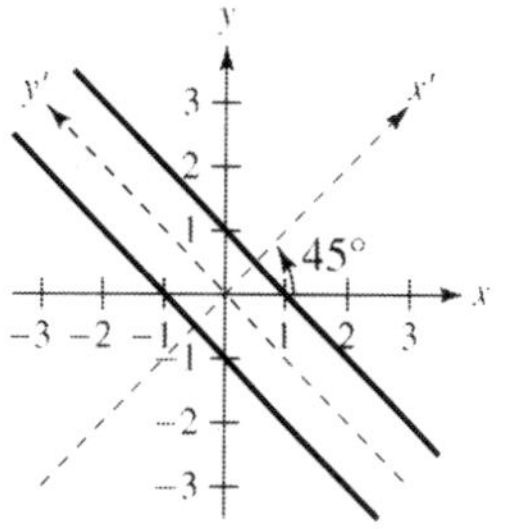

81. If $\theta = \dfrac{\pi}{4}$, then $\sin\theta = \dfrac{1}{\sqrt{2}}$ and $\cos\theta = \dfrac{1}{\sqrt{2}}$. So,

$$x = x'\cos\theta - y'\sin\theta = \frac{1}{\sqrt{2}}(x' - y')$$

and

$$y = x'\sin\theta - y'\cos\theta = \frac{1}{\sqrt{2}}(x' + y').$$

Substituting these expressions for x and y into $ax^2 + bxy + ay^2 + dx + ey + f = 0$, you obtain,

$$a\frac{1}{2}(x' - y')^2 + b\frac{1}{2}(x' - y')(x' + y') + a\frac{1}{2}(x' - y')^2 + d\frac{1}{\sqrt{2}}(x' - y') + e\frac{1}{\sqrt{2}}(x' + y') + f = 0.$$

Expanding out the first three terms, you see that $x'y'$-term has been eliminated.

© 2013 Cengage Learning. All Rights Reserved. May not be scanned, copied or duplicated, or posted to a publicly accessible website, in whole or in part.

83. (a) Let $A = \begin{bmatrix} a & \frac{b}{2} \\ \frac{b}{2} & c \end{bmatrix}$ and assume $|A| = ac - \dfrac{b^2}{4} \neq 0$. If $a = 0$, then

$ax^2 + bxy + cy^2 = bxy + cy^2 = y(cy + bx) = 0$, which implies that $y = 0$ or $y = \dfrac{-bx}{c}$,

the equations of two intersecting lines.

On the other hand, if $a \neq 0$, then you can divide $ax^2 + bxy + cy^2 = 0$ through by a to obtain

$$x^2 + \frac{b}{a}xy + \frac{c}{a}y^2 = x^2 + \frac{b}{a}xy + \left(\frac{b}{2a}\right)^2 y^2 + \frac{c}{a}y^2 - \left(\frac{b}{2a}\right)^2 y^2 = 0 \Rightarrow \left(x + \frac{b}{2a}y\right)^2 = \left(\left(\frac{b}{2a}\right)^2 - \frac{c}{a}\right)y^2.$$

Because $4ac \neq b^2$, you can see that this last equation represents two intersecting lines.

(b) If $|A| = \begin{vmatrix} a & \frac{b}{2} \\ \frac{b}{2} & c \end{vmatrix} = 0$, then $b^2 = 4ac$, and you have

$$ax^2 + bxy + cy^2 = acx^2 + bcxy + c^2y^2 = 0 \Rightarrow b^2x^2 + 4bcxy + 4c^2y^2 = 0 \Rightarrow (bx + 2cy)^2 = 0$$

$\Rightarrow bx + 2cy = 0$, which is the equation of a line.

Review Exercises for Chapter 4

1. (a) $\mathbf{u} + \mathbf{v} = (-1, 2, 3) + (1, 0, 2) = (-1 + 1, 2 + 0, 3 + 2) = (0, 2, 5)$

(b) $2\mathbf{v} = 2(1, 0, 2) = (2, 0, 4)$

(c) $\mathbf{u} - \mathbf{v} = (-1, 2, 3) - (1, 0, 2) = (-1 - 1, 2 - 0, 3 - 2) = (-2, 2, 1)$

(d) $3\mathbf{u} - 2\mathbf{v} = 3(-1, 2, 3) - 2(1, 0, 2) = (-3, 6, 9) - (2, 0, 4) = (-3 - 2, 6 - 0, 9 - 4) = (-5, 6, 5)$

3. (a) $\mathbf{u} + \mathbf{v} = (3, -1, 2, 3) + (0, 2, 2, 1) = (3, 1, 4, 4)$

(b) $2\mathbf{v} = 2(0, 2, 2, 1) = (0, 4, 4, 2)$

(c) $\mathbf{u} - \mathbf{v} = (3, -1, 2, 3) - (0, 2, 2, 1) = (3, -3, 0, 2)$

(d) $3\mathbf{u} - 2\mathbf{v} = 3(3, -1, 2, 3) - 2(0, 2, 2, 1) = (9, -3, 6, 9) - (0, 4, 4, 2) = (9, -7, 2, 7)$

5. $\mathbf{x} = \frac{1}{2}\mathbf{u} - \frac{3}{2}\mathbf{v} - \frac{1}{2}\mathbf{w}$

$= \frac{1}{2}(1, -1, 2) - \frac{3}{2}(0, 2, 3) - \frac{1}{2}(0, 1, 1)$

$= \left(\frac{1}{2}, -4, -4\right)$

7. $5\mathbf{u} - 2\mathbf{x} = 3\mathbf{v} + \mathbf{w}$

$-2\mathbf{x} = -5\mathbf{u} + 3\mathbf{v} + \mathbf{w}$

$\mathbf{x} = \frac{5}{2}\mathbf{u} - \frac{3}{2}\mathbf{v} - \frac{1}{2}\mathbf{w}$

$= \frac{5}{2}(1, -1, 2) - \frac{3}{2}(0, 2, 3) - \frac{1}{2}(0, 1, 1)$

$= \left(\frac{5}{2}, -\frac{5}{2}, 5\right) - \left(0, 3, \frac{9}{2}\right) - \left(0, \frac{1}{2}, \frac{1}{2}\right)$

$= \left(\frac{5}{2} - 0 - 0, -\frac{5}{2} - 3 - \frac{1}{2}, 5 - \frac{9}{2} - \frac{1}{2}\right)$

$= \left(\frac{5}{2}, -6, 0\right)$

9. To write $\mathbf{v}$ as a linear combination of $\mathbf{u}_1$, $\mathbf{u}_2$, and $\mathbf{u}_3$, solve the equation

$c_1\mathbf{u}_1 + c_2\mathbf{u}_2 + c_3\mathbf{u}_3 = \mathbf{v}$

for c_1, c_2, and c_3. This vector equation corresponds to the system of linear equations

$$\begin{aligned} c_1 + 2c_2 + c_3 &= 3 \\ -c_1 + 4c_2 + 2c_3 &= 0 \\ 2c_1 - 2c_2 - 4c_3 &= -6. \end{aligned}$$

The solution of this system is $c_1 = 2$, $c_2 = -1$, and $c_3 = 3$. So, $\mathbf{v} = 2\mathbf{u}_1 - \mathbf{u}_2 + 3\mathbf{u}_3$.

© 2013 Cengage Learning. All Rights Reserved. May not be scanned, copied or duplicated, or posted to a publicly accessible website, in whole or in part.

11. To write $\mathbf{v}$ as a linear combination of $\mathbf{u}_1$, $\mathbf{u}_2$, and $\mathbf{u}_3$, solve the equation

$$c_1\mathbf{u}_1 + c_2\mathbf{u}_2 + c_3\mathbf{u}_3 = \mathbf{v}$$

for c_1, c_2, and c_3. This vector equation corresponds to the system of linear equations

$$\begin{aligned} c_1 - c_2 &= 1 \\ 2c_1 - 2c_2 &= 2 \\ 3c_1 - 3c_2 + c_3 &= 3 \\ 4c_1 + 4c_2 + c_3 &= 5. \end{aligned}$$

The solution of this system is $c_1 = \frac{9}{8}$, $c_2 = \frac{1}{8}$, and $c_3 = 0$. So, $\mathbf{v} = \frac{9}{8}\mathbf{u}_1 + \frac{1}{8}\mathbf{u}_2$.

13. The zero vector is $\begin{bmatrix} 0 & 0 & 0 & 0 \\ 0 & 0 & 0 & 0 \\ 0 & 0 & 0 & 0 \end{bmatrix}$.

The additive inverse of $\begin{bmatrix} a_{11} & a_{12} & a_{13} & a_{14} \\ a_{21} & a_{22} & a_{23} & a_{24} \\ a_{31} & a_{32} & a_{33} & a_{34} \end{bmatrix}$ is

$$\begin{bmatrix} -a_{11} & -a_{12} & -a_{13} & -a_{14} \\ -a_{21} & -a_{22} & -a_{23} & -a_{24} \\ -a_{31} & -a_{32} & -a_{33} & -a_{34} \end{bmatrix}.$$

15. The zero vector is $(0, 0, 0)$. The additive inverse of a vector in R^3 is $(-a_1, -a_2, -a_3)$.

17. Because $W = \{(x, y) : x = 2y\}$ is nonempty and $W \subset R^2$, you need only check that W is closed under addition and scalar multiplication. Because

$$(2x_1, x_1) + (2x_2, x_2) = (2(x_1 + x_2), x_1 + x_2) \in W$$

and

$$c(2x_1, x_1) = (2cx_1, cx_1) \in W$$

conclude that W is a subspace of R^2.

19. W is not a space of R^2.

Because $W = \{(x, y) : y = ax, a \text{ is an integer}\}$ is nonempty and $W \subset R^2$, you need only check that W is closed under addition and scalar multiplication. Because

$$(x_1, ax_1) + (x_2, ax_2) = (x_1 + x_2, a(x_1 + x_2)) \in W$$

and

$c(x_1, ax_1) = (cx_1, acx_1)$ is not in W, ac is not necessarily an integer. So, W is not closed under scalar multiplication.

21. Because $W = \{(x, 2x, 3x) : x \text{ is a real number}\}$ is nonempty and $W \subset R^2$, you need only check that W is closed under addition and scalar multiplication. Because

$$(x_1, 2x_1, 3x_1) + (x_2, 2x_2, 3x_2) = (x_1 + x_2, 2(x_1 + x_2), 3(x_1 + x_2)) \in W$$

and

$$c(x_1, 2x_1, 3x_1) = (cx_2, 2(cx_1), 3(cx_1)) \in W$$

conclude that W is a subspace of R^3.

23. W is not a subspace of $C[-1, 1]$. For instance, $f(x) = x - 1$ and $g(x) = -1$ are in W, but their sum $(f + g)(x) = x - 2$ is not in W, because $(f + g)(0) = -2 \neq -1$. So, W is closed under addition (nor scalar multiplication).

25. (a) The only vector in W is the zero vector. So, W is nonempty and $W \subset R^3$. Furthermore, because W is closed under addition and scalar multiplication, it is a subspace of R^3.

(b) W is not closed under addition or scalar multiplication, so it is not a subspace of R^3. For example, $(1, 0, 0) \in W$, and yet

$$2(1, 0, 0) = (2, 0, 0) \notin W.$$

27. (a) To find out whether S spans R^3, form the vector equation

$$c_1(1, -5, 4) + c_2(11, 6, -1) + c_3(2, 3, 5) = (u_1, u_2, u_3).$$

This yields the system of linear equations

$$\begin{aligned} c_1 + 11c_2 + 2c_3 &= u_1 \\ -5c_1 + 6c_2 + 3c_3 &= u_2 \\ 4c_1 - c_2 + 5c_3 &= u_3. \end{aligned}$$

This system has a unique solution for every (u_1, u_2, u_3) because the determinant of the coefficient matrix is not zero. So, S spans R^3.

(b) Solving the same system in (a) with $(u_1, u_2, u_3) = (0, 0, 0)$ yields the trivial solution. So, S is linearly independent.

(c) Because S is linearly independent and S spans R^3, it is a basis for R^3.

© 2013 Cengage Learning. All Rights Reserved. May not be scanned, copied or duplicated, or posted to a publicly accessible website, in whole or in part.

29. (a) To find out whether S spans R^3, form the vector equation

$$c_1\left(-\tfrac{1}{2}, \tfrac{3}{4}, -1\right) + c_2(5, 2, 3) + c_3(-4, 6, -8) = (u_1, u_2, u_3).$$

This yields the system

$$\begin{aligned} -\tfrac{1}{2}c_1 + 5c_2 - 4c_3 &= u_1 \\ \tfrac{3}{4}c_1 + 2c_2 + 6c_3 &= u_2 \\ -c_1 + 3c_2 - 8c_3 &= u_3. \end{aligned}$$

which is equivalent to the system

$$\begin{aligned} c_1 - 10c_2 + 8c_3 &= -2u_1 \\ 7c_2 &= 2u_2 - u_3 \\ 0 &= -34u_1 + 28u_2 + 38u_3. \end{aligned}$$

So, there are vectors (u_1, u_2, u_3) not spanned by S. For instance, $(0, 0, 1) \notin \text{span}(S)$.

(b) Solving the same system in (a) for $(u_1, u_2, u_3) = (0, 0, 0)$ yields nontrivial solutions. For instance, $c_1 = -8, c_2 = 0$, and $c_3 = 1$.

So, $-8\left(-\tfrac{1}{2}, \tfrac{3}{4}, -1\right) + 0(5, 2, 3) + 1(-4, 6, -8) = (0, 0, 0)$ and S is linearly dependent.

(c) S is not a basis because it does not span R^3 nor is it linearly independent.

31. (a) S span R^3 because the first three vectors in the set form the standard basis of R^3.

(b) S is linearly dependent because the fourth vector is a linear combination of the first three

$$(-1, 2, -3) = -1(1, 0, 0) + 2(0, 1, 0) - 3(0, 0, 1).$$

(c) S is not a basis because it is not linearly independent.

33. S has four vectors, so you need only check that S is linearly independent. Form the vector equation

$$c_1(1 - t) + c_2(2t + 3t^2) + c_3(t^2 - 2t^3) + c_4(2 + t^3) = 0 + 0t + 0t^2 + 0t^3$$

which yields the homogenous system of linear equations

$$\begin{aligned} c_1 \qquad\qquad + 2c_4 &= 0 \\ -c_1 + 2c_2 \qquad\qquad &= 0 \\ 3c_2 + c_3 \qquad &= 0 \\ -2c_3 + c_4 &= 0. \end{aligned}$$

This system has only the trivial solution. So, S is linearly independent and S is a basis for P_3.

35. S has four vectors, so you need only check that S is linearly independent. Form the vector equation

$$c_1\begin{bmatrix} 1 & 0 \\ 2 & 3 \end{bmatrix} + c_2\begin{bmatrix} -2 & 1 \\ -1 & 0 \end{bmatrix} + c_3\begin{bmatrix} 3 & 4 \\ 2 & 3 \end{bmatrix} + c_4\begin{bmatrix} -3 & -3 \\ 1 & 3 \end{bmatrix} = \begin{bmatrix} 0 & 0 \\ 0 & 0 \end{bmatrix}$$

which yields the homogeneous system of linear equations

$$\begin{aligned} c_1 - 2c_2 + 3c_3 - 3c_4 &= 0 \\ c_2 + 4c_3 - 3c_4 &= 0 \\ 2c_1 - c_2 + 2c_3 + c_4 &= 0 \\ 3c_1 \qquad + 3c_3 + 3c_4 &= 0. \end{aligned}$$

Because this system has nontrivial solutions, S is not a basis. For example, one solution is $c_1 = 2, c_2 = 1, c_3 = -1$, and $c_4 = -1$.

$$2\begin{bmatrix} 1 & 0 \\ 2 & 3 \end{bmatrix} + \begin{bmatrix} -2 & 1 \\ -1 & 0 \end{bmatrix} - \begin{bmatrix} 3 & 4 \\ 2 & 3 \end{bmatrix} - \begin{bmatrix} -3 & -3 \\ 1 & 3 \end{bmatrix} = \begin{bmatrix} 0 & 0 \\ 0 & 0 \end{bmatrix}$$

© 2013 Cengage Learning. All Rights Reserved. May not be scanned, copied or duplicated, or posted to a publicly accessible website, in whole or in part.

37. (a) The system given by $A\mathbf{x} = \mathbf{0}$ has solutions of the form $(8t, 5t)$, where t is any real number. So, a basis for the solution space is $\{(8, 5)\}$.

(b) The nullity is 1.

Note that $\text{rank}(A) + \text{nullity}(A) = 1 + 1 = 2 = n$.

(c) The rank of A is 1 (the number of nonzero row vectors in the reduced row-echelon matrix)

39. (a) The system given by $A\mathbf{x} = \mathbf{0}$ has solutions of the form $(3s - t, -2t, s, t)$, where s and t are any real numbers. So, a basis for the solution space of $A\mathbf{x} = \mathbf{0}$ is $\{(3, 0, 1, 0), (-1, -2, 0, 1)\}$.

(b) The nullity of A is 2.

Note that $\text{rank}(A) + \text{nullity}(A) = 2 + 2 = 4 = n$.

(c) The rank of A is 2 (the number of nonzero row vectors in the reduced row-echelon matrix)

41. (a) The system given by $A\mathbf{x} = \mathbf{0}$ has solutions of the form $(4t, -2t, t)$, where t is any real number. So, a basis for the solution space is $\{(4, -2, 1)\}$.

(b) The nullity is 1.

Note that $\text{rank}(A) + \text{nullity}(A) = 2 + 1 = 3 = n$.

(c) The rank of A is 2 (the number of nonzero row vectors in the reduced row-echelon matrix)

43. (a) This system has solutions of the form $(-2s - 3t, s, 4t, t)$, where s and t are any real numbers. A basis for the solution space is $\{(-2, 1, 0, 0), (-3, 0, 4, 1)\}$.

(b) The dimension of the solution space is 2—the number of vectors in a basis for the solution space.

45. (a) This system has solutions of the form $\left(\frac{2}{7}s - t, \frac{3}{7}s, s, t\right)$, where s and t are any real numbers. A basis for the solution space is $\{(2, 3, 7, 0), (-1, 0, 0, 1)\}$.

(b) The dimension of the solution space is 2—the number of vectors in a basis for the solution space.

47. (a) Using Gauss-Jordan elimination, the matrix reduces to

$$\begin{bmatrix} 1 & 0 \\ 0 & 1 \\ 0 & 0 \end{bmatrix}.$$ So, the rank is 2.

(b) A basis for the row space is $\{(1, 0), (0, 1)\}$.

49. (a) Because the matrix is already row-reduced, its rank is 1.

(b) A basis for the row space is $\{(1, -4, 0, 4)\}$.

51. (a) Using Gauss-Jordan elimination, the matrix reduces to

$$\begin{bmatrix} 1 & 0 & 0 \\ 0 & 1 & 0 \\ 0 & 0 & 1 \end{bmatrix}.$$ So, the rank is 3.

(b) A basis for the row space is $\{(1, 0, 0), (0, 1, 0), (0, 0, 1)\}$.

53. Because $[\mathbf{x}]_B = \begin{bmatrix} 3 \\ 5 \end{bmatrix}$, write $\mathbf{x}$ as

$$\mathbf{x} = 3(1, 1) + 5(-1, 1) = (-2, 8).$$

Because $(-2, 8) = -2(1, 0) + 8(0, 1)$, the coordinate vector of $\mathbf{x}$ relative to the standard basis is

$$[\mathbf{x}]_S = \begin{bmatrix} -2 \\ 8 \end{bmatrix}.$$

55. Because $[\mathbf{x}]_B = \begin{bmatrix} \frac{1}{2} \\ \frac{1}{2} \end{bmatrix}$, write $\mathbf{x}$ as

$$\mathbf{x} = \tfrac{1}{2}\left(\tfrac{1}{2}, \tfrac{1}{2}\right) + \tfrac{1}{2}(1, 0) = \left(\tfrac{3}{4}, \tfrac{1}{4}\right).$$

Because $\left(\frac{3}{4}, \frac{1}{4}\right) = \frac{3}{4}(1, 0) + \frac{1}{4}(0, 1)$, the coordinate vector of $\mathbf{x}$ relative to the standard basis is

$$[\mathbf{x}]_S = \begin{bmatrix} \frac{3}{4} \\ \frac{1}{4} \end{bmatrix}.$$

57. Because $[\mathbf{x}]_B = \begin{bmatrix} 2 \\ 0 \\ -1 \end{bmatrix}$, write $\mathbf{x}$ as

$$\mathbf{x} = 2(1, 0, 0) + 0(1, 1, 0) - 1(0, 1, 1) = (2, -1, -1).$$

Because $(-2, -1, -1) = 2(1, 0, 0) - 1(0, 1, 0) - 1(0, 0, 1)$, the coordinate vector of $\mathbf{x}$ relative to the standard basis is

$$[\mathbf{x}]_S = \begin{bmatrix} 2 \\ -1 \\ -1 \end{bmatrix}.$$

59. To find $[\mathbf{x}]_{B'} = \begin{bmatrix} c_1 \\ c_2 \end{bmatrix}$, solve the equation

$$c_1(5, 0) + c_2(0, -8) = (2, 2).$$

The resulting system of linear equations is

$$\begin{aligned} 5c_1 \quad &= 2 \\ -8c_2 &= 2. \end{aligned}$$

So, $c_1 = \frac{2}{5}$, $c_2 = -\frac{1}{4}$, and $[\mathbf{x}]_{B'} = \begin{bmatrix} \frac{2}{5} \\ -\frac{1}{4} \end{bmatrix}$.

© 2013 Cengage Learning. All Rights Reserved. May not be scanned, copied or duplicated, or posted to a publicly accessible website, in whole or in part.

61. To find $[\mathbf{x}]_{B'} = \begin{bmatrix} c_1 \\ c_2 \\ c_3 \end{bmatrix}$ solve the equation

$$c_1(1, 2, 3) + c_2(1, 2, 0) + c_3(0, -6, 2) = (3, -3, 0).$$

The resulting system of linear equations is

$$\begin{aligned} c_1 + c_2 &= 3 \\ 2c_1 + 2c_2 - 6c_3 &= -3 \\ 3c_1 + 2c_3 &= 0. \end{aligned}$$

The solution of this system is $c_1 = -1$, $c_2 = 4$, $c_3 = \frac{3}{2}$, and

$$[\mathbf{x}]_{B'} = \begin{bmatrix} -1 \\ 4 \\ \frac{3}{2} \end{bmatrix}.$$

63. To find $[\mathbf{x}]_{B'} = \begin{bmatrix} c_1 \\ c_2 \\ c_3 \\ c_4 \end{bmatrix}$, solve the equation

$$c_1(9, -3, 15, 4) + c_2(-3, 0, 0, -1) + c_3(0, -5, 6, 8) + c_4(-3, 4, -2, 3) = (21, -5, 43, 14).$$

The resulting system of linear equations is

$$\begin{aligned} 9c_1 - 3c_2 - 3c_4 &= 21 \\ -3c_1 - 5c_3 + 4c_4 &= -5 \\ 15c_1 + 6c_3 - 2c_4 &= 43 \\ 4c_1 - c_2 + 8c_3 + 3c_4 &= 14. \end{aligned}$$

The solution of this system is

$c_1 = 3$, $c_2 = 1$, $c_3 = 0$, and $c_4 = 1$. So,

$$[\mathbf{x}]_{B'} = \begin{bmatrix} 3 \\ 1 \\ 0 \\ 1 \end{bmatrix}.$$

65. Begin by forming

$$[B' \;\; B] = \begin{bmatrix} 1 & 0 & 1 & 3 \\ 0 & 1 & -1 & 1 \end{bmatrix}.$$

Because this matrix is already in the form $[I_2 \;\; P^{-1}]$, you have

$$P^{-1} = \begin{bmatrix} 1 & 3 \\ -1 & 1 \end{bmatrix}.$$

67. Begin by forming.

$$[B' \;\; B] = \begin{bmatrix} 0 & 0 & 1 & 1 & 0 & 0 \\ 0 & 1 & 0 & 0 & 1 & 0 \\ 1 & 0 & 0 & 0 & 0 & 1 \end{bmatrix}.$$

Then use Gauss-Jordan elimination to obtain

$$[I_3 \;\; P^{-1}] = \begin{bmatrix} 1 & 0 & 0 & 0 & 0 & 1 \\ 0 & 1 & 0 & 0 & 1 & 0 \\ 0 & 0 & 1 & 1 & 0 & 0 \end{bmatrix}.$$

So, you have

$$P^{-1} = \begin{bmatrix} 0 & 0 & 1 \\ 0 & 1 & 0 \\ 1 & 0 & 0 \end{bmatrix}.$$

© 2013 Cengage Learning. All Rights Reserved. May not be scanned, copied or duplicated, or posted to a publicly accessible website, in whole or in part.

69. (a) $[B'\ B] = \begin{bmatrix} 0 & 1 & 1 & -1 \\ 1 & 2 & 1 & 1 \end{bmatrix} \Rightarrow \begin{bmatrix} 1 & 0 & -1 & 3 \\ 0 & 1 & 1 & -1 \end{bmatrix} = [I\ P^{-1}]$

(b) $[B\ B'] = \begin{bmatrix} 1 & -1 & 0 & 1 \\ 1 & 1 & 1 & 2 \end{bmatrix} \Rightarrow \begin{bmatrix} 1 & 0 & \frac{1}{2} & \frac{3}{2} \\ 0 & 1 & \frac{1}{2} & \frac{1}{2} \end{bmatrix} = [I\ P]$

(c) $P^{-1}P = \begin{bmatrix} -1 & 3 \\ 1 & -1 \end{bmatrix}\begin{bmatrix} \frac{1}{2} & \frac{3}{2} \\ \frac{1}{2} & \frac{1}{2} \end{bmatrix} = \begin{bmatrix} 1 & 0 \\ 0 & 1 \end{bmatrix}$

(d) $[\mathbf{x}]_{B'} = P^{-1}[\mathbf{x}]_B = \begin{bmatrix} -1 & 3 \\ 1 & -1 \end{bmatrix}\begin{bmatrix} 3 \\ -3 \end{bmatrix} = \begin{bmatrix} -12 \\ 6 \end{bmatrix}$

71. (a) $[B'\ B] = \begin{bmatrix} 0 & 0 & 1 & 1 & 1 & 1 \\ 0 & 1 & 1 & 0 & 1 & 1 \\ 1 & 1 & 1 & 0 & 0 & 1 \end{bmatrix} \Rightarrow \begin{bmatrix} 1 & 0 & 0 & 0 & -1 & 0 \\ 0 & 1 & 0 & -1 & 0 & 0 \\ 0 & 0 & 1 & 1 & 1 & 1 \end{bmatrix} = [I\ P^{-1}]$

(b) $[B\ B'] = \begin{bmatrix} 1 & 1 & 1 & 0 & 0 & 1 \\ 0 & 1 & 1 & 0 & 1 & 1 \\ 0 & 0 & 1 & 1 & 1 & 1 \end{bmatrix} \Rightarrow \begin{bmatrix} 1 & 0 & 0 & 0 & -1 & 0 \\ 0 & 1 & 0 & -1 & 0 & 0 \\ 0 & 0 & 1 & 1 & 1 & 1 \end{bmatrix} = [I\ P]$

(c) $P^{-1}P = \begin{bmatrix} 0 & -1 & 0 \\ -1 & 0 & 0 \\ 1 & 1 & 1 \end{bmatrix}\begin{bmatrix} 0 & -1 & 0 \\ -1 & 0 & 0 \\ 1 & 1 & 1 \end{bmatrix} = \begin{bmatrix} 1 & 0 & 0 \\ 0 & 1 & 0 \\ 0 & 0 & 1 \end{bmatrix}$

(d) $[\mathbf{x}]_{B'} = P^{-1}[\mathbf{x}]_B = \begin{bmatrix} 0 & -1 & 0 \\ -1 & 0 & 0 \\ 1 & 1 & 1 \end{bmatrix}\begin{bmatrix} -1 \\ 2 \\ -3 \end{bmatrix} = \begin{bmatrix} -2 \\ 1 \\ -2 \end{bmatrix}$

73. Begin by finding a basis for W. The polynomials in W must have x as a factor. Consequently, a polynomial in W is of the form

$p = x(c_1 + c_2x + c_3x^2) = c_1x + c_2x^2 + c_3x^3.$

A basis for W is $\{x, x^2, x^3\}$. Similarly, the polynomials in U must have $(x - 1)$ as a factor. A polynomial in U is of the form

$$\begin{aligned} p &= (x - 1)(c_1 + c_2x + c_3x^2) \\ &= c_1(x - 1) + c_2(x^2 - x) + c_3(x^3 - x^2). \end{aligned}$$

So, a basis for U is $\{x - 1, x^2 - x, x^3 - x^2\}$. The intersection of W and U contains polynomials with x and $(x - 1)$ as a factor. A polynomial in $W \cap M$ is of the form

$p = x(x - 1)(c_1 + c_2x) = c_1(x^2 - x) + c_2(x^3 - x^2).$

So, a basis for $W \cap U$ is $\{x^2 - x, x^3 - x^2\}$.

75. No. For example, the set $\{x^2 + x, x^2 - x, 1\}$ is a basis for P_2.

77. Because W is a nonempty subset of V, you need only show that W is closed under addition and scalar multiplication. If $(x^3 + x)p(x)$ and $(x^3 + x)q(x)$ are in W, then $(x^3 + x)p(x) + (x^3 + x)q(x) = (x^3 + x)(p(x) + q(x)) \in W$. Finally, $c(x^3 + x)p(x) = (x^3 + x)(cp(x)) \in W$. So, W is a subspace of $P_5 = V$.

79. The row vectors of A are linearly dependent if and only if the rank of A is less than n, which is equivalent to the column vectors of A being linearly dependent.

© 2013 Cengage Learning. All Rights Reserved. May not be scanned, copied or duplicated, or posted to a publicly accessible website, in whole or in part.

81. (a) Consider the equation

$c_1f + c_2g = c_1x + c_2|x| = 0$. If $x = \frac{1}{2}$, then

$\frac{1}{2}c_1 + \frac{1}{2}c_2 = 0$, while if $x = -\frac{1}{2}$, you obtain

$-\frac{1}{2}c_1 + \frac{1}{2}c_2 = 0$. This implies that

$c_1 = c_2 = 0$, and f and g are linearly independent.

(b) On the interval $[0, 1]$, $f = g = x$, and so they are linearly dependent.

83. (a) True. See discussion above "Definitions of Vector Addition and Scalar Multiplication in R^n," page 149.

(b) False. See Theorem 4.3, part 2, page 151.

(c) True. See "Definition of a Vector Space" and the discussion following, page 155.

85. (a) True. See discussion under "Vectors in R^n," page 149.

(b) False. See "Definition of a Vector Space," part 4, page 155.

(c) True. See discussion following "Summary of Important Vector Spaces," page 158.

87. (a) Because $y' = 3e^{3x}$ and $y'' = 9e^{3x}$, you have

$$y'' - y' - 6y = 9e^{3x} - 3e^{3x} - 6(e^{3x}) = 0.$$

Therefore, e^{3x} is a solution.

(b) Because $y' = 2e^{2x}$ and $y'' = 4e^{2x}$, you have

$$\begin{aligned} y'' - y' - 6y &= 4e^{2x} - 2e^{2x} - 6(e^{2x}) \\ &= -4e^{2x} \\ &\neq 0. \end{aligned}$$

Therefore, e^{2x} is *not* a solution.

(c) Because $y' = -3e^{-3x}$ and $y'' = 9e^{-3x}$, you have

$$\begin{aligned} y'' - y' - 6y &= 9e^{-3x} - (-3e^{-3x}) - 6(e^{-3x}) \\ &= 6e^{-3x} \\ &\neq 0. \end{aligned}$$

Therefore, e^{-3x} is *not* a solution.

(d) Because $y' = -2e^{-2x}$ and $y'' = 4e^{-2x}$, you have

$$y'' - y' - 6y = 4e^{-2x} - (-2e^{-2x}) - 6(e^{-2x}) = 0.$$

Therefore, e^{-2x} is a solution.

89. (a) Because $y' = -2e^{-2x}$, you have $y' + 2y = -2e^{-2x} + 2e^{-2x} = 0$.

Therefore, e^{-2x} is a solution.

(b) Because $y' = e^{-2x} - 2xe^{-2x}$, you have $y' + 2y = e^{-2x} - 2xe^{-2x} + 2xe^{-2x} = e^{-2x} \neq 0$.

Therefore, xe^{-2x} is *not* a solution.

(c) Because $y' = 2xe^{-x} - x^2e^{-x}$, you have $y' + 2y = 2xe^{-x} - x^2e^{-x} + 2x^2e^{-x} \neq 0$.

Therefore, x^2e^{-x} is *not* a solution.

(d) Because $y' = 2e^{-2x} - 4xe^{-2x}$, you have

$$y' + 2y = 2e^{-2x} - 4xe^{-2x} + 2(2xe^{-2x}) = 2e^{-2x} \neq 0.$$

Therefore, $2xe^{-2x}$ is *not* a solution.

91. $W(1, x, e^x) = \begin{bmatrix} 1 & x & e^x \\ 0 & 1 & e^x \\ 0 & 0 & e^x \end{bmatrix} = e^x$

93. $W(1, \sin 2x, \cos 2x) = \begin{vmatrix} 1 & \sin 2x & \cos 2x \\ 0 & 2\cos 2x & -2\sin 2x \\ 0 & -4\sin 2x & -4\cos 2x \end{vmatrix} = -8$

95. (a) $y = e^{-3y} \Rightarrow y' = -3e^{-3x}, y'' = 9e^{-3x} \Rightarrow y'' + 6y' + 9y = 0$

$y = xe^{-3x} \Rightarrow y' = (1 - 3x)e^{-3x}, y'' = (9x - 6)e^{-3x} \Rightarrow y'' + 6y' + 9y = 0$

(b) The Wronskian of this set is

$$W(e^{-3x}, xe^{-3x}) = \begin{vmatrix} e^{-3x} & xe^{-3x} \\ -3e^{-3x} & (1 - 3x)e^{-3x} \end{vmatrix} = (1 - 3x)e^{-6x} + 3xe^{-6x} = e^{-6x}.$$

Because $W(e^{-3x}, xe^{-3x}) = e^{-6x} \neq 0$, the set is linearly independent.

(c) $y = C_1e^{-3x} + C_2xe^{-3x}$

© 2013 Cengage Learning. All Rights Reserved. May not be scanned, copied or duplicated, or posted to a publicly accessible website, in whole or in part.

97. (a) $y = e^x \Rightarrow y' = y'' = y''' = e^x \Rightarrow y''' - 6y'' + 11y' - 6y = 0$

$y = e^{2x} \Rightarrow y' = 2e^{2x}, y'' = 4e^{2x}, y''' = 8e^{2x} \Rightarrow y''' - 6y'' + 11y' - 6y = 0$

$y = e^x = e^{2x} \Rightarrow y' = e^x - 2e^{2x}, y'' = e^x - 4e^{2x}, y''' = e^x - 8e^{2x} \Rightarrow y''' - 6y'' + 11y' - 6y = 0$

(b) The Wronskian of this set is

$$W\left(e^x, e^{2x}, e^x - e^{2x}\right) = \begin{vmatrix} e^x & e^{2x} & e^x - e^{2x} \\ e^x & 2e^{2x} & e^x - 2e^{2x} \\ e^x & 4e^{2x} & e^x - 4e^{2x} \end{vmatrix} = 0.$$

Because the third column is the difference of the first two columns, the set is linearly dependent.

(c) Not applicable

99. Begin by completing the square.

$$\left(x^2 - 4x + 4\right) + \left(y^2 + 2y + 1\right) = 4 + 4 + 1$$

$$(x - 2)^2 + (y + 1)^2 = 9$$

This is the equation of a circle of radius $\sqrt{9} = 3$, centered at $(2, -1)$.

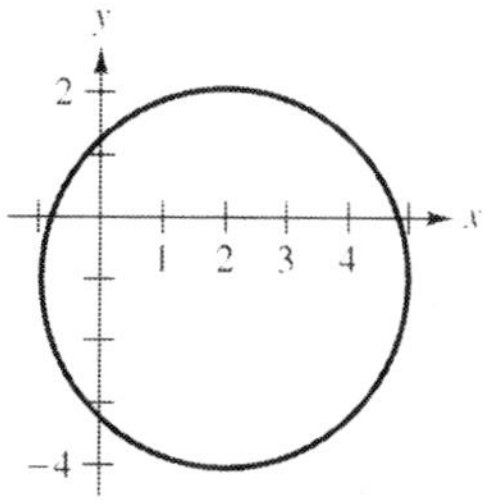

$x^2 + y^2 - 4x + 2y - 4 = 0$

101. Begin by completing the square.

$$x^2 - y^2 + 2x - 3 = 0$$

$$\left(x^2 + 2x + 1\right) - y^2 = 3 + 1$$

$$(x + 1)^2 - y^2 = 4$$

$$\frac{(x + 1)^2}{2^2} - \frac{y^2}{2^2} = 1$$

This is the equation of a hyperbola with center $(-1, 0)$.

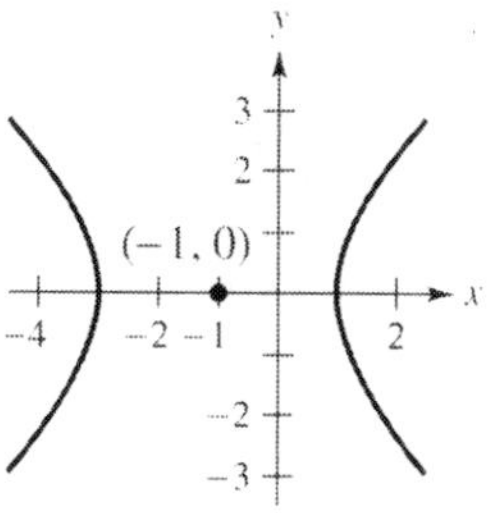

$x^2 - y^2 + 2x - 3 = 0$

103. Begin by completing the square.

$$2x^2 - 20x - y + 46 = 0$$

$$2\left(x^2 - 10x + 25\right) = y - 46 + 50$$

$$2(x - 5)^2 = y + 4$$

This is the equation of a parabola with vertex $(5, -4)$.

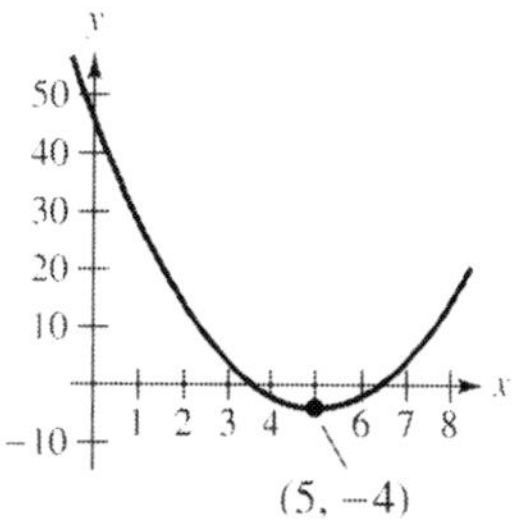

$2x^2 - 20x - y + 46 = 0$

105. Begin by completing the square.

$$4x^2 + y^2 + 32x + 4y + 63 = 0$$

$$4\left(x^2 + 8x + 16\right) + \left(y^2 + 4y + 4\right) = -63 + 64 + 4$$

$$4(x + 4)^2 + (y + 2)^2 = 5$$

This is the equation of an ellipse with center $(-4, -2)$.

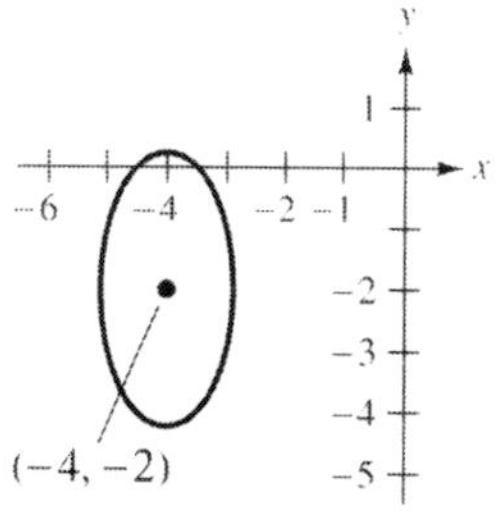

$4x^2 + y^2 + 32x + 4y + 63 = 0$

© 2013 Cengage Learning. All Rights Reserved. May not be scanned, copied or duplicated, or posted to a publicly accessible website, in whole or in part.

107. From the equation

$$\cot 2\theta = \frac{a-c}{b} = \frac{0-0}{1} = 0$$

you find that the angle of rotation is $\theta = \frac{\pi}{4}$. Therefore,

$\sin\theta = \frac{1}{\sqrt{2}}$ and $\cos\theta = \frac{1}{\sqrt{2}}$. By substituting

$$x = x'\cos\theta - y'\sin\theta = \frac{1}{\sqrt{2}} = (x' - y')$$

and

$$y = x'\sin\theta + y'\cos\theta = \frac{1}{\sqrt{2}} = (x' + y')$$

into $xy = 3$, you obtain $\frac{1}{2}(x')^2 - \frac{1}{2}(y')^2 = 3$.

In standard form,

$$\frac{(x')^2}{6} - \frac{(y')^2}{6} = 1$$

you can recognize this to be the equation of a hyperbola whose transverse axis is the x'-axis.

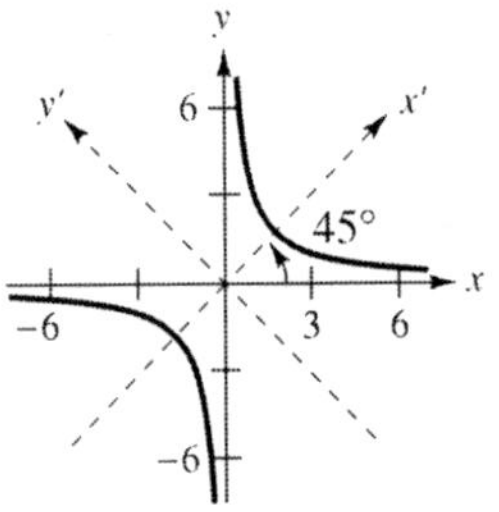

109. From the equation

$$\cos 2\theta = \frac{a-c}{b} = \frac{16-9}{-24} = -\frac{7}{24}$$

you find that the angle of rotation is $\theta \approx -36.87°$. Therefore, $\sin\theta \approx -0.6$ and $\cos\theta \approx 0.8$.

By substituting

$$x = x'\cos\theta - y'\sin\theta = 0.8x' + 0.6y'$$

and

$$y = x'\sin\theta + y'\cos\theta = -0.6x' + 0.8y'$$

into $16x^2 - 24xy + 9y^2 - 60x - 80y + 100 = 0$, you obtain $25(x')^2 - 100y' = -100$. In standard form,

$$(x')^2 = 4(y' - 1)$$

you can recognize this to be the equation of a parabola with vertex at $(x', y') = (0, 1)$.

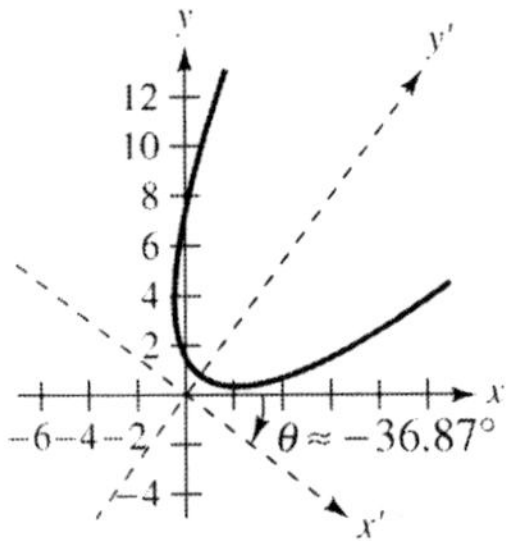

© 2013 Cengage Learning. All Rights Reserved. May not be scanned, copied or duplicated, or posted to a publicly accessible website, in whole or in part.

CHAPTER 5
Inner Product Spaces

© 2013 Cengage Learning. All Rights Reserved. May not be scanned, copied or duplicated, or posted to a publicly accessible website, in whole or in part.

CHAPTER 5
Inner Product Spaces

Section 5.1 Length and Dot Product in R^n

1. $\|\mathbf{v}\| = \sqrt{4^2 + 3^2} = \sqrt{25} = 5$

3. $\|\mathbf{v}\| = \sqrt{1^2 + 2^2 + 2^2} = \sqrt{9} = 3$

5. (a) $\|\mathbf{u}\| = \sqrt{(-1)^2 + \left(\frac{1}{4}\right)^2} = \sqrt{\frac{17}{16}} = \frac{\sqrt{17}}{4}$

(b) $\|\mathbf{v}\| = \sqrt{4^2 + \left(-\frac{1}{8}\right)^2} = \sqrt{\frac{1025}{64}} = \frac{5\sqrt{41}}{8}$

(c) $\|\mathbf{u} + \mathbf{v}\| = \left\|\left(3, \frac{1}{8}\right)\right\|$

$= \sqrt{3^2 + \left(\frac{1}{8}\right)^2} = \sqrt{\frac{577}{64}} = \frac{\sqrt{577}}{8}$

7. (a) $\|\mathbf{u}\| = \sqrt{1^2 + 2^2 + 1^2} = \sqrt{6}$

(b) $\|\mathbf{v}\| = \sqrt{0^2 + 2^2 + (-2)^2} = \sqrt{8} = 2\sqrt{2}$

(c) $\|\mathbf{u} + \mathbf{v}\| = \|(1, 4, -1)\|$

$= \sqrt{1^2 + 4^2 + (-1)^2} = \sqrt{18} = 3\sqrt{2}$

9. (a) A unit vector $\mathbf{v}$ in the direction of $\mathbf{u}$ is given by

$$\mathbf{v} = \frac{\mathbf{u}}{\|\mathbf{u}\|} = \frac{1}{\sqrt{(-5)^2 + 12^2}}(-5, 12)$$

$$= \frac{1}{13}(-5, 12) = \left(-\frac{5}{13}, \frac{12}{13}\right).$$

(b) A unit vector in the direction opposite that of $\mathbf{u}$ is given by

$$-\mathbf{v} = -\left(-\frac{5}{13}, \frac{12}{13}\right) = \left(\frac{5}{13}, -\frac{12}{13}\right).$$

11. (a) A unit vector $\mathbf{v}$ in the direction of $\mathbf{u}$ is given by

$$\mathbf{v} = \frac{\mathbf{u}}{\|\mathbf{u}\|} = \frac{1}{\sqrt{3^2 + 2^2 + (-5)^2}}(3, 2, -5)$$

$$= \frac{1}{\sqrt{38}}(3, 2, -5) = \left(\frac{3}{\sqrt{38}}, \frac{2}{\sqrt{38}}, -\frac{5}{\sqrt{38}}\right).$$

(b) A unit vector in the direction opposite that of $\mathbf{u}$ is given by

$$-\mathbf{v} = -\left(\frac{3}{\sqrt{38}}, \frac{2}{\sqrt{38}}, -\frac{5}{\sqrt{38}}\right)$$

$$= \left(-\frac{3}{\sqrt{38}}, -\frac{2}{\sqrt{38}}, \frac{5}{\sqrt{38}}\right).$$

13. First find a unit vector in the direction of $\mathbf{u}$.

$$\frac{\mathbf{u}}{\|\mathbf{u}\|} = \frac{1}{\sqrt{1^2 + 1^2}}(1, 1) = \left(\frac{1}{\sqrt{2}}, \frac{1}{\sqrt{2}}\right)$$

Then $\mathbf{v}$ is four times this vector.

$$\mathbf{v} = 4\frac{\mathbf{u}}{\|\mathbf{u}\|} = 4\left(\frac{1}{\sqrt{2}}, \frac{1}{\sqrt{2}}\right) = \left(2\sqrt{2}, 2\sqrt{2}\right)$$

15. First find a unit vector in the direction of $\mathbf{u}$.

$$\frac{\mathbf{u}}{\|\mathbf{u}\|} = \frac{1}{\sqrt{3 + 9 + 0}}\left(\sqrt{3}, 3, 0\right)$$

$$= \frac{1}{2\sqrt{3}}\left(\sqrt{3}, 3, 0\right) = \left(\frac{1}{2}, \frac{\sqrt{3}}{2}, 0\right)$$

Then $\mathbf{v}$ is twice this vector.

$$\mathbf{v} = 2\left(\frac{1}{2}, \frac{\sqrt{3}}{2}, 0\right) = \left(1, \sqrt{3}, 0\right)$$

17. (a) Because $\frac{\mathbf{v}}{\|\mathbf{v}\|}$ is a unit vector in the direction of $\mathbf{v}$, you have

$$\mathbf{u} = \frac{\|\mathbf{v}\|}{2}\frac{\mathbf{v}}{\|\mathbf{v}\|} = \frac{1}{2}\mathbf{v}$$

$$= \frac{1}{2}(-1, 3, 0, 4) = \left(-\frac{1}{2}, \frac{3}{2}, 0, 2\right).$$

(b) Because $-\frac{\mathbf{v}}{\|\mathbf{v}\|}$ is a unit vector with direction opposite that of $\mathbf{v}$,

$$\mathbf{u} = 2\|\mathbf{v}\|\left(-\frac{\mathbf{v}}{\|\mathbf{v}\|}\right)$$

$$= -2\mathbf{v} = -2(-1, 3, 0, 4) = (2, -6, 0, -8).$$

19. $d(\mathbf{u}, \mathbf{v}) = \|\mathbf{u} - \mathbf{v}\| = \|(2, -2)\| = \sqrt{4 + 4} = 2\sqrt{2}$

21. $d(\mathbf{u}, \mathbf{v}) = \|\mathbf{u} - \mathbf{v}\|$

$= \|(2, -2, -1)\|$

$= \sqrt{2^2 + (-2)^2 + (-1)^2} = \sqrt{9} = 3$

23. (a) $\mathbf{u} \cdot \mathbf{v} = 3(2) + 4(-3) = 6 - 12 = -6$

(b) $\mathbf{v} \cdot \mathbf{v} = 2(2) + (-3)(-3) = 4 + 9 = 13$

(c) $\|\mathbf{u}\|^2 = \mathbf{u} \cdot \mathbf{u} = 3(3) + 4(4) = 9 + 16 = 25$

(d) $(\mathbf{u} \cdot \mathbf{v})\mathbf{v} = -6(2, -3) = (-12, 18)$

(e) $\mathbf{u} \cdot (5\mathbf{v}) = 5(\mathbf{u} \cdot \mathbf{v}) = 5(-6) = -30$

© 2013 Cengage Learning. All Rights Reserved. May not be scanned, copied or duplicated, or posted to a publicly accessible website, in whole or in part.

25. (a) $\mathbf{u} \cdot \mathbf{v} = (-1)(1) + (1)(-3) + (-2)(-2) = 0$

(b) $\mathbf{v} \cdot \mathbf{v} = 1(1) + (-3)(-3) + (-2)(-2)$

$= 1 + 9 + 4$

$= 14$

(c) $\|\mathbf{u}\|^2 = \mathbf{u} \cdot \mathbf{u} = (-1)(-1) + 1(1) + (-2)(-2)$

$= 1 + 1 + 4$

$= 6$

(d) $(\mathbf{u} \cdot \mathbf{v})\mathbf{v} = 0(1, -3, -2)$

$= (0, 0, 0)$

$= \mathbf{0}$

(e) $\mathbf{u} \cdot (5\mathbf{v}) = 5(\mathbf{u} \cdot \mathbf{v}) = 5 \cdot 0 = 0$

27. $(\mathbf{u} + \mathbf{v}) \cdot (2\mathbf{u} - \mathbf{v}) = \mathbf{u} \cdot (2\mathbf{u} - \mathbf{v}) + \mathbf{v} \cdot (2\mathbf{u} - \mathbf{v})$

$= 2\mathbf{u} \cdot \mathbf{u} - \mathbf{u} \cdot \mathbf{v} + 2\mathbf{v} \cdot \mathbf{u} - \mathbf{v} \cdot \mathbf{v}$

$= 2(\mathbf{u} \cdot \mathbf{u}) + \mathbf{u} \cdot \mathbf{v} - \mathbf{v} \cdot \mathbf{v}$

$= 2(4) + (-5) - 10 = -7$

29. (a) $\|\mathbf{u}\| = 1.0843$, $\|\mathbf{v}\| = 0.3202$

(b) $\dfrac{\mathbf{v}}{\|\mathbf{v}\|} = (0, 0.7809, 0.6247)$

(c) $-\dfrac{\mathbf{u}}{\|\mathbf{u}\|} = (-0.9223, -0.1153, -0.3689)$

(d) $\mathbf{u} \cdot \mathbf{v} = 0.1113$

(e) $\mathbf{u} \cdot \mathbf{u} = 1.1756$

(f) $\mathbf{v} \cdot \mathbf{v} = 0.1025$

31. (a) $\|\mathbf{u}\| = 1.7321$, $\|\mathbf{v}\| = 2$

(b) $\dfrac{\mathbf{v}}{\|\mathbf{v}\|} = (-0.5, 0.7071, -0.5)$

(c) $-\dfrac{\mathbf{u}}{\|\mathbf{u}\|} = (0, -0.5774, -0.8165)$

(d) $\mathbf{u} \cdot \mathbf{v} = 0$

(e) $\mathbf{u} \cdot \mathbf{u} = 3$

(f) $\mathbf{v} \cdot \mathbf{v} = 4$

33. You have

$\mathbf{u} \cdot \mathbf{v} = 3(2) + 4(-3) = -6,$

$\|\mathbf{u}\| = \sqrt{3^2 + 4^2} = \sqrt{25} = 5$, and

$\|\mathbf{v}\| = \sqrt{2^2 + (-3)^2} = \sqrt{13}$. So,

$|\mathbf{u} \cdot \mathbf{v}| \le \|\mathbf{u}\|\|\mathbf{v}\|$

$|-6| \le 5(\sqrt{13})$

$6 \le 5\sqrt{3} \approx 8.66.$

35. You have

$\mathbf{u} \cdot \mathbf{v} = (1) + 1(-3) + (-2)(-2) = 2,$

$\|\mathbf{u}\| = \sqrt{1^2 + 1^2 + (-2)^2} = \sqrt{6}$, and

$\|\mathbf{v}\| = \sqrt{1^2 + (-3)^2 + (-2)^2} = \sqrt{14}$. So,

$|\mathbf{u} \cdot \mathbf{v}| \le \|\mathbf{u}\|\|\mathbf{v}\|$

$|2| \le \sqrt{6}\sqrt{14}$

$2 \le 2\sqrt{21} \approx 9.17.$

37. The cosine of the angle θ between **u** and **v** is given by

$$\cos\theta = \frac{\mathbf{u} \cdot \mathbf{v}}{\|\mathbf{u}\|\|\mathbf{v}\|} = \frac{3(-2) + 1(4)}{\sqrt{3^2 + 1^2}\sqrt{(-2)^2 + 4^2}}$$

$$= -\frac{2}{10\sqrt{2}} = -\frac{\sqrt{2}}{10}.$$

So, $\theta = \cos^{-1}\left(-\dfrac{\sqrt{2}}{10}\right) \approx 1.713$ radians $(98.13°)$.

39. The cosine of the angle θ between **u** and **v** is given by

$$\cos\theta = \frac{\mathbf{u} \cdot \mathbf{v}}{\|\mathbf{u}\|\|\mathbf{v}\|}$$

$$= \frac{\cos\frac{\pi}{6}\cos\frac{3\pi}{4} + \sin\frac{\pi}{6}\sin\frac{3\pi}{4}}{\sqrt{\cos^2\frac{\pi}{6} + \sin^2\frac{\pi}{6}} \cdot \sqrt{\cos^2\frac{3\pi}{4} + \sin^2\frac{3\pi}{4}}}$$

$$= \frac{\cos\left(\frac{\pi}{6} - \frac{3\pi}{4}\right)}{1 \cdot 1}$$

$$= \cos\left(-\frac{7\pi}{12}\right)$$

$$= \cos\left(\frac{7\pi}{12}\right).$$

So, $\theta = \dfrac{7\pi}{12}$ radians $(105°)$.

41. The cosine of the angle θ between **u** and **v** is given by

$$\cos\theta = \frac{\mathbf{u} \cdot \mathbf{v}}{\|\mathbf{u}\|\|\mathbf{v}\|} = \frac{1(2) + 1(1) + 1(-1)}{\sqrt{1^2 + 1^2 + 1^2}\sqrt{2^2 + 1^2 + (-1)^2}}$$

$$= \frac{2}{3\sqrt{2}} = \frac{\sqrt{2}}{3}.$$

So, $\theta = \cos^{-1}\left(\dfrac{\sqrt{2}}{3}\right) \approx 1.080$ radians $(61.87°)$.

© 2013 Cengage Learning. All Rights Reserved. May not be scanned, copied or duplicated, or posted to a publicly accessible website, in whole or in part.

43. The cosine of the angle θ between $\mathbf{u}$ and $\mathbf{v}$ is given by

$$\cos\theta = \frac{\mathbf{u}\cdot\mathbf{v}}{\|\mathbf{u}\|\|\mathbf{v}\|} = \frac{0(3) + 1(3) + 0(3) + 1(3)}{\sqrt{0^2 + 1^2 + 0^2 + 1^2}\sqrt{3^2 + 3^2 + 3^2 + 3^2}} = \frac{6}{6\sqrt{2}} = \frac{\sqrt{2}}{2}.$$

So, $\theta = \cos^{-1}\left(\frac{\sqrt{2}}{2}\right) = \frac{\pi}{4}.$

45. Because $\mathbf{u}\cdot\mathbf{v} = 2\left(\frac{3}{2}\right) + 18\left(-\frac{1}{6}\right) = 0$, the vectors $\mathbf{u}$ and $\mathbf{v}$ are orthogonal.

47. Because $\mathbf{v} = -6\mathbf{u}$, the vectors are parallel.

49. Because $\mathbf{u}\cdot\mathbf{v} = 0(1) + 1(-2) + 0(0) = -2 \neq 0$, the vectors $\mathbf{u}$ and $\mathbf{v}$ are not orthogonal. Moreover, because one is not a scalar multiple of the other, they are not parallel.

51. Because

$\mathbf{u}\cdot\mathbf{v} = -2\left(\frac{1}{4}\right) + 5\left(-\frac{5}{4}\right) + 1(0) + 0(1) = -\frac{27}{4} \neq 0$, the vectors $\mathbf{u}$ and $\mathbf{v}$ are not orthogonal. Moreover, because one is not a scalar multiple of the other, they are not parallel.

53.

$$\begin{aligned} \mathbf{u}\cdot\mathbf{v} &= 0 \\ (0, 5)\cdot(v_1, v_2) &= 0 \\ 0v_1 + 5v_2 &= 0 \\ v_2 &= 0 \end{aligned}$$

So, $\mathbf{v} = (t, 0)$, where t is any real number.

55.

$$\begin{aligned} \mathbf{u}\cdot\mathbf{v} &= 0 \\ (2, -1, 1)\cdot(\mathbf{v}_1, \mathbf{v}_2, \mathbf{v}_3) &= 0 \\ 2\mathbf{v}_1 - \mathbf{v}_2 + \mathbf{v}_3 &= 0 \end{aligned}$$

So, $\mathbf{v} = (t, s, -2t + s)$, where s and t are any real numbers.

57. Because $\mathbf{u} + \mathbf{v} = (4, 0) + (1, 1) = (5, 1)$, you have

$$\begin{aligned} \|\mathbf{u} + \mathbf{v}\| &\le \|\mathbf{u}\| + \|\mathbf{v}\| \\ \|(5, 1)\| &\le \|(4, 0)\| + \|(1, 1)\| \\ \sqrt{26} &\le 4 + \sqrt{2}. \end{aligned}$$

59. Because $\mathbf{u} + \mathbf{v} = (1, 1, 1) + (0, -1, 2) = (1, 0, 3)$, you have

$$\begin{aligned} \|\mathbf{u} + \mathbf{v}\| &\le \|\mathbf{u}\| + \|\mathbf{v}\| \\ \|(1, 0, 3)\| &\le \|(1, 1, 1)\| + \|(0, -1, 2)\| \\ \sqrt{10} &\le \sqrt{3} + \sqrt{5}. \end{aligned}$$

61. First note that $\mathbf{u}$ and $\mathbf{v}$ are orthogonal, because $\mathbf{u}\cdot\mathbf{v} = (1, -1)\cdot(1, 1) = 0$. Then note

$$\begin{aligned} \|\mathbf{u} + \mathbf{v}\|^2 &= \|\mathbf{u}\|^2 + \|\mathbf{v}\|^2 \\ \|(2, 0)\|^2 &= \|(1, -1)\|^2 + \|(1, 1)\|^2 \\ 4 &= 2 + 2. \end{aligned}$$

63. First note that $\mathbf{u}$ and $\mathbf{v}$ are orthogonal, because $\mathbf{u}\cdot\mathbf{v} = (3, 4, -2)\cdot(4, -3, 0) = 0$. Then note

$$\begin{aligned} \|\mathbf{u} + \mathbf{v}\|^2 &= \|\mathbf{u}\|^2 + \|\mathbf{v}\|^2 \\ \|(7, 1, -2)\|^2 &= \|(3, 4, -2)\|^2 + \|(4, -3, 0)\|^2 \\ 54 &= 29 + 25. \end{aligned}$$

65. (a) $$\mathbf{u}\cdot\mathbf{v} = \mathbf{u}^T\mathbf{v} = \begin{bmatrix} 3 & 4 \end{bmatrix}\begin{bmatrix} 2 \\ -3 \end{bmatrix} = \begin{bmatrix} 3(2) + 4(-3) \end{bmatrix} = 6 - 12 = -6$$

(b) $$\mathbf{v}\cdot\mathbf{v} = \mathbf{v}^T\mathbf{v} = \begin{bmatrix} 2 & -3 \end{bmatrix}\begin{bmatrix} 2 \\ -3 \end{bmatrix} = \begin{bmatrix} 2(2) + (-3)(-3) \end{bmatrix} = \begin{bmatrix} 4 + 9 \end{bmatrix} = 13$$

(c) $$\|\mathbf{u}\|^2 = \mathbf{u}^T\mathbf{u} = \begin{bmatrix} 3 & 4 \end{bmatrix}\begin{bmatrix} 3 \\ 4 \end{bmatrix} = \begin{bmatrix} 3(3) + 4(4) \end{bmatrix} = \begin{bmatrix} 9 + 16 \end{bmatrix} = 25$$

(d) $$(\mathbf{u}\cdot\mathbf{v})\mathbf{v} = (\mathbf{u}^T\mathbf{v})\mathbf{v} = \left(\begin{bmatrix} 3 & 4 \end{bmatrix}\begin{bmatrix} 2 \\ -3 \end{bmatrix}\right)\begin{bmatrix} 2 \\ -3 \end{bmatrix} = -6\begin{bmatrix} 2 \\ -3 \end{bmatrix} = \begin{bmatrix} -12 \\ 18 \end{bmatrix}$$

(e) $$\mathbf{u}\cdot(5\mathbf{v}) = 5(\mathbf{u}^T\cdot\mathbf{v}) = 5\left(\begin{bmatrix} 3 & 4 \end{bmatrix}\begin{bmatrix} 2 \\ -3 \end{bmatrix}\right) = 5(-6) = -30$$

© 2013 Cengage Learning. All Rights Reserved. May not be scanned, copied or duplicated, or posted to a publicly accessible website, in whole or in part.

67. (a) $\mathbf{u} \cdot \mathbf{v} = \mathbf{u}^T \mathbf{v} = \begin{bmatrix} -1 & 1 & -2 \end{bmatrix} \begin{bmatrix} 1 \\ -3 \\ -2 \end{bmatrix} = \left[(-1)(1) + 1(-3) + (-2)(-2)\right] = \left[-1 + (-3) + 4\right] = 0$

(b) $\mathbf{v} \cdot \mathbf{v} = \mathbf{v}^T \mathbf{v} = \begin{bmatrix} 1 & -3 & -2 \end{bmatrix} \begin{bmatrix} 1 \\ -3 \\ -2 \end{bmatrix} = \left[1(1) + (-3)(-3) + (-2)(-2)\right] = \left[1 + 9 + 4\right] = 14$

(c) $\|\mathbf{u}\|^2 = \mathbf{u} \cdot \mathbf{u} = \mathbf{u}^T \mathbf{v} = \begin{bmatrix} -1 & 1 & -2 \end{bmatrix} \begin{bmatrix} -1 \\ 1 \\ -2 \end{bmatrix} = \left[(-1)(-1) + 1(1) + (-2)(-2)\right] = \left[1 + 1 + 4\right] = 6$

(d) $(\mathbf{u} \cdot \mathbf{v})\mathbf{v} = \left(\mathbf{u}^T \cdot \mathbf{v}\right)\mathbf{v} = \left(\begin{bmatrix} -1 & 1 & -2 \end{bmatrix} \begin{bmatrix} 1 \\ -3 \\ -2 \end{bmatrix}\right) \begin{bmatrix} 1 \\ -3 \\ -2 \end{bmatrix} = 0 \begin{bmatrix} 1 \\ -3 \\ -2 \end{bmatrix} = \begin{bmatrix} 0 \\ 0 \\ 0 \end{bmatrix}$

(e) $\mathbf{u} \cdot (5\mathbf{v}) = 5\left(\mathbf{u}^T \cdot \mathbf{v}\right) = 5\left(\begin{bmatrix} -1 & 1 & -2 \end{bmatrix} \begin{bmatrix} 1 \\ -3 \\ -2 \end{bmatrix}\right) = 5(0) = 0$

69. Because $\mathbf{u} \cdot \mathbf{v} = \cos\theta \sin\theta + \sin\theta(-\cos\theta) - 1(0)$

$= \cos\theta \sin\theta - \cos\theta \sin\theta = 0,$

the vectors **u** and **v** are orthogonal.

71. (a) False. See "Definition of Length of a Vector in R^n," page 226.

(b) False. See "Definition of Dot Product in R^n," page 229.

73. (a) $(\mathbf{u} \cdot \mathbf{v}) - \mathbf{v}$ is meaningless because **v** is a vector and $\mathbf{u} \cdot \mathbf{v}$ is a scalar.

(b) $\mathbf{u} + (\mathbf{u} \cdot \mathbf{v})$ is meaningless because **u** is a vector and $\mathbf{u} \cdot \mathbf{v}$ is a scalar.

75. $\mathbf{v} = (\mathbf{v}_1, \mathbf{v}_2) = (12, 5), (\mathbf{v}_2, -\mathbf{v}_1) = (5, -12)$

$(12, 5) \cdot (5, -12) = 12(5) + 5(-12) = 60 - 60 = 0$

So, $(\mathbf{v}_2, -\mathbf{v}_1)$ is orthogonal to **v**.

Two unit vectors orthogonal to **v**: $-1(5, -12) = (-5, 12)$: $(12, 5) \cdot (-5, 12) = 12(-5) + 5(12)$
$= -60 + 60$
$= 0$

$3(5, -12) = (15, -36)$: $(12, 5) \cdot (15, -36) = 12(15) + 5(-36)$
$= 180 - 180$
$= 0$

(Answer is not unique.)

77. $\mathbf{u} = \langle 3140, 2750 \rangle, \mathbf{v} = \langle 2.25, 1.75 \rangle$

$\mathbf{u} \cdot \mathbf{v} = 3140(2.25) + 2750(1.75) = 11{,}877.5$

The total revenue earned from selling the hamburgers and hot dogs is $11,877.50.

© 2013 Cengage Learning. All Rights Reserved. May not be scanned, copied or duplicated, or posted to a publicly accessible website, in whole or in part.

79. Let t = length of side of cube. The diagonal of the cube can be represented by the vector $\mathbf{v} = (t, t, t)$, and one side by the vector $\mathbf{u} = (t, 0, 0)$. So,

$$\cos\theta = \frac{\mathbf{u}\cdot\mathbf{v}}{\|\mathbf{u}\|\|\mathbf{v}\|} = \frac{t^2}{t(t\sqrt{3})} = \frac{1}{\sqrt{3}} \Rightarrow \theta = \cos^{-1}\left(\frac{1}{\sqrt{3}}\right) \approx 54.7°.$$

81. Given $\mathbf{u}\cdot\mathbf{v} = 0$ and $\mathbf{u}\cdot\mathbf{w} = 0$,

$$\begin{aligned}\mathbf{u}\cdot(c\mathbf{v} + d\mathbf{w}) &= \mathbf{u}\cdot(c\mathbf{v}) + \mathbf{u}\cdot(d\mathbf{w})\\ &= c(\mathbf{u}\cdot\mathbf{v}) + d(\mathbf{u}\cdot\mathbf{w})\\ &= c(0) + d(0)\\ &= 0.\end{aligned}$$

So, $\mathbf{u}$ is orthogonal to $c\mathbf{v} + d\mathbf{w}$.

83. Let $\mathbf{u} = (\cos\theta)\mathbf{i} - (\sin\theta)\mathbf{j}$ and $\mathbf{v} = (\sin\theta)\mathbf{i} + (\cos\theta)\mathbf{j}$. Then

$$\|\mathbf{u}\| = \sqrt{\cos^2\theta + \sin^2\theta} = 1, \qquad \|\mathbf{v}\| = \sqrt{\sin^2\theta + \cos^2\theta} = 1,$$

and $\mathbf{u}\cdot\mathbf{v} = \cos\theta\sin\theta - \sin\theta\cos\theta = 0$. So, $\mathbf{u}$ and $\mathbf{v}$ are orthogonal unit vectors for any value of θ. If $\theta = \frac{\pi}{3}$, you have the following graph.

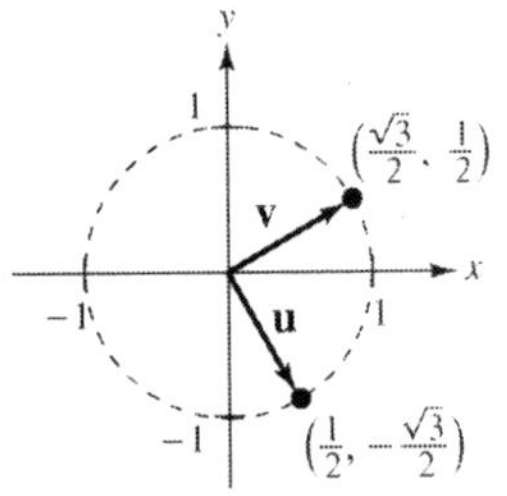

85. Property 1: $\mathbf{u}\cdot\mathbf{v} = \mathbf{u}^T\mathbf{v} = (\mathbf{u}^T\mathbf{v})^T = \mathbf{v}^T\mathbf{u} = \mathbf{v}\cdot\mathbf{u}$

Property 2: $\mathbf{u}\cdot(\mathbf{v} + \mathbf{w}) = \mathbf{u}^T(\mathbf{v} + \mathbf{w}) = \mathbf{u}^T\mathbf{v} + \mathbf{u}^T\mathbf{w} = \mathbf{u}\cdot\mathbf{v} + \mathbf{u}\cdot\mathbf{w}$

Property 3: $c(\mathbf{u}\cdot\mathbf{v}) = c(\mathbf{u}^T\mathbf{v}) = (c\mathbf{u})^T\mathbf{v} = (c\mathbf{u})\cdot\mathbf{v}$ and $c(\mathbf{u}\cdot\mathbf{v}) = c(\mathbf{u}^T\mathbf{v}) = \mathbf{u}^T(c\mathbf{v}) = \mathbf{u}\cdot(c\mathbf{v})$

87. $A\mathbf{x} = \mathbf{0}$ means that the dot product of each row of A with the column vector $\mathbf{x}$ is zero. So, $\mathbf{x}$ is orthogonal to the row vectors of A.

Section 5.2 Inner Product Spaces

1. 1. Since the product of real numbers is commutative,

$$\begin{aligned}\langle\mathbf{u}, \mathbf{v}\rangle &= 3u_1v_1 + u_2v_2 = 3v_1u_1 + v_2u_2\\ &= \langle\mathbf{v}, \mathbf{u}\rangle.\end{aligned}$$

2. Let $\mathbf{w} = (w_1, w_2)$. Then

$$\begin{aligned}\langle\mathbf{u}, \mathbf{v} + \mathbf{w}\rangle &= 3u_1(v_1 + w_1) + u_2(v_2 + w_2)\\ &= 3u_1v_1 + 3u_1w_1 + u_2v_2 + u_2w_2\\ &= 3u_1v_1 + u_2v_2 + 3u_1w_1 + u_2w_2\\ &= \langle\mathbf{u}, \mathbf{v}\rangle + \langle\mathbf{u}, \mathbf{w}\rangle.\end{aligned}$$

3. If c is any scalar, then

$$\begin{aligned}c\langle\mathbf{u}, \mathbf{v}\rangle &= c(3u_1v_1 + u_2v_2)\\ &= 3(cu_1)v_1 + (cu_2)v_2\\ &= \langle c\mathbf{u}, \mathbf{v}\rangle.\end{aligned}$$

4. Since the square of a real number is nonnegative, $\langle\mathbf{v}, \mathbf{v}\rangle = 3v_1^2 + v_2^2 \ge 0$. Moreover, this expression is equal to zero if and only if $\mathbf{v} = \mathbf{0}$ (that is, if and only if $v_1 = v_2 = 0$).

© 2013 Cengage Learning. All Rights Reserved. May not be scanned, copied or duplicated, or posted to a publicly accessible website, in whole or in part.

3. 1. Since the product of real numbers is commutative,

$$\langle \mathbf{u}, \mathbf{v} \rangle = \tfrac{1}{2}u_1v_1 + \tfrac{1}{4}u_2v_2 = \tfrac{1}{2}v_1u_1 + \tfrac{1}{4}v_2u_2$$
$$= \langle \mathbf{v}, \mathbf{u} \rangle.$$

2. Let $\mathbf{w} = (w_1, w_2)$. Then,

$$\begin{aligned}\langle \mathbf{u}, \mathbf{v} + \mathbf{w} \rangle &= \tfrac{1}{2}u_1(v_1 + w_1) + \tfrac{1}{4}u_2(v_2 + w_2)\\ &= \tfrac{1}{2}u_1v_1 + \tfrac{1}{2}u_1w_1 + \tfrac{1}{4}u_2v_2 + \tfrac{1}{4}u_2w_2\\ &= \tfrac{1}{2}u_1v_1 + \tfrac{1}{4}u_2v_2 + \tfrac{1}{2}u_1w_1 + \tfrac{1}{4}u_2w_2\\ &= \langle \mathbf{u}, \mathbf{v} \rangle + \langle \mathbf{u}, \mathbf{w} \rangle.\end{aligned}$$

3. If c is any scalar, then

$$\begin{aligned}c\langle \mathbf{u}, \mathbf{v} \rangle &= c\left(\tfrac{1}{2}u_1v_1 + \tfrac{1}{4}u_2v_2\right)\\ &= \tfrac{1}{2}(cu_1)v_1 + \tfrac{1}{4}(cu_2)v_2\\ &= \langle c\mathbf{u}, \mathbf{v} \rangle.\end{aligned}$$

4. Since the square of a real number is nonnegative, $\langle \mathbf{v}, \mathbf{v} \rangle = \tfrac{1}{2}v_1^2 + \tfrac{1}{4}v_2^2 \geq 0$. Moreover, this expression is equal to zero if and only if $\mathbf{v} = \mathbf{0}$ (that is, if and only if $v_1 = v_2 = 0$).

5. 1. Since the product of real numbers is commutative,

$$\begin{aligned}\langle \mathbf{u}, \mathbf{v} \rangle &= 2u_1v_1 + 3u_2v_2 + u_3v_3\\ &= 2v_1u_1 + 3v_2u_2 + v_3u_3\\ &= \langle \mathbf{v}, \mathbf{u} \rangle.\end{aligned}$$

2. Let $\mathbf{w} = (w_1, w_2, w_3)$. Then,

$$\begin{aligned}\langle \mathbf{u}, \mathbf{v} + \mathbf{w} \rangle &= 2u_1(v_1 + w_1) + 3u_2(v_2 + w_2) + u_3(v_3 + w_3)\\ &= 2u_1v_1 + 2u_1w_1 + 3u_2v_2 + 3u_2w_2 + u_3v_3 + u_3w_3\\ &= 2u_1v_1 + 3u_2v_2 + u_3v_3 + 2u_1w_1 + 3u_2w_2 + u_3w_3\\ &= \langle \mathbf{u}, \mathbf{v} \rangle + \langle \mathbf{u}, \mathbf{w} \rangle.\end{aligned}$$

3. If c is any scalar, then

$$\begin{aligned}c\langle \mathbf{u}, \mathbf{v} \rangle &= c(2u_1v_1 + 3u_2v_2 + u_3v_3)\\ &= 2(cu_1)v_1 + 3(cu_2)v_2 + (cu_3)v_3\\ &= \langle c\mathbf{u}, \mathbf{v} \rangle.\end{aligned}$$

4. Since the square of a real number is nonnegative, $\langle \mathbf{v}, \mathbf{v} \rangle = 2v_1^2 + 3v_2^2 + v_3^2 \geq 0$. Moreover, this expression is equal to zero if and only if $\mathbf{v} = \mathbf{0}$ (that is, if and only if $v_1 = v_2 = v_3 = 0$).

© 2013 Cengage Learning. All Rights Reserved. May not be scanned, copied or duplicated, or posted to a publicly accessible website, in whole or in part.

7. 1. Since the product of real numbers is commutative,

$$\begin{aligned}\langle \mathbf{u}, \mathbf{v}\rangle &= 2u_1v_1 + u_2v_2 + 2u_3v_3\\ &= 2v_1u_1 + v_2u_2 + 2v_3u_3\\ &= \langle \mathbf{v}, \mathbf{u}\rangle.\end{aligned}$$

2. Let $\mathbf{w} = (w_1, w_2, w_3)$. Then

$$\begin{aligned}\langle \mathbf{u}, \mathbf{v} + \mathbf{w}\rangle &= 2u_1(v_1 + w_1) + u_2(v_2 + w_2) + 2u_3(v_3 + w_3)\\ &= 2u_1v_1 + 2u_1w_1 + u_2v_2 + u_2w_2 + 2u_3v_3 + 2u_3w_3\\ &= 2u_1v_1 + u_2v_2 + 2u_3v_3 + 2u_1w_1 + u_2w_2 + 2u_3w_3\\ &= \langle \mathbf{u}, \mathbf{v}\rangle + \langle \mathbf{u}, \mathbf{w}\rangle.\end{aligned}$$

3. If c is any scalar, then

$$\begin{aligned}c\langle \mathbf{u}, \mathbf{v}\rangle &= c(2u_1v_1 + u_2v_2 + 2u_3v_3)\\ &= 2(cu_1)v_1 + (cu_2)v_2 + 2(cu_3)v_3\\ &= \langle c\mathbf{u}, \mathbf{v}\rangle.\end{aligned}$$

4. Since the square of a real number is nonnegative, $\langle \mathbf{v}, \mathbf{v}\rangle = 2v_1^2 + v_2^2 + 2v_3^2 \geq 0$. Moreover, this expression is equal to zero, if and only if $\mathbf{v} = \mathbf{0}$ (that is, if and only if $v_1 = v_2 = v_3 = 0$).

9. The product $\langle \mathbf{u}, \mathbf{v}\rangle$ is not an inner product because nonzero vectors can have a norm of zero. For example, let $\mathbf{v} = (0, 1)$, then, $\langle \mathbf{v}, \mathbf{v}\rangle = 0^2 = 0$.

11. The product $\langle \mathbf{u}, \mathbf{v}\rangle$ is not an inner product because Axiom 2, Axiom 3, Axiom 4 are not satisfied. For example, let $\mathbf{u} = (1, 1)$, $\mathbf{v} = (1, 2)$, $\mathbf{w} = (2, 0)$, and $c = 2$.

Axiom 2: Then $\langle \mathbf{u}, \mathbf{v} + \mathbf{w}\rangle = (1)^2(3)^2 - (1)^2(2)^2 = 5$ and

$\langle \mathbf{u}, \mathbf{v}\rangle + \langle \mathbf{u}, \mathbf{w}\rangle = \left[(1)^2(1)^2 - (1)^2(2)^2\right] + \left[(1)^2(2)^2 - (1)^2(0)^2\right] = 1$, which are not equal.

Axiom 3: Then $c\langle \mathbf{u}, \mathbf{v}\rangle = 2\left[(1)^2(1)^2 - (1)^2(2)^2\right] = -6$ and $\langle c\mathbf{u}, \mathbf{v}\rangle = (2)^2(1)^2 - (2)^2(2)^2 = -12$, which are not equal.

Axiom 4: Then $\langle \mathbf{v}, \mathbf{v}\rangle = (1)^2(1)^2 - (2)^2(2)^2 = -15$, which is less than zero.

13. The product $\langle \mathbf{u}, \mathbf{v}\rangle$ is not an inner product because Axiom 4 is not satisfied. For example, let $\mathbf{v} = (1, 1, 1)$. Then, $\langle \mathbf{v}, \mathbf{v}\rangle = -(1)(1)(1) = -1$, which is less than zero.

15. The product $\langle \mathbf{u}, \mathbf{v}\rangle$ is not an inner product because Axiom 2 is not satisfied. For example, if $\mathbf{u} = (1, 0, 0)$, $\mathbf{v} = (1, 0, 0)$, and $\mathbf{w} = (1, 0, 0)$, then $\langle \mathbf{u}, \mathbf{v} + \mathbf{w}\rangle = 1^2(2)^2 + 0^2(0)^2 + 0^2(0)^2 = 4$ and

$\langle \mathbf{u}, \mathbf{v}\rangle + \langle \mathbf{u}, \mathbf{w}\rangle = 1^2(1)^2 + 0^2(0)^2 + 0^2(0)^2 + 1^2(1)^2 + 0^2(0)^2 + 0^2(0)^2 = 2$.

So, $\langle \mathbf{u}, \mathbf{v} + \mathbf{w}\rangle \neq \langle \mathbf{u}, \mathbf{v}\rangle + \langle \mathbf{u}, \mathbf{w}\rangle$.

17. (a) $\langle \mathbf{u}, \mathbf{v}\rangle = \mathbf{u} \cdot \mathbf{v} = 3(5) + 4(-12) = -33$

(b) $\|\mathbf{u}\| = \sqrt{\langle \mathbf{u}, \mathbf{u}\rangle} = \sqrt{\mathbf{u} \cdot \mathbf{u}} = \sqrt{3(3) + 4(4)} = 5$

(c) $\|\mathbf{v}\| = \sqrt{\langle \mathbf{v}, \mathbf{v}\rangle} = \sqrt{\mathbf{v} \cdot \mathbf{v}} = \sqrt{5(5) + (-12)(-12)} = \sqrt{169} = 13$

(d) $d(\mathbf{u}, \mathbf{v}) = \|\mathbf{u} - \mathbf{v}\| = \sqrt{\langle \mathbf{u} - \mathbf{v}, \mathbf{u} - \mathbf{v}\rangle} = \sqrt{(\mathbf{u} - \mathbf{v}) \cdot (\mathbf{u} - \mathbf{v})} = \sqrt{(-2)(-2) + 16(16)} = 2\sqrt{65}$

© 2013 Cengage Learning. All Rights Reserved. May not be scanned, copied or duplicated, or posted to a publicly accessible website, in whole or in part.

19. (a) $\langle \mathbf{u}, \mathbf{v}\rangle = 3u_1v_1 + u_2v_2 = 3(-4)(0) + 3(5) = 15$

(b) $\|\mathbf{u}\| = \sqrt{\langle \mathbf{u}, \mathbf{u}\rangle} = \sqrt{3(-4)^2 + 3^2} = \sqrt{57}$

(c) $\|\mathbf{v}\| = \sqrt{\langle \mathbf{v}, \mathbf{v}\rangle} = \sqrt{3 \cdot 0^2 + 5^2} = 5$

(d) $d(\mathbf{u}, \mathbf{v}) = \|\mathbf{u} - \mathbf{v}\| = \sqrt{\langle \mathbf{u} - \mathbf{v}, \mathbf{u} - \mathbf{v}\rangle} = \sqrt{3(-4)^2 + (-2)^2} = 2\sqrt{13}$

21. (a) $\langle \mathbf{u}, \mathbf{v}\rangle = \mathbf{u} \cdot \mathbf{v} = 0(9) + 9(-2) + 4(-4) = -34$

(b) $\|\mathbf{u}\| = \sqrt{\langle \mathbf{u}, \mathbf{u}\rangle} = \sqrt{\mathbf{u} \cdot \mathbf{u}} = \sqrt{0 + 9^2 + 4^2} = \sqrt{97}$

(c) $\|\mathbf{v}\| = \sqrt{\langle \mathbf{v}, \mathbf{v}\rangle} = \sqrt{\mathbf{v} \cdot \mathbf{v}} = \sqrt{9^2 + (-2)^2 + (-4)^2} = \sqrt{101}$

(d) $d(\mathbf{u}, \mathbf{v}) = \|\mathbf{u} - \mathbf{v}\| = \|(-9, 11, 8)\| = \sqrt{9^2 + 11^2 + 8^2} = \sqrt{266}$

23. (a) $\langle \mathbf{u}, \mathbf{v}\rangle = 2u_1v_1 + 3u_2v_2 + u_3v_3 = 2 \cdot 8 \cdot 8 + 3 \cdot 0 \cdot 3 + (-8) \cdot 16 = 0.$

(b) $\|\mathbf{u}\| = \sqrt{2 \cdot 8 \cdot 8 + 3 \cdot 0 \cdot 0 + (-8)^2} = 8\sqrt{3}.$

(c) $\|\mathbf{v}\| = \sqrt{\langle \mathbf{v}, \mathbf{v}\rangle} = \sqrt{2 \cdot 8^2 + 3 \cdot 3^2 + 16^2} = \sqrt{411}$

(d) $d(\mathbf{u}, \mathbf{v}) = \|\mathbf{u} - \mathbf{v}\| = \|(0, -3, -24)\| = 3\sqrt{67}.$

25. (a) $\langle \mathbf{u}, \mathbf{v}\rangle = \mathbf{u} \cdot \mathbf{v} = 2(2) + 0(2) + 1(0) + (-1)(1) = 3$

(b) $\|\mathbf{u}\| = \sqrt{\langle \mathbf{u}, \mathbf{u}\rangle} = \sqrt{\mathbf{u} \cdot \mathbf{u}} = \sqrt{2^2 + 0^2 + 1^2 + (-1)^2} = \sqrt{6}$

(c) $\|\mathbf{v}\| = \sqrt{\langle \mathbf{v}, \mathbf{v}\rangle} = \sqrt{\mathbf{v} \cdot \mathbf{v}} = \sqrt{2^2 + 2^2 + 0^2 + 1^2} = 3$

(d) $d(\mathbf{u}, \mathbf{v}) = \|\mathbf{u} - \mathbf{v}\| = \|(0, -2, 1, -2)\| = \sqrt{0^2 + (-2)^2 + 1^2 + (-2)^2} = 3$

27. 1. Since the product of real numbers is commutative,

$$\begin{aligned}\langle A, B\rangle &= a_{11}b_{11} + a_{21}b_{21} + a_{12}b_{12} + a_{22}b_{22}\\ &= b_{11}a_{11} + b_{21}a_{21} + b_{12}a_{12} + b_{22}a_{22}\\ &= \langle B, A\rangle.\end{aligned}$$

(*Note:* Multiplication of matrices is *not* commutative!)

2. Let $W = \begin{bmatrix} w_{11} & w_{12} \\ w_{21} & w_{22}\end{bmatrix}$. Then,

$$\begin{aligned}\langle A, B + W\rangle &= a_{11}(b_{11} + w_{11}) + a_{21}(b_{21} + w_{21}) + a_{12}(b_{12} + w_{12}) + a_{22}(b_{22} + w_{22})\\ &= a_{11}b_{11} + a_{11}w_{11} + a_{21}b_{21} + a_{21}w_{21} + a_{12}b_{12} + a_{12}w_{12} + a_{22}b_{22} + a_{22}w_{22}\\ &= a_{11}b_{11} + a_{21}b_{21} + a_{12}b_{12} + a_{22}b_{22} + a_{11}w_{11} + a_{21}w_{21} + a_{12}w_{12} + a_{22}w_{22}\\ &= \langle A, B\rangle + \langle A, W\rangle.\end{aligned}$$

3. If c is any scalar, then

$$\begin{aligned}c\langle A, B\rangle &= c(a_{11}b_{11}) + c(a_{21}b_{21}) + c(a_{12}b_{12}) + c(a_{22}b_{22})\\ &= (ca_{11})b_{11} + (ca_{21})b_{21} + (ca_{12})b_{12} + (ca_{22})b_{22}\\ &= \langle CA, B\rangle\end{aligned}$$

4. Since the square of a real number is nonnegative, $\langle B, B\rangle = b_{11}^2 + b_{21}^2 + b_{12}^2 + b_{22}^2 \ge 0$. Moreover, this expression is equal to zero if and only if $B = O$ (that is, if and only if $b_{11} = b_{12} = b_{21} = b_{22} = 0$).

© 2013 Cengage Learning. All Rights Reserved. May not be scanned, copied or duplicated, or posted to a publicly accessible website, in whole or in part.

29. (a) $\langle A, B\rangle = 2(-1)(0) + 4(1) + 3(-2) + 2(-2)(1) = -6$

(b) $\|A\|^2 = \langle A, A\rangle = 2(-1)^2 + 4^2 + 3^2 + 2(-2)^2 = 35$

$\|A\| = \sqrt{\langle A, A\rangle} = \sqrt{35}$

(c) $\|B\|^2 = \langle B, B\rangle = 2 \cdot 0^2 + 1^2 + (-2)^2 + 2 \cdot 1^2 = 7$

$\|B\| = \sqrt{\langle B, B\rangle} = \sqrt{7}$

(d) Use the fact that $d(A, B) = \|A - B\|$.

$$\langle A - B, A - B\rangle = 2(-1)^2 + 3^2 + 5^2 + 2(-3)^2 = 54$$

$$d(A, B) = \sqrt{\langle A - B, A - B\rangle} = 3\sqrt{6}$$

31. (a) $\langle A, B\rangle = 2(1)(0) + (2)(-2) + (-1)(1) + 2(4)(0) = -5$

(b) $\|A\|^2 = \langle A, A\rangle = 2(1)^2 + (2)^2 + (-1)^2 + 2(4)^2 = 39$

$\|A\| = \sqrt{\langle A, A\rangle} = \sqrt{39}$

(c) $\|B\|^2 = \langle B, B\rangle = 2(0)^2 + (-2)^2 + 1^2 + 0^2 = 5$

$\|B\| = \sqrt{\langle B, B\rangle} = \sqrt{5}$

(d) Use the fact that $d(A, B) = \|A - B\|$.

$$\langle A - B, A - B\rangle = 2(1)^2 + 4^2 + (-2)^2 + 2(4)^2 = 54$$

$$d(A, B) = \sqrt{\langle A - B, A - B\rangle} = \sqrt{54} = 3\sqrt{6}$$

33. 1. Because the product of real numbers is commutative,

$$\begin{aligned}\langle p, q\rangle &= a_0b_0 + 2a_1b_1 + a_2b_2\\ &= b_0a_0 + 2b_1a_1 + b_2a_2\\ &= \langle q, p\rangle.\end{aligned}$$

2. Let $W(x) = w_0 + w_1x + w_2x^2$, then

$$\begin{aligned}\langle p, q + w\rangle &= a_0(b_0 + w_0) + 2a_1(b_1 + w_1) + a_2(b_2 + w_2)\\ &= a_0b_0 + a_0w_0 + 2a_1b_1 + 2a_1w_1 + a_2b_2 + a_2w_2\\ &= a_0b_0 + 2a_1b_1 + a_2b_2 + a_0w_0 + 2a_1w_1 + a_2w_2\\ &= \langle p, q\rangle + \langle p, w\rangle.\end{aligned}$$

3. If c is any scalar, then

$$\begin{aligned}c\langle p, q\rangle &= c(a_0b_0 + 2a_1b_1 + a_2b_2)\\ &= (ca_0)b_0 + (2ca_1)b_1 + (ca_2)b_2\\ &= \langle cp, q\rangle.\end{aligned}$$

4. Since the square of a real number is nonnegative, $\langle q, q\rangle = b_0^2 + 2b_1^2 + b_2^2 \geq 0$.

Moreover, this expression is equal to zero if and only if $q = 0$ (that is, if and only if $q_0 = q_1 = q_2 = 0$).

35. (a) $\langle p, q\rangle = 1(0) + (-1)(1) + 3(-1) = -4$

(b) $\|p\| = \sqrt{\langle p, p\rangle} = \sqrt{1^2 + (-1)^2 + 3^2} = \sqrt{11}$

(c) $\|q\| = \sqrt{\langle q, q\rangle} = \sqrt{0^2 + 1^2 + (-1)^2} = \sqrt{2}$

(d) $d(p, q) = \|p - q\| = \sqrt{\langle p - q, p - q\rangle} = \sqrt{1^2 + (-2)^2 + 4^2} = \sqrt{21}$

© 2013 Cengage Learning. All Rights Reserved. May not be scanned, copied or duplicated, or posted to a publicly accessible website, in whole or in part.

37. (a) $\langle p, q\rangle = 1(1) + 0(0) + 1(-1) = 0$

(b) $\|p\| = \sqrt{\langle p, p\rangle} = \sqrt{1^2 + 0^2 + 1^2} = \sqrt{2}$

(c) $\|q\| = \sqrt{\langle q, q\rangle} = \sqrt{1^2 + 0^2 + (-1)^2} = \sqrt{2}$

(d) $d(p, q) = \|p - q\| = \sqrt{\langle p - q, p - q\rangle} = \sqrt{0^2 + 0^2 + 2^2} = 2$

39. (a) $\langle f, g\rangle = \int_{-1}^{1} f(x)g(x)dx = \int_{-1}^{1} x^2(x^2 + 1)dx = \int_{-1}^{1} (x^4 + x^2)dx = \left[\frac{x^5}{5} + \frac{x^3}{3}\right]_{-1}^{1} = \frac{16}{15}$

(b) $\|f\|^2 = \langle f, f\rangle = \int_{-1}^{1} (x^2)^2 dx = \frac{x^5}{5}\Big]_{-1}^{1} = \frac{2}{5}$

$\|f\| = \sqrt{\frac{2}{5}} = \frac{\sqrt{10}}{5}$

(c) $\|g\|^2 = \langle g, g\rangle = \int_{-1}^{1} (x^2 + 1)^2 dx = \int_{-1}^{1} (x^4 + 2x^2 + 1)dx = \left[\frac{x^5}{5} + \frac{2x^3}{3} + x\right]_{-1}^{1} = \frac{56}{15}$

$\|g\| = \sqrt{\frac{56}{15}} = \frac{2\sqrt{210}}{15}$

(d) Use the fact that $d(f, g) = \|f - g\|$. Because $f - g = x^2 - (x^2 + 1) = -1$, you have

$\langle f - g, f - g\rangle = \int_{-1}^{1} (-1)(-1)\, dx = x\Big]_{-1}^{1} = 2.$

So, $d(f, g) = \sqrt{\langle f - g, f - g\rangle} = \sqrt{2}.$

41. (a) $\langle f, g\rangle = \int_{-1}^{1} xe^x\, dx = (x - 1)e^x\Big]_{-1}^{1} = \frac{2}{e}$

(b) $\|f\|^2 = \langle f, f\rangle = \int_{-1}^{1} x^2 dx = \frac{x^3}{3}\Big]_{-1}^{1} = \frac{2}{3}$

$\|f\| = \frac{\sqrt{6}}{3}$

(c) $\|g\|^2 = \langle g, g\rangle = \int_{-1}^{1} e^{2x} dx = \frac{e^{2x}}{2}\Big]_{-1}^{1} = \frac{1}{2}(e^2 - e^{-2})$

$\|g\| = \sqrt{\frac{1}{2}(e^2 - e^{-2})}$

(d) Use the fact that $d(f, g) = \|f - g\|$. Because $f - g = x - e^x$, you have

$$\begin{aligned}\langle f - g, f - g\rangle &= \int_{-1}^{1} (x - e^x)^2 dx \\ &= \int_{-1}^{1} (x^2 - 2xe^x + e^{2x})dx \\ &= \left[\frac{x^3}{3} - 2(x - 1)e^x + \frac{e^{2x}}{2}\right]_{-1}^{1} \\ &= \frac{2}{3} - \frac{4}{e} + \frac{e^2 - e^{-2}}{2} \\ &= \frac{2}{3} - \frac{4}{e} + \frac{e^2}{2} - \frac{1}{2e^2}.\end{aligned}$$

So, $d(f, g) = \sqrt{\langle f - g, f - g\rangle} = \sqrt{\frac{2}{3} - \frac{4}{e} + \frac{e^2}{2} - \frac{1}{2e^2}}.$

© 2013 Cengage Learning. All Rights Reserved. May not be scanned, copied or duplicated, or posted to a publicly accessible website, in whole or in part.

43. Because

$$\frac{\langle \mathbf{u}, \mathbf{v}\rangle}{\|\mathbf{u}\|\|\mathbf{v}\|} = \frac{3(5) + 4(-12)}{\sqrt{3^2 + 4^2}\sqrt{5^2 + (-12)^2}} = \frac{-33}{5 \cdot 13} = \frac{-33}{65},$$

the angle between **u** and **v** is

$$\cos^{-1}\left(\frac{-33}{65}\right) \approx 2.103 \text{ radians } (120.51°).$$

45. Because

$$\frac{\langle \mathbf{u}, \mathbf{v}\rangle}{\|\mathbf{u}\|\|\mathbf{v}\|} = \frac{3(-4)(0) + (3)(5)}{\sqrt{3(-4)^2 + 3^2}\sqrt{3(0)^2 + 5^2}} = \frac{15}{\sqrt{57} \cdot 5} = \frac{3}{\sqrt{57}},$$

the angle between **u** and **v** is

$$\cos^{-1}\left(\frac{3}{\sqrt{57}}\right) \approx 1.16 \text{ radians } (66.59°).$$

47. Because

$$\langle \mathbf{u}, \mathbf{v}\rangle = 1(2) + 2(1)(-2) + 1(2) = 0,$$

the angle between **u** and **v** is $\cos^{-1}(0) = \frac{\pi}{2}$.

49. Because

$$\frac{\langle p, q\rangle}{\|p\|\|q\|} = \frac{1 - 1 + 1}{\sqrt{3}\sqrt{3}} = \frac{1}{3},$$

the angle between p and q is

$$\cos^{-1}\left(\frac{1}{3}\right) \approx 1.23 \text{ radians } (70.53°).$$

51. Because

$$\langle f, g\rangle = \int_{-1}^{1} x^3\, dx = \frac{x^4}{4}\Bigg]_{-1}^{1} = 0,$$

the angle between f and g is $\cos^{-1}(0) = \frac{\pi}{2}$.

53. (a) To verify the Cauchy-Schwarz Inequality, observe

$$|\langle \mathbf{u}, \mathbf{v}\rangle| \le \|\mathbf{u}\|\|\mathbf{v}\|$$
$$|(5, 12) \cdot (3, 4)| \le \|(5, 12)\|\|(3, 4)\|$$
$$63 \le (13)(5) = 65.$$

(b) To verify the Triangle Inequality, observe

$$\|\mathbf{u} + \mathbf{v}\| \le \|\mathbf{u}\| + \|\mathbf{v}\|$$
$$\|(8, 16)\| \le \|(5, 12)\| + \|(3, 4)\|$$
$$\sqrt{320} \le 13 + 5$$
$$8\sqrt{5} \le 18.$$

55. (a) To verify the Cauchy-Schwarz Inequality, observe

$$|\langle \mathbf{u}, \mathbf{v}\rangle| \le \|\mathbf{u}\|\|\mathbf{v}\|$$
$$|(1, 0, 4) \cdot (-5, 4, 1)| \le \|(1, 0, 4)\|\|(-5, 4, 1)\|$$
$$1 \le \sqrt{17}\sqrt{42}$$
$$1 \le \sqrt{714}.$$

(b) To verify the Triangle Inequality, observe

$$\|\mathbf{u} + \mathbf{v}\| \le \|\mathbf{u}\| + \|\mathbf{v}\|$$
$$\|(-4, 4, 5)\| \le \|(1, 0, 4)\| + \|(-5, 4, 1)\|$$
$$\sqrt{57} \le \sqrt{17} + \sqrt{42}$$
$$7.5498 \le 10.6038.$$

57. (a) To verify the Cauchy-Schwarz Inequality, observe

$$|\langle p, q\rangle| \le \|p\|\|q\|$$
$$|0(1) + 2(0) + 0(3)| \le (2)\sqrt{10}$$
$$0 \le 2\sqrt{10}.$$

(b) To verify the Triangle Inequality, observe

$$\|p + q\| \le \|p\| + \|q\|$$
$$\|1 + 2x + 3x^2\| \le 2 + \sqrt{10}$$
$$\sqrt{14} \le 2 + \sqrt{10}$$
$$3.742 \le 5.162.$$

59. (a) To verify the Cauchy-Schwarz Inequality, observe

$$|\langle A, B\rangle| \le \|A\|\|B\|$$
$$|0(-3) + 3(1) + 2(4) + 1(3)| \le \sqrt{14}\sqrt{35}$$
$$14 \le \sqrt{14}\sqrt{35}.$$

(b) To verify the Triangle Inequality, observe

$$\|A + B\| \le \|A\| + \|B\|$$
$$\left\|\begin{bmatrix} -3 & 4 \\ 6 & 4 \end{bmatrix}\right\| \le \sqrt{14} + \sqrt{35}$$
$$\sqrt{77} \le \sqrt{14} + \sqrt{35}$$
$$8.775 \le 9.658.$$

© 2013 Cengage Learning. All Rights Reserved. May not be scanned, copied or duplicated, or posted to a publicly accessible website, in whole or in part.

61. (a) To verify the Cauchy-Schwarz Inequality, compute each part of the inequality.

$$\begin{aligned}\langle f, g\rangle &= \langle \sin x, \cos x\rangle \\ &= \int_0^{\pi/4} \sin x \cos x\, dx \\ &= \left[\frac{\sin^2 x}{2}\right]_0^{\pi/4} \\ &= \frac{\sin^2(\pi/4)}{2} - \frac{\sin^2 0}{2} = \frac{1}{4}\end{aligned}$$

$$\begin{aligned}\|f\|^2 &= \langle \sin x, \sin x\rangle \\ &= \int_0^{\pi/4} \sin^2 x\, dx = \int_0^{\pi/4} \frac{1 - \cos 2x}{2} dx \\ &= \left[\frac{1}{2}x - \frac{\sin 2x}{4}\right]_0^{\pi/4} = \left[\frac{1}{2}\left(\frac{\pi}{4}\right) - \frac{\sin 2(\pi/4)}{4}\right] - 0 \\ &= \frac{\pi}{8} - \frac{1}{4}\end{aligned}$$

So, $\|f\| = \sqrt{\frac{\pi}{8} - \frac{1}{4}}$.

$$\begin{aligned}\|g\|^2 &= \langle \cos x, \cos x\rangle \\ &= \int_0^{\pi/4} \cos^2 x\, dx = \int_0^{\pi/4} \frac{1 + \cos 2x}{2} dx \\ &= \left[\frac{1}{2x} + \frac{\sin 2x}{4}\right]_0^{\pi/4} = \frac{\pi}{8} + \frac{1}{4}\end{aligned}$$

So, $\|g\| = \sqrt{\frac{\pi}{8} + \frac{1}{4}}$. Therefore, observe that $|\langle f, g\rangle| \le \|f\|\|g\|$, $\frac{1}{4} \le \left(\sqrt{\frac{\pi}{8} - \frac{1}{4}}\right)\left(\sqrt{\frac{\pi}{8} + \frac{1}{4}}\right)$.

(b) From part (a), $\|f\| = \sqrt{\frac{\pi}{8} - \frac{1}{4}}$ and $\|g\| = \sqrt{\frac{\pi}{8} + \frac{1}{4}}$.

$$\|f\| + \|g\| = \sqrt{\frac{\pi}{8} - \frac{1}{4}} + \sqrt{\frac{\pi}{8} + \frac{1}{4}}$$

$$\begin{aligned}\|f + g\|^2 &= \langle \sin x + \cos x, \sin x + \cos x\rangle \\ &= \int_0^{\pi/4} \left(\sin^2 x + 2 \sin x \cos x + \cos^2 x\right) dx \\ &= \int_0^{\pi/4} \frac{1 - \cos 2x\, dx}{2} + 2\int_0^{\pi/4} \sin x \cos x\, dx + \int_0^{\pi/4} \frac{1 + \cos 2x\, dx}{2} \\ &= \frac{\pi}{8} - \frac{1}{4} + \frac{1}{2} + \frac{\pi}{8} + \frac{1}{4} \\ &= \frac{\pi}{4} + \frac{1}{2}\end{aligned}$$

So, $\|f + g\| = \sqrt{\frac{\pi}{4} + \frac{1}{2}}$. Therefore, observe that $\|f + g\| \le \|f\| + \|g\|$.

© 2013 Cengage Learning. All Rights Reserved. May not be scanned, copied or duplicated, or posted to a publicly accessible website, in whole or in part.

63. (a) To verify the Cauchy-Schwarz Inequality, compute

$$\langle f, g\rangle = \langle x, e^x\rangle = \int_0^1 x\,e^x\,dx = \left[e^x(x-1)\right]_0^1 = 1$$

$$\|f\|^2 = \langle x, x\rangle = \int_0^1 x^2\,dx = \left.\frac{x^3}{3}\right]_0^1 = \frac{1}{3} \Rightarrow \|f\| = \frac{\sqrt{3}}{3}$$

$$\|g\|^2 = \langle e^x, e^x\rangle = \int_0^1 e^{2x}\,dx = \left.\frac{e^{2x}}{2}\right]_0^1 = \frac{e^2}{2} - \frac{1}{2} \Rightarrow \|g\| = \sqrt{\frac{e^2}{2} - \frac{1}{2}}$$

and observe that

$$|\langle f, g\rangle| \le \|f\|\|g\|$$

$$1 \le \left(\sqrt{\frac{3}{3}}\right)\left(\sqrt{\frac{e^2}{2} - \frac{1}{2}}\right)$$

$$1 \le 1.032.$$

(b) To verify the Triangle Inequality, compute

$$\|f + g\|^2 = \langle x + e^x, x + e^x\rangle = \int_0^1 (x + e^x)^2\,dx = \left[\frac{e^{2x}}{2} + 2e^x(x-1) + \frac{x^3}{3}\right]_0^1$$

$$= \left[\frac{e^2}{2} + 2e(0) + \frac{1}{3}\right] - \left[\frac{1}{2} + 2(1)(-1) + 0\right]$$

$$= \frac{e^2}{2} + \frac{11}{6} \Rightarrow \|f + g\| = \sqrt{\frac{e^2}{2} + \frac{11}{6}}$$

and observe that

$$\|f + g\| \le \|f\| + \|g\|$$

$$\sqrt{\frac{e^2}{2} + \frac{11}{6}} \le \frac{\sqrt{3}}{3} + \sqrt{\frac{e^2}{2} - \frac{1}{2}}$$

$$2.351 \le 2.364.$$

65. Because $\langle f, g\rangle = \int_{-\pi/2}^{\pi/2} \cos x \sin x = \left[\frac{1}{2}\sin^2 x\right]_{-\pi/2}^{\pi/2} = 0$, f and g are orthogonal.

67. Because

$$\langle f, g\rangle = \int_{-1}^1 x\frac{1}{2}(5x^3 - 3x)\,dx = \frac{1}{2}\int_{-1}^1 (5x^4 - 3x^2)\,dx = \left.\frac{1}{2}(x^5 - x^3)\right]_{-1}^1 = 0,$$

f and g are orthogonal.

69. (a) $\text{proj}_{\mathbf{v}}\mathbf{u} = \frac{\langle \mathbf{u}, \mathbf{v}\rangle}{\langle \mathbf{v}, \mathbf{v}\rangle}\mathbf{v} = \frac{1(2) + 2(1)}{2^2 + 1^2}(2, 1) = \frac{4}{5}(2, 1) = \left(\frac{8}{5}, \frac{4}{5}\right)$

(b) $\text{proj}_{\mathbf{u}}\mathbf{v} = \frac{\langle \mathbf{v}, \mathbf{u}\rangle}{\langle \mathbf{u}, \mathbf{u}\rangle}\mathbf{u} = \frac{2(1) + 1(2)}{1^2 + 2^2}(1, 2) = \frac{4}{5}(1, 2) = \left(\frac{4}{5}, \frac{8}{5}\right)$

(c)

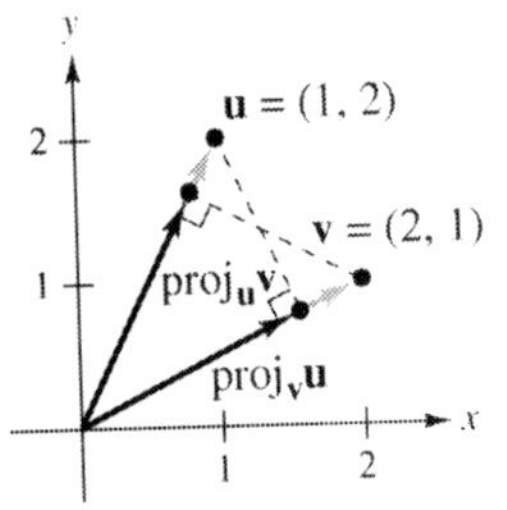

© 2013 Cengage Learning. All Rights Reserved. May not be scanned, copied or duplicated, or posted to a publicly accessible website, in whole or in part.

71. (a) $\text{proj}_{\mathbf{v}}\mathbf{u} = \dfrac{\langle \mathbf{u}, \mathbf{v}\rangle}{\langle \mathbf{v}, \mathbf{v}\rangle}\mathbf{v} = \dfrac{(-1)(4) + 3(4)}{4(4) + 4(4)}(4, 4) = \dfrac{1}{4}(4, 4) = (1, 1)$

(b) $\text{proj}_{\mathbf{u}}\mathbf{v} = \dfrac{\langle \mathbf{v}, \mathbf{u}\rangle}{\langle \mathbf{u}, \mathbf{u}\rangle}\mathbf{u} = \dfrac{4(-1) + 4(3)}{(-1)(-1) + 3(3)}(-1, 3) = \dfrac{4}{5}(-1, 3) = \left(-\dfrac{4}{5}, \dfrac{12}{5}\right)$

(c)

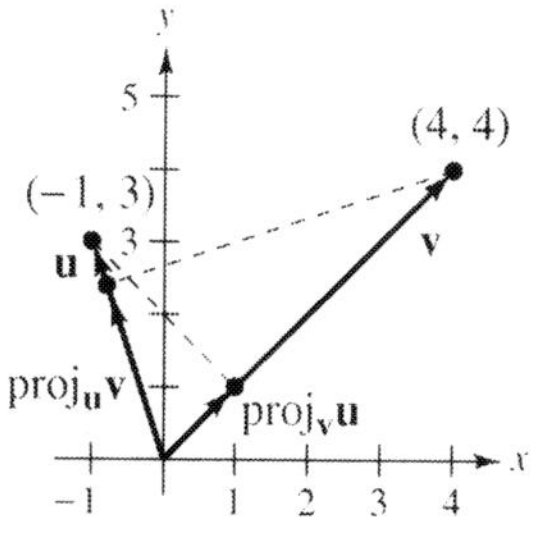

73. (a) $\text{proj}_{\mathbf{v}}\mathbf{u} = \dfrac{\langle \mathbf{u}, \mathbf{v}\rangle}{\langle \mathbf{v}, \mathbf{v}\rangle}\mathbf{v} = \dfrac{1(0) + 3(-1) + (-2)(1)}{0^2 + (-1)^2 + 1^2}(0, -1, 1) = \dfrac{-5}{2}(0, -1, 1) = \left(0, \dfrac{5}{2}, -\dfrac{5}{2}\right)$

(b) $\text{proj}_{\mathbf{u}}\mathbf{v} = \dfrac{\langle \mathbf{v}, \mathbf{u}\rangle}{\langle \mathbf{u}, \mathbf{u}\rangle}\mathbf{u} = \dfrac{0(1) + (-1)(3) + 1(-2)}{1^2 + 3^2 + (-2)^2}(1, 3, -2) = \dfrac{-5}{14}(1, 3, -2) = \left(-\dfrac{5}{14}, -\dfrac{15}{14}, \dfrac{5}{7}\right)$

75. (a) $\text{proj}_{\mathbf{v}}\mathbf{u} = \dfrac{\langle \mathbf{u}, \mathbf{v}\rangle}{\langle \mathbf{v}, \mathbf{v}\rangle}\mathbf{v} = \dfrac{0(-1) + 1(1) + 3(2) + (-6)(2)}{(-1)(-1) + 1(1) + 2(2) + 2(2)}\mathbf{v} = -\dfrac{5}{10}(-1, 1, 2, 2) = \left(\dfrac{1}{2}, -\dfrac{1}{2}, -1, -1\right)$

(b) $\text{proj}_{\mathbf{u}}\mathbf{v} = \dfrac{\langle \mathbf{v}, \mathbf{u}\rangle}{\langle \mathbf{u}, \mathbf{u}\rangle}\mathbf{u} = \dfrac{(-1)(0) + 1(1) + 2(3) + 2(-6)}{0 + 1(1) + 3(3) + (-6)(-6)}\mathbf{u} = -\dfrac{5}{46}(0, 1, 3, -6) = \left(0, -\dfrac{5}{46}, -\dfrac{15}{46}, \dfrac{15}{23}\right)$

77. The inner products $\langle f, g\rangle$ and $\langle g, g\rangle$ are as follows.

$$\langle f, g\rangle = \int_{-1}^{1} x\,dx = \left.\frac{x^2}{2}\right]_{-1}^{1} = 0$$

$$\langle g, g\rangle = \int_{-1}^{1} dx = \left.x\right]_{-1}^{1} = 2$$

So, the projection of f onto g is

$$\text{proj}_g f = \frac{\langle f, g\rangle}{\langle g, g\rangle}g = \frac{0}{2}(1) = 0.$$

79. The inner products $\langle f, g\rangle$ and $\langle g, g\rangle$ are as follows.

$$\langle f, g\rangle = \int_0^1 xe^x\,dx = \Big[(x - 1)e^x\Big]_0^1 = 0 + 1 = 1$$

$$\langle g, g\rangle = \int_0^1 e^{2x}\,dx = \left.\frac{1}{2}e^{2x}\right]_0^1 = \frac{e^2 - 1}{2}$$

So, the projection of f onto g is

$$\text{proj}_g f = \frac{\langle f, g\rangle}{\langle g, g\rangle}g = \frac{1}{(e^2 - 1)/2}e^x = \frac{2e^x}{e^2 - 1}.$$

81. The inner product $\langle f, g\rangle$ is

$$\langle f, g\rangle = \int_{-\pi}^{\pi} \sin x \cos x\,dx = \left.\frac{\sin^2 x}{2}\right]_{-\pi}^{\pi} = 0.$$

So, the projection of f onto g is

$$\text{proj}_g f = \frac{\langle f, g\rangle}{\langle g, g\rangle}g = \frac{0}{\langle g, g\rangle}g = 0.$$

83. The inner products $\langle f, g\rangle$ and $\langle g, g\rangle$ are as follows.

$$\langle f, g\rangle = \int_{-\pi}^{\pi} x \sin 2x\,dx = \left[\frac{\sin 2x}{4} - \frac{x\cos 2x}{2}\right]_{-\pi}^{\pi} = -\pi$$

$$\langle g, g\rangle = \int_{-\pi}^{\pi} (\sin 2x)^2\,dx = \left[\frac{x}{2} - \frac{\sin 4x}{8}\right]_{-\pi}^{\pi} = \pi$$

So, the projection of f onto g is

$$\text{proj}_g f = \frac{\langle f, g\rangle}{\langle g, g\rangle}g = \frac{-\pi}{\pi}(\sin 2x) = -\sin 2x.$$

85. (a) False. See the introduction to this section, page, 237.

(b) False. $\|\mathbf{v}\| = \mathbf{0}$ if and only if $\mathbf{v} = \mathbf{0}$.

© 2013 Cengage Learning. All Rights Reserved. May not be scanned, copied or duplicated, or posted to a publicly accessible website, in whole or in part.

87. (a) $\langle \mathbf{u}, \mathbf{v} \rangle = 4(2) + 2(2)(-2) = 0 \Rightarrow \mathbf{u}$ and $\mathbf{v}$ are orthogonal.

(b) The vectors are not orthogonal in the Euclidean sense.

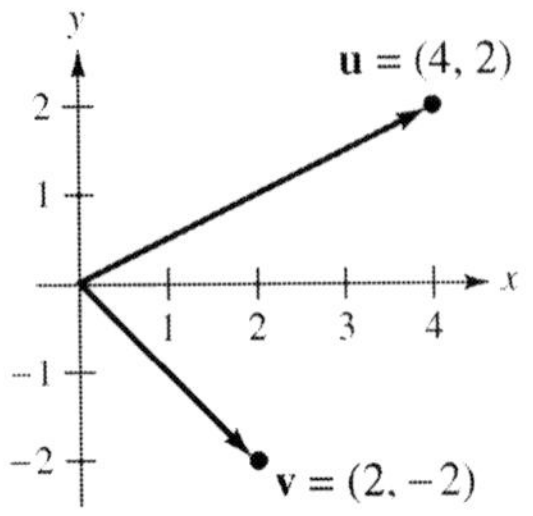

89. Verify the four parts of the definition for the function

$$\langle \mathbf{u}, \mathbf{v} \rangle = c_1u_1v_1 + \cdots + c_nu_nv_n = \sum_{i=1}^{n} c_iu_iv_i.$$

1. $\langle \mathbf{u}, \mathbf{v} \rangle = \sum_{i=1}^{n} c_iu_iv_i = \sum_{i=1}^{n} c_iv_iu_i = \langle \mathbf{v}, \mathbf{u} \rangle$

2. $\langle \mathbf{u}, \mathbf{v} + \mathbf{w} \rangle = \sum_{i=1}^{n} c_iu_i(v_i + w_i) = \sum_{i=1}^{n} c_iu_iv_i + \sum_{i=1}^{n} c_iu_iw_i$
$= \langle \mathbf{u}, \mathbf{v} \rangle + \langle \mathbf{u}, \mathbf{w} \rangle$

3. $d\langle \mathbf{u}, \mathbf{v} \rangle = d\sum_{i=1}^{n} c_iu_iv_i = \sum_{i=1}^{n} c_i(du_i)v_i = \langle d\mathbf{u}, \mathbf{v} \rangle$

4. $\langle \mathbf{v}, \mathbf{v} \rangle = \sum_{i=1}^{n} c_iv_i^2 \geq 0$, and $\langle \mathbf{v}, \mathbf{v} \rangle = 0$ if and only if $\mathbf{v} = \mathbf{0}$.

91. From the definition of inner product,

$$\langle \mathbf{u} + \mathbf{v}, \mathbf{w} \rangle = \langle \mathbf{w}, \mathbf{u} + \mathbf{v} \rangle$$
$$= \langle \mathbf{w}, \mathbf{u} \rangle + \langle \mathbf{w}, \mathbf{v} \rangle = \langle \mathbf{u}, \mathbf{w} \rangle + \langle \mathbf{v}, \mathbf{w} \rangle.$$

93. (i) $W^{\perp} = \{\mathbf{v} \in V: \langle \mathbf{v}, \mathbf{w} \rangle = 0 \text{ for all } \mathbf{w} \in W\}$ is nonempty because $\mathbf{0} \in W^{\perp}$.

(ii) Let $\mathbf{v}_1, \mathbf{v}_2 \in W^{\perp}$. Then $\langle \mathbf{v}_1, \mathbf{w} \rangle = \langle \mathbf{v}_2, \mathbf{w} \rangle = 0$ for all $\mathbf{w} \in W$.

So,

$\langle \mathbf{v}_1 + \mathbf{v}_2, \mathbf{w} \rangle = \langle \mathbf{v}_1, \mathbf{w} \rangle + \langle \mathbf{v}_2, \mathbf{w} \rangle = 0 + 0 = 0$

for all $\mathbf{w} \in W \Rightarrow \mathbf{v}_1 + \mathbf{v}_2 \in W^{\perp}$.

(iii) Let $\mathbf{v} \in W^{\perp}$ and $c \in R$. Then, $\langle \mathbf{v}, \mathbf{w} \rangle = 0$ for all $\mathbf{w} \in W$, and $\langle c\mathbf{v}, \mathbf{w} \rangle = c\langle \mathbf{v}, \mathbf{w} \rangle = c0 = 0$ for all $\mathbf{w} \in W \Rightarrow c\mathbf{v} \in W^{\perp}$.

95. (a) Let $\langle \mathbf{u}, \mathbf{v} \rangle$ be the Euclidean inner product on R^n. Because $\langle \mathbf{u}, \mathbf{v} \rangle = \mathbf{u}^T\mathbf{v}$, it follows that

$$\langle A^T\mathbf{u}, \mathbf{v} \rangle = (A^T\mathbf{u})^T\mathbf{v} = \mathbf{u}^T(A^T)^T\mathbf{v} = \mathbf{u}^TA\mathbf{v} = \langle \mathbf{u}, A\mathbf{v} \rangle.$$

(b) $\langle A^TA\mathbf{u}, \mathbf{u} \rangle = (A^TA\mathbf{u})^T\mathbf{u} = \mathbf{u}^TA^T(A^T)^T\mathbf{u} = \mathbf{u}^TA^TA\mathbf{u} = (A\mathbf{u})^TA\mathbf{u} = \langle A\mathbf{u}, A\mathbf{u} \rangle = \|A\mathbf{u}\|^2$

97. Let $\mathbf{u} = (x, y)$. Then $\|\mathbf{u}\| = \sqrt{c_1x^2 + c_2y^2} = 1$. Since the equation of the graph is $\frac{1}{4}x^2 + y^2 = 1$, $c_1 = \frac{1}{4}$ and $c_2 = 1$.

99. Let $\mathbf{u} = (x, y)$. Then $\|\mathbf{u}\| = \sqrt{c_1x^2 + c_2y^2} = 1$. Since the equation of the graph is $\frac{1}{4}x^2 + \frac{1}{16}y^2 = 1$, $c_1 = \frac{1}{4}$ and $c_2 = \frac{1}{16}$.

101. From example 10, you have $\text{proj}_{\mathbf{v}}\mathbf{u} = (2, 4, 0)$. So,

$$d(\mathbf{u}, \text{proj}_{\mathbf{v}}\mathbf{u}) = \|\mathbf{u} - \text{proj}_{\mathbf{v}}\mathbf{u}\| = \|(4, -2, 4)\| = \sqrt{36} = 6.$$

Let x be any real number different from $2 = \dfrac{\langle \mathbf{u}, \mathbf{v} \rangle}{\langle \mathbf{v}, \mathbf{v} \rangle}$, so that $x\mathbf{v} \neq \text{proj}_{\mathbf{v}}\mathbf{u}$. You want to show that

$$d(\mathbf{u}, x\mathbf{v}) > d(\mathbf{u}, \text{proj}_{\mathbf{v}}\mathbf{u}).$$

$$d(\mathbf{u}, x\mathbf{v}) = \sqrt{(6 - x)^2 + 4(1 - x)^2 + 16} = \sqrt{36 + 5x^2 - 20x + 20}$$
$$= \sqrt{36 + 5(x - 2)^2} > \sqrt{36} = d(\mathbf{u}, \text{proj}_{\mathbf{v}}\mathbf{u}),$$

if

$$x \neq 2 = \frac{\langle \mathbf{u}, \mathbf{v} \rangle}{\langle \mathbf{v}, \mathbf{v} \rangle}.$$

© 2013 Cengage Learning. All Rights Reserved. May not be scanned, copied or duplicated, or posted to a publicly accessible website, in whole or in part.

Section 5.3 Orthonormal Bases: Gram-Schmidt Process

1. (a) The set is orthogonal since
$(2, -4)\cdot(2, 1) = 2(2) - 4(1) = 0.$

(b) This set is *not* orthonormal since
$\|(2, -4)\| = \sqrt{2^2 + (-4)^2} = \sqrt{20} \neq 1.$

(c) Because the two vectors are not scalar multiples of each other, by the Corollary to Theorem 4.8 they are linearly independent. By Theorem 4.12, they are a basis for R^2.

3. (a) The set is *not* orthogonal since
$(-4, 6)\cdot(5, 0) = (-4)(5) + 6(0) = -20 \neq 0.$

(b) The set is *not* orthonormal since it is *not* orthogonal.

(c) Because the two vectors are not scalar multiples of each other, by the Corollary to Theorem 4.8 they are linearly independent. By Theorem 4.12, they are a basis for R^2.

5. (a) The set is orthogonal since
$\left(\frac{3}{5}, \frac{4}{5}\right)\cdot\left(-\frac{4}{5}, \frac{3}{5}\right) = \frac{3}{5}\left(-\frac{4}{5}\right) + \frac{4}{5}\left(\frac{3}{5}\right) = -\frac{12}{25} + \frac{12}{25} = 0.$

(b) The set is orthonormal since
$\left\|\left(\frac{3}{5}, \frac{4}{5}\right)\right\| = \sqrt{\left(\frac{3}{5}\right)^2 + \left(\frac{4}{5}\right)^2} = 1$ and
$\left\|\left(-\frac{4}{5}, \frac{3}{5}\right)\right\| = \sqrt{\left(-\frac{4}{5}\right)^2 + \left(\frac{3}{5}\right)^2} = 1.$

(c) Because the two vectors are not scalar multiples of each other, by the Corollary to Theorem 4.8 they are linearly independent. By Theorem 4.12, they are a basis for R^2.

7. (a) The set is orthogonal since
$(4, -1, 1)\cdot(-1, 0, 4) = -4 + 4 = 0,$
$(4, -1, 1)\cdot(-4, -17, -1) = -16 + 17 - 1 = 0,$ and
$(-1, 0, 4)\cdot(-4, -17, -1) = 4 - 4 = 0.$

(b) The set is *not* orthonormal since
$\|(4, -1, 1)\| = \sqrt{4^2 + (-1)^2 + 1^2} = \sqrt{18} \neq 1.$

(c) Because the three vectors do not lie in the same plane, they span R^3. By Theorem 4.12, they form a basis for R^3.

9. (a) The set is orthogonal since $\left(\frac{\sqrt{2}}{2}, 0, \frac{\sqrt{2}}{2}\right)\cdot\left(\frac{-\sqrt{6}}{6}, \frac{\sqrt{6}}{3}, \frac{\sqrt{6}}{6}\right) = 0,$ $\left(\frac{\sqrt{2}}{2}, 0, \frac{\sqrt{2}}{2}\right)\cdot\left(\frac{\sqrt{3}}{3}, \frac{\sqrt{3}}{3}, \frac{-\sqrt{3}}{3}\right) = 0,$ and
$\left(\frac{-\sqrt{6}}{6}, \frac{\sqrt{6}}{3}, \frac{\sqrt{6}}{6}\right)\cdot\left(\frac{\sqrt{3}}{3}, \frac{\sqrt{3}}{3}, \frac{-\sqrt{3}}{3}\right) = 0.$

(b) The set is orthonormal since $\left\|\left(\frac{\sqrt{2}}{2}, 0, \frac{\sqrt{2}}{2}\right)\right\| = \sqrt{\frac{1}{2} + 0 + \frac{1}{2}} = 1,$ $\left\|\left(\frac{-\sqrt{6}}{6}, \frac{\sqrt{6}}{3}, \frac{\sqrt{6}}{6}\right)\right\| = \sqrt{\frac{1}{6} + \frac{2}{3} + \frac{1}{6}} = 1,$ and
$\left\|\left(\frac{\sqrt{3}}{3}, \frac{\sqrt{3}}{3}, \frac{-\sqrt{3}}{3}\right)\right\| = \sqrt{\frac{1}{3} + \frac{1}{3} + \frac{1}{3}} = 1.$

(c) Because the three vectors do not lie in the same plane, they span R^3. By Theorem 4.12, they form a basis for R^3.

11. (a) The set is orthogonal since $(2, 5, -3)\cdot(4, 2, 6) = 8 + 10 - 18 = 0.$

(b) The set is *not* orthonormal since $\|(2, 5, -3)\| = \sqrt{2^2 + 5^2 + (-3)^2} = \sqrt{4 + 25 + 9} = \sqrt{38} \neq 1.$

(c) Since there are not enough vectors, the set is *not* a basis for R^3.

13. (a) The set is orthonormal since $\left(\frac{\sqrt{2}}{2}, 0, 0, \frac{\sqrt{2}}{2}\right)\cdot\left(0, \frac{\sqrt{2}}{2}, \frac{\sqrt{2}}{2}, 0\right) = 0,$ $\left(\frac{\sqrt{2}}{2}, 0, 0, \frac{\sqrt{2}}{2}\right)\cdot\left(\frac{-1}{2}, \frac{1}{2}, \frac{-1}{2}, \frac{1}{2}\right) = 0,$ and
$\left(0, \frac{\sqrt{2}}{2}, \frac{\sqrt{2}}{2}, 0\right)\cdot\left(\frac{-1}{2}, \frac{1}{2}, \frac{-1}{2}, \frac{1}{2}\right) = 0.$

(b) The set is orthonormal since $\left\|\left(\frac{\sqrt{2}}{2}, 0, 0, \frac{\sqrt{2}}{2}\right)\right\| = \sqrt{\frac{1}{2} + 0 + 0 + \frac{1}{2}} = 1,$ $\left\|\left(0, \frac{\sqrt{2}}{2}, \frac{\sqrt{2}}{2}, 0\right)\right\| = \sqrt{0 + \frac{1}{2} + \frac{1}{2} + 0} = 1,$
and $\left\|\left(\frac{-1}{2}, \frac{1}{2}, \frac{-1}{2}, \frac{1}{2}\right)\right\| = \sqrt{\frac{1}{4} + \frac{1}{4} + \frac{1}{4} + \frac{1}{4}} = 1.$

(c) Since there are not enough vectors, the set is *not* a basis for R^4.

© 2013 Cengage Learning. All Rights Reserved. May not be scanned, copied or duplicated, or posted to a publicly accessible website, in whole or in part.

15. (a) The set is orthogonal since $(-1, 4) \cdot (8, 2) = -8 + 8 = 0$.

(b) Since $\|(-1, 4)\| = \sqrt{(-1)^2 + 4^2} = \sqrt{17}$, normalizing the set produces an orthonormal set.

$$\mathbf{u}_1 = \frac{\mathbf{v}_1}{\|\mathbf{v}_1\|} = \frac{1}{\sqrt{17}}(-1, 4) = \left(\frac{-\sqrt{17}}{17}, \frac{4\sqrt{17}}{17}\right)$$

$$\mathbf{u}_2 = \frac{\mathbf{v}_2}{\|\mathbf{v}_2\|} = \frac{1}{2\sqrt{17}}(8, 2) = \left(\frac{4\sqrt{17}}{17}, \frac{\sqrt{17}}{17}\right)$$

17. (a) The set is orthogonal since $\left(\sqrt{3}, \sqrt{3}, \sqrt{3}\right) \cdot \left(-\sqrt{2}, 0, \sqrt{2}\right) = -\sqrt{6} + 0 + \sqrt{6} = 0$.

(b) Since $\left\|\left(\sqrt{3}, \sqrt{3}, \sqrt{3}\right)\right\| = \sqrt{\left(\sqrt{3}\right)^2 + \left(\sqrt{3}\right)^2 + \left(\sqrt{3}\right)^2} = \sqrt{9} = 3$, normalizing the set produces an orthonormal set.

$$\mathbf{u}_1 = \frac{\mathbf{v}_1}{\|\mathbf{v}_1\|} = \frac{1}{3}\left(\sqrt{3}, \sqrt{3}, \sqrt{3}\right) = \left(\frac{\sqrt{3}}{3}, \frac{\sqrt{3}}{3}, \frac{\sqrt{3}}{3}\right)$$

$$\mathbf{u}_2 = \frac{\mathbf{v}_2}{\|\mathbf{v}_2\|} = \frac{1}{2}\left(-\sqrt{2}, 0, \sqrt{2}\right) = \left(\frac{-\sqrt{2}}{2}, 0, \frac{\sqrt{2}}{2}\right)$$

19. The set $\{1, x, x^2, x^3\}$ is orthogonal because

$$\langle 1, x\rangle = 0, \langle 1, x^2\rangle = 0, \langle 1, x^3\rangle = 0,$$

$$\langle x, x^2\rangle = 0, \langle x, x^3\rangle = 0, \text{ and } \langle x^2, x^3\rangle = 0.$$

Furthermore, the set is orthonormal because

$$\|1\| = 1, \|x\| = 1, \|x^2\| = 1 \text{ and } \|x^3\| = 1.$$

So, $\{1, x, x^2, x^3\}$ is an orthonormal basis for P_3.

21. Use Theorem 5.11 to find the coordinates of $\mathbf{x} = (1, 2)$ relative to B.

$$(1, 2) \cdot \left(-\frac{2\sqrt{13}}{13}, \frac{3\sqrt{13}}{13}\right) = -\frac{2\sqrt{13}}{13} + \frac{6\sqrt{13}}{13} = \frac{4\sqrt{13}}{13}$$

$$(1, 2) \cdot \left(\frac{3\sqrt{13}}{13}, \frac{2\sqrt{13}}{13}\right) = \frac{3\sqrt{13}}{13} + \frac{4\sqrt{13}}{13} = \frac{7\sqrt{13}}{13}$$

So, $[\mathbf{x}]_B = \begin{bmatrix} \frac{4\sqrt{13}}{13} \\ \frac{7\sqrt{13}}{13} \end{bmatrix}$.

23. Use Theorem 5.11 to find the coordinates of $\mathbf{x} = (2, -2, 1)$ relative to B.

$$(2, -2, 1) \cdot \left(\frac{\sqrt{10}}{10}, 0, \frac{3\sqrt{10}}{10}\right) = \frac{2\sqrt{10}}{10} + \frac{3\sqrt{10}}{10} = \frac{\sqrt{10}}{2}$$

$$(2, -2, 1) \cdot (0, 1, 0) = -2$$

$$(2, -2, 1) \cdot \left(-\frac{3\sqrt{10}}{10}, 0, \frac{\sqrt{10}}{10}\right) = -\frac{6\sqrt{10}}{10} + \frac{\sqrt{10}}{10} = -\frac{\sqrt{10}}{2}$$

So, $[\mathbf{x}]_B = \begin{bmatrix} \frac{\sqrt{10}}{2} \\ -2 \\ -\frac{\sqrt{10}}{2} \end{bmatrix}$.

25. Use Theorem 5.11 to find the coordinates of $\mathbf{x} = (5, 10, 15)$ relative to B.

$$(5, 10, 15) \cdot \left(\tfrac{3}{5}, \tfrac{4}{5}, 0\right) = 3 + 8 = 11$$

$$(5, 10, 15) \cdot \left(-\tfrac{4}{5}, \tfrac{3}{5}, 0\right) = -4 + 6 = 2$$

$$(5, 10, 15) \cdot (0, 0, 1) = 15$$

So, $[\mathbf{x}]_B = \begin{bmatrix} 11 \\ 2 \\ 15 \end{bmatrix}$.

© 2013 Cengage Learning. All Rights Reserved. May not be scanned, copied or duplicated, or posted to a publicly accessible website, in whole or in part.

27. First, orthogonalize each vector in B.

$$\mathbf{w}_1 = \mathbf{v}_1 = (3, 4)$$

$$\mathbf{w}_2 = \mathbf{v}_2 - \frac{\langle \mathbf{v}_2, \mathbf{w}_1 \rangle}{\langle \mathbf{w}_1, \mathbf{w}_1 \rangle}\mathbf{w}_1 = (1, 0) - \frac{1(3) + 0(4)}{3^2 + 4^2}(3, 4) = (1, 0) - \frac{3}{25}(3, 4) = \left(\frac{16}{25}, -\frac{12}{25}\right)$$

Then, normalize the vectors.

$$\mathbf{u}_1 = \frac{\mathbf{w}_1}{\|\mathbf{w}_1\|} = \frac{1}{\sqrt{3^2 + 4^2}}(3, 4) = \left(\frac{3}{5}, \frac{4}{5}\right)$$

$$\mathbf{u}_2 = \frac{\mathbf{w}_2}{\|\mathbf{w}_2\|} = \frac{1}{\sqrt{\left(\frac{16}{25}\right)^2 + \left(-\frac{12}{25}\right)^2}}\left(\frac{16}{25}, -\frac{12}{25}\right) = \left(\frac{4}{5}, -\frac{3}{5}\right)$$

So, the orthonormal basis is $\left\{\left(\frac{3}{5}, \frac{4}{5}\right), \left(\frac{4}{5}, -\frac{3}{5}\right)\right\}$.

29. First, orthogonalize each vector in B.

$$\mathbf{w}_1 = \mathbf{v}_1 = (0, 1)$$

$$\mathbf{w}_2 = \mathbf{v}_2 - \frac{\langle \mathbf{v}_2, \mathbf{w}_1 \rangle}{\langle \mathbf{w}_1, \mathbf{w}_1 \rangle}\mathbf{w}_1 = (2, 5) - \frac{2(0) + 5(1)}{0^2 + 1^2}(0, 1) = (2, 5) - 5(0, 1) = (2, 0)$$

Then, normalize the vectors.

$$\mathbf{u}_1 = \frac{\mathbf{w}_1}{\|\mathbf{w}_1\|} = \mathbf{w}_1 = (0, 1)$$

$$\mathbf{u}_2 = \frac{\mathbf{w}_2}{\|\mathbf{w}_2\|} = \frac{1}{2}(2, 0) = (1, 0)$$

So, the orthonormal basis is $\{(0, 1), (1, 0)\}$.

31. Because $\mathbf{v}_i \cdot \mathbf{v}_j = 0$ for $i \neq j$, the given vectors are orthogonal. Normalize the vectors.

$$\mathbf{u}_1 = \frac{\mathbf{v}_1}{\|\mathbf{v}_1\|} = \frac{1}{3}(1, -2, 2) = \left(\frac{1}{3}, -\frac{2}{3}, \frac{2}{3}\right)$$

$$\mathbf{u}_2 = \frac{\mathbf{v}_2}{\|\mathbf{v}_2\|} = \frac{1}{3}(2, 2, 1) = \left(\frac{2}{3}, \frac{2}{3}, \frac{1}{3}\right)$$

$$\mathbf{u}_3 = \frac{\mathbf{v}_3}{\|\mathbf{v}_3\|} = \frac{1}{3}(2, -1, -2) = \left(\frac{2}{3}, -\frac{1}{3}, -\frac{2}{3}\right)$$

So, the orthonormal basis is

$$\left\{\left(\frac{1}{3}, -\frac{2}{3}, \frac{2}{3}\right), \left(\frac{2}{3}, \frac{2}{3}, \frac{1}{3}\right), \left(\frac{2}{3}, -\frac{1}{3}, -\frac{2}{3}\right)\right\}.$$

33. First, orthogonalize each vector in B.

$$\mathbf{w}_1 = \mathbf{v}_1 = (4, -3, 0)$$

$$\mathbf{w}_2 = \mathbf{v}_2 - \frac{\langle \mathbf{v}_2, \mathbf{w}_1 \rangle}{\langle \mathbf{w}_1, \mathbf{w}_1 \rangle}\mathbf{w}_1$$

$$= (1, 2, 0) - \frac{-2}{25}(4, -3, 0) = \left(\frac{33}{25}, \frac{44}{25}, 0\right)$$

$$\mathbf{w}_3 = \mathbf{v}_3 - \frac{\langle \mathbf{v}_3, \mathbf{w}_1 \rangle}{\langle \mathbf{w}_1, \mathbf{w}_1 \rangle}\mathbf{w}_1 - \frac{\langle \mathbf{v}_3, \mathbf{w}_2 \rangle}{\langle \mathbf{w}_2, \mathbf{w}_2 \rangle}\mathbf{w}_2$$

$$= (0, 0, 4) - 0(4, -3, 0) - 0\left(\frac{33}{25}, \frac{44}{25}, 0\right) = (0, 0, 4)$$

Then, normalize the vectors.

$$\mathbf{u}_1 = \frac{\mathbf{w}_1}{\|\mathbf{w}_1\|} = \frac{1}{5}(4, -3, 0) = \left(\frac{4}{5}, -\frac{3}{5}, 0\right)$$

$$\mathbf{u}_2 = \frac{\mathbf{w}_2}{\|\mathbf{w}_2\|} = \frac{5}{11}\left(\frac{33}{25}, \frac{44}{25}, 0\right) = \left(\frac{3}{5}, \frac{4}{5}, 0\right)$$

$$\mathbf{u}_3 = \frac{\mathbf{w}_3}{\|\mathbf{w}_3\|} = \frac{1}{4}(0, 0, 4) = (0, 0, 1)$$

So, the orthonormal basis is

$$\left\{\left(\frac{4}{5}, -\frac{3}{5}, 0\right), \left(\frac{3}{5}, \frac{4}{5}, 0\right), (0, 0, 1)\right\}.$$

© 2013 Cengage Learning. All Rights Reserved. May not be scanned, copied or duplicated, or posted to a publicly accessible website, in whole or in part.

35. First, orthogonalize each vector in B.

$$\mathbf{w}_1 = \mathbf{v}_1 = (0, 1, 1)$$

$$\mathbf{w}_2 = \mathbf{v}_2 - \frac{\langle \mathbf{v}_2, \mathbf{w}_1 \rangle}{\langle \mathbf{w}_1, \mathbf{w}_1 \rangle}\mathbf{w}_1 = (1, 1, 0) - \frac{1(0) + 1(1) + 0(1)}{0^2 + 1^2 + 1^2}(0, 1, 1) = (1, 1, 0) - \frac{1}{2}(0, 1, 1) = \left(1, \frac{1}{2}, -\frac{1}{2}\right)$$

$$\mathbf{w}_3 = \mathbf{v}_3 - \frac{\langle \mathbf{v}_3, \mathbf{w}_2 \rangle}{\langle \mathbf{w}_2, \mathbf{w}_2 \rangle}\mathbf{w}_2 - \frac{\langle \mathbf{v}_3, \mathbf{w}_1 \rangle}{\langle \mathbf{w}_1, \mathbf{w}_1 \rangle}\mathbf{w}_1$$

$$= (1, 0, 1) - \frac{1(1) + 0\left(\frac{1}{2}\right) + 1\left(-\frac{1}{2}\right)}{1^2 + \left(\frac{1}{2}\right)^2 + \left(-\frac{1}{2}\right)^2}\left(1, \frac{1}{2}, -\frac{1}{2}\right) - \frac{1(0) + 0(1) + 1(1)}{0^2 + 1^2 + 1^2}(0, 1, 1)$$

$$= (1, 0, 1) - \frac{1}{3}\left(1, \frac{1}{2}, -\frac{1}{2}\right) - \frac{1}{2}(0, 1, 1) = \left(\frac{2}{3}, -\frac{2}{3}, \frac{2}{3}\right)$$

Then, normalize the vectors.

$$\mathbf{u}_1 = \frac{\mathbf{w}_1}{\|\mathbf{w}_1\|} = \frac{1}{\sqrt{0^2 + 1^2 + 1^2}}(0, 1, 1) = \left(0, \frac{\sqrt{2}}{2}, \frac{\sqrt{2}}{2}\right)$$

$$\mathbf{u}_2 = \frac{\mathbf{w}_2}{\|\mathbf{w}_2\|} = \frac{1}{\sqrt{1^2 + \left(\frac{1}{2}\right)^2 + \left(-\frac{1}{2}\right)^2}}\left(1, \frac{1}{2}, -\frac{1}{2}\right) = \left(\frac{\sqrt{6}}{3}, \frac{\sqrt{6}}{6}, -\frac{\sqrt{6}}{6}\right)$$

$$\mathbf{u}_3 = \frac{\mathbf{w}_3}{\|\mathbf{w}_3\|} = \frac{1}{\sqrt{\left(\frac{2}{3}\right)^2 + \left(-\frac{2}{3}\right)^2 + \left(\frac{2}{3}\right)^2}}\left(\frac{2}{3}, -\frac{2}{3}, \frac{2}{3}\right) = \left(\frac{\sqrt{3}}{3}, -\frac{\sqrt{3}}{3}, \frac{\sqrt{3}}{3}\right)$$

So, the orthonormal basis is $\left\{\left(0, \frac{\sqrt{2}}{2}, \frac{\sqrt{2}}{2}\right), \left(\frac{\sqrt{6}}{3}, \frac{\sqrt{6}}{6}, -\frac{\sqrt{6}}{6}\right), \left(\frac{\sqrt{3}}{3}, -\frac{\sqrt{3}}{3}, \frac{\sqrt{3}}{3}\right)\right\}$.

37. Because there is just one vector, you simply need to normalize it.

$$\mathbf{u}_1 = \frac{1}{\sqrt{(-8)^2 + 3^2 + 5^2}}(-8, 3, 5)$$

$$= \left(-\frac{4\sqrt{2}}{7}, \frac{3\sqrt{2}}{14}, \frac{5\sqrt{2}}{14}\right)$$

So, the orthonormal basis is $\left\{\left(-\frac{4\sqrt{2}}{7}, \frac{3\sqrt{2}}{14}, \frac{5\sqrt{2}}{14}\right)\right\}$.

39. First, orthogonalize each vector in B.

$$\mathbf{w}_1 = \mathbf{v}_1 = (3, 4, 0)$$

$$\mathbf{w}_2 = \mathbf{v}_2 - \frac{\mathbf{v}_2 \cdot \mathbf{w}_1}{\mathbf{w}_1 \cdot \mathbf{w}_1}\mathbf{w}_1$$

$$= (2,0,0) - \frac{6}{25}(3,4,0) = \left(\frac{32}{25}, -\frac{24}{25}, 0\right)$$

Then, normalize the vectors.

$$\mathbf{u}_1 = \frac{\mathbf{w}_1}{\|\mathbf{w}_1\|} = \frac{1}{\sqrt{3^2 + 4^2 + 0^2}}(3, 4, 0) = \left(\frac{3}{5}, \frac{4}{5}, 0\right)$$

$$\mathbf{u}_2 = \frac{\mathbf{w}_2}{\|\mathbf{w}_2\|} = \frac{1}{\sqrt{\left(\frac{32}{25}\right)^2 + \left(\frac{-24}{25}\right)^2 + 0^2}}\left(\frac{32}{25}, -\frac{24}{25}, 0\right)$$

$$= \frac{1}{8/5}\left(\frac{32}{25}, -\frac{24}{25}, 0\right)$$

$$= \frac{5}{8}\left(\frac{32}{25}, -\frac{24}{25}, 0\right)$$

$$= \left(\frac{4}{5}, -\frac{3}{5}, 0\right)$$

So, the orthonormal basis is $\left\{\left(\frac{3}{5}, \frac{4}{5}, 0\right), \left(\frac{4}{5}, -\frac{3}{5}, 0\right)\right\}$.

© 2013 Cengage Learning. All Rights Reserved. May not be scanned, copied or duplicated, or posted to a publicly accessible website, in whole or in part.

41. First, orthogonalize each vector in B.

$$\mathbf{w}_1 = \mathbf{v}_1 = (1, 2, -1, 0)$$

$$\mathbf{w}_2 = \mathbf{v}_2 - \frac{\langle \mathbf{v}_2, \mathbf{w}_1 \rangle}{\langle \mathbf{w}_1, \mathbf{w}_1 \rangle}\mathbf{w}_1 = (2, 2, 0, 1) - \frac{2(1) + 2(2) + 0(-1) + 1(0)}{1^2 + 2^2 + (-1)^2 + 0^2}(1, 2, -1, 0) = (2, 2, 0, 1) - (1, 2, -1, 0) = (1, 0, 1, 1)$$

$$\mathbf{w}_3 = \mathbf{v}_3 - \frac{\langle \mathbf{v}_3, \mathbf{w}_2 \rangle}{\langle \mathbf{w}_2, \mathbf{w}_2 \rangle}\mathbf{w}_2 - \frac{\langle \mathbf{v}_3, \mathbf{w}_1 \rangle}{\langle \mathbf{w}_1, \mathbf{w}_1 \rangle}\mathbf{w}_1$$

$$= (1, 1, -1, 0) - \frac{1(1) + 1(0) - 1(1) + 0(1)}{1^2 + 0^2 + 1^2 + 1^2}(1, 0, 1, 1) - \frac{1(1) + 1(2) - 1(-1) + 0(0)}{1^2 + 2^2 + (-1)^2 + 0^2}(1, 2, -1, 0)$$

$$= (1, 1, -1, 0) - \frac{0}{3}(1, 0, 1, 1) - \frac{2}{3}(1, 2, -1, 0) = \left(\frac{1}{3}, -\frac{1}{3}, -\frac{1}{3}, 0\right)$$

Then, normalize the vectors.

$$\mathbf{u}_1 = \frac{\mathbf{w}_1}{\|\mathbf{w}_1\|} = \frac{1}{\sqrt{1^2 + 2^2 + (-1)^2 + 0^2}}(1, 2, -1, 0) = \left(\frac{\sqrt{6}}{6}, \frac{\sqrt{6}}{3}, -\frac{\sqrt{6}}{6}, 0\right)$$

$$\mathbf{u}_2 = \frac{\mathbf{w}_2}{\|\mathbf{w}_2\|} = \frac{1}{\sqrt{1^2 + 0^2 + 1^2 + 1^2}}(1, 0, 1, 1) = \left(\frac{\sqrt{3}}{3}, 0, \frac{\sqrt{3}}{3}, \frac{\sqrt{3}}{3}\right)$$

$$\mathbf{u}_3 = \frac{\mathbf{w}_3}{\|\mathbf{w}_3\|} = \frac{1}{\sqrt{\left(\frac{1}{3}\right)^2 + \left(-\frac{1}{3}\right)^2 + \left(-\frac{1}{3}\right)^2 + 0^2}}\left(\frac{1}{3}, -\frac{1}{3}, -\frac{1}{3}, 0\right) = \left(\frac{\sqrt{3}}{3}, -\frac{\sqrt{3}}{3}, -\frac{\sqrt{3}}{3}, 0\right)$$

So, the orthonormal basis is $\left\{\left(\frac{\sqrt{6}}{6}, \frac{\sqrt{6}}{3}, -\frac{\sqrt{6}}{6}, 0\right), \left(\frac{\sqrt{3}}{3}, 0, \frac{\sqrt{3}}{3}, \frac{\sqrt{3}}{3}\right), \left(\frac{\sqrt{3}}{3}, -\frac{\sqrt{3}}{3}, -\frac{\sqrt{3}}{3}, 0\right)\right\}$.

43. Begin by orthogonalizing the set.

$$\mathbf{w}_1 = \mathbf{v}_1 = (2, -1)$$

$$\mathbf{w}_2 = \mathbf{v}_2 - \frac{\langle \mathbf{v}_2, \mathbf{w}_1 \rangle}{\langle \mathbf{w}_1, \mathbf{w}_1 \rangle}\mathbf{w}_1 = (-2, 10) - \frac{2(-2)(2) + 10(-1)}{2(2)^2 + (-1)^2}(2, -1) = (-2, 10) + 2(2, -1) = (2, 8)$$

Then, normalize each vector.

$$\mathbf{u}_1 = \frac{\mathbf{w}_1}{\|\mathbf{w}_1\|} = \frac{1}{\sqrt{2(2)^2 + (-1)^2}}(2, -1) = \left(\frac{2}{3}, -\frac{1}{3}\right)$$

$$\mathbf{u}_2 = \frac{\mathbf{w}_2}{\|\mathbf{w}_2\|} = \frac{1}{\sqrt{2(2)^2 + 8^2}}(2, 8) = \left(\frac{\sqrt{2}}{6}, \frac{2\sqrt{2}}{3}\right)$$

So, an orthonormal basis, using the given inner product is $\left\{\left(\frac{2}{3}, -\frac{1}{3}\right), \left(\frac{\sqrt{2}}{6}, \frac{2\sqrt{2}}{3}\right)\right\}$.

45. $\langle x, 1 \rangle = \int_{-1}^{1} x \, dx = \left.\frac{x^2}{2}\right]_{-1}^{1} = 0$

47. $\langle x^2, 1 \rangle = \int_{-1}^{1} x^2 \, dx = \left.\frac{x^3}{3}\right]_{-1}^{1} = \frac{2}{3}$

49. $\langle x, x \rangle = \int_{-1}^{1} x^2 dx = \left.\frac{x^3}{3}\right]_{-1}^{1} = \frac{2}{3}$

© 2013 Cengage Learning. All Rights Reserved. May not be scanned, copied or duplicated, or posted to a publicly accessible website, in whole or in part.

51. The solutions of the homogeneous system are of the form $(3s, -2t, s, t)$, where s and t are any real numbers. So, a basis for the solution space is $\{(3, 0, 1, 0), (0, -2, 0, 1)\}$.

Orthogonalize this basis as follows.

$$\mathbf{w}_1 = \mathbf{v}_1 = (3, 0, 1, 0)$$

$$\mathbf{w}_2 = \mathbf{v}_2 - \frac{\langle \mathbf{v}_2, \mathbf{w}_1 \rangle}{\langle \mathbf{w}_1, \mathbf{w}_1 \rangle}\mathbf{w}_1 = (0, -2, 0, 1) - \frac{0(3) + (-2)(0) + 0(1) + 1(0)}{3^2 + 0^2 + 1^2 + 0^2}(3, 0, 1, 0) = (0, -2, 0, 1)$$

Then, normalize these vectors.

$$\mathbf{u}_1 = \frac{\mathbf{w}_1}{\|\mathbf{w}_1\|} = \frac{1}{\sqrt{3^2 + 0^2 + 1^2 + 0^2}}(3, 0, 1, 0) = \left(\frac{3\sqrt{10}}{10}, 0, \frac{\sqrt{10}}{10}, 0\right)$$

$$\mathbf{u}_2 = \frac{\mathbf{w}_2}{\|\mathbf{w}_2\|} = \frac{1}{\sqrt{0^2 + (-2)^2 + 0^2 + 1^2}}(0, -2, 0, 1) = \left(0, -\frac{2\sqrt{5}}{5}, 0, \frac{\sqrt{5}}{5}\right)$$

So, the orthonormal basis for the solution set is $\left\{\left(\frac{3\sqrt{10}}{10}, 0, \frac{\sqrt{10}}{10}, 0\right), \left(0, -\frac{2\sqrt{5}}{5}, 0, \frac{\sqrt{5}}{5}\right)\right\}$.

53. The solutions of the homogeneous system are of the form $(-s - t, 0, s, t)$, where s and t are any real numbers. So, a basis for the solution space is $\{(-1, 0, 1, 0), (-1, 0, 0, 1)\}$.

Orthogonalize this basis as follows.

$$\mathbf{w}_1 = \mathbf{v}_1 = (-1, 0, 1, 0)$$

$$\mathbf{w}_2 = \mathbf{v}_2 - \frac{\langle \mathbf{v}_2, \mathbf{w}_1 \rangle}{\langle \mathbf{w}_1, \mathbf{w}_1 \rangle}\mathbf{w}_1 = (-1, 0, 0, 1) - \frac{1}{2}(-1, 0, 1, 0) = \left(-\frac{1}{2}, 0, -\frac{1}{2}, 1\right)$$

Then, normalize these vectors.

$$\mathbf{u}_1 = \frac{\mathbf{w}_1}{\|\mathbf{w}_1\|} = \frac{1}{\sqrt{2}}(-1, 0, 1, 0) = \left(-\frac{\sqrt{2}}{2}, 0, \frac{\sqrt{2}}{2}, 0\right)$$

$$\mathbf{u}_2 = \frac{\mathbf{w}_2}{\|\mathbf{w}_2\|} = \frac{1}{\sqrt{3/2}}\left(-\frac{1}{2}, 0, \frac{1}{2}, 1\right) = \left(-\frac{\sqrt{6}}{6}, 0, \frac{\sqrt{6}}{6}, \frac{\sqrt{6}}{3}\right)$$

So, an orthonormal basis for the solution space is $\left\{\left(-\frac{\sqrt{2}}{2}, 0, \frac{\sqrt{2}}{2}, 0\right), \left(-\frac{\sqrt{6}}{6}, 0, \frac{\sqrt{6}}{6}, \frac{\sqrt{6}}{3}\right)\right\}$.

55. The solutions of the homogeneous system are of the form $(2s - t, s, t)$, where s and t are any real numbers. So, a basis for the solution space is $\{(2, 1, 0), (-1, 0, 1)\}$.

Orthogonalize this basis as follows.

$$\mathbf{w}_1 = \mathbf{v}_1 = (2, 1, 0)$$

$$\mathbf{w}_2 = \mathbf{v}_2 - \frac{\langle \mathbf{v}_2, \mathbf{w}_1 \rangle}{\langle \mathbf{w}_1, \mathbf{w}_1 \rangle}\mathbf{w}_1 = (-1, 0, 1) + \frac{2}{5}(2, 1, 0) = \left(-\frac{1}{5}, \frac{2}{5}, 1\right)$$

Then, normalize these vectors.

$$\mathbf{u}_1 = \frac{\mathbf{w}_1}{\|\mathbf{w}_1\|} = \frac{1}{\sqrt{5}}(2, 1, 0) = \left(\frac{2\sqrt{5}}{5}, \frac{\sqrt{5}}{5}, 0\right)$$

$$\mathbf{u}_2 = \frac{\mathbf{w}_2}{\|\mathbf{w}_2\|} = \frac{1}{\sqrt{6/5}}\left(-\frac{1}{5}, \frac{2}{5}, 1\right) = \left(-\frac{\sqrt{30}}{30}, \frac{\sqrt{30}}{15}, \frac{\sqrt{30}}{6}\right)$$

So, an orthonormal basis for the solution space is $\left\{\left(\frac{2\sqrt{5}}{5}, \frac{\sqrt{5}}{5}, 0\right), \left(-\frac{\sqrt{30}}{30}, \frac{\sqrt{30}}{15}, \frac{\sqrt{30}}{6}\right)\right\}$.

© 2013 Cengage Learning. All Rights Reserved. May not be scanned, copied or duplicated, or posted to a publicly accessible website, in whole or in part.

57. (a) True. See "Definition of Orthogonal and Orthornormal Sets," page 248.

(b) False. See "Remark," page 254.

59. Let

$$p(x) = \frac{x^2 + 1}{\sqrt{2}} \quad \text{and} \quad q(x) = \frac{x^2 + x - 1}{\sqrt{3}}.$$

Then

$$\langle p, q\rangle = \frac{1}{\sqrt{2}}\left(-\frac{1}{\sqrt{3}}\right) + 0\left(\frac{1}{\sqrt{3}}\right) + \frac{1}{\sqrt{2}}\left(\frac{1}{\sqrt{3}}\right) = 0. \text{ Furthermore,}$$

$$\|p\| = \sqrt{\left(\frac{1}{\sqrt{2}}\right)^2 + 0^2 + \left(\frac{1}{\sqrt{2}}\right)^2} = 1$$

$$\|q\| = \sqrt{\left(-\frac{1}{\sqrt{3}}\right)^2 + \left(\frac{1}{\sqrt{3}}\right)^2 + \left(\frac{1}{\sqrt{3}}\right)^2} = 1.$$

So, $\{p, q\}$ is an orthonormal set.

61. Let $p_1(x) = x^2$, $p_2(x) = x^2 + 2x$, and $p_3(x) = x^2 + 2x + 1$.

Then, because $\langle p_1, p_2\rangle = 0(0) + 0(2) + 1(1) = 1 \neq 0$, the set is not orthogonal. Orthogonalize the set as follows.

$$\mathbf{w}_1 = p_1 = x^2$$

$$\mathbf{w}_2 = p_2 - \frac{\langle p_2, \mathbf{w}_1\rangle}{\langle \mathbf{w}_1, \mathbf{w}_1\rangle}\mathbf{w}_1 = x^2 + 2x - \frac{0(0) + 2(0) + 1(1)}{0^2 + 0^2 + 1^2}x^2 = 2x$$

$$\mathbf{w}_3 = p_3 - \frac{\langle p_3, \mathbf{w}_2\rangle}{\langle \mathbf{w}_2, \mathbf{w}_2\rangle}\mathbf{w}_2 - \frac{\langle p_3, \mathbf{w}_1\rangle}{\langle \mathbf{w}_1, \mathbf{w}_1\rangle}\mathbf{w}_1$$

$$= x^2 + 2x + 1 - \frac{1(0) + 2(2) + 1(0)}{0^2 + 2^2 + 0^2}(2x) - \frac{1(0) + 2(0) + 1(1)}{0^2 + 0^2 + 1^2}x^2$$

$$= x^2 + 2x + 1 - 2x - x^2 = 1$$

Then, normalize the vectors.

$$\mathbf{u}_1 = \frac{\mathbf{w}_1}{\|\mathbf{w}_1\|} = \frac{1}{\sqrt{0^2 + 0^2 + 1^2}}x^2 = x^2$$

$$\mathbf{u}_2 = \frac{\mathbf{w}_2}{\|\mathbf{w}_2\|} = \frac{1}{\sqrt{0^2 + 2^2 + 0^2}}(2x) = x$$

$$\mathbf{u}_3 = \frac{\mathbf{w}_3}{\|\mathbf{w}_3\|} = \frac{1}{\sqrt{1^2 + 0^2 + 0^2}}(1) = 1$$

So, the orthonormal set is $\{x^2, x, 1\}$.

63. Let $p_1(x) = \dfrac{3x^2 + 4x}{5}$, $p_2(x) = \dfrac{-4x^2 + 3x}{5}$, $p_3(x) = 1$. Then,

$$\langle p_1, p_2\rangle = -\frac{12}{25} + \frac{12}{25} = 0, \langle p_1, p_3\rangle = 0 \text{ and } \langle p_2, p_3\rangle = 0$$

Furthermore,

$$\|p_1\| = \sqrt{\frac{9 + 16}{25}} = 1, \|p_2\| = \sqrt{\frac{16 + 9}{25}} = 1 \text{ and } \|p_3\| = 1.$$

So, $\{p_1, p_2, p_3\}$ is an orthonormal set.

© 2013 Cengage Learning. All Rights Reserved. May not be scanned, copied or duplicated, or posted to a publicly accessible website, in whole or in part.

65. For $\{\mathbf{u}_1, \mathbf{u}_2, \ldots, \mathbf{u}_n\}$ an orthonormal basis for R^n and $\mathbf{v}$ any vector in R^n,

$$\mathbf{v} = \langle \mathbf{v}, \mathbf{u}_1\rangle \mathbf{u}_1 + \langle \mathbf{v}, \mathbf{u}_2\rangle \mathbf{u}_2 + \ldots + \langle \mathbf{v}, \mathbf{u}_n\rangle \mathbf{u}_n$$

$$\|\mathbf{v}\|^2 = \|\langle \mathbf{v}, \mathbf{u}_1\rangle \mathbf{u}_1 + \langle \mathbf{v}, \mathbf{u}_2\rangle \mathbf{u}_2 + \ldots + \langle \mathbf{v}, \mathbf{u}_n\rangle \mathbf{u}_n\|^2.$$

Because $\mathbf{u}_i \cdot \mathbf{u}_j = 0$ for $i \neq j$, it follows that

$$\|\mathbf{v}\|^2 = (\langle \mathbf{v}, \mathbf{u}_1\rangle)^2 \|\mathbf{u}_1\|^2 + (\langle \mathbf{v}, \mathbf{u}_2\rangle)^2 \|\mathbf{u}_2\|^2 + \ldots + (\langle \mathbf{v}, \mathbf{u}_n\rangle)^2 \|\mathbf{u}_n\|^2$$

$$\|\mathbf{v}\|^2 = (\langle \mathbf{v}, \mathbf{u}_1\rangle)^2 (1) + (\langle \mathbf{v}, \mathbf{u}_2\rangle)^2 (1) + \ldots + (\langle \mathbf{v}, \mathbf{u}_n\rangle)^2 (1)$$

$$\|\mathbf{v}\|^2 = |\mathbf{v} \cdot \mathbf{u}_1|^2 + |\mathbf{v} \cdot \mathbf{u}_2|^2 + \ldots + |\mathbf{v} \cdot \mathbf{u}_n|^2.$$

67. First prove that condition (a) implies (b). If $P^{-1} = P^T$, consider $\mathbf{p}_i$ the ith row vector of P. Because $P P^T = I_n$, you have $\mathbf{p}_i \cdot \mathbf{p}_i = 1$ and $\mathbf{p}_i \cdot \mathbf{p}_j = 0$, for $i \neq j$. So the row vectors of P form an orthonormal basis for R^n.

(b) implies (c) if the row vectors of P form an orthonormal basis, then $P P^T = I_n \Rightarrow P^T P = I_n$, which implies that the column vectors of P form an orthonormal basis.

(c) implies (a) because the column vectors of P form an orthonormal basis, you have $P^T P = I_n$, which implies that $P^{-1} = P^T$.

69. $$A = \begin{bmatrix} 1 & 1 & -1 \\ 0 & 2 & 1 \\ 1 & 3 & 0 \end{bmatrix} \Rightarrow \begin{bmatrix} 2 & 0 & -3 \\ 0 & 2 & 1 \\ 0 & 0 & 0 \end{bmatrix}$$

$$A^T = \begin{bmatrix} 1 & 0 & 1 \\ 1 & 2 & 3 \\ -1 & 1 & 0 \end{bmatrix} \Rightarrow \begin{bmatrix} 1 & 0 & 1 \\ 0 & 1 & 1 \\ 0 & 0 & 0 \end{bmatrix}$$

$N(A)$-basis: $\left\{\begin{bmatrix} 3 \\ -1 \\ 2 \end{bmatrix}\right\}$

$N(A^T)$-basis: $\left\{\begin{bmatrix} -1 \\ -1 \\ 1 \end{bmatrix}\right\}$

$R(A)$-basis: $\left\{\begin{bmatrix} 1 \\ 0 \\ 1 \end{bmatrix}, \begin{bmatrix} 1 \\ 2 \\ 3 \end{bmatrix}\right\}$

$R(A^T)$-basis: $\left\{\begin{bmatrix} 1 \\ 1 \\ -1 \end{bmatrix}, \begin{bmatrix} 0 \\ 2 \\ 1 \end{bmatrix}\right\}$

$N(A) = R(A^T)^\perp$ and $N(A^T) = R(A)^\perp$

71. (a) The row space of A is the column space of A^T, $R(A^T)$.

(b) Let $\mathbf{x} \in N(A) \Rightarrow A\mathbf{x} = \mathbf{0} \Rightarrow \mathbf{x}$ is orthogonal to all the rows of $A \Rightarrow \mathbf{x}$ is orthogonal to all the columns of $A^T \Rightarrow \mathbf{x} \in R(A^T)^\perp$.

(c) Let $\mathbf{x} \in R(A^T)^\perp \Rightarrow \mathbf{x}$ is orthogonal to each column vector of $A^T \Rightarrow A\mathbf{x} = \mathbf{0} \Rightarrow \mathbf{x} \in N(A)$. Combining this with part (b), $N(A) = R(A^T)^\perp$.

(d) Substitute A^T for A in part (c).

Section 5.4 Mathematical Models and Least Squares Analysis

1. The system

$$\begin{aligned} c_0 &= 1 \\ c_0 + c_1 &= 3 \\ c_0 + 2c_1 &= 5 \end{aligned}$$

has the solution $c_0 = 1$ and $c_1 = 2$.

So, the three points are collinear. The equation that models the points is $y = 1 + 2x$.

3. The system

$$\begin{aligned} c_0 - c_1 &= 0 \\ c_0 &= 1 \\ c_0 + c_1 &= 1 \end{aligned}$$

has no solution.

The points are not collinear.

© 2013 Cengage Learning. All Rights Reserved. May not be scanned, copied or duplicated, or posted to a publicly accessible website, in whole or in part.

5. Not orthogonal: $\begin{bmatrix}0\\1\\1\end{bmatrix}\cdot\begin{bmatrix}-1\\2\\0\end{bmatrix} = 2 \neq 0$

7. Orthogonal: $\begin{bmatrix}1\\1\\1\\1\end{bmatrix}\cdot\begin{bmatrix}-1\\1\\-1\\1\end{bmatrix} = \begin{bmatrix}1\\1\\1\\1\end{bmatrix}\cdot\begin{bmatrix}0\\2\\-2\\0\end{bmatrix} = 0$

9. (a) $S = \text{span}\left\{\begin{bmatrix}1\\0\\0\end{bmatrix}, \begin{bmatrix}0\\0\\1\end{bmatrix}\right\}$

$\Rightarrow S^{\perp} = \text{span}\left\{\begin{bmatrix}0\\1\\0\end{bmatrix}\right\}$ (the y-axis)

(b) $S \oplus S^{\perp} = R^3$

11. (a) $A^T = \begin{bmatrix}1 & 2 & 0 & 0\\0 & 1 & 0 & 1\end{bmatrix} \Rightarrow \begin{bmatrix}1 & 0 & 0 & -2\\0 & 1 & 0 & 1\end{bmatrix}$

$\Rightarrow S^{\perp} = \text{span}\left\{\begin{bmatrix}0\\0\\1\\0\end{bmatrix}, \begin{bmatrix}2\\-1\\0\\1\end{bmatrix}\right\}$

(b) From part (a), since

$S^{\perp} = \text{span}\left\{\begin{bmatrix}0\\0\\1\\0\end{bmatrix}, \begin{bmatrix}2\\-1\\0\\1\end{bmatrix}\right\}$ and $S = \text{span}\left\{\begin{bmatrix}1\\2\\0\\0\end{bmatrix}, \begin{bmatrix}0\\1\\0\\1\end{bmatrix}\right\}$

it follows that $R^4 = S \oplus S^{\perp}$.

13. The orthogonal complement of span $\left\{\begin{bmatrix}0\\0\\1\\0\end{bmatrix}, \begin{bmatrix}2\\-1\\0\\1\end{bmatrix}\right\} = S^{\perp}$ is

$(S^{\perp})^{\perp} = S = \text{span}\left\{\begin{bmatrix}1\\2\\0\\0\end{bmatrix}, \begin{bmatrix}0\\1\\0\\1\end{bmatrix}\right\}$.

15. An orthonormal basis for S is $\left\{\begin{bmatrix}0\\0\\-\frac{1}{\sqrt{2}}\\\frac{1}{\sqrt{2}}\end{bmatrix}, \begin{bmatrix}0\\\frac{1}{\sqrt{3}}\\\frac{1}{\sqrt{3}}\\\frac{1}{\sqrt{3}}\end{bmatrix}\right\}$.

$\text{proj}_S \mathbf{v} = (\mathbf{v}\cdot\mathbf{u}_1)\mathbf{u}_1 + (\mathbf{v}\cdot\mathbf{u}_2)\mathbf{u}_2$

$= 0\mathbf{u}_1 + \frac{2}{\sqrt{3}}\begin{bmatrix}0\\\frac{1}{\sqrt{3}}\\\frac{1}{\sqrt{3}}\\\frac{1}{\sqrt{3}}\end{bmatrix} = \begin{bmatrix}0\\\frac{2}{3}\\\frac{2}{3}\\\frac{2}{3}\end{bmatrix}$

17. Use Gram-Schmidt to construct an orthonormal basis for S.

$\begin{bmatrix}0\\1\\1\end{bmatrix} - \frac{1}{2}\begin{bmatrix}1\\0\\1\end{bmatrix} = \begin{bmatrix}-\frac{1}{2}\\1\\\frac{1}{2}\end{bmatrix}$

orthonormal basis: $\left\{\begin{bmatrix}\frac{1}{\sqrt{2}}\\0\\\frac{1}{\sqrt{2}}\end{bmatrix}, \begin{bmatrix}-\frac{1}{\sqrt{6}}\\\frac{2}{\sqrt{6}}\\\frac{1}{\sqrt{6}}\end{bmatrix}\right\}$

$\text{proj}_S \mathbf{v} = (\mathbf{u}_1\cdot\mathbf{v})\mathbf{u}_1 + (\mathbf{u}_2\cdot\mathbf{v})\mathbf{u}_2$

$= \frac{6}{\sqrt{2}}\begin{bmatrix}\frac{1}{\sqrt{2}}\\0\\\frac{1}{\sqrt{2}}\end{bmatrix} + \frac{8}{\sqrt{6}}\begin{bmatrix}-\frac{1}{\sqrt{6}}\\\frac{2}{\sqrt{6}}\\\frac{1}{\sqrt{6}}\end{bmatrix}$

$= \begin{bmatrix}3\\0\\3\end{bmatrix} + \begin{bmatrix}-\frac{4}{3}\\\frac{8}{3}\\\frac{4}{3}\end{bmatrix} = \begin{bmatrix}\frac{5}{3}\\\frac{8}{3}\\\frac{13}{3}\end{bmatrix}$

© 2013 Cengage Learning. All Rights Reserved. May not be scanned, copied or duplicated, or posted to a publicly accessible website, in whole or in part.

19. $A = \begin{bmatrix} 1 & 2 & 3 \\ 0 & 1 & 0 \end{bmatrix} \Rightarrow \begin{bmatrix} 1 & 0 & 3 \\ 0 & 1 & 0 \end{bmatrix}$

$A^T = \begin{bmatrix} 1 & 0 \\ 2 & 1 \\ 3 & 0 \end{bmatrix} \Rightarrow \begin{bmatrix} 1 & 0 \\ 0 & 1 \\ 0 & 0 \end{bmatrix}$

$N(A)$-basis: $\left\{\begin{bmatrix} -3 \\ 0 \\ 1 \end{bmatrix}\right\}$

$N(A^T) = \left\{\begin{bmatrix} 0 \\ 0 \end{bmatrix}\right\}$

$R(A)$-basis: $\left\{\begin{bmatrix} 1 \\ 0 \end{bmatrix}, \begin{bmatrix} 2 \\ 1 \end{bmatrix}\right\}$ $\quad (R(A) = R^2)$

$R(A^T)$-basis: $\left\{\begin{bmatrix} 1 \\ 2 \\ 3 \end{bmatrix}, \begin{bmatrix} 0 \\ 1 \\ 0 \end{bmatrix}\right\}$

21. $A = \begin{bmatrix} 1 & 0 & 0 & 1 \\ 0 & 1 & 1 & 1 \\ 1 & 1 & 1 & 2 \\ 1 & 2 & 2 & 3 \end{bmatrix} \Rightarrow \begin{bmatrix} 1 & 0 & 0 & 1 \\ 0 & 1 & 1 & 1 \\ 0 & 0 & 0 & 0 \\ 0 & 0 & 0 & 0 \end{bmatrix}$

$A^T = \begin{bmatrix} 1 & 0 & 1 & 1 \\ 0 & 1 & 1 & 2 \\ 0 & 1 & 1 & 2 \\ 1 & 1 & 2 & 3 \end{bmatrix} \Rightarrow \begin{bmatrix} 1 & 0 & 1 & 1 \\ 0 & 1 & 1 & 2 \\ 0 & 0 & 0 & 0 \\ 0 & 0 & 0 & 0 \end{bmatrix}$

$N(A)$-basis: $\left\{\begin{bmatrix} -1 \\ -1 \\ 0 \\ 1 \end{bmatrix}, \begin{bmatrix} 0 \\ -1 \\ 1 \\ 0 \end{bmatrix}\right\}$

$N(A^T)$-basis: $\left\{\begin{bmatrix} -1 \\ -1 \\ 1 \\ 0 \end{bmatrix}, \begin{bmatrix} -1 \\ -2 \\ 0 \\ 1 \end{bmatrix}\right\}$

$R(A)$-basis: $\left\{\begin{bmatrix} 1 \\ 0 \\ 1 \\ 1 \end{bmatrix}, \begin{bmatrix} 0 \\ 1 \\ 1 \\ 2 \end{bmatrix}\right\}$

$R(A^T)$-basis: $\left\{\begin{bmatrix} 1 \\ 0 \\ 0 \\ 1 \end{bmatrix}, \begin{bmatrix} 0 \\ 1 \\ 1 \\ 1 \end{bmatrix}\right\}$

23. $A^T A = \begin{bmatrix} 2 & 1 & 1 \\ 1 & 2 & 1 \end{bmatrix}\begin{bmatrix} 2 & 1 \\ 1 & 2 \\ 1 & 1 \end{bmatrix} = \begin{bmatrix} 6 & 5 \\ 5 & 6 \end{bmatrix}$

$A^T\mathbf{b} = \begin{bmatrix} 2 & 1 & 1 \\ 1 & 2 & 1 \end{bmatrix}\begin{bmatrix} 2 \\ 0 \\ -3 \end{bmatrix} = \begin{bmatrix} 1 \\ -1 \end{bmatrix}$

$\begin{bmatrix} 6 & 5 & 1 \\ 5 & 6 & -1 \end{bmatrix} \Rightarrow \begin{bmatrix} 1 & 0 & 1 \\ 0 & 1 & -1 \end{bmatrix} \Rightarrow \mathbf{x} = \begin{bmatrix} 1 \\ -1 \end{bmatrix}$

25. $A^T A = \begin{bmatrix} 1 & 1 & 0 & 1 \\ 0 & 1 & 1 & 1 \\ 1 & 1 & 1 & 0 \end{bmatrix}\begin{bmatrix} 1 & 0 & 1 \\ 1 & 1 & 1 \\ 0 & 1 & 1 \\ 1 & 1 & 0 \end{bmatrix} = \begin{bmatrix} 3 & 2 & 2 \\ 2 & 3 & 2 \\ 2 & 2 & 3 \end{bmatrix}$

$A^T\mathbf{b} = \begin{bmatrix} 1 & 1 & 0 & 1 \\ 0 & 1 & 1 & 1 \\ 1 & 1 & 1 & 0 \end{bmatrix}\begin{bmatrix} 4 \\ -1 \\ 0 \\ 1 \end{bmatrix} = \begin{bmatrix} 4 \\ 0 \\ 3 \end{bmatrix}$

$\begin{bmatrix} 3 & 2 & 2 & 4 \\ 2 & 3 & 2 & 0 \\ 2 & 2 & 3 & 3 \end{bmatrix} \Rightarrow \begin{bmatrix} 1 & 0 & 0 & 2 \\ 0 & 1 & 0 & -2 \\ 0 & 0 & 1 & 1 \end{bmatrix} \Rightarrow \mathbf{x} = \begin{bmatrix} 2 \\ -2 \\ 1 \end{bmatrix}$

27. $A^T A = \begin{bmatrix} 1 & 0 & 1 \\ 2 & 1 & 1 \end{bmatrix}\begin{bmatrix} 1 & 2 \\ 0 & 1 \\ 1 & 1 \end{bmatrix} = \begin{bmatrix} 2 & 3 \\ 3 & 6 \end{bmatrix}$

$A^T b = \begin{bmatrix} 1 & 0 & 1 \\ 2 & 1 & 1 \end{bmatrix}\begin{bmatrix} 2 \\ -2 \\ 1 \end{bmatrix} = \begin{bmatrix} 3 \\ 3 \end{bmatrix}$

The normal equations are

$$A^T A\mathbf{x} = A^T\mathbf{b}$$

$$\begin{bmatrix} 2 & 3 \\ 3 & 6 \end{bmatrix}\begin{bmatrix} x_1 \\ x_2 \end{bmatrix} = \begin{bmatrix} 3 \\ 3 \end{bmatrix}.$$

The solution of this system is

$$\mathbf{x} = \begin{bmatrix} x_1 \\ x_2 \end{bmatrix} = \begin{bmatrix} 3 \\ -1 \end{bmatrix}.$$

Finally, the projection of **b** onto S is

$$A\mathbf{x} = \begin{bmatrix} 1 & 2 \\ 0 & 1 \\ 1 & 1 \end{bmatrix}\begin{bmatrix} 3 \\ -1 \end{bmatrix} = \begin{bmatrix} 1 \\ -1 \\ 2 \end{bmatrix}.$$

© 2013 Cengage Learning. All Rights Reserved. May not be scanned, copied or duplicated, or posted to a publicly accessible website, in whole or in part.

29. $A^TA = \begin{bmatrix} 1 & 1 & 1 \\ -1 & 1 & 3 \end{bmatrix}\begin{bmatrix} 1 & -1 \\ 1 & 1 \\ 1 & 3 \end{bmatrix} = \begin{bmatrix} 3 & 3 \\ 3 & 11 \end{bmatrix}$

$A^T\mathbf{b} = \begin{bmatrix} 1 & 1 & 1 \\ -1 & 1 & 3 \end{bmatrix}\begin{bmatrix} 1 \\ 0 \\ -3 \end{bmatrix} = \begin{bmatrix} -2 \\ -10 \end{bmatrix}$

$\begin{bmatrix} 3 & 3 & -2 \\ 3 & 11 & -10 \end{bmatrix} \Rightarrow \begin{bmatrix} 1 & 0 & \frac{1}{3} \\ 0 & 1 & -1 \end{bmatrix} \Rightarrow \mathbf{x} = \begin{bmatrix} \frac{1}{3} \\ -1 \end{bmatrix}$

line: $y = \frac{1}{3} - x$

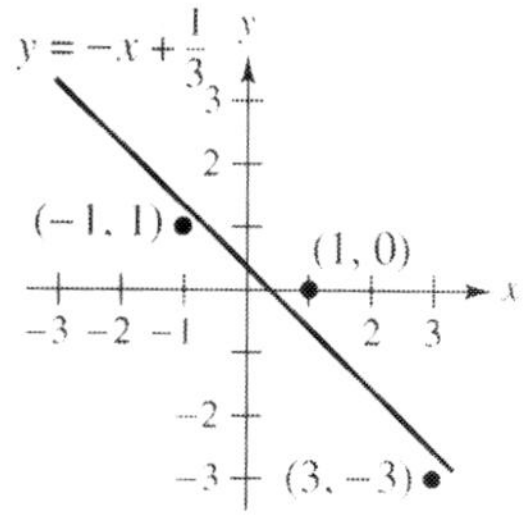

31. $A^TA = \begin{bmatrix} 1 & 1 & 1 & 1 & 1 \\ -2 & -1 & 0 & 1 & 2 \end{bmatrix}\begin{bmatrix} 1 & -2 \\ 1 & -1 \\ 1 & 0 \\ 1 & 1 \\ 1 & 2 \end{bmatrix} = \begin{bmatrix} 5 & 0 \\ 0 & 10 \end{bmatrix}$

$A^T\mathbf{b} = \begin{bmatrix} 1 & 1 & 1 & 1 & 1 \\ -2 & -1 & 0 & 1 & 2 \end{bmatrix}\begin{bmatrix} 1 \\ 2 \\ 1 \\ 2 \\ 1 \end{bmatrix} = \begin{bmatrix} 7 \\ 0 \end{bmatrix}$

$\begin{bmatrix} 5 & 0 & 7 \\ 0 & 10 & 0 \end{bmatrix} \Rightarrow \begin{bmatrix} 1 & 0 & \frac{7}{5} \\ 0 & 1 & 0 \end{bmatrix} \Rightarrow \mathbf{x} = \begin{bmatrix} \frac{7}{5} \\ 0 \end{bmatrix}$

line: $y = \frac{7}{5}$

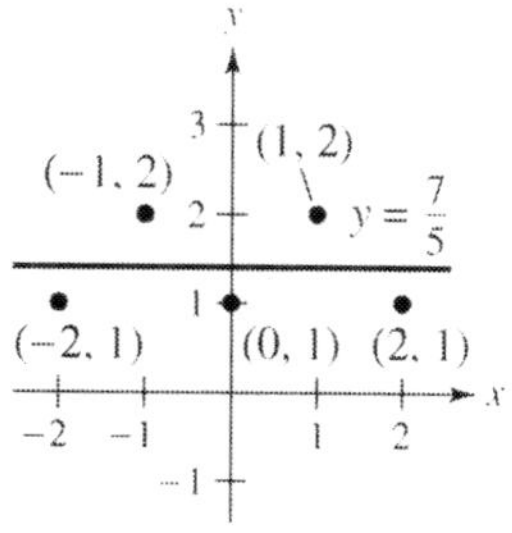

33. $A^TA = \begin{bmatrix} 1 & 1 & 1 & 1 \\ 0 & 2 & 3 & 4 \\ 0 & 4 & 9 & 16 \end{bmatrix}\begin{bmatrix} 1 & 0 & 0 \\ 1 & 2 & 4 \\ 1 & 3 & 9 \\ 1 & 4 & 16 \end{bmatrix} = \begin{bmatrix} 4 & 9 & 29 \\ 9 & 29 & 99 \\ 29 & 99 & 353 \end{bmatrix}$

$A^T\mathbf{b} = \begin{bmatrix} 1 & 1 & 1 & 1 \\ 0 & 2 & 3 & 4 \\ 0 & 4 & 9 & 16 \end{bmatrix}\begin{bmatrix} 0 \\ 2 \\ 6 \\ 12 \end{bmatrix} = \begin{bmatrix} 20 \\ 70 \\ 254 \end{bmatrix}$

$\begin{bmatrix} 4 & 9 & 29 & 20 \\ 9 & 29 & 99 & 70 \\ 29 & 99 & 353 & 254 \end{bmatrix} \Rightarrow \begin{bmatrix} 1 & 0 & 0 & 0 \\ 0 & 1 & 0 & -1 \\ 0 & 0 & 1 & 1 \end{bmatrix} \Rightarrow \mathbf{x} = \begin{bmatrix} 0 \\ -1 \\ 1 \end{bmatrix}$

Quadratic Polynomial: $y = x^2 - x$

35. $A^TA = \begin{bmatrix} 1 & 1 & 1 & 1 & 1 \\ -2 & -1 & 0 & 1 & 2 \\ 4 & 1 & 0 & 1 & 4 \end{bmatrix}\begin{bmatrix} 1 & -2 & 4 \\ 1 & -1 & 1 \\ 1 & 0 & 0 \\ 1 & 1 & 1 \\ 1 & 2 & 4 \end{bmatrix} = \begin{bmatrix} 5 & 0 & 10 \\ 0 & 10 & 0 \\ 10 & 0 & 34 \end{bmatrix}$

$A^T\mathbf{b} = \begin{bmatrix} 1 & 1 & 1 & 1 & 1 \\ -2 & -1 & 0 & 1 & 2 \\ 4 & 1 & 0 & 1 & 4 \end{bmatrix}\begin{bmatrix} 0 \\ 0 \\ 1 \\ 2 \\ 5 \end{bmatrix} = \begin{bmatrix} 8 \\ 12 \\ 22 \end{bmatrix}$

$\begin{bmatrix} 5 & 0 & 10 & 8 \\ 0 & 10 & 0 & 12 \\ 10 & 0 & 34 & 22 \end{bmatrix} \Rightarrow \begin{bmatrix} 1 & 0 & 0 & \frac{26}{35} \\ 0 & 1 & 0 & \frac{6}{5} \\ 0 & 0 & 1 & \frac{3}{7} \end{bmatrix} \Rightarrow \mathbf{x} = \begin{bmatrix} \frac{26}{35} \\ \frac{6}{5} \\ \frac{3}{7} \end{bmatrix}$

Quadratic Polynomial: $y = \frac{26}{35} + \frac{6}{5}x + \frac{3}{7}x^2$

© 2013 Cengage Learning. All Rights Reserved. May not be scanned, copied or duplicated, or posted to a publicly accessible website, in whole or in part.

37. Substitute the data points $(5, 52.6), (6, 56.1), (7, 60.6)$, and $(8, 63.7)$ into the equation $y = c_0 + c_1 t$ to obtain the following system.

$$c_0 + c_1(5) = 52.6$$
$$c_0 + c_1(6) = 56.1$$
$$c_0 + c_1(7) = 60.6$$
$$c_0 + c_1(8) = 63.7$$

This produces the least squares problem

$$A\mathbf{x} = \mathbf{b}$$

$$\begin{bmatrix} 1 & 5 \\ 1 & 6 \\ 1 & 7 \\ 1 & 8 \end{bmatrix} \begin{bmatrix} c_0 \\ c_1 \end{bmatrix} = \begin{bmatrix} 52.6 \\ 56.1 \\ 60.6 \\ 63.7 \end{bmatrix}.$$

The normal equations are

$$A^T A\mathbf{x} = A^T\mathbf{b}$$

$$\begin{bmatrix} 4 & 26 \\ 26 & 174 \end{bmatrix} \begin{bmatrix} c_0 \\ c_1 \end{bmatrix} = \begin{bmatrix} 233 \\ 1533.4 \end{bmatrix}$$

and the solution is

$$\mathbf{x} = \begin{bmatrix} c_0 \\ c_1 \end{bmatrix} = \begin{bmatrix} 33.68 \\ 3.78 \end{bmatrix}.$$

So, the least squares regression line is $y = 33.68 + 3.78t$. In 2015, there were $33.68 + 3.78(15) = 90.38$, or 90,380 degrees.

39. When you plot the data as given, they do not lie in a straight line. By taking the natural logarithm of each coordinate, however, you obtain points of the form $(\ln x, \ln y)$ as follows.

$\ln x$	3.219	3.555	3.912	4.317	6.215	6.908
$\ln y$	5.255	5.208	5.158	5.101	4.835	4.738

Substitute these data points into the equation $y = c_0 + c_1 x$ to obtain the following system.

$$c_0 + 3.219c_1 = 5.255$$
$$c_0 + 3.555c_1 = 5.208$$
$$c_0 + 3.912c_1 = 5.158$$
$$c_0 + 4.317c_1 = 5.101$$
$$c_0 + 6.215c_1 = 4.835$$
$$c_0 + 6.908c_1 = 4.738$$

This produces the least squares problem

$$A\mathbf{x} = \mathbf{b}$$

$$\begin{bmatrix} 1 & 3.219 \\ 1 & 3.555 \\ 1 & 3.912 \\ 1 & 4.317 \\ 1 & 6.215 \\ 1 & 6.908 \end{bmatrix} \begin{bmatrix} c_0 \\ c_1 \end{bmatrix} = \begin{bmatrix} 5.255 \\ 5.208 \\ 5.158 \\ 5.101 \\ 4.835 \\ 4.738 \end{bmatrix}.$$

The normal equations are

$$A^T A\mathbf{x} = A^T\mathbf{b}$$

$$\begin{bmatrix} 6 & 28.126 \\ 28.126 & 143.286908 \end{bmatrix} \begin{bmatrix} c_0 \\ c_1 \end{bmatrix} = \begin{bmatrix} 30.295 \\ 140.409027 \end{bmatrix}$$

and the solution is

$$\mathbf{x} = \begin{bmatrix} c_0 \\ c_1 \end{bmatrix} = \begin{bmatrix} 5.706264046 \\ -0.1401757902 \end{bmatrix}.$$

So, $\ln y = 5.706 - 0.140 \ln x$. Exponentiating each side produces $y = e^{5.706} \cdot x^{-0.14} \approx 300.7x^{-0.14}$.

$\left(\textit{Note: } e^{x+y} = e^x \cdot e^y\right)$

41. (a) False. The orthogonal complement of R^n is $\{\mathbf{0}\}$.

(b) True. See "Definition of Direct Sum," page 261.

43. Let $\mathbf{v} \in S_1 \cap S_2$. Because $\mathbf{v} \in S_1$, $\mathbf{v} \cdot \mathbf{x}_2 = 0$ for all $\mathbf{x}_2 \in S_2 \Rightarrow \mathbf{v} \cdot \mathbf{v} = 0$, because $\mathbf{v} \in S_2 \Rightarrow \mathbf{v} = \mathbf{0}$.

45. Let $\mathbf{v} \in R^n$, $\mathbf{v} = \mathbf{v}_1 + \mathbf{v}_2$, $\mathbf{v}_1 \in S$, $\mathbf{v}_2 \in S^\perp$.

Let $\{\mathbf{u}_1, \ldots, \mathbf{u}_t\}$ be an orthonormal basis for S.

Then

$$\mathbf{v} = \mathbf{v}_1 + \mathbf{v}_2 = c_1\mathbf{u}_1 + \cdots + c_t\mathbf{u}_t + \mathbf{v}_2, c_i \in R$$

and

$$\mathbf{v} \cdot \mathbf{u}_i = \left(c_1\mathbf{u}_1 + \cdots + c_t\mathbf{u}_t + \mathbf{v}_2\right) \cdot \mathbf{u}_i = c_i\left(\mathbf{u}_i \cdot \mathbf{u}_i\right) = c_i$$

which shows that $\mathbf{v}_1 = \text{proj}_S\mathbf{v} = (\mathbf{v} \cdot \mathbf{u}_1)\mathbf{u}_1 + \cdots + (\mathbf{v} \cdot \mathbf{u}_t)\mathbf{u}_t$.

© 2013 Cengage Learning. All Rights Reserved. May not be scanned, copied or duplicated, or posted to a publicly accessible website, in whole or in part.

Section 5.5 Applications of Inner Product Spaces

1. $\mathbf{j} \times \mathbf{i} = \begin{vmatrix} \mathbf{i} & \mathbf{j} & \mathbf{k} \\ 0 & 1 & 0 \\ 1 & 0 & 0 \end{vmatrix}$

$= \begin{vmatrix} 1 & 0 \\ 0 & 0 \end{vmatrix}\mathbf{i} - \begin{vmatrix} 0 & 0 \\ 1 & 0 \end{vmatrix}\mathbf{j} + \begin{vmatrix} 0 & 1 \\ 1 & 0 \end{vmatrix}\mathbf{k}$

$= 0\mathbf{i} - 0\mathbf{j} - \mathbf{k} = -\mathbf{k}$

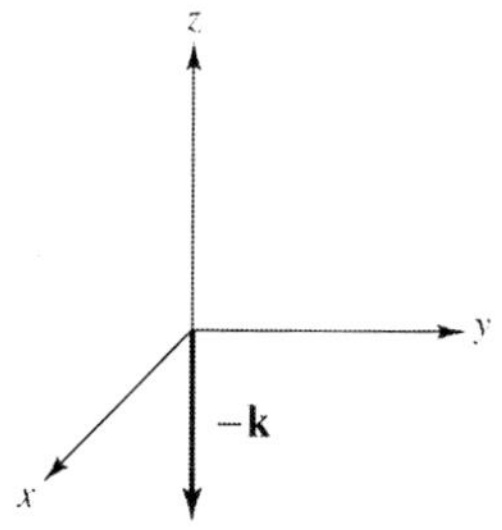

3. $\mathbf{j} \times \mathbf{k} = \begin{vmatrix} \mathbf{i} & \mathbf{j} & \mathbf{k} \\ 0 & 1 & 0 \\ 0 & 0 & 1 \end{vmatrix}$

$= \begin{vmatrix} 1 & 0 \\ 0 & 1 \end{vmatrix}\mathbf{i} - \begin{vmatrix} 0 & 0 \\ 0 & 1 \end{vmatrix}\mathbf{j} + \begin{vmatrix} 0 & 1 \\ 0 & 0 \end{vmatrix}\mathbf{k}$

$= \mathbf{i} - 0\mathbf{j} + 0\mathbf{k} = \mathbf{i}$

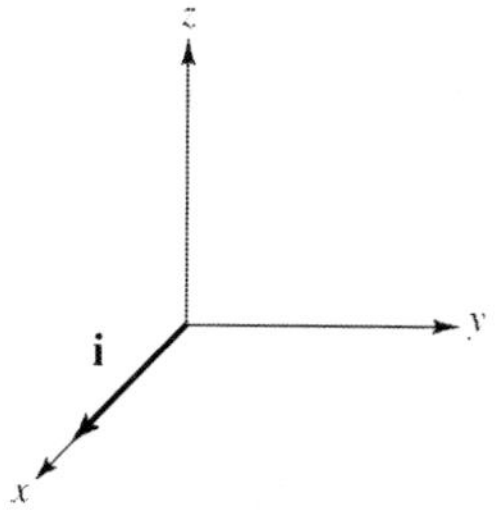

5. $\mathbf{i} \times \mathbf{k} = \begin{vmatrix} \mathbf{i} & \mathbf{j} & \mathbf{k} \\ 1 & 0 & 0 \\ 0 & 0 & 1 \end{vmatrix}$

$= \begin{vmatrix} 0 & 0 \\ 0 & 1 \end{vmatrix}\mathbf{i} - \begin{vmatrix} 1 & 0 \\ 0 & 1 \end{vmatrix}\mathbf{j} + \begin{vmatrix} 1 & 0 \\ 0 & 0 \end{vmatrix}\mathbf{k}$

$= 0\mathbf{i} - \mathbf{j} + 0\mathbf{k} = -\mathbf{j}$

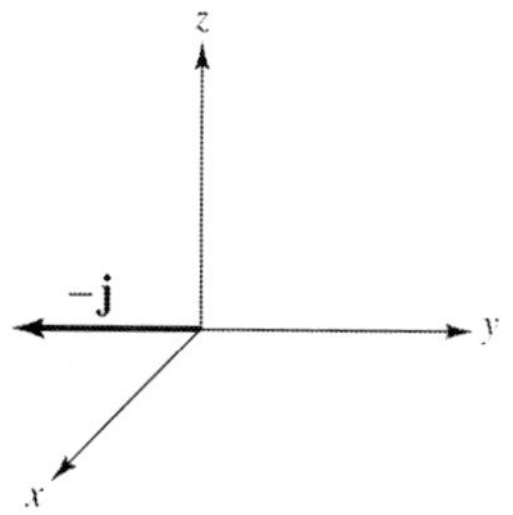

7. (a) $\mathbf{u} \times \mathbf{v} = \begin{vmatrix} \mathbf{i} & \mathbf{j} & \mathbf{k} \\ 1 & -1 & 0 \\ 0 & 1 & 1 \end{vmatrix}$

$= \mathbf{i}\begin{vmatrix} -1 & 0 \\ 1 & 1 \end{vmatrix} - \mathbf{j}\begin{vmatrix} 1 & 0 \\ 0 & 1 \end{vmatrix} + \mathbf{k}\begin{vmatrix} 1 & -1 \\ 0 & 1 \end{vmatrix}$

$= \mathbf{i}(-1 - 0) - \mathbf{j}(1 - 0) + \mathbf{k}(1 + 0)$

$= -\mathbf{i} - \mathbf{j} + \mathbf{k}$

(b) $\mathbf{v} \times \mathbf{u} = \begin{vmatrix} \mathbf{i} & \mathbf{j} & \mathbf{k} \\ 0 & 1 & 1 \\ 1 & -1 & 0 \end{vmatrix}$

$= \mathbf{i}\begin{vmatrix} 1 & 1 \\ -1 & 0 \end{vmatrix} - \mathbf{j}\begin{vmatrix} 0 & 1 \\ 1 & 0 \end{vmatrix} + \mathbf{k}\begin{vmatrix} 0 & 1 \\ 1 & -1 \end{vmatrix}$

$= \mathbf{i}(0 + 1) - \mathbf{j}(0 - 1) + \mathbf{k}(0 - 1)$

$= \mathbf{i} + \mathbf{j} - \mathbf{k}$

(c) $\mathbf{v} \times \mathbf{v} = \begin{vmatrix} \mathbf{i} & \mathbf{j} & \mathbf{k} \\ 0 & 1 & 1 \\ 0 & 1 & 1 \end{vmatrix}$

$= \mathbf{i}\begin{vmatrix} 1 & 1 \\ 1 & 1 \end{vmatrix} + \mathbf{j}\begin{vmatrix} 0 & 1 \\ 0 & 1 \end{vmatrix} + \mathbf{k}\begin{vmatrix} 0 & 1 \\ 0 & 1 \end{vmatrix}$

$= \mathbf{i}(1 - 1) - \mathbf{j}(0 - 0) + \mathbf{k}(0 - 0)$

$= \mathbf{0}$

9. (a) $\mathbf{u} \times \mathbf{v} = \begin{vmatrix} \mathbf{i} & \mathbf{j} & \mathbf{k} \\ 1 & 2 & -1 \\ 1 & 1 & 2 \end{vmatrix}$

$= \mathbf{i}\begin{vmatrix} 2 & -1 \\ 1 & 2 \end{vmatrix} - \mathbf{j}\begin{vmatrix} 1 & -1 \\ 1 & 2 \end{vmatrix} + \mathbf{k}\begin{vmatrix} 1 & 2 \\ 1 & 1 \end{vmatrix}$

$= \mathbf{i}(4 + 1) - \mathbf{j}(2 + 1) + \mathbf{k}(1 - 2)$

$= 5\mathbf{i} - 3\mathbf{j} - \mathbf{k}$

(b) $\mathbf{v} \times \mathbf{u} = \begin{vmatrix} \mathbf{i} & \mathbf{j} & \mathbf{k} \\ 1 & 1 & 2 \\ 1 & 2 & -1 \end{vmatrix}$

$= \mathbf{i}\begin{vmatrix} 1 & 2 \\ 2 & -1 \end{vmatrix} - \mathbf{j}\begin{vmatrix} 1 & 2 \\ 1 & -1 \end{vmatrix} + \mathbf{k}\begin{vmatrix} 1 & 1 \\ 1 & 2 \end{vmatrix}$

$= \mathbf{i}(-1 - 4) - \mathbf{j}(-1 - 2) + \mathbf{k}(2 - 1)$

$= -5\mathbf{i} + 3\mathbf{j} + \mathbf{k}$

(c) $\mathbf{v} \times \mathbf{v} = \begin{vmatrix} \mathbf{i} & \mathbf{j} & \mathbf{k} \\ 1 & 1 & 2 \\ 1 & 1 & 2 \end{vmatrix}$

$= \mathbf{i}\begin{vmatrix} 1 & 2 \\ 1 & 2 \end{vmatrix} - \mathbf{j}\begin{vmatrix} 1 & 2 \\ 1 & 2 \end{vmatrix} + \mathbf{k}\begin{vmatrix} 1 & 1 \\ 1 & 1 \end{vmatrix}$

$= \mathbf{i}(2 - 2) - \mathbf{j}(2 - 2) + \mathbf{k}(1 - 1)$

$= \mathbf{0}$

© 2013 Cengage Learning. All Rights Reserved. May not be scanned, copied or duplicated, or posted to a publicly accessible website, in whole or in part.

11. (a) $\mathbf{u} \times \mathbf{v} = \begin{vmatrix} \mathbf{i} & \mathbf{j} & \mathbf{k} \\ 3 & -2 & 4 \\ 1 & 5 & -3 \end{vmatrix}$

$= \mathbf{i}\begin{vmatrix} -2 & 4 \\ 5 & -3 \end{vmatrix} - \mathbf{j}\begin{vmatrix} 3 & 4 \\ 1 & -3 \end{vmatrix} + \mathbf{k}\begin{vmatrix} 3 & -2 \\ 1 & 5 \end{vmatrix}$

$= \mathbf{i}(6 - 20) - \mathbf{j}(-9 - 4) + \mathbf{k}(15 + 2)$

$= -14\mathbf{i} + 13\mathbf{j} + 17\mathbf{k}$

(b) $\mathbf{v} \times \mathbf{u} = \begin{vmatrix} \mathbf{i} & \mathbf{j} & \mathbf{k} \\ 1 & 5 & -3 \\ 3 & -2 & 4 \end{vmatrix}$

$= \mathbf{i}\begin{vmatrix} 5 & -3 \\ -2 & 4 \end{vmatrix} - \mathbf{j}\begin{vmatrix} 1 & -3 \\ 3 & 4 \end{vmatrix} + \mathbf{k}\begin{vmatrix} 1 & 5 \\ 3 & -2 \end{vmatrix}$

$= \mathbf{i}(20 - 6) - \mathbf{j}(4 + 9) + \mathbf{k}(-2 - 15)$

$= 14\mathbf{i} - 13\mathbf{j} - 17\mathbf{k}$

(c) $\mathbf{v} \times \mathbf{v} = \begin{vmatrix} \mathbf{i} & \mathbf{j} & \mathbf{k} \\ 1 & 5 & -3 \\ 1 & 5 & -3 \end{vmatrix}$

$= \mathbf{i}\begin{vmatrix} 5 & -3 \\ 5 & -3 \end{vmatrix} - \mathbf{j}\begin{vmatrix} 1 & -3 \\ 1 & -3 \end{vmatrix} + \mathbf{k}\begin{vmatrix} 1 & 5 \\ 1 & 5 \end{vmatrix}$

$= \mathbf{i}(-15 + 15) - \mathbf{j}(-3 + 3) + \mathbf{k}(5 - 5)$

$= \mathbf{0}$

13. $\mathbf{u} \times \mathbf{v} = \begin{vmatrix} \mathbf{i} & \mathbf{j} & \mathbf{k} \\ 0 & 1 & -2 \\ 1 & -1 & 0 \end{vmatrix} = -2\mathbf{i} - 2\mathbf{j} - \mathbf{k} = (-2, -2, -1)$

Furthermore, $\mathbf{u} \times \mathbf{v} = (-2, -2, -1)$ is orthogonal to both $(0, 1, -2)$ and $(1, -1, 0)$ because

$(-2, -2, -1) \cdot (0, 1, -2) = 0$ and

$(-2, -2, -1) \cdot (1, -1, 0) = 0.$

15. $\mathbf{u} \times \mathbf{v} = \begin{vmatrix} \mathbf{i} & \mathbf{j} & \mathbf{k} \\ 12 & -3 & 1 \\ -2 & 5 & 1 \end{vmatrix} = -8\mathbf{i} - 14\mathbf{j} + 54\mathbf{k} = (-8, -14, 54)$

Furthermore, $\mathbf{u} \times \mathbf{v} = (-8, -14, 54)$ is orthogonal to both $(12, -3, 1)$ and $(-2, 5, 1)$ because

$(-8, -14, 54) \cdot (12, -3, 1) = 0$ and

$(-8, -14, 54) \cdot (-2, 5, 1) = 0.$

17. $\mathbf{u} \times \mathbf{v} = \begin{vmatrix} \mathbf{i} & \mathbf{j} & \mathbf{k} \\ 2 & -3 & 1 \\ 1 & -2 & 1 \end{vmatrix} = -\mathbf{i} - \mathbf{j} - \mathbf{k} = (-1, -1, -1)$

Furthermore, $\mathbf{u} \times \mathbf{v} = (-1, -1, -1)$ is orthogonal to both $(2, -3, 1)$ and $(1, -2, 1)$ because

$(-1, -1, -1) \cdot (2, -3, 1) = 0$

and $(-1, -1, -1) \cdot (1, -2, 1) = 0.$

19. $\mathbf{u} \times \mathbf{v} = \begin{vmatrix} \mathbf{i} & \mathbf{j} & \mathbf{k} \\ 0 & 1 & 6 \\ 2 & 0 & -1 \end{vmatrix} = -\mathbf{i} + 12\mathbf{j} - 2\mathbf{k} = (-1, 12, -2)$

Furthermore, $\mathbf{u} \times \mathbf{v} = (-1, 12, -2)$ is orthogonal to both $(0, 1, 6)$ and $(2, 0, -1)$ because

$(-1, 12, -2) \cdot (0, 1, 6) = 0$ and

$(-1, 12, -2) \cdot (2, 0, -1) = 0.$

21. $\mathbf{u} \times \mathbf{v} = \begin{vmatrix} \mathbf{i} & \mathbf{j} & \mathbf{k} \\ 1 & 1 & 1 \\ 2 & 1 & -1 \end{vmatrix} = -2\mathbf{i} + 3\mathbf{j} - \mathbf{k} = (-2, 3, -1).$

Furthermore, $\mathbf{u} \times \mathbf{v} = (-2, 3, -1)$ is orthogonal to both $(1, 1, 1)$ and $(2, 1, -1)$ because

$(-2, 3, -1) \cdot (1, 1, 1) = 0$ and

$(-2, 3, -1) \cdot (2, 1, -1) = 0.$

23. Using a graphing utility:

$\mathbf{w} = \mathbf{u} \times \mathbf{v} = (5, -4, -3)$

Check if $\mathbf{w}$ is orthogonal to both $\mathbf{u}$ and $\mathbf{v}$:

$\mathbf{w} \cdot \mathbf{u} = (5, -4, -3) \cdot (1, 2, -1) = 5 - 8 + 3 = 0$

$\mathbf{w} \cdot \mathbf{v} = (5, -4, -3) \cdot (2, 1, 2) = 10 - 4 - 6 = 0$

25. Using a graphing utility:

$\mathbf{w} = \mathbf{u} \times \mathbf{v} = (2, -1, -1)$

Check if $\mathbf{w}$ is orthogonal to both $\mathbf{u}$ and $\mathbf{v}$:

$\mathbf{w} \cdot \mathbf{u} = (2, -1, -1) \cdot (0, 1, -1) = -1 + 1 = 0$

$\mathbf{w} \cdot \mathbf{v} = (2, -1, -1) \cdot (1, 2, 0) = 2 - 2 = 0$

27. Using a graphing utility:

$\mathbf{w} = \mathbf{u} \times \mathbf{v} = (1, -1, -3)$

Check if $\mathbf{w}$ is orthogonal to both $\mathbf{u}$ and $\mathbf{v}$:

$\mathbf{w} \cdot \mathbf{u} = (1, -1, -3) \cdot (2, -1, 1) = 2 + 1 - 3 = 0$

$\mathbf{w} \cdot \mathbf{v} = (1, -1, -3) \cdot (1, -2, 1) = 1 + 2 - 3 = 0$

© 2013 Cengage Learning. All Rights Reserved. May not be scanned, copied or duplicated, or posted to a publicly accessible website, in whole or in part.

29. Using a graphing utility:

$\mathbf{w} = \mathbf{u} \times \mathbf{v} = (1, -5, -3)$

Check if **w** is orthogonal to both **u** and **v**:

$\mathbf{w} \cdot \mathbf{u} = (1, -5, -3) \cdot (2, 1, -1) = 2 - 5 + 3 = 0$

$\mathbf{w} \cdot \mathbf{v} = (1, -5, -3) \cdot (1, -1, 2) = 1 + 5 - 6 = 0$

31. $\mathbf{u} \times \mathbf{v} = \begin{vmatrix} \mathbf{i} & \mathbf{j} & \mathbf{k} \\ 2 & -3 & 4 \\ 0 & -1 & 1 \end{vmatrix} = \mathbf{i} - 2\mathbf{j} - 2\mathbf{k}$

$\|\mathbf{u} \times \mathbf{v}\| = \sqrt{(1)^2 + (-2)^2 + (-2)^2} = \sqrt{9} = 3$

$$\text{Unit vector} = \frac{u \times v}{\|u \times v\|}$$

$$= \frac{1}{3}(i - 2j - 2k) = \frac{1}{3}i - \frac{2}{3}j - \frac{2}{3}k$$

33. $\mathbf{u} \times \mathbf{v} = \begin{vmatrix} \mathbf{i} & \mathbf{j} & \mathbf{k} \\ 3 & 1 & 0 \\ 0 & 1 & 1 \end{vmatrix} = \mathbf{i} - 3\mathbf{j} + 3\mathbf{k}$

$\|\mathbf{u} \times \mathbf{v}\| = \sqrt{19}$

$$\text{Unit vector} = \frac{\mathbf{u} \times \mathbf{v}}{\|\mathbf{u} \times \mathbf{v}\|}$$

$$= \frac{1}{\sqrt{19}}(\mathbf{i} - 3\mathbf{j} - 3\mathbf{k}) = \frac{\sqrt{19}}{19}(1, -3, 3)$$

35. $\mathbf{u} \times \mathbf{v} = \begin{vmatrix} \mathbf{i} & \mathbf{j} & \mathbf{k} \\ -3 & 2 & -5 \\ \frac{1}{2} & -\frac{3}{4} & \frac{1}{10} \end{vmatrix} = \left\langle -\frac{71}{20}, -\frac{11}{5}, \frac{5}{4} \right\rangle$

Consider the parallel vector $\langle -71, -44, 25 \rangle = w$.

$$\|\mathbf{w}\| = \sqrt{71^2 + 44^2 + 25^2} = \sqrt{7602}$$

$$\text{Unit vector} = \frac{\mathbf{u} \times \mathbf{v}}{\|\mathbf{w}\|} = \frac{1}{\sqrt{7602}}\langle -71, -44, 25 \rangle$$

$$= \frac{\sqrt{7602}}{7602}\langle -71, -44, 25 \rangle$$

37. $\mathbf{u} \times \mathbf{v} = \begin{vmatrix} \mathbf{i} & \mathbf{j} & \mathbf{k} \\ 1 & 1 & -1 \\ 1 & 1 & 1 \end{vmatrix} = 2\mathbf{i} - 2\mathbf{j}$

$\|\mathbf{u} \times \mathbf{v}\| = 2\sqrt{2}$

$$\text{Unit vector} = \frac{\mathbf{u} \times \mathbf{v}}{\|\mathbf{u} \times \mathbf{v}\|} = \frac{1}{2\sqrt{2}}(2\mathbf{i} - 2\mathbf{j})$$

$$= \frac{1}{\sqrt{2}}\mathbf{i} - \frac{1}{\sqrt{2}}\mathbf{j} = \frac{\sqrt{2}}{2}\mathbf{i} - \frac{\sqrt{2}}{2}\mathbf{j}$$

39. Because

$$\mathbf{u} \times \mathbf{v} = \begin{vmatrix} \mathbf{i} & \mathbf{j} & \mathbf{k} \\ 0 & 1 & 0 \\ 0 & 1 & 1 \end{vmatrix} = \mathbf{i} = (1, 0, 0)$$

the area of the parallelogram is

$\|\mathbf{u} \times \mathbf{v}\| = \|\mathbf{i}\| = 1.$

41. Because

$$\mathbf{u} \times \mathbf{v} = \begin{vmatrix} \mathbf{i} & \mathbf{j} & \mathbf{k} \\ 3 & 2 & -1 \\ 1 & 2 & 3 \end{vmatrix} = 8\mathbf{i} - 10\mathbf{j} + 4\mathbf{k} = (8, -10, 4),$$

the area of the parallelogram is

$\|(8, -10, 4)\| = \sqrt{8^2 + (-10)^2 + 4^2} = \sqrt{180} = 6\sqrt{5}.$

43. $(2, 3, 4) - (1, 1, 1) = (1, 2, 3)$

$(7, 7, 5) - (6, 5, 2) = (1, 2, 3)$

$(7, 7, 5) - (2, 3, 4) = (5, 4, 1)$

$(6, 5, 2) - (1, 1, 1) = (5, 4, 1)$

$\mathbf{u} = (1, 2, 3)$ and $\mathbf{v} = (5, 4, 1)$

Because

$$\mathbf{u} \times \mathbf{v} = \begin{vmatrix} \mathbf{i} & \mathbf{j} & \mathbf{k} \\ 1 & 2 & 3 \\ 5 & 4 & 1 \end{vmatrix} = -10\mathbf{i} + 14\mathbf{j} - 6\mathbf{k} = (-10, 14, -6),$$

the area of the parallelogram is

$\|\mathbf{u} \times \mathbf{v}\| = \sqrt{(-10)^2 + 14^2 + (-6)^2} = \sqrt{332} = 2\sqrt{83}.$

45. Because

$$\mathbf{v} \times \mathbf{w} = \begin{vmatrix} \mathbf{i} & \mathbf{j} & \mathbf{k} \\ 0 & 1 & 0 \\ 0 & 0 & 1 \end{vmatrix} = \mathbf{i} = (1, 0, 0),$$

the triple scalar product of **u**, **v**, and **w** is

$\mathbf{u} \cdot (\mathbf{v} \times \mathbf{w}) = (1, 0, 0) \cdot (1, 0, 0) = 1.$

47. Because

$$\mathbf{v} \times \mathbf{w} = \begin{vmatrix} \mathbf{i} & \mathbf{j} & \mathbf{k} \\ 2 & 1 & 0 \\ 0 & 0 & 1 \end{vmatrix} = \mathbf{i} - 2\mathbf{j} = (1, -2, 0),$$

the triple scalar product of **u**, **v**, and **w** is

$\mathbf{u} \cdot (\mathbf{v} \times \mathbf{w}) = (1, 1, 1) \cdot (1, -2, 0) = -1.$

© 2013 Cengage Learning. All Rights Reserved. May not be scanned, copied or duplicated, or posted to a publicly accessible website, in whole or in part.

49. The area of the base of the parallelogram is $\|\mathbf{v} \times \mathbf{w}\|$.

The height is $|\cos\theta|\,\|\mathbf{u}\|$, where

$$|\cos\theta| = \frac{|\mathbf{u} \cdot (\mathbf{v} \times \mathbf{w})|}{\|\mathbf{u}\|\,\|\mathbf{v} \times \mathbf{w}\|}.$$

So,

$$\text{volume} = \text{base} \times \text{height} = \|\mathbf{v} \times \mathbf{w}\| \frac{|\mathbf{u} \cdot (\mathbf{v} \times \mathbf{w})|}{\|\mathbf{u}\|\,\|\mathbf{v} \times \mathbf{w}\|}\|\mathbf{u}\|$$

$$= |\mathbf{u} \cdot (\mathbf{v} \times \mathbf{w})|.$$

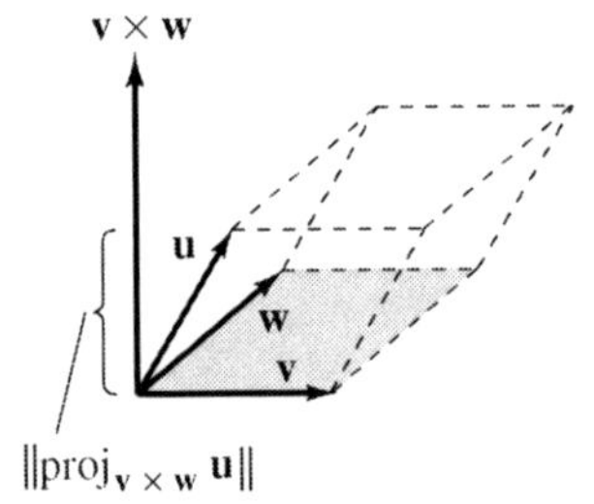

51. $(3, 3, 0) - (1, 3, 5) = (2, 0, -5)$

$(3, 3, 0) - (-2, 0, 5) = (5, 3, -5)$

Because

$$\mathbf{u} \times \mathbf{v} = \begin{vmatrix} \mathbf{i} & \mathbf{j} & \mathbf{k} \\ 2 & 0 & -5 \\ 5 & 3 & -5 \end{vmatrix} = 15\mathbf{i} - 15\mathbf{j} + 6\mathbf{k} = (15, -15, 6),$$

the area of the triangle is

$$A = \frac{1}{2}\|\mathbf{u} \times \mathbf{v}\|$$

$$= \frac{1}{2}\sqrt{15^2 + (-15)^2 + 6^2} = \frac{1}{2}\sqrt{486} = \frac{9\sqrt{6}}{2}.$$

53. $\mathbf{u} \times (\mathbf{v} + \mathbf{w})$

$$= \begin{vmatrix} \mathbf{i} & \mathbf{j} & \mathbf{k} \\ u_1 & u_2 & u_3 \\ v_1 + w_1 & v_2 + w_2 & v_3 + w_3 \end{vmatrix}$$

$$= \left[u_2(v_3 + w_3) - u_3(v_2 + w_2)\right]\mathbf{i} - \left[u_1(v_3 + w_3) - u_3(v_1 + w_1)\right]\mathbf{j} + \left[u_1(v_2 + w_2) - u_2(v_1 + w_1)\right]\mathbf{k}$$

$$= (u_2v_3 - v_2u_3)\mathbf{i} - (u_1v_3 - u_3v_1)\mathbf{j} + (u_1v_2 - u_2v_1)\mathbf{k} + (u_2w_3 - u_3w_2)\mathbf{i} - (u_1w_3 - u_3w_1)\mathbf{j} + (u_1w_2 - u_2w_1)\mathbf{k}$$

$$= \begin{vmatrix} \mathbf{i} & \mathbf{j} & \mathbf{k} \\ u_1 & u_2 & u_3 \\ v_1 & v_2 & v_3 \end{vmatrix} + \begin{vmatrix} \mathbf{i} & \mathbf{j} & \mathbf{k} \\ u_1 & u_2 & u_3 \\ w_1 & w_2 & w_3 \end{vmatrix}$$

$$= (\mathbf{u} \times \mathbf{v}) + (\mathbf{u} \times \mathbf{w})$$

55. $$\mathbf{u} \times \mathbf{0} = \begin{vmatrix} \mathbf{i} & \mathbf{j} & \mathbf{k} \\ u_1 & u_2 & u_3 \\ 0 & 0 & 0 \end{vmatrix} = \mathbf{0} = \begin{vmatrix} \mathbf{i} & \mathbf{j} & \mathbf{k} \\ 0 & 0 & 0 \\ u_1 & u_2 & u_3 \end{vmatrix} = \mathbf{0} \times \mathbf{u}$$

57. $$\mathbf{u} \cdot (\mathbf{v} \times \mathbf{w}) = \mathbf{u} \cdot \begin{vmatrix} \mathbf{i} & \mathbf{j} & \mathbf{k} \\ v_1 & v_2 & v_3 \\ w_1 & w_2 & w_3 \end{vmatrix}$$

$$= (u_1, u_2, u_3) \cdot \left[(v_2w_3 - v_3w_2)\mathbf{i} - (v_1w_3 - v_3w_1)\mathbf{j} + (v_1w_2 - v_2w_1)\mathbf{k}\right]$$

$$= u_1(v_2w_3 - v_3w_2) - u_2(v_1w_3 - v_3w_1) + u_3(v_1w_2 - v_2w_1)$$

$$= (u_2v_3 - u_3v_2)w_1 - (u_1v_3 - v_1u_3)w_2 + (u_1v_2 - v_1u_2)w_3$$

$$= \begin{vmatrix} \mathbf{i} & \mathbf{j} & \mathbf{k} \\ u_1 & u_2 & u_3 \\ v_1 & v_2 & v_3 \end{vmatrix} \cdot (w_1, w_2, w_3)$$

$$= (\mathbf{u} \times \mathbf{v}) \cdot \mathbf{w}$$

© 2013 Cengage Learning. All Rights Reserved. May not be scanned, copied or duplicated, or posted to a publicly accessible website, in whole or in part.

59. $\|\mathbf{u}\|\|\mathbf{v}\|\sin\theta = \|\mathbf{u}\|\|\mathbf{v}\|\sqrt{1-\cos^2\theta}$

$$= \|\mathbf{u}\|\|\mathbf{v}\|\sqrt{1-\frac{(\mathbf{u}\cdot\mathbf{v})^2}{\|\mathbf{u}\|^2\|\mathbf{v}\|^2}}$$

$$= \sqrt{\|\mathbf{u}\|^2\|\mathbf{v}\|^2 - (\mathbf{u}\cdot\mathbf{v})^2}$$

$$= \sqrt{\left(u_1^2+u_2^2+u_3^2\right)\left(v_1^2+v_2^2+v_3^2\right) - \left(u_1v_1+u_2v_2+u_3v_3\right)^2}$$

$$= \sqrt{\left(u_2v_3-u_3v_2\right)^2+\left(u_3v_1-u_1v_3\right)^2+\left(u_1v_2-u_2v_1\right)^2}$$

$$= \|\mathbf{u}\times\mathbf{v}\|$$

61. $\|\mathbf{u}\times\mathbf{v}\|^2 = \|\mathbf{u}\|^2\|\mathbf{v}\|^2\sin^2\theta = \|\mathbf{u}\|^2\|\mathbf{v}\|^2\left(1-\cos^2\theta\right) = \|\mathbf{u}\|^2\|\mathbf{v}\|^2\left(1-\frac{(\mathbf{u}\cdot\mathbf{v})^2}{\|\mathbf{u}\|^2\|\mathbf{v}\|^2}\right) = \|\mathbf{u}\|^2\|\mathbf{v}\|^2-(\mathbf{u}\cdot\mathbf{v})^2$

63. (a) The standard basis for P_1 is $\{1, x\}$. Applying the Gram-Schmidt orthonormalization process produces the orthonormal basis

$$B = \{\mathbf{w}_1, \mathbf{w}_2\} = \left\{1, \sqrt{3}(2x-1)\right\}.$$

The least squares approximating function is then given by $g(x) = \langle f, \mathbf{w}_1\rangle\mathbf{w}_1 + \langle f, \mathbf{w}_2\rangle\mathbf{w}_2$.

Find the inner products

$$\langle f, \mathbf{w}_1\rangle = \int_0^1 \left(x^2\right)(1)\,dx = \frac{1}{3}x^3\Big]_0^1 = \frac{1}{3}$$

$$\langle f, \mathbf{w}_2\rangle = \int_0^1 \left(x^2\right)\left[\sqrt{3}(2x-1)\right]dx$$

$$= \sqrt{3}\left(\frac{1}{2}x^4-\frac{1}{3}x^3\right)\Big]_0^1 = \frac{\sqrt{3}}{6}$$

and conclude that g is given by

$$g(x) = \langle f, \mathbf{w}_1\rangle\mathbf{w}_1 + \langle f, \mathbf{w}_2\rangle\mathbf{w}_2$$

$$= \frac{1}{3}(1) + \frac{\sqrt{3}}{6}\left[\sqrt{3}(2x-1)\right]$$

$$= \frac{1}{3} + x - \frac{1}{2} = x - \frac{1}{6}.$$

(b)

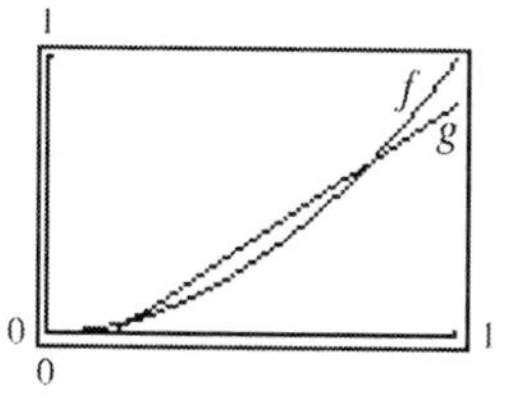

65. (a) The standard basis for P_1 is $\{1, x\}$. Applying the Gram-Schmidt orthonormalization process produces the orthonormal basis

$$B = \{\mathbf{w}_1, \mathbf{w}_2\} = \left\{1, \sqrt{3}(2x-1)\right\}.$$

The least squares approximating function is then given by $g(x) = \langle f, \mathbf{w}_1\rangle\mathbf{w}_1 + \langle f, \mathbf{w}_2\rangle\mathbf{w}_2$.

Find the inner products

$$\langle f, \mathbf{w}_1\rangle = \int_0^1 e^{2x}\,dx = \frac{1}{2}e^{2x}\Big]_0^1 = \frac{1}{2}\left(e^2-1\right)$$

$$\langle f, \mathbf{w}_2\rangle = \int_0^1 e^{2x}\sqrt{3}(2x-1)\,dx$$

$$= \sqrt{3}(x-1)e^{2x}\Big]_0^1 = \sqrt{3}$$

and conclude that

$$g(x) = \langle f, \mathbf{w}_1\rangle\mathbf{w}_1 + \langle f, \mathbf{w}_2\rangle\mathbf{w}_2$$

$$= \frac{1}{2}\left(e^2-1\right) + \sqrt{3}\left(\sqrt{3}(2x-1)\right)$$

$$= 6x + \frac{1}{2}\left(e^2-7\right)(\approx 6x + 0.1945)$$

(b)

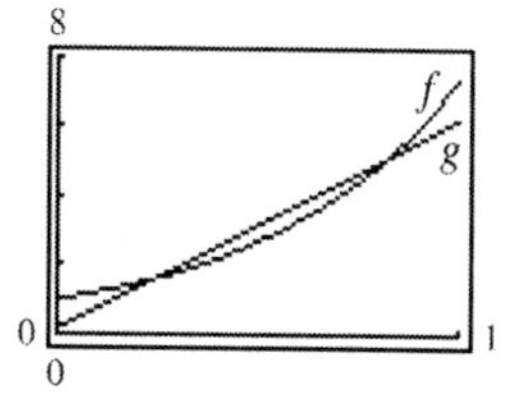

© 2013 Cengage Learning. All Rights Reserved. May not be scanned, copied or duplicated, or posted to a publicly accessible website, in whole or in part.

67. (a) The standard basis for P_1 is $\{1, x\}$. Applying the Gram-Schmidt orthonormalization process produces the orthonormal basis.

$$B = \{\mathbf{w}_1, \mathbf{w}_2\} = \left\{\frac{1}{\sqrt{\pi}}, \frac{\sqrt{3}}{\pi^{3/2}}(2x - \pi)\right\}$$

The least squares approximating function is given by

$g(x) = \langle f, \mathbf{w}_1\rangle\mathbf{w}_1 + \langle f, \mathbf{w}_2\rangle\mathbf{w}_2$.

Find the inner products

$$\langle f, \mathbf{w}_1\rangle = \int_0^\pi \frac{1}{\sqrt{\pi}}\cos x\, dx = \frac{1}{\sqrt{\pi}}\sin x\Big]_0^\pi = 0$$

$$\langle f, \mathbf{w}_2\rangle = \int_0^\pi \frac{\sqrt{3}}{\pi^{3/2}}(2x - \pi)\cos x\, dx = \frac{\sqrt{3}}{\pi^{3/2}}\Big[(2x - \pi)\sin x + 2\cos x\Big]_0^\pi = -\frac{4\sqrt{3}}{\pi^{3/2}}$$

and conclude that

$$g(x) = \langle f, \mathbf{w}_1\rangle\mathbf{w}_1 + \langle f, \mathbf{w}_2\rangle\mathbf{w}_2 = 0 + \frac{-4\sqrt{3}}{\pi^{3/2}}\left(\frac{\sqrt{3}}{\pi^{3/2}}\right)(2x - \pi) = \frac{12}{\pi^3}(\pi - 2x).$$

(b)

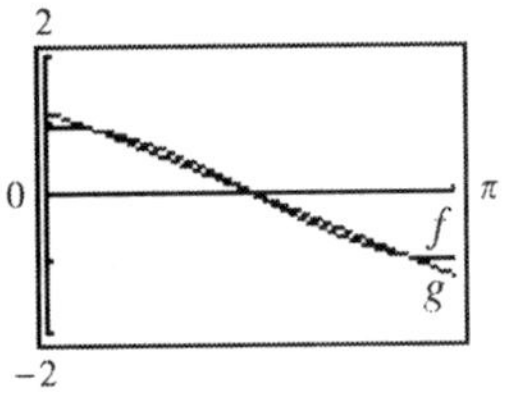

69. (a) The standard basis for P_2 is $\{1, x, x^2\}$. Applying the Gram-Schmidt orthonormalization process produces the orthonormal basis $B = \{\mathbf{w}_1, \mathbf{w}_2, \mathbf{w}_3\} = \left\{1, \sqrt{3}(2x - 1), \sqrt{5}(6x^2 - 6x + 1)\right\}$.

The least squares approximating function for f is given by $g(x) = \langle f, \mathbf{w}_1\rangle\mathbf{w}_1 + \langle f, \mathbf{w}_2\rangle\mathbf{w}_2 + \langle f, \mathbf{w}_3\rangle\mathbf{w}_3$.

Find the inner products

$$\langle f, \mathbf{w}_1\rangle = \int_0^1 x^3(1)dx = \frac{1}{4}x^4\Big]_0^1 = \frac{1}{4}$$

$$\langle f, \mathbf{w}_2\rangle = \int_0^1 x^3\left[\sqrt{3}(2x - 1)\right]dx = \sqrt{3}\left[\frac{2}{5}x^5 - \frac{1}{4}x^4\right]_0^1 = \frac{3\sqrt{3}}{20}$$

$$\langle f, \mathbf{w}_3\rangle = \int_0^1 x^3\left[\sqrt{5}(6x^2 - 6x + 1)\right]dx = \sqrt{5}\left[x^6 - \frac{6}{5}x^5 + \frac{1}{4}x^4\right]_0^1 = \frac{\sqrt{5}}{20}$$

and conclude that

$$g(x) = \langle f, \mathbf{w}_1\rangle\mathbf{w}_1 + \langle f, \mathbf{w}_2\rangle\mathbf{w}_2 + \langle f, \mathbf{w}_3\rangle\mathbf{w}_3 = \frac{1}{4}(1) + \frac{3\sqrt{3}}{20}\left[\sqrt{3}(2x - 1)\right] + \frac{\sqrt{5}}{20}\left[\sqrt{5}(6x^2 - 6x + 1)\right] = \frac{3}{2}x^2 - \frac{3}{5}x + \frac{1}{20}.$$

(b)

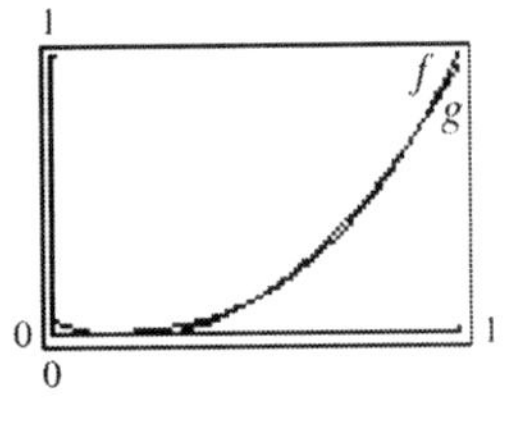

© 2013 Cengage Learning. All Rights Reserved. May not be scanned, copied or duplicated, or posted to a publicly accessible website, in whole or in part.

71. (a) The standard basis for P_2 is $\{1, x, x^2\}$. Applying the Gram-Schmidt orthonormalization process produces the orthonormal basis

$$B = \{\mathbf{w}_1, \mathbf{w}_2, \mathbf{w}_3\} = \left\{\frac{1}{\sqrt{\pi}}, \frac{2\sqrt{3}}{\pi^{3/2}}x, \frac{6\sqrt{5}}{\pi^{5/2}}\left(x^2 - \frac{\pi^2}{12}\right)\right\}.$$

The least squares approximating function is given by $g(x) = \langle f, \mathbf{w}_1\rangle\mathbf{w}_1 + \langle f, \mathbf{w}_2\rangle\mathbf{w}_2 + \langle f, \mathbf{w}_3\rangle\mathbf{w}_3$.

Find the inner products

$$\langle f, \mathbf{w}_1\rangle = \int_{-\pi/2}^{\pi/2} \frac{1}{\sqrt{\pi}} \sin x \, dx = 0 \text{ (see Exercise 54)}$$

$$\langle f, \mathbf{w}_2\rangle = \int_{-\pi/2}^{\pi/2} \frac{2\sqrt{3}}{\pi^{3/2}} x \sin x \, dx = \frac{4\sqrt{3}}{\pi^{3/2}} \text{ (See Exercise 54)}$$

$$\langle f, \mathbf{w}_3\rangle = \int_{-\pi/2}^{\pi/2} \frac{6\sqrt{5}}{\pi^{5/2}}\left(x^2 - \frac{x^2}{12}\right)\sin x \, dx = \frac{6\sqrt{5}}{\pi^{3/2}}\left[2\cos x - x^2\cos x + 2x\sin x + \frac{\pi^2}{12}\cos x\right]_{-\pi/2}^{\pi/2} = 0$$

and conclude that $g(x) = 0 + \frac{4\sqrt{3}}{\pi^{3/2}}\left(\frac{2\sqrt{3}}{\pi^{3/2}}x\right) + 0 = \frac{24}{\pi^3}x.$

(b)

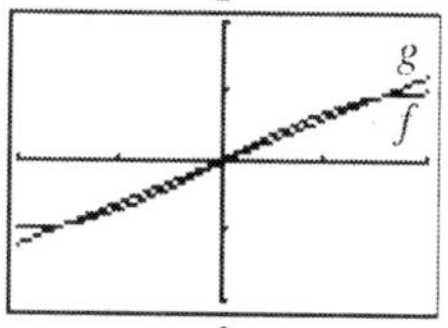

73. The third order Fourier approximation of $f(x) = \pi - x$ is of the form

$$g(x) = \frac{a_0}{2} + a_1 \cos x + b_1 \sin x + a_2 \cos 2x + b_2 \sin 2x + a_3 \cos 3x + b_3 \sin 3x.$$

Find the coefficients as follows.

$$a_0 = \frac{1}{\pi}\int_0^{2\pi} f(x)\,dx = \frac{1}{\pi}\int_0^{2\pi} (\pi - x)\,dx = -\frac{1}{2\pi}(\pi - x)^2\Big]_0^{2\pi} = 0$$

$$a_j = \frac{1}{\pi}\int_0^{2\pi} f(x)\cos jx\,dx = \frac{1}{\pi}\int_0^{2\pi} (\pi - x)\cos jx\,dx = \left[-\frac{1}{j\pi}x\sin jx - \frac{1}{j^2\pi}\cos jx + \frac{1}{j}\sin jx\right]_0^{2\pi} = 0,\ j = 1, 2, 3$$

$$b_j = \frac{1}{\pi}\int_0^{2\pi} f(x)\sin x\,jx\,dx = \frac{1}{\pi}\int_0^{2\pi} (\pi - x)\sin jx\,dx = \left[\frac{1}{j\pi}x\cos jx - \frac{1}{j^2\pi}\sin jx - \frac{1}{j}\cos jx\right]_0^{2\pi} = \frac{2}{j},\ j = 1, 2, 3$$

So, the approximation is $g(x) = 2\sin x + \sin 2x + \frac{2}{3}\sin 3x.$

© 2013 Cengage Learning. All Rights Reserved. May not be scanned, copied or duplicated, or posted to a publicly accessible website, in whole or in part.

75. The third order Fourier approximation of $f(x) = (x - \pi)^2$ is of the form

$$g(x) = \frac{a_0}{2} + a_1 \cos x + b_1 \sin x + a_2 \cos 2x + b_2 \sin 2x + a_3 \cos 3x + b_3 \sin 3x.$$

Find the coefficients as follows.

$$a_0 = \frac{1}{\pi}\int_0^{2\pi} f(x)\,dx = \frac{1}{\pi}\int_0^{2\pi} (x - \pi)^2\,dx = \frac{1}{3\pi}(x - \pi)^3\Big]_0^{2\pi} = \frac{2\pi^2}{3}$$

$$a_j = \frac{1}{\pi}\int_0^{2\pi} f(x)\cos jx\,dx$$

$$= \frac{1}{\pi}\int_0^{2\pi} (x - \pi)^2 \cos jx\,dx$$

$$= \left[\frac{\pi}{j}\sin jx - \frac{2}{j}x\sin jx + \frac{2}{j^2\pi}x\cos jx + \frac{1}{j\pi}x^2\sin jx - \frac{2}{j^3\pi}\sin jx - \frac{2}{j^2}\cos jx\right]_0^{2\pi}$$

$$= \frac{4}{j^2},\ j = 1, 2, 3$$

$$b_j = \frac{1}{\pi}\int_0^{2\pi} f(x)\sin jx\,dx$$

$$= \frac{1}{\pi}\int_0^{2\pi} (x - \pi)^2 \sin jx\,dx$$

$$= \left[-\frac{\pi}{j}\cos jx + \frac{2}{j}x\cos jx + \frac{2}{j^2\pi}x\sin jx - \frac{1}{j\pi}x^2\cos jx + \frac{2}{j^3\pi}\cos jx - \frac{2}{j^2}\sin jx\right]_0^{2\pi}$$

$$= 0,\ j = 1, 2, 3$$

So, the approximation is $g(x) = \dfrac{\pi^2}{3} + 4\cos x + \cos 2x + \dfrac{4}{9}\cos 3x.$

77. The first order Fourier approximation of $f(x) = e^{-x}$ is of the form $g(x) = \dfrac{a_0}{2} + a_1\cos x + b_1 \sin x.$

Find the coefficients as follows.

$$a_0 = \frac{1}{\pi}\int_0^{2\pi} f(x)\,dx = \frac{1}{\pi}\int_0^{2\pi} e^{-x}\,dx = -\frac{1}{\pi}e^{-x}\Big]_0^{2\pi} = \frac{1 - e^{-2\pi}}{\pi}$$

$$a_1 = \frac{1}{\pi}\int_0^{2\pi} f(x)\cos x\,dx = \frac{1}{\pi}\int_0^{2\pi} e^{-x}\cos x\,dx = \left[-\frac{1}{2\pi}e^{-x}\cos x + \frac{1}{2\pi}e^{-x}\sin x\right]_0^{2\pi} = \frac{1 - e^{-2\pi}}{2\pi}$$

$$b_1 = \frac{1}{\pi}\int_0^{2\pi} f(x)\sin x\,dx = \frac{1}{\pi}\int_0^{2\pi} e^{-x}\sin x\,dx = \left[-\frac{1}{2\pi}e^{-x}\cos x - \frac{1}{2\pi}e^{-x}\sin x\right]_0^{2\pi} = \frac{1 - e^{-2\pi}}{2\pi}$$

So, the approximation is $g(x) = \dfrac{1}{2\pi}\left(1 - e^{-2\pi}\right)\left(1 + \cos x + \sin x\right).$

© 2013 Cengage Learning. All Rights Reserved. May not be scanned, copied or duplicated, or posted to a publicly accessible website, in whole or in part.

79. The first order Fourier approximation of $f(x) = e^{-2x}$ is of the form $g(x) = \frac{a_0}{2} + a_1 \cos x + b_1 \sin x$.

Find the coefficients as follows.

$$a_0 = \frac{1}{\pi}\int_0^{2\pi} f(x)\,dx = \frac{1}{\pi}\int_0^{2\pi} e^{-2x}dx = \left.\frac{1 - e^{-2x}}{2\pi}\right]_0^{2\pi} = \frac{1 - e^{-4\pi}}{2\pi}$$

$$a_1 = \frac{1}{\pi}\int_0^{2\pi} f(x)\cos x\,dx = \frac{1}{\pi}\int_0^{2\pi} e^{-2x}\cos x\,dx = \left[-\frac{2}{5\pi}e^{-2x}\cos x + \frac{1}{5\pi}e^{-2x}\sin x\right]_0^{2\pi} = 2\left(\frac{1 - e^{-4\pi}}{5\pi}\right)$$

$$b_1 = \frac{1}{\pi}\int_0^{2\pi} f(x)\sin x\,dx = \frac{1}{\pi}\int_0^{2\pi} e^{-2x}\sin x\,dx = \left[-\frac{2}{5\pi}e^{-2x}\sin x - \frac{1}{5\pi}e^{-2x}\cos x\right]_0^{2\pi} = \frac{1 - e^{-4\pi}}{5\pi}$$

So, the approximation is

$$\begin{aligned} g(x) &= \frac{1 - e^{-4\pi}}{4\pi} + 2\left(\frac{1 - e^{-4\pi}}{5\pi}\right)\cos x + \frac{1 - e^{-4\pi}}{5\pi}\sin x \\ &= 5\left(\frac{1 - e^{-4\pi}}{20\pi}\right) + 8\left(\frac{1 - e^{-4\pi}}{20\pi}\right)\cos x + 4\left(\frac{1 - e^{-4\pi}}{20\pi}\right)\sin x \\ &= \left(\frac{1 - e^{-4\pi}}{20\pi}\right)(5 + 8\cos x + 4\sin x). \end{aligned}$$

81. The third order Fourier approximation of $f(x) = 1 + x$ is of the form

$$g(x) = \frac{a_0}{2} + a_1\cos x + b_1\sin x + a_2\cos 2x + b_2\sin 2x + a_3\cos 3x + b_3\sin 3x.$$

Find the coefficients as follows.

$$a_0 = \frac{1}{\pi}\int_0^{2\pi} f(x)\,dx = \frac{1}{\pi}\int_0^{2\pi}(1 + x)\,dx = \frac{1}{\pi}\left(x + \frac{x^2}{2}\right)\Bigg]_0^{2\pi} = 2 + 2\pi$$

$$a_j = \frac{1}{\pi}\int_0^{2\pi} f(x)\cos(jx)\,dx = \frac{1}{\pi}\int_0^{2\pi}(1 + x)\cos(jx)\,dx = \frac{1}{\pi}\left[\frac{1 + x}{j}\sin(jx) + \frac{1}{j^2}\cos(jx)\right]_0^{2\pi} = 0$$

$$b_j = \frac{1}{\pi}\int_0^{2\pi} f(x)\sin(jx)\,dx = \frac{1}{\pi}\int_0^{2\pi}(1 + x)\sin(jx)\,dx = \frac{1}{\pi}\left[\frac{-(1 + x)}{j}\cos(jx) + \frac{1}{j^2}\sin(jx)\right]_0^{2\pi} = \frac{-2}{j}$$

So, the approximation is $g(x) = (1 + \pi) - 2\sin x - \sin 2x - \frac{2}{3}\sin 3x$.

83. Because $f(x) = 2\sin x\cos x = \sin 2x$, you see that the fourth order Fourier approximation is simply $g(x) = \sin 2x$.

85. Because $a_0 = 0, a_j = 0\,(j = 1, 2, 3, \ldots, n)$, and $b_j = \frac{2}{j}\,(j = 1, 2, 3, \ldots, n)$ the nth-order Fourier approximation is

$$g(x) = 2\sin x + \sin 2x + \frac{2}{3}\sin 3x + \ldots + \frac{2}{n}\sin nx = \sum_{j=1}^{n}\frac{2}{j}\sin jx.$$

87. Since $a_0 = \frac{1 - e^{-2\pi}}{\pi}$, $a_j = \frac{1 - e^{-2\pi}}{(1 + j^2)\pi}$, and $b_j = \frac{j(1 - e^{-2\pi})}{(1 + j^2)\pi}$, the nth-order Fourier approximation is

$$f(x) = \frac{1 - e^{-2\pi}}{2\pi} + \frac{1 - e^{-2\pi}}{\pi}\sum_{j=1}^{n}\left(\frac{1}{j^2 + 1}\cos jx + \frac{2}{j^2 + 1}\sin jx\right)$$

© 2013 Cengage Learning. All Rights Reserved. May not be scanned, copied or duplicated, or posted to a publicly accessible website, in whole or in part.

Review Exercises for Chapter 5

1. (a) $\|\mathbf{u}\| = \sqrt{1^2 + 2^2} = \sqrt{5}$

(b) $\|\mathbf{v}\| = \sqrt{4^2 + 1^2} = \sqrt{17}$

(c) $\mathbf{u} \cdot \mathbf{v} = 1(4) + 2(1) = 6$

(d) $d(\mathbf{u}, \mathbf{v}) = \|\mathbf{u} - \mathbf{v}\| = \|(-3, 1)\|$

$= \sqrt{(-3)^2 + 1^2} = \sqrt{10}$

3. (a) $\|\mathbf{u}\| = \sqrt{2^2 + 1^2 + 1^2} = \sqrt{6}$

(b) $\|\mathbf{v}\| = \sqrt{3^2 + 2^2 + (-1)^2} = \sqrt{14}$

(c) $\mathbf{u} \cdot \mathbf{v} = 2(3) + 1(2) + 1(-1) = 7$

(d) $d(\mathbf{u} \cdot \mathbf{v}) = \|\mathbf{u} - \mathbf{v}\| = \|(-1, -1, -2)\|$

$= \sqrt{(-1)^2 + (-1)^2 + (-2)^2}$

$= \sqrt{6}$

5. (a) $\|\mathbf{u}\| = \sqrt{1^2 + (-2)^2 + 0^2 + 1^2} = \sqrt{6}$

(b) $\|\mathbf{v}\| = \sqrt{1^2 + 1^2 + (-1)^2 + 0^2} = \sqrt{3}$

(c) $\mathbf{u} \cdot \mathbf{v} = 1(1) + (-2)(1) + 0(-1) + 1(0) = -1$

(d) $d(\mathbf{u} \cdot \mathbf{v}) = \|\mathbf{u} - \mathbf{v}\| = \|(0, -3, 1, 1)\|$

$= \sqrt{0^2 + (-3)^2 + 1^2 + 1^2}$

$= \sqrt{11}$

7. (a) $\|\mathbf{u}\| = \sqrt{0^2 + 1^2 + (-1)^2 + 1^2 + 2^2} = \sqrt{7}$

(b) $\|\mathbf{v}\| = \sqrt{0^2 + 1^2 + (-2)^2 + 1^2 + 1^2} = \sqrt{7}$

(c) $\mathbf{u} \cdot \mathbf{v} = 0(0) + 1(1) + (-1)(-2) + 1(1) + 2(1) = 6$

(d) $d(\mathbf{u} \cdot \mathbf{v}) = \|\mathbf{u} - \mathbf{v}\| = \|(0, 0, 1, 0, 1)\|$

$= \sqrt{0^2 + 0^2 + 1^2 + 0^2 + 1^2}$

$= \sqrt{2}$

9. The norm of $\mathbf{v}$ is

$$\|\mathbf{v}\| = \sqrt{5^2 + 3^2 + (-2)^2} = \sqrt{38}.$$

So, a unit vector in the direction of $\mathbf{v}$ is

$$\mathbf{u} = \frac{1}{\|\mathbf{v}\|}\mathbf{v} = \frac{1}{\sqrt{38}}(5, 3, -2) = \left(\frac{5}{\sqrt{38}}, \frac{3}{\sqrt{38}}, -\frac{2}{\sqrt{38}}\right).$$

11. The norm of $\mathbf{v}$ is

$$\|\mathbf{v}\| = \sqrt{(-1)^2 + 1^2 + 2^2} = \sqrt{6}.$$

So, a unit vector in the direction of $\mathbf{v}$ is given by

$$\mathbf{u} = \frac{1}{\|\mathbf{v}\|}\mathbf{v} = \frac{1}{\sqrt{6}}(-1, 1, 2) = \left(\frac{-1}{\sqrt{6}}, \frac{1}{\sqrt{6}}, \frac{2}{\sqrt{6}}\right).$$

13. (a) Because $\mathbf{v}/\|\mathbf{v}\|$ is a unit vector in the direction of $\mathbf{v}$,

$$\mathbf{u} = \frac{\|\mathbf{v}\|}{2}\frac{\mathbf{v}}{\|\mathbf{v}\|} = \frac{1}{2}\mathbf{v} = \frac{1}{2}(8, 8, 6) = (4, 4, 3).$$

(b) Because $-\mathbf{v}/\|\mathbf{v}\|$ is a unit vector with direction opposite that of $\mathbf{v}$,

$$\mathbf{u} = \frac{\|\mathbf{v}\|}{4}\left(-\frac{\mathbf{v}}{\|\mathbf{v}\|}\right)$$

$$= \frac{-1}{4}\mathbf{v} = -\frac{1}{4}(8, 8, 6) = \left(-2, -2, -\frac{3}{2}\right).$$

(c) Because $-\mathbf{v}/\|\mathbf{v}\|$ is a unit vector with direction opposite that of $\mathbf{v}$,

$$\mathbf{u} = 2\|\mathbf{v}\|\left(-\frac{\mathbf{v}}{\|\mathbf{v}\|}\right)$$

$$= -2\mathbf{v} = -2(8, 8, 6) = (-16, -16, -12).$$

15. The cosine of the angle θ between $\mathbf{u}$ and $\mathbf{v}$ is given by

$$\cos\theta = \frac{\mathbf{u} \cdot \mathbf{v}}{\|\mathbf{u}\|\|\mathbf{v}\|} = \frac{2(-3) + 2(3)}{\sqrt{2^2 + 2^2}\sqrt{(-3)^2 + 3^2}} = 0.$$

So, $\theta = \dfrac{\pi}{12}$ radians (90°).

17. The cosine of the angle θ between $\mathbf{u}$ and $\mathbf{v}$ is given by

$$\cos\theta = \frac{\mathbf{u} \cdot \mathbf{v}}{\|\mathbf{u}\|\|\mathbf{v}\|} = \frac{\cos\frac{3\pi}{4}\cos\frac{2\pi}{3} + \sin\frac{3\pi}{4}\sin\frac{2\pi}{3}}{\sqrt{\cos^2\frac{3\pi}{4} + \sin^2\frac{3\pi}{4}} \cdot \sqrt{\cos^2\frac{2\pi}{3} + \sin^2\frac{2\pi}{3}}} = \frac{\cos\left(\frac{3\pi}{4} - \frac{2\pi}{3}\right)}{1 \cdot 1} = \cos\frac{\pi}{12}$$

So, $\theta = \dfrac{\pi}{12}$ radians (15°).

© 2013 Cengage Learning. All Rights Reserved. May not be scanned, copied or duplicated, or posted to a publicly accessible website, in whole or in part.

19. The cosine of the angle θ between **u** and **v** is given by

$$\cos\theta = \frac{\mathbf{u}\cdot\mathbf{v}}{\|\mathbf{u}\|\|\mathbf{v}\|} = \frac{10(-2) + (-5)(1) + 15(-3)}{\sqrt{10^2 + (-5)^2 + 15^2}\sqrt{(-2)^2 + 1^2 + (-3)^2}} = -1.$$

So, $\theta = \pi$ radians (180°).

21. A vector $\mathbf{v} = (v_1, v_2, v_3)$ that is orthogonal to $\mathbf{u} = (0, -4, 3)$ must satisfy the equation

$$\mathbf{u}\cdot\mathbf{v} = (0, -4, 3)\cdot(v_1, v_2, v_3) = 0v_1 - 4v_2 + 3v_3 = 0.$$

This equation has solutions of the form $\mathbf{v} = (s, 3t, 4t)$, where s and t are any real numbers.

23. A vector $\mathbf{v} = (v_1, v_2, v_3, v_4)$ that is orthogonal to $\mathbf{u} = (1, -2, 2, 1)$ must satisfy the equation

$$\mathbf{u}\cdot\mathbf{v} = (1, -2, 2, 1)\cdot(v_1, v_2, v_3, v_4) = v_1 - 2v_2 + 2v_3 + v_4 = 0.$$

This equation has solutions of the form $(2r - 2s - t, r, s, t)$, where r, s, and t are any real numbers.

25. (a) $\langle\mathbf{u}, \mathbf{v}\rangle = 2\left(\frac{3}{2}\right) + 2\left(-\frac{1}{2}\right)(2) + 3(1)(-1) = -2$

(b) $d(\mathbf{u}, \mathbf{v}) = \|\mathbf{u} - \mathbf{v}\| = \sqrt{\langle\mathbf{u} - \mathbf{v}, \mathbf{u} - \mathbf{v}\rangle} = \sqrt{\left(2 - \frac{3}{2}\right)^2 + 2\left(-\frac{1}{2} - 2\right)^2 + 3(1 - (-1))^2} = \frac{3}{2}\sqrt{11}$

27. Verify the Triangle Inequality as follows.

$$\|\mathbf{u} + \mathbf{v}\| \le \|\mathbf{u}\| + \|\mathbf{v}\|$$

$$\left\|\left(\frac{7}{2}, \frac{3}{2}, 0\right)\right\| \le \sqrt{2^2 + 2\left(-\frac{1}{2}\right)^2 + 3(1)^2} + \sqrt{\left(\frac{3}{2}\right)^2 + 2(2)^2 + 3(-1)^2}$$

$$\sqrt{\left(\frac{7}{2}\right)^2 + 2\left(\frac{3}{2}\right)^2 + 0} \le \sqrt{\frac{15}{2}} + \frac{\sqrt{53}}{2}$$

$$\frac{\sqrt{67}}{2} \le \frac{\sqrt{30}}{2} + \frac{\sqrt{53}}{2}$$

$$4.093 \le 6.379$$

Verify the Cauchy-Schwarz Inequality as follows.

$$|\langle\mathbf{u}, \mathbf{v}\rangle| \le \|\mathbf{u}\|\|\mathbf{v}\|$$

$$\left|2\left(\frac{3}{2}\right) + 2\left(-\frac{1}{2}\right)(2) + 3(1)(-1)\right| \le \sqrt{\frac{15}{2}}\frac{\sqrt{53}}{2}$$

$$2 \le \frac{\sqrt{30}}{2}\frac{\sqrt{53}}{2}$$

$$2 \le 9.969$$

29. (a) $\langle f, g\rangle = \int_{-1}^{1} x\frac{1}{x^2 + 1}\,dx = \frac{1}{2}\ln(x^2 + 1)\Big]_{-1}^{1} = \frac{1}{2}\ln 2 - \frac{1}{2}\ln 2 = 0$

(b) The vectors are orthogonal.

(c) Because $\langle f, g\rangle = 0$, it follows that $|\langle f, g\rangle| \le \|f\|\,\|g\|$.

31. The projection of **u** onto **v** is given by

$$\text{proj}_{\mathbf{v}}\mathbf{u} = \frac{\mathbf{u}\cdot\mathbf{v}}{\mathbf{v}\cdot\mathbf{v}}\mathbf{v} = \frac{2(1) + 4(-5)}{1^2 + (-5)^2}(1, -5) = -\frac{9}{13}(1, -5) = \left(-\frac{9}{13}, \frac{45}{13}\right).$$

© 2013 Cengage Learning. All Rights Reserved. May not be scanned, copied or duplicated, or posted to a publicly accessible website, in whole or in part.

33. The projection of **u** onto **v** is given by

$$\text{proj}_{\mathbf{v}}\mathbf{u} = \frac{\mathbf{u}\cdot\mathbf{v}}{\mathbf{v}\cdot\mathbf{v}}\mathbf{v} = \frac{1(2)+2(5)}{2^2+5^2}(2, 5) = \frac{12}{29}(2, 5) = \left(\frac{24}{29}, \frac{60}{29}\right).$$

35. The projection of **u** onto **v** is given by

$$\text{proj}_{\mathbf{v}}\mathbf{u} = \frac{\mathbf{u}\cdot\mathbf{v}}{\mathbf{v}\cdot\mathbf{v}}\mathbf{v} = \frac{0(3)+(-1)(2)+2(4)}{3^2+2^2+4^2}(3, 2, 4) = \left(\frac{18}{29}, \frac{12}{29}, \frac{24}{29}\right).$$

37. First, orthogonalize each vector in B.

$$\mathbf{w}_1 = \mathbf{v}_1 = (1, 1)$$

$$\mathbf{w}_2 = \mathbf{v}_2 - \frac{\mathbf{v}_2\cdot\mathbf{w}_1}{\mathbf{w}_1\cdot\mathbf{w}_1}\mathbf{w}_1 = (0, 2) - \frac{2}{2}(1, 1) = (-1, 1)$$

Then, normalize the vectors.

$$\mathbf{u}_1 = \frac{\mathbf{w}_1}{\|\mathbf{w}_1\|} = \frac{1}{\sqrt{1^2+1^2}}(1, 1) = \left(\frac{1}{\sqrt{2}}, \frac{1}{\sqrt{2}}\right)$$

$$\mathbf{u}_2 = \frac{\mathbf{w}_2}{\|\mathbf{w}_2\|} = \frac{1}{\sqrt{(-1)^2+1^2}}(-1, 1) = \left(-\frac{1}{\sqrt{2}}, \frac{1}{\sqrt{2}}\right)$$

So, the orthonormal basis is $\left\{\left(\frac{1}{\sqrt{2}}, \frac{1}{\sqrt{2}}\right), \left(-\frac{1}{\sqrt{2}}, \frac{1}{\sqrt{2}}\right)\right\}$.

39. $\mathbf{w}_1 = (0, 3, 4)$

$$\mathbf{w}_2 = (1, 0, 0) - \frac{1(0)+0(3)+0(4)}{0^2+3^2+4^2}(0, 3, 4) = (1, 0, 0)$$

$$\mathbf{w}_3 = (1, 1, 0) - \frac{1(1)+1(0)+0(0)}{1^2+0^2+0^2}(1, 0, 0) - \frac{1(0)+1(3)+0(4)}{0^2+3^2+4^2}(0, 3, 4) = \left(0, \frac{16}{25}, -\frac{12}{25}\right)$$

Then, normalize each vector.

$$\mathbf{u}_1 = \frac{1}{\|\mathbf{w}_1\|}\mathbf{w}_1 = \frac{1}{5}(0, 3, 4) = \left(0, \frac{3}{5}, \frac{4}{5}\right)$$

$$\mathbf{u}_2 = \frac{1}{\|\mathbf{w}_2\|}\mathbf{w}_2 = 1(1, 0, 0) = (1, 0, 0)$$

$$\mathbf{u}_3 = \frac{1}{\|\mathbf{w}_3\|}\mathbf{w}_3 = \frac{5}{4}\left(0, \frac{16}{25}, -\frac{12}{25}\right) = \left(0, \frac{4}{5}, -\frac{3}{5}\right)$$

So, an orthonormal basis for R^3 is $\left\{\left(0, \frac{3}{5}, \frac{4}{5}\right), (1, 0, 0), \left(0, \frac{4}{5} - \frac{3}{5}\right)\right\}$.

41. (a) To find **x** as a linear combination of the vectors in B, solve the vector equation $c_1(0, 2, -2) + c_2(1, 0, -2) = (-1, 4, -2)$.

This produces the system of linear equations

$$\begin{aligned} c_2 &= -1 \\ 2c_1 \quad &= 4 \\ -2c_1 - 2c_2 &= -2 \end{aligned}$$

which has the solution $c_1 = 2$ and $c_2 = -1$. So, $[\mathbf{x}]_B = (2, -1)$, and you can write $(-1, 4, -2) = 2(0, 2, -2) - (1, 0, -2)$.

© 2013 Cengage Learning. All Rights Reserved. May not be scanned, copied or duplicated, or posted to a publicly accessible website, in whole or in part.

(b) To apply the Gram-Schmidt orthonormalization process, first orthogonalize each vector in B.

$$\mathbf{w}_1 = (0, 2, -2)$$

$$\mathbf{w}_2 = (1, 0, -2) - \frac{1(0) + 0(2) + (-2)(-2)}{0^2 + 2^2 + (-2)^2}(0, 2, -2) = (1, -1, -1).$$

Then normalize $\mathbf{w}_1$ and $\mathbf{w}_2$ as follows.

$$\mathbf{u}_1 = \frac{1}{\|\mathbf{w}_1\|}\mathbf{w}_1 = \frac{1}{2\sqrt{2}}(0, 2, -2) = \left(0, \frac{1}{\sqrt{2}}, -\frac{1}{\sqrt{2}}\right)$$

$$\mathbf{u}_2 = \frac{1}{\|\mathbf{w}_2\|}\mathbf{w}_2 = \frac{1}{\sqrt{3}}(1, -1, -1) = \left(\frac{1}{\sqrt{3}}, -\frac{1}{\sqrt{3}}, -\frac{1}{\sqrt{3}}\right)$$

So, $B' = \left\{\left(0, \frac{1}{\sqrt{2}}, -\frac{1}{\sqrt{2}}\right), \left(\frac{1}{\sqrt{3}}, -\frac{1}{\sqrt{3}}, -\frac{1}{\sqrt{3}}\right)\right\}.$

(c) To find $\mathbf{x}$ as a linear combination of the vectors in B', solve the vector equation

$$c_1\left(0, \frac{1}{\sqrt{2}}, -\frac{1}{\sqrt{2}}\right) + c_2\left(\frac{1}{\sqrt{3}}, -\frac{1}{\sqrt{3}}, -\frac{1}{\sqrt{3}}\right) = (-1, 4, -2).$$

This produces the system of linear equations

$$\begin{aligned} \frac{1}{\sqrt{3}}c_2 &= -1 \\ \frac{1}{\sqrt{2}}c_1 - \frac{1}{\sqrt{3}}c_2 &= 4 \\ -\frac{1}{\sqrt{2}}c_1 - \frac{1}{\sqrt{3}}c_2 &= -2 \end{aligned}$$

which has the solution $c_1 = 3\sqrt{2}$ and $c_2 = -\sqrt{3}$. So, $[\mathbf{x}]_{B'} = (3\sqrt{2}, -\sqrt{3})$, and you can write

$$(-1, 4, -2) = 3\sqrt{2}\left(0, \frac{1}{\sqrt{2}}, -\frac{1}{\sqrt{2}}\right) - \sqrt{3}\left(\frac{1}{\sqrt{3}}, -\frac{1}{\sqrt{3}}, -\frac{1}{\sqrt{3}}\right).$$

43. These functions are orthogonal since

$$\langle f, g\rangle = \int_0^\pi \sin x \cos x \, dx = \frac{1}{2}\sin^2 x\Big]_0^\pi = \left[\frac{1}{2}\sin^2\pi\right] - \left[\frac{1}{2}\sin^2 0\right] = 0$$

45. (a) $\langle f, g\rangle = \int_0^1 x^4 dx = \frac{1}{5}x^5\Big]_0^1 = \frac{1}{5}$

(b) Since $\langle g, g\rangle = \int_0^1 g(x)g(x)dx = \int_0^1 x^6 dx = \frac{1}{7}x^7\Big]_0^1 = \frac{1}{7}$, the norm of g is $\|g\| = \sqrt{\langle g, g\rangle} = \sqrt{\frac{1}{7}} = \frac{1}{\sqrt{7}}$.

(c) Since

$$\langle f - g, f - g\rangle = \int_0^1 (x - x^3)^2 dx = \int_0^1 (x^2 - 2x^4 + x^6)dx = \frac{1}{3}x^3 - \frac{2}{5}x^5 + \frac{1}{7}x^7\Big]_0^1 = \frac{8}{105},$$

the distance between f and g is

$$d(f, g) = \|f - g\| = \sqrt{\langle f - g, f - g\rangle} = \sqrt{\frac{8}{105}} = \frac{2\sqrt{2}}{\sqrt{105}}.$$

© 2013 Cengage Learning. All Rights Reserved. May not be scanned, copied or duplicated, or posted to a publicly accessible website, in whole or in part.

(d) First, orthogonalize the vectors.

$$\mathbf{w}_1 = f = x$$

$$\mathbf{w}_2 = g - \frac{\langle g_1 \mathbf{w}_1 \rangle}{\langle \mathbf{w}_1 \mathbf{w}_1 \rangle}\mathbf{w}_1 = x^3 - \frac{\int_0^1 x^4 dx}{\int_0^1 x^2 dx}x = x^3 - \frac{\frac{1}{5}}{\frac{1}{3}}x = x^3 - \frac{3}{5}x$$

Then, normalize each vector.

$$\langle \mathbf{w}_1, \mathbf{w}_1 \rangle = \int_0^1 x^2 dx = \frac{1}{3}x^3\Big]_0^1 = \frac{1}{3}$$

$$\langle \mathbf{w}_2, \mathbf{w}_2 \rangle = \int_0^1 \left(x^3 - \frac{3}{5}x\right)^2 dx = \int_0^1 \left(x^6 - \frac{6}{5}x^4 + \frac{9}{25}x^2\right)dx = \frac{1}{7}x^7 - \frac{6}{25}x^5 + \frac{9}{75}x^3\Big]_0^1 = \frac{4}{175}$$

$$\mathbf{u}_1 = \frac{1}{\|\mathbf{w}_1\|}\mathbf{w}_1 = \frac{1}{\sqrt{\frac{1}{3}}}x = \sqrt{3}x$$

$$\mathbf{u}_2 = \frac{1}{\|\mathbf{w}_2\|}\mathbf{w}_2 = \frac{1}{\sqrt{\frac{4}{175}}}\left(x^3 - \frac{3}{5}x\right) = \frac{1}{\frac{2}{5\sqrt{7}}}\left(x^3 - \frac{3}{5}x\right)$$

The orthonormal set is $B' = \left\{\sqrt{3}x, \frac{\sqrt{7}}{2}(5x^3 - 3x)\right\}$.

47. Vectors in W are of the form $(-s - t, s, t)$ where s and t are any real numbers. So, a basis for W is $\{(-1, 0, 1), (-1, 1, 0)\}$.

Orthogonalize these vectors as follows.

$$\mathbf{w}_1 = (-1, 0, 1)$$

$$\mathbf{w}_2 = (-1, 1, 0) - \frac{-1(-1) + 1(0) + 0(1)}{(-1)^2 + 0^2 + 1^2}(-1, 0, 1) = \left(-\frac{1}{2}, 1, -\frac{1}{2}\right)$$

Finally, normalize $\mathbf{w}_1$ and $\mathbf{w}_2$ to obtain

$$\mathbf{u}_1 = \frac{1}{\|\mathbf{w}_1\|}\mathbf{w}_1 = \frac{1}{\sqrt{2}}(-1, 0, 1) = \left(-\frac{1}{\sqrt{2}}, 0, \frac{1}{\sqrt{2}}\right)$$

$$\mathbf{u}_2 = \frac{1}{\|\mathbf{w}_2\|}\mathbf{w}_2 = \frac{2}{\sqrt{6}}\left(-\frac{1}{2}, 1, -\frac{1}{2}\right) = \left(-\frac{1}{\sqrt{6}}, \frac{2}{\sqrt{6}}, -\frac{1}{\sqrt{6}}\right).$$

So, $W' = \left\{\left(-\frac{1}{\sqrt{2}}, 0, \frac{1}{\sqrt{2}}\right), \left(-\frac{1}{\sqrt{6}}, \frac{2}{\sqrt{6}}, -\frac{1}{\sqrt{6}}\right)\right\}$.

49. $(\mathbf{u} + \mathbf{v}) \cdot \mathbf{w} = \mathbf{w} \cdot (\mathbf{u} + \mathbf{v})$ (Theorem 5.3, part 1)

$= \mathbf{w} \cdot \mathbf{u} + \mathbf{w} \cdot \mathbf{v}$ (Theorem 5.3, part 2)

$= \mathbf{u} \cdot \mathbf{w} + \mathbf{v} \cdot \mathbf{w}$ (Theorem 5.3, part 1)

51. If $\|\mathbf{u}\| \le 1$ and $\|\mathbf{v}\| \le 1$, then the Cauchy-Schwarz Inequality implies that $|\langle \mathbf{u}, \mathbf{v} \rangle| \le \|\mathbf{u}\|\,\|\mathbf{v}\| \le 1$.

53. Let $\{\mathbf{v}_1, \ldots, \mathbf{v}_m\}$ be a basis for V. You can extend this basis to one for R^n.

$$B = \{\mathbf{v}_1, \ldots, \mathbf{v}_m, \mathbf{w}_{m+1}, \ldots, \mathbf{w}_n\}$$

Now apply the Gram-Schmidt orthonormalization process to this basis, which results in the following basis for R^n.

$$B' = \{\mathbf{u}_1, \ldots, \mathbf{u}_m, \mathbf{z}_{m+1}, \ldots, \mathbf{z}_n\}$$

The first m vectors of B' still span V. Therefore, any vector $\mathbf{u} \in R^n$ is of the form

$$\mathbf{u} = c_1\mathbf{u}_1 + \cdots + c_m\mathbf{u}_m + c_{m+1}\mathbf{z}_{m+1} + \cdots + c_n\mathbf{z}_n = \mathbf{v} + \mathbf{w}$$

where $\mathbf{v} \in V$ and $\mathbf{w}$ is orthogonal to every vector in V.

© 2013 Cengage Learning. All Rights Reserved. May not be scanned, copied or duplicated, or posted to a publicly accessible website, in whole or in part.

55. First extend the set $\{\mathbf{u}_1, \ldots, \mathbf{u}_m\}$ to an orthonormal basis for R^n.

$B = \{\mathbf{u}_1, \ldots, \mathbf{u}_m, \mathbf{u}_{m+1}, \ldots, \mathbf{u}_n\}$

If $\mathbf{v}$ is any vector in R^n, you have

$\mathbf{v} = \sum_{i=1}^{n}(\mathbf{v} \cdot \mathbf{u}_i)\mathbf{u}_i$ which implies that

$$\|\mathbf{v}\|^2 = \langle \mathbf{v}, \mathbf{v} \rangle = \sum_{i=1}^{n}(\mathbf{v} \cdot \mathbf{u}_i)^2 \geq \sum_{i=1}^{m}(\mathbf{v} \cdot \mathbf{u}_i)^2.$$

57. If $\mathbf{u}$ and $\mathbf{v}$ are orthogonal, then

$\|\mathbf{u}\|^2 + \|\mathbf{v}\|^2 = \|\mathbf{u} + \mathbf{v}\|^2$ by the Pythagorean Theorem.

Furthermore, $\|\mathbf{u}\|^2 + \|-\mathbf{v}\|^2 = \|\mathbf{u}\|^2 + \|\mathbf{v}\|^2 = \|\mathbf{u} - \mathbf{v}\|^2$, which gives

$\|\mathbf{u} + \mathbf{v}\|^2 = \|\mathbf{u} - \mathbf{v}\|^2 \Rightarrow \|\mathbf{u} + \mathbf{v}\| = \|\mathbf{u} - \mathbf{v}\|$.

On the other hand, if $\|\mathbf{u} + \mathbf{v}\| = \|\mathbf{u} - \mathbf{v}\|$, then

$\langle \mathbf{u} + \mathbf{v}, \mathbf{u} + \mathbf{v} \rangle^2 = \langle \mathbf{u} - \mathbf{v}, \mathbf{u} - \mathbf{v} \rangle^2$, which implies that

$\|\mathbf{u}\|^2 + \|\mathbf{v}\|^2 + 2\langle \mathbf{u}, \mathbf{v} \rangle = \|\mathbf{u}\|^2 + \|\mathbf{v}\|^2 - 2\langle \mathbf{u}, \mathbf{v} \rangle$, or

$\langle \mathbf{u}, \mathbf{v} \rangle = 0$, and $\mathbf{u}$ and $\mathbf{v}$ are orthogonal.

59. $S^{\perp} = N(A^T)$, the orthogonal complement of S is the nullspace of A^T.

$$A = \begin{bmatrix} 1 & 2 & 0 \\ 2 & 1 & -1 \end{bmatrix} \Rightarrow \begin{bmatrix} 1 & 2 & 0 \\ 0 & -3 & -1 \end{bmatrix} \Rightarrow \begin{bmatrix} 1 & 0 & -\frac{2}{3} \\ 0 & 1 & \frac{1}{3} \end{bmatrix}$$

So, $S^{\perp}$ is spanned by $\mathbf{u} = \begin{bmatrix} 2 \\ -1 \\ 3 \end{bmatrix}$.

61. $$A = \begin{bmatrix} 0 & 1 & 0 \\ 0 & -3 & 0 \\ 1 & 0 & 1 \end{bmatrix} \Rightarrow \begin{bmatrix} 1 & 0 & 1 \\ 0 & 1 & 0 \\ 0 & 0 & 0 \end{bmatrix}$$

$$A^T = \begin{bmatrix} 0 & 0 & 1 \\ 1 & -3 & 0 \\ 0 & 0 & 1 \end{bmatrix} \Rightarrow \begin{bmatrix} 1 & -3 & 0 \\ 0 & 0 & 1 \\ 0 & 0 & 0 \end{bmatrix}$$

$R(A)$-basis: $\left\{\begin{bmatrix} 0 \\ 0 \\ 1 \end{bmatrix}, \begin{bmatrix} 1 \\ -3 \\ 0 \end{bmatrix}\right\}$ $\quad R(A^T)$-basis: $\left\{\begin{bmatrix} 0 \\ 1 \\ 0 \end{bmatrix}, \begin{bmatrix} 1 \\ 0 \\ 1 \end{bmatrix}\right\}$

$N(A)$-basis: $\left\{\begin{bmatrix} 1 \\ 0 \\ -1 \end{bmatrix}\right\}$ $\quad N(A^T)$-basis: $\left\{\begin{bmatrix} 3 \\ 1 \\ 0 \end{bmatrix}\right\}$

63. Substitute the data points $(1, 400.5)$, $(2, 410.7)$, $(3, 425.7)$, $(4, 448.9)$, $(5, 461.6)$, $(6, 470.9)$, $(7, 482.3)$, $(8, 493.0)$ into the equation $y = c_0 + c_1 t$ to obtain the following system.

$$\begin{aligned} c_0 + c_1(1) &= 400.5 \\ c_0 + c_1(2) &= 410.7 \\ c_0 + c_1(3) &= 425.7 \\ c_0 + c_1(4) &= 448.9 \\ c_0 + c_1(5) &= 461.6 \\ c_0 + c_1(6) &= 470.9 \\ c_0 + c_1(7) &= 482.3 \\ c_0 + c_1(8) &= 493.0 \end{aligned}$$

This produces the least squares problem:

$A\mathbf{x} = \mathbf{b}$

$$\begin{bmatrix} 1 & 1 \\ 1 & 2 \\ 1 & 3 \\ 1 & 4 \\ 1 & 5 \\ 1 & 6 \\ 1 & 7 \\ 1 & 8 \end{bmatrix} \begin{bmatrix} c_0 \\ c_1 \end{bmatrix} = \begin{bmatrix} 400.5 \\ 410.7 \\ 425.7 \\ 448.9 \\ 461.6 \\ 470.9 \\ 482.3 \\ 493.0 \end{bmatrix}.$$

The normal equations are

$$A^T A\mathbf{x} = A^T\mathbf{b}$$

$$\begin{bmatrix} 8 & 36 \\ 36 & 204 \end{bmatrix} \begin{bmatrix} c_0 \\ c_1 \end{bmatrix} = \begin{bmatrix} 3593.6 \\ 16748.1 \end{bmatrix}$$

and the solution is

$$\mathbf{x} = \begin{bmatrix} c_0 \\ c_1 \end{bmatrix} = \begin{bmatrix} 387.389285714 \\ 13.7357142857 \end{bmatrix}.$$

So, the least squares regression line is $y = 387.389 + 13.736t$. Energy consumption in 2015 will be

$387.389 + 13.736(15) \approx 593.414$ quadrillion Btu.

65. $$\mathbf{u} \times \mathbf{v} = \begin{vmatrix} \mathbf{i} & \mathbf{j} & \mathbf{k} \\ 1 & 1 & 1 \\ 1 & 0 & 0 \end{vmatrix} = \mathbf{j} - \mathbf{k} = (0, 1, -1)$$

$\mathbf{u} \times \mathbf{v}$ is orthogonal to both $\mathbf{u}$ and $\mathbf{v}$ because,

$\mathbf{u} \cdot (\mathbf{u} \times \mathbf{v}) = 1(0) + 1(1) + 1(-1) = 0$

and

$\mathbf{v} \cdot (\mathbf{u} \times \mathbf{v}) = 1(0) + 0(1) + 0(-1) = 0.$

© 2013 Cengage Learning. All Rights Reserved. May not be scanned, copied or duplicated, or posted to a publicly accessible website, in whole or in part.

67. $\mathbf{u} \times \mathbf{v} = \begin{bmatrix} \mathbf{i} & \mathbf{j} & \mathbf{k} \\ 0 & 1 & 6 \\ 1 & -2 & 1 \end{bmatrix} = 13\mathbf{i} + 6\mathbf{j} - \mathbf{k} = (13, 6, -1)$

$\mathbf{u} \times \mathbf{v}$ is orthogonal to both $\mathbf{u}$ and $\mathbf{v}$ because

$\mathbf{u} \cdot (\mathbf{u} \times \mathbf{v}) = (0, 1, 6) \cdot (13, 6, -1) = 6 - 6 = 0$

and

$\mathbf{v} \cdot (\mathbf{u} \times \mathbf{v}) = (1, -2, 1) \cdot (13, 6, -1) = 13 - 12 - 1 = 0.$

69. Because

$$\mathbf{v} \times \mathbf{w} = \begin{bmatrix} \mathbf{i} & \mathbf{j} & \mathbf{k} \\ 0 & 0 & 1 \\ 0 & 1 & 0 \end{bmatrix} = -\mathbf{i} = (-1, 0, 0),$$

the volume is

$|\mathbf{u} \cdot (\mathbf{v} \times \mathbf{w})| = |1(-1) + 0(0) + 0(0)| = 1.$

71. $\mathbf{v} \times \mathbf{w} = \begin{vmatrix} \mathbf{i} & \mathbf{j} & \mathbf{k} \\ 3 & -2 & 1 \\ 2 & -3 & -2 \end{vmatrix}$

$$= \mathbf{i}\begin{vmatrix} -2 & 1 \\ -3 & -2 \end{vmatrix} - \mathbf{j}\begin{vmatrix} 3 & 1 \\ 2 & -2 \end{vmatrix} + \mathbf{k}\begin{vmatrix} 3 & -2 \\ 2 & -3 \end{vmatrix}$$

$$= \mathbf{i}(4 + 3) - \mathbf{j}(-6 - 2) + \mathbf{k}(-9 + 4)$$

$$= 7\vec{\mathbf{i}} + 8\vec{\mathbf{j}} - 5\vec{\mathbf{k}}$$

The volume is given by

$$|\mathbf{u} \cdot (\mathbf{v} \times \mathbf{w})| = |(-2, 1, 0) \cdot (7, 8, -5)|$$
$$= |(-2)(7) + (1)(8) + (0)(-5)|$$
$$= |-14 + 8|$$
$$= 6 \text{ units}^3.$$

73. Because

$$\mathbf{u} \times \mathbf{v} = \begin{bmatrix} \mathbf{i} & \mathbf{j} & \mathbf{k} \\ 1 & 3 & 0 \\ -1 & 0 & 2 \end{bmatrix} = 6\mathbf{i} - 2\mathbf{j} + 3\mathbf{k} = (6, -2, 3),$$

the area of the parallelogram is

$\|\mathbf{u} \times \mathbf{v}\| = \sqrt{6^2 + (-2)^2 + 3^2} = 7.$

75. (a) The standard basis for P_1 is $\{1, x\}$. In the interval $c[-1, 1]$, the Gram-Schmidt orthonormalization process yields the orthonormal basis $\left\{\frac{\sqrt{2}}{2}, \frac{\sqrt{6}}{2}x\right\}$.

The linear least squares approximating function is given by

$$g(x) = \langle f, \mathbf{w}_1 \rangle \mathbf{w}_1 + \langle f, \mathbf{w}_2 \rangle \mathbf{w}_2.$$

Because

$$\langle f, \mathbf{w}_1 \rangle = \int_{-1}^{1} \frac{\sqrt{2}}{2}x^3\,dx = \frac{\sqrt{2}}{8}x^4\Big]_{-1}^{1} = 0$$

$$\langle f, \mathbf{w}_2 \rangle = \int_{-1}^{1} \frac{\sqrt{6}}{2}x^4\,dx = \frac{\sqrt{6}}{10}x^5\Big]_{-1}^{1} = \frac{\sqrt{6}}{5},$$

g is given by

$$g(x) = 0\left(\frac{\sqrt{2}}{2}\right) + \frac{\sqrt{6}}{5}\left(\frac{\sqrt{6}}{2}x\right) = \frac{3}{5}x.$$

(b)

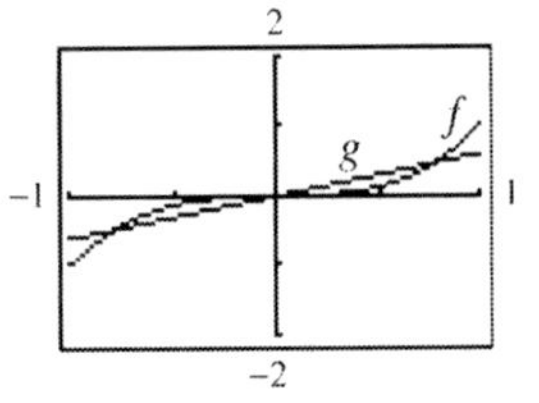

77. (a) The standard basis for P_1 is $\{1, x\}$. In the interval $c\left[0, \frac{\pi}{2}\right]$, the Gram-Schmidt orthonormalization process yields the orthonormal basis

$$\left\{\sqrt{\frac{2}{\pi}}, \frac{\sqrt{6\pi}}{\pi^2}(4x - \pi)\right\}.$$

Because

$$\langle f, \mathbf{w}_1 \rangle = \int_0^{\frac{\pi}{2}} \sin(2x)\sqrt{\frac{2}{\pi}}\,dx = \sqrt{\frac{2}{\pi}}$$

$$\langle f, \mathbf{w}_2 \rangle = \int_0^{\frac{\pi}{2}} \sin(2x)\left(\frac{\sqrt{6\pi}}{\pi^2}\right)(4x - \pi)dx = 0,$$

g is given by

$$g(x) = \langle f, \mathbf{w}_1 \rangle \mathbf{w}_1 + \langle f, \mathbf{w}_2 \rangle \mathbf{w}_2 = \sqrt{\frac{2}{\pi}}\sqrt{\frac{2}{\pi}} = \frac{2}{\pi}.$$

(b)

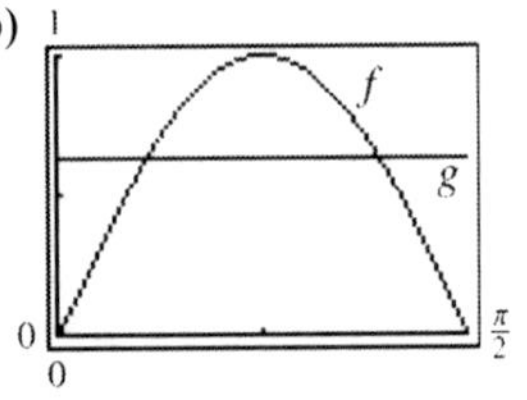

© 2013 Cengage Learning. All Rights Reserved. May not be scanned, copied or duplicated, or posted to a publicly accessible website, in whole or in part.

79. (a) The standard basis for P_2 is $\{1, x, x^2\}$. In the interval $c[0, 1]$, the Gram-Schmidt orthonormalization process yields the orthonormal basis

$$\left\{1, \sqrt{3}(2x - 1), \sqrt{5}\left(6x^2 - 6x + 1\right)\right\}.$$

Because

$$\langle f, \mathbf{w}_1 \rangle = \int_0^1 \sqrt{x}\, dx = \frac{2}{3}$$

$$\langle f, \mathbf{w}_2 \rangle = \int_0^1 \sqrt{x}\sqrt{3}(2x - 1)dx = \sqrt{3}\int_0^1\left(2x^{3/2} - x^{1/2}\right)dx = \sqrt{3}\left(\frac{4}{5}x^{5/2} - \frac{2}{3}x^{3/2}\right)\Bigg]_0^1 = \frac{2}{15}\sqrt{3}$$

$$\langle f, \mathbf{w}_3 \rangle = \int_0^1 \sqrt{x}\sqrt{5}\left(6x^2 - 6x + 1\right) dx = \sqrt{5}\int_0^1\left(6x^{5/2} - 6x^{3/2} + x^{1/2}\right) dx = \sqrt{5}\left(\frac{12}{7}x^{7/2} - \frac{12}{5}x^{5/2} + \frac{2}{3}x^{3/2}\right)\Bigg]_0^1 = \frac{-2\sqrt{5}}{105},$$

g is given by

$$\begin{aligned} g(x) &= \langle f, \mathbf{w}_1\rangle\mathbf{w}_1 + \langle f, \mathbf{w}_2\rangle\mathbf{w}_2 + \langle f, \mathbf{w}_3\rangle\mathbf{w}_3 \\ &= \frac{2}{3}(1) + \frac{2}{15}\sqrt{3}\left(\sqrt{3}\right)(2x - 1) + \frac{-2\sqrt{5}}{105}\sqrt{5}\left(6x^2 - 6x + 1\right) \\ &= -\frac{4}{7}x^2 + \frac{48}{35}x + \frac{6}{35} = \frac{2}{35}\left(-10x^2 + 24x + 3\right) \end{aligned}$$

(b)

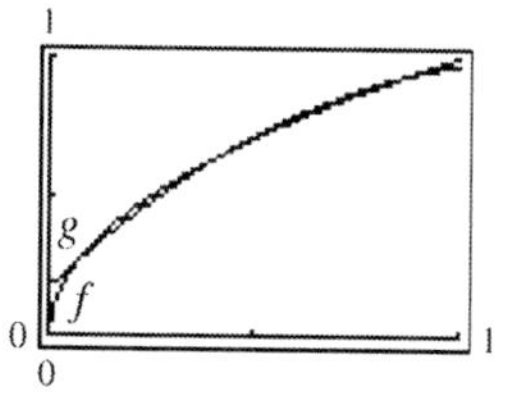

81. Find the coefficients as follows.

$$a_0 = \frac{1}{\pi}\int_{-\pi}^{\pi} f(x)\, dx = \frac{1}{\pi}\int_{-\pi}^{\pi} x^2 dx = \frac{1}{3\pi}x^3\Bigg]_{-\pi}^{\pi} = \frac{2\pi^2}{3}$$

$$a_1 = \frac{1}{\pi}\int_{-\pi}^{\pi} f(x)\cos x\, dx = \frac{1}{\pi}\int_{-\pi}^{\pi} x^2 \cos x\, dx = \frac{1}{\pi}\left(x^2 \sin x + 2x\cos x - 2\sin x\right)\Bigg]_{-\pi}^{\pi} = -4$$

$$b_1 = \frac{1}{\pi}\int_{-\pi}^{\pi} f(x)\sin x\, dx = \frac{1}{\pi}\int_{-\pi}^{\pi} x^2 \sin x\, dx = \frac{1}{\pi}\left(-x^2 \cos x + 2x\sin x + 2\cos x\right)\Bigg]_{-\pi}^{\pi} = 0$$

So, the approximation is $g(x) = \dfrac{a_0}{2} + a_1 \cos x + b_1 \sin x = \dfrac{\pi^2}{3} - 4\cos x.$

83. (a) True. See Theorem 5.18, part 1, page 273.

(b) False. See Theorem 5.17, part 1, page 272.

(c) True. See discussion before Theorem 5.19, page 277.

Cumulative Test for Chapters 4 and 5

1. (a) $(1, -2) + (2, -5) = (3, -7)$

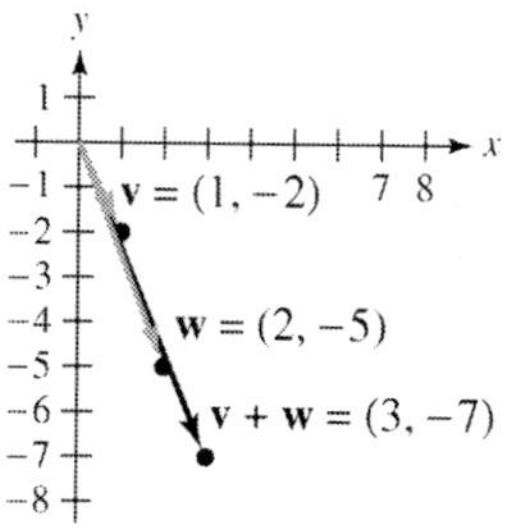

(b) $3(1, -2) = (3, -6)$

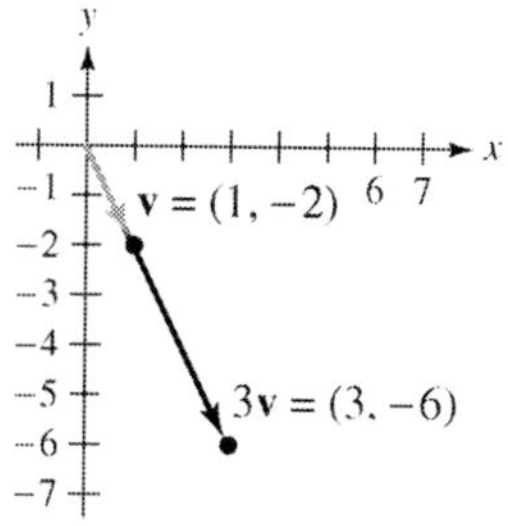

(c) $2(1, -2) - 4(2, -5) = (2, -4) - (8, -20)$
$= (-6, 16)$

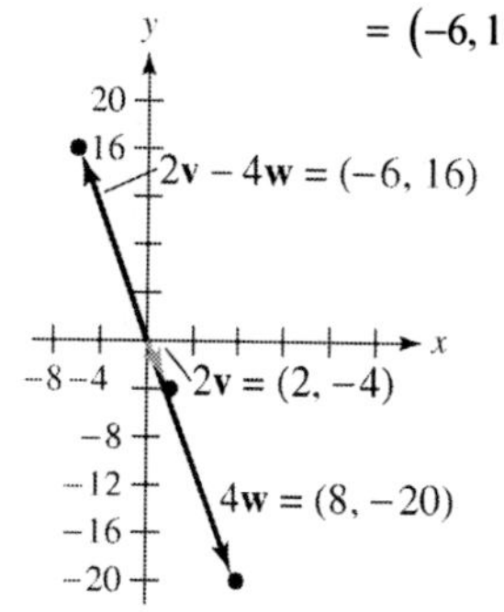

© 2013 Cengage Learning. All Rights Reserved. May not be scanned, copied or duplicated, or posted to a publicly accessible website, in whole or in part.

2. $\begin{bmatrix} 1 & -1 & 0 & 2 \\ 2 & 0 & 3 & 4 \\ 0 & 1 & 0 & 1 \end{bmatrix} \Rightarrow \begin{bmatrix} 1 & 0 & 0 & 3 \\ 0 & 1 & 0 & 1 \\ 0 & 0 & 1 & -\frac{2}{3} \end{bmatrix}$

$3(1, 2, 0) + (-1, 0, 1) - \frac{2}{3}(0, 3, 0) = (2, 4, 1)$

3. $\begin{bmatrix} 1 & 2 & 3 \\ 7 & 8 & 9 \\ 4 & 5 & 7 \end{bmatrix} \Rightarrow \begin{bmatrix} 1 & 0 & 0 \\ 0 & 1 & 0 \\ 0 & 0 & 1 \end{bmatrix}$

No.

4. Write a matrix using the given $\mathbf{u}_1, \mathbf{u}_2, \ldots, \mathbf{u}_6$ as columns and augment this matrix with $\mathbf{v}$ as a column.

$$A = \begin{bmatrix} 1 & 1 & 0 & 1 & 1 & 3 & 10 \\ 2 & -2 & 2 & 0 & -2 & 2 & 30 \\ -3 & 1 & -1 & 3 & 1 & 1 & -13 \\ 4 & -1 & 2 & -4 & -1 & -2 & 14 \\ -1 & 2 & -1 & 1 & 2 & 3 & -7 \\ 2 & 1 & -1 & 2 & -3 & 0 & 27 \end{bmatrix}$$

The reduced row-echelon form for A is

$$A = \begin{bmatrix} 1 & 0 & 0 & 0 & 0 & 0 & 5 \\ 0 & 1 & 0 & 0 & 0 & 0 & -1 \\ 0 & 0 & 1 & 0 & 0 & 0 & 1 \\ 0 & 0 & 0 & 1 & 0 & 0 & 2 \\ 0 & 0 & 0 & 0 & 1 & 0 & -5 \\ 0 & 0 & 0 & 0 & 0 & 1 & 3 \end{bmatrix}.$$

So, $\mathbf{v} = 5\mathbf{u}_1 - \mathbf{u}_2 + \mathbf{u}_3 + 2\mathbf{u}_4 - 5\mathbf{u}_5 + 3\mathbf{u}_6$. Verify the solution by showing that

$5(1, 2, -3, 4, -1, 2) - (1, -2, 1, -1, 2, 1) + (0, 2, -1, 2, -1, -1) + 2(1, 0, 3, -4, 1, 2) - 5(1, -2, 1, -1, 2, 3) + 3(3, 2, 1, -2, 3, 0).$

equals $(10, 30, -13, 14, -7, 27)$.

5. Not closed under addition: $\begin{bmatrix} 1 & 0 & 0 \\ 0 & 1 & 0 \\ 0 & 0 & 0 \end{bmatrix} + \begin{bmatrix} 0 & 0 & 0 \\ 0 & 0 & 0 \\ 0 & 0 & 1 \end{bmatrix} = \begin{bmatrix} 1 & 0 & 0 \\ 0 & 1 & 0 \\ 0 & 0 & 1 \end{bmatrix}$

6. Let $\mathbf{v} = (v_1, v_1 + v_2, v_2, v_2)$ and $\mathbf{u} = (u_1, u_1 + u_2, u_2, u_2)$ be two vectors in W.

$\mathbf{v} + \mathbf{u} = (v_1 + u_1, (v_1 + v_2) + (u_1 + u_2), v_2 + u_2, v_2 + u_2) = (v_1 + u_1, (v_1 + u_1) + (v_2 + u_2), v_2 + u_2, v_2 + u_2)$

$= (x_1, x_1 + x_2, x_2, x_2)$ where $x_1 = v_1 + u_1$ and $x_2 = v_2 + u_2$. So, $\mathbf{v} + \mathbf{u}$ is in W.

$c\mathbf{v} = c(v_1, v_1 + v_2, v_2, v_2) = (cv_1, c(v_1 + v_2), cv_2, cv_2) = (cv_1, cv_1 + cv_2, cv_2, cv_2) = (x_1, x_1 + x_2, x_2, x_2)$ where $x_1 = cv_1$ and $v_2 = cv_2$. So, $c\mathbf{v}$ is in W. So, you can conclude that it is a subspace of R^4.

7. No: $(1, 1, 1) + (1, 1, 1) = (2, 2, 2)$

8. Yes, because $\begin{bmatrix} 1 & 2 & -1 & 0 \\ 1 & 3 & 0 & 2 \\ 0 & 0 & 1 & -1 \\ 1 & 0 & 0 & 1 \end{bmatrix}$ row reduces to I.

9. (a) See definition page 213.

(b) Linearly dependent

10. (a) A set of vectors $\{\mathbf{v}_1, \ldots, \mathbf{v}_n\}$ in a vector space V is a basis for $\mathbf{V}$ if the set is linearly independent and spans V.

(b) $\mathbf{v}_1 = (1, 1), \mathbf{v}_2 = (-1, 1)$

$c_1(1, 1) + c_2(-1, 1) = (c_1 - c_2, c_1 + c_2)$

$\begin{bmatrix} 1 & -1 \\ 1 & 1 \end{bmatrix} \sim \begin{bmatrix} 1 & 0 \\ 0 & 1 \end{bmatrix}$

Since the matrix reduces to I_2, $\mathbf{v}_1$ and $\mathbf{v}_2$ form a basis in IR^2.

(c) Yes. Because the set is linearly independent.

© 2013 Cengage Learning. All Rights Reserved. May not be scanned, copied or duplicated, or posted to a publicly accessible website, in whole or in part.

11. $\begin{bmatrix} 1 & 1 & 0 & 0 \\ -2 & -2 & 0 & 0 \\ 0 & 0 & 1 & 1 \\ 1 & 1 & 0 & 0 \end{bmatrix} \Rightarrow \begin{bmatrix} 1 & 1 & 0 & 0 \\ 0 & 0 & 1 & 1 \\ 0 & 0 & 0 & 0 \\ 0 & 0 & 0 & 0 \end{bmatrix} \begin{matrix} x_1 = -s \\ x_2 = s \\ x_3 = -t \\ x_4 = t \end{matrix}$ basis $\left\{ \begin{bmatrix} -1 \\ 1 \\ 0 \\ 0 \end{bmatrix}, \begin{bmatrix} 0 \\ 0 \\ -1 \\ 1 \end{bmatrix} \right\}$

12. $\begin{bmatrix} 0 & 1 & 1 & 1 \\ 1 & 1 & 0 & 2 \\ 1 & 1 & 1 & -3 \end{bmatrix} \Rightarrow \begin{bmatrix} 1 & 0 & 0 & -4 \\ 0 & 1 & 0 & 6 \\ 0 & 0 & 1 & -5 \end{bmatrix}; [\mathbf{v}]_B = \begin{bmatrix} -4 \\ 6 \\ -5 \end{bmatrix}$

13. $[B' \vdots B] \Rightarrow [I \vdots P^{-1}]; \left[\begin{array}{ccc|ccc} 1 & 1 & 0 & 2 & 1 & 0 \\ 1 & 1 & 1 & 1 & 0 & 1 \\ 2 & 1 & 2 & 0 & 0 & 1 \end{array}\right] \Rightarrow \left[\begin{array}{ccc|ccc} 1 & 0 & 0 & 0 & 1 & -1 \\ 0 & 1 & 0 & 2 & 0 & 1 \\ 0 & 0 & 1 & -1 & -1 & 1 \end{array}\right]$

14. (a) $\|\mathbf{u}\| = \sqrt{1^2 + 0^2 + 2^2} = \sqrt{5}$

(b) $\|\mathbf{u} - \mathbf{v}\| = \|(3, -1, -1)\|$

$= \sqrt{3^2 + (-1)^2 + (-1)^2} = \sqrt{11}$

(c) $\mathbf{u} \cdot \mathbf{v} = 1(-2) + 0(1) + 2(3) = 4$

(d) $\cos\theta = \dfrac{\mathbf{u} \cdot \mathbf{v}}{\|\mathbf{u}\|\|\mathbf{v}\|} = \dfrac{4}{\sqrt{5} \cdot \sqrt{14}} = \dfrac{4}{\sqrt{70}}$

$\theta = \cos^{-1}\dfrac{4}{\sqrt{70}} \approx 1.0723$ radians $(61.45°)$

15. $\int_0^1 x^2(x + 2)\, dx = \left[\dfrac{x^4}{4} + \dfrac{2x^3}{3}\right]_0^1 = \dfrac{11}{12}$

16. $\mathbf{w}_1 = (2, 0, 0)$

$\mathbf{w}_2 = (1, 1, 1) - \frac{1}{2}(2, 0, 0) = (0, 1, 1)$

$\mathbf{w}_3 = (0, 1, 2) - 0(2, 0, 0) - \frac{3}{2}(0, 1, 1) = \left(0, -\frac{1}{2}, \frac{1}{2}\right)$

Normalize each vector.

$\mathbf{u}_1 = \dfrac{\mathbf{w}_1}{\|\mathbf{w}_1\|} = \frac{1}{2}(2, 0, 0) = (1, 0, 0)$

$\mathbf{u}_2 = \dfrac{\mathbf{w}_2}{\|\mathbf{w}_2\|} = \frac{\sqrt{2}}{2}(0, 1, 1) = \left(0, \frac{\sqrt{2}}{2}, \frac{\sqrt{2}}{2}\right)$

$\mathbf{u}_3 = \dfrac{\mathbf{w}_3}{\|\mathbf{w}_3\|} = \sqrt{2}\left(0, -\frac{1}{2}, \frac{1}{2}\right) = \left(0, -\frac{\sqrt{2}}{2}, \frac{\sqrt{2}}{2}\right)$

So, an orthonormal basis for R^3 is

$\left\{(1, 0, 0), \left(0, \frac{\sqrt{2}}{2}, \frac{\sqrt{2}}{2}\right), \left(0, -\frac{\sqrt{2}}{2}, \frac{\sqrt{2}}{2}\right)\right\}.$

17. $\text{proj}_{\mathbf{v}}\mathbf{u} = \dfrac{\mathbf{u} \cdot \mathbf{v}}{\mathbf{v} \cdot \mathbf{v}}\mathbf{v} = \dfrac{1}{13}(-3, 2)$

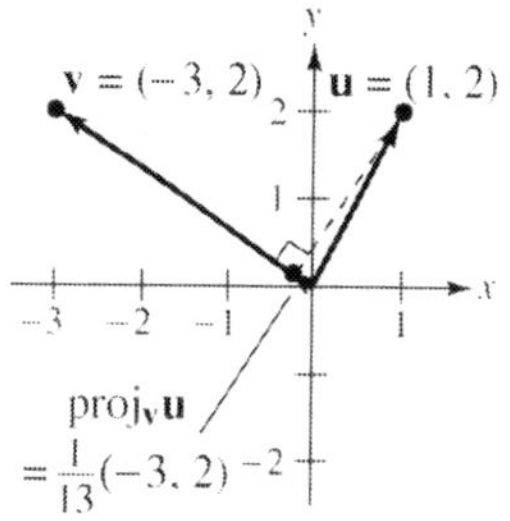

18. $A = \begin{bmatrix} 0 & 1 & 1 & 0 \\ -1 & 0 & 0 & 1 \\ 1 & 1 & 1 & 1 \end{bmatrix} \Rightarrow \begin{bmatrix} 1 & 0 & 0 & 0 \\ 0 & 1 & 1 & 0 \\ 0 & 0 & 0 & 1 \end{bmatrix}$

$A^T = \begin{bmatrix} 0 & -1 & 1 \\ 1 & 0 & 1 \\ 1 & 0 & 1 \\ 0 & 1 & 1 \end{bmatrix} \Rightarrow \begin{bmatrix} 1 & 0 & 0 \\ 0 & 1 & 0 \\ 0 & 0 & 1 \\ 0 & 0 & 0 \end{bmatrix}$

$R(A)$ = column space of $A = R^3$

$N(A)$-basis: $\left\{ \begin{bmatrix} 0 \\ 1 \\ -1 \\ 0 \end{bmatrix} \right\}$

$R(A^T)$-basis: $\left\{ \begin{bmatrix} 0 \\ 1 \\ 1 \\ 0 \end{bmatrix}, \begin{bmatrix} -1 \\ 0 \\ 0 \\ 1 \end{bmatrix}, \begin{bmatrix} 1 \\ 1 \\ 1 \\ 1 \end{bmatrix} \right\}$

$N(A^T) = \{\mathbf{0}\}$

19. $S^\perp = N(A^T)$;

$\begin{bmatrix} 1 & 0 & 1 \\ -1 & 1 & 0 \end{bmatrix} \Rightarrow \begin{bmatrix} 1 & 0 & 1 \\ 0 & 1 & 1 \end{bmatrix} \Rightarrow S^\perp = \text{span}\left\{ \begin{bmatrix} -1 \\ -1 \\ 1 \end{bmatrix} \right\}$

© 2013 Cengage Learning. All Rights Reserved. May not be scanned, copied or duplicated, or posted to a publicly accessible website, in whole or in part.

20. Suppose $c_1\mathbf{x}_1 + \cdots + c_n\mathbf{x}_n + c\mathbf{y} = \mathbf{0}$

If $c = 0$, then $c_1\mathbf{x}_1 + \ldots + c_n\mathbf{x}_n = \mathbf{0}$ and

$\mathbf{x}_i$ independent $\Rightarrow c_i = 0$

If $c \neq 0$, then $\mathbf{y} = -(c_1/c)\mathbf{x}_1 + \cdots + -(c_n/c)\mathbf{x}_n$, a contradiction.

21. Substitute the points $(1, 1)$, $(2, 0)$, and $(5, -5)$ into the equation $y = c_0 + c_1x$ to obtain the following system.

$c_0 + c_1(1) = 1$

$c_0 + c_1(2) = 0$

$c_0 + c_1(5) = -5$

This produces the least squares problem

$$A\mathbf{x} = \mathbf{b}$$

$$\begin{bmatrix} 1 & 1 \\ 1 & 2 \\ 1 & 5 \end{bmatrix}\begin{bmatrix} c_0 \\ c_1 \end{bmatrix} = \begin{bmatrix} 1 \\ 0 \\ -5 \end{bmatrix}.$$

The normal equations are

$$A^TA\mathbf{x} = A^T\mathbf{b}$$

$$\begin{bmatrix} 3 & 8 \\ 8 & 30 \end{bmatrix}\begin{bmatrix} c_0 \\ c_1 \end{bmatrix} = \begin{bmatrix} -4 \\ -24 \end{bmatrix}$$

and the solution is $\mathbf{x} = \begin{bmatrix} c_0 \\ c_1 \end{bmatrix} = \begin{bmatrix} \frac{36}{13} \\ -\frac{20}{13} \end{bmatrix}$.

So, the least squares regression line is $y = \frac{36}{13} - \frac{20}{13}x$.

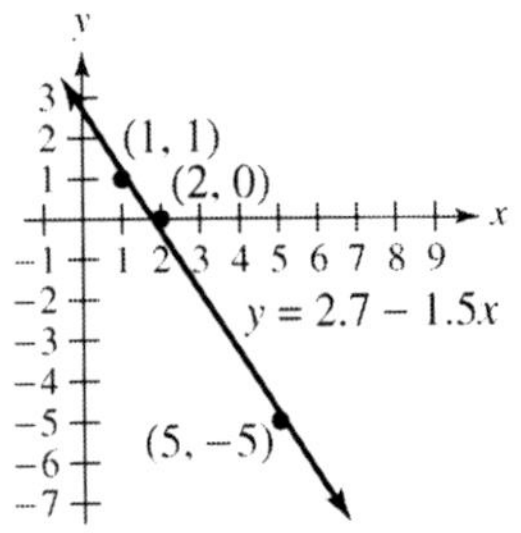

22. (a) rank $A = 3$

(b) first 3 rows of A

(c) columns 1, 3, 4 of A

(d) $x_1 = 2r - 3s - 2t$

$x_2 = r$

$x_3 = 5s + 3t$

$x_4 = -s - 7t$

$x_5 = s$

$x_6 = t$

$$\left\{\begin{bmatrix} 2 \\ 1 \\ 0 \\ 0 \\ 0 \\ 0 \end{bmatrix}, \begin{bmatrix} -3 \\ 0 \\ 5 \\ -1 \\ 1 \\ 0 \end{bmatrix}, \begin{bmatrix} -2 \\ 0 \\ 3 \\ -7 \\ 0 \\ 1 \end{bmatrix}\right\}$$

(e) no

(f) no

(g) yes

(h) no

23. No. Two planes can intersect in a line, but not a single point.

S_1–basis:	$\{(1, 0, 0), (1, 1, 0)\}$	$\dim(S_1) = 2$
S_2–basis:	$\{(0, 0, 1), (0, 1, 0)\}$	$\dim(S_2) = 2$
$S_1 \cap S_2$–basis:	$\{(0, 1, 0)\}$	$\dim(S_1 \cap S_2) = 1$
$S_1 + S_2$–basis:	$\{(1, 0, 0), (0, 1, 0), (0, 0, 1)\}$	$\dim(S_1 + S_2) = 3$

(Answers are not unique.)

24. If a set $\{\mathbf{v}_1, \ldots, \mathbf{v}_m\}$ spans V where $m < n$, then by Exercise 84, you could reduce this set to an independent spanning set of less than n vectors. But then $\dim V \neq n$.

© 2013 Cengage Learning. All Rights Reserved. May not be scanned, copied or duplicated, or posted to a publicly accessible website, in whole or in part.

CHAPTER 6
Linear Transformations

© 2013 Cengage Learning. All Rights Reserved. May not be scanned, copied or duplicated, or posted to a publicly accessible website, in whole or in part.

CHAPTER 6
Linear Transformations

Section 6.1 Introduction to Linear Transformations

1. (a) The image of **v** is

$$T(3, -4) = (3 + (-4), 3 - (-4)) = (-1, 7).$$

(b) If $T(v_1, v_2) = (v_1 + v_2, v_1 - v_2) = (3, 19)$, then

$$v_1 + v_2 = 3$$
$$v_1 - v_2 = 19,$$

which implies that $v_1 = 11$ and $v_2 = -8$. So, the preimage of **w** is $(11, -8)$.

3. (a) The image of **v** is

$$T(2, 3, 0) = (3 - 2, 2 + 3, 2(2)) = (1, 5, 4).$$

(b) If $T(v_1, v_2, v_3) = (v_2 - v_1, v_1 + v_2, 2v_1) = (-11, -1, 10)$,

then

$$v_2 - v_1 = -11$$
$$v_1 + v_2 = -1$$
$$2v_1 = 10$$

which implies that $v_1 = 5$ and $v_2 = -6$. So, the preimage of **w** is $\{(5, -6, t) : t \text{ is any real number}\}$.

5. (a) The image of **v** is

$$T(2, -3, -1) = (4(-3) - 2, 4(2) + 5(-3)) = (-14, -7).$$

(b) If $T(v_1, v_2, v_3) = (4v_2 - v_1, 4v_1 + 5v_2) = (3, 9)$, then

$$-v_1 + 4v_2 = 3$$
$$4v_1 + 5v_2 = 9,$$

which implies that $v_1 = 1$, $v_2 = 1$, and $v_3 = t$, where t is any real number. So, the preimage of **w** is $\{(1, 1, t) : t \text{ is any real number}\}$.

7. (a) The image of **v** is

$$T(1, 1) = \left(\frac{\sqrt{2}}{2}(1) - \frac{\sqrt{2}}{2}(1), 1 + 1, 2(1) - 1\right) = (0, 2, 1).$$

(b) If $T(v_1, v_2) = \left(\frac{\sqrt{2}}{2}v_1 - \frac{\sqrt{2}}{2}v_2, v_1 + v_2, 2v_1 - v_2\right) = (-5\sqrt{2}, -2, -16)$,

then

$$\frac{\sqrt{2}}{2}v_1 - \frac{\sqrt{2}}{2}v_2 = -5\sqrt{2}$$
$$v_1 + v_2 = -2$$
$$2v_1 - v_2 = -16$$

which implies that $v_1 = -6$ and $v_2 = 4$. So, the preimage of **w** is $(-6, 4)$.

9. T is *not* a linear transformation because it does not preserve addition nor scalar multiplication. For example,

$$T(1, 1) + T(1, 1) = (1, 1) + (1, 1) = (2, 2) \neq (2, 1) = T(2, 2).$$

11. T preserves addition.

$$\begin{aligned} T(x_1, y_1, z_1) + T(x_2, y_2, z_2) &= (x_1 + y_1, x_1 - y_1, z_1) + (x_2 + y_2, x_2 - y_2, z_2) \\ &= (x_1 + y_1 + x_2 + y_2, x_1 - y_1 + x_2 - y_2, z_1 + z_2) \\ &= ((x_1 + x_2) + (y_1 + y_2), (x_1 + x_2) - (y_1 + y_2), z_1 + z_2) \\ &= T(x_1 + x_2, y_1 + y_2, z_1 + z_2) \end{aligned}$$

T preserves scalar multiplication.

$$T(c(x, y, z)) = T(cx, cy, cz) = (cx + cy, cx - cy, cz) = c(x + y, x - y, z) = cT(x, y, z)$$

Therefore, T *is* a linear transformation.

© 2013 Cengage Learning. All Rights Reserved. May not be scanned, copied or duplicated, or posted to a publicly accessible website, in whole or in part.

13. T is *not* a linear transformation because it does not preserve addition nor scalar multiplication. For example,

$T(0, 1) + T(1, 0) = (0, 0, 1) + (1, 0\ 0) = (1, 0, 1) \neq (1, 1, 1) = T(1, 1).$

15. T is *not* a linear transformation because it does not preserve addition nor scalar multiplication. For example, $T(I_2) = 1$ but $T(2I_2) = 4 \neq 2T(I_2)$.

17. T preserves addition.

$$\begin{aligned} T(A_1) + T(A_2) &= T\left(\begin{bmatrix} a_1 & b_1 \\ c_1 & d_1 \end{bmatrix}\right) + T\left(\begin{bmatrix} a_2 & b_2 \\ c_2 & d_2 \end{bmatrix}\right) \\ &= a_1 + b_1 - c_1 + d_1 + a_2 + b_2 - c_2 + d_2 \\ &= (a_1 + a_2) + (b_1 + b_2) - (c_1 + c_2) + (d_1 + d_2) \end{aligned}$$

T preserves scalar multiplication.

$T(kA) = ka + kb - kc + kd = k(a + b - c + d) = kT(A)$

Therefore, T *is* a linear transformation.

19. Let A and B be two elements of $M_{3,3}$—two 3×3 matrices—and let c be a scalar. First

$$T(A + B) = \begin{bmatrix} 0 & 0 & 1 \\ 0 & 1 & 0 \\ 1 & 0 & 0 \end{bmatrix}(A + B) = \begin{bmatrix} 0 & 0 & 1 \\ 0 & 1 & 0 \\ 1 & 0 & 0 \end{bmatrix}A + \begin{bmatrix} 0 & 0 & 1 \\ 0 & 1 & 0 \\ 1 & 0 & 0 \end{bmatrix}B = T(A) + T(B)$$

by Theorem 2.3, part 2. And

$$T(cA) = \begin{bmatrix} 0 & 0 & 1 \\ 0 & 1 & 0 \\ 1 & 0 & 0 \end{bmatrix}(cA) = c\begin{bmatrix} 0 & 0 & 1 \\ 0 & 1 & 0 \\ 1 & 0 & 0 \end{bmatrix}A = cT(A)$$

by Theorem 2.3, part 4. So, T *is* a linear transformation.

21. Let $\mathbf{u} = a_0 + a_1x + a_2x^2$, $\mathbf{v} = b_0 + b_1x + b_2x^2$.

$$\begin{aligned} \text{Then } T(\mathbf{u} + \mathbf{v}) &= (a_0 + b_0) + (a_1 + b_1) + (a_2 + b_2) + [(a_1 + b_1) + (a_2 + b_2)]x + (a_2 + b_2)x^2 \\ &= (a_0 + a_1 + a_2) + (b_0 + b_1 + b_2) + [(a_1 + a_2) + (b_1 + b_2)]x + (a_2 + b_2)x^2 \\ &= T(\mathbf{u}) + T(\mathbf{v}), \text{ and} \end{aligned}$$

$T(c\mathbf{u}) = ca_0 + ca_1 + ca_2 + (ca_1 + ca_2)x + ca_2x^2 = cT(\mathbf{u})$.

T *is* a linear transformation.

23. Because $(1, 4) = 1(1, 0) + 4(0, 1)$, you have

$$\begin{aligned} T(1, 4) &= T[(1, 0) + 4(0, 1)] \\ &= T(1, 0) + 4T(0, 1) \\ &= (1, 1) + 4(-1, 1) \\ &= (-3, 5). \end{aligned}$$

Similarly, $(-2, 1) = -2(1, 0) + 1(0, 1)$, which gives

$$\begin{aligned} T(-2, 1) &= T[-2(1, 0) + (0, 1)] \\ &= -2T(1, 0) + T(0, 1) \\ &= -2(1, 1) + (-1, 1) \\ &= (-3, -1). \end{aligned}$$

25. Because $(0, 3, -1)$ can be written as $(0, 3, -1) = 0(1, 0, 0) + 3(0, 1, 0) - (0, 0, 1)$, you can use Property 4 of Theorem 6.1 to write

$$\begin{aligned} T(0, 3, -1) &= 0 \cdot T(1, 0, 0) + 3T(0, 1, 0) - T(0, 0, 1) \\ &= (0, 0, 0) + 3(1, 3, -2) - (0, -2, 2) \\ &= (3, 11, -8). \end{aligned}$$

© 2013 Cengage Learning. All Rights Reserved. May not be scanned, copied or duplicated, or posted to a publicly accessible website, in whole or in part.

27. Because $(2, -4, 1)$ can be written as

$(2, -4, 1) = 2(1, 0, 0) - 4(0, 1, 0) + (0, 0, 1)$, you can use Property 4 of Theorem 6.1 to write

$$\begin{aligned} T(2, -4, 1) &= 2T(1, 0, 0) - 4T(0, 1, 0) + T(0, 0, 1) \\ &= 2(2, 4, -1) - 4(1, 3, -2) + (0, -2, 2) \\ &= (0, -6, 8). \end{aligned}$$

29. Because $(2, 1, 0)$ can be written as

$(2, 1, 0) = 0(1, 1, 1) - (0, -1, 2) + 2(1, 0, 1)$, you can use Property 4 of Theorem 6.1 to write

$$\begin{aligned} T(2, 1, 0) &= 0 \cdot T(1, 1, 1) - T(0, -1, 2) + 2T(1, 0, 1) \\ &= (0, 0, 0) - (-3, 2, -1) + 2(1, 1, 0) \\ &= (5, 0, 1). \end{aligned}$$

31. Because $(2, -1, 1)$ can be written as

$(2, -1, 1) = -\frac{3}{2}(1, 1, 1) - \frac{1}{2}(0, -1, 2) + \frac{7}{2}(1, 0, 1)$, you can use Property 4 of Theorem 6.1 to write

$$\begin{aligned} T(2, -1, 1) &= -\tfrac{3}{2}T(1, 1, 1) - \tfrac{1}{2}T(0, -1, 2) + \tfrac{7}{2}T(1, 0, 1) \\ &= -\tfrac{3}{2}(2, 0, -1) - \tfrac{1}{2}(-3, 2, -1) + \tfrac{7}{2}(1, 1, 0) \\ &= \left(2, \tfrac{5}{2}, 2\right). \end{aligned}$$

33. Because the matrix has two columns, the dimension of R^n is 2. Because the matrix has two rows, the dimension of R^m is 2. So, $T: R^2 \to R^2$.

35. Because the matrix has 4 columns and 4 rows it defines a linear transformation from R^4 to R^4.

37. Because the matrix has four columns, the dimension of R^n is 4. Because the matrix has three rows, the dimension of R^m is 3. So, $T: R^4 \to R^3$.

39. (a) $T(1, 1) = \begin{bmatrix} 0 & -1 \\ -1 & 0 \end{bmatrix}\begin{bmatrix} 1 \\ 1 \end{bmatrix} = \begin{bmatrix} -1 \\ -1 \end{bmatrix} = (-1, -1).$

(b) The preimage of $(1, 1)$ is determined by solving the equation

$$T(v_1, v_2) = \begin{bmatrix} 0 & -1 \\ -1 & 0 \end{bmatrix}\begin{bmatrix} v_1 \\ v_2 \end{bmatrix} = \begin{bmatrix} 1 \\ 1 \end{bmatrix}.$$

The equivalent system of linear equations has the solution $v_1 = -1$ where $v_2 = -1$. So, the preimage is $(-1, -1)$.

(c) The preimage of $(0, 0)$ is determined by solving the equation

$$T(v_1, v_2) = \begin{bmatrix} 0 & -1 \\ -1 & 0 \end{bmatrix}\begin{bmatrix} v_1 \\ v_2 \end{bmatrix} = \begin{bmatrix} 0 \\ 0 \end{bmatrix}.$$

The equivalent system of linear equations has the solution $v_1 = 0$ and $v_2 = 0$. So, the preimage is $(0, 0)$.

41. (a) $T(1, 1, 1, 1) = \begin{bmatrix} -1 & 0 & 0 & 0 \\ 0 & 1 & 0 & 0 \\ 0 & 0 & 2 & 0 \\ 0 & 0 & 0 & 1 \end{bmatrix}\begin{bmatrix} 1 \\ 1 \\ 1 \\ 1 \end{bmatrix} = \begin{bmatrix} -1 \\ 1 \\ 2 \\ 1 \end{bmatrix} = (-1, 1, 2, 1).$

(b) The preimage of $(1, 1, 1, 1)$ is determined by solving the equation

$$T(v_1, v_2, v_3, v_4) = \begin{bmatrix} -1 & 0 & 0 & 0 \\ 0 & 1 & 0 & 0 \\ 0 & 0 & 2 & 0 \\ 0 & 0 & 0 & 1 \end{bmatrix}\begin{bmatrix} v_1 \\ v_2 \\ v_3 \\ v_4 \end{bmatrix} = \begin{bmatrix} 1 \\ 1 \\ 1 \\ 1 \end{bmatrix}$$

for $\mathbf{v} = (v_1, v_2, v_3, v_4)$. The equivalent system of linear equations has solution $v_1 = -1, v_2 = 1, v_3 = \frac{1}{2}, v_4 = 1$.

So, the preimage is $\left(-1, 1, \frac{1}{2}, 1\right)$.

© 2013 Cengage Learning. All Rights Reserved. May not be scanned, copied or duplicated, or posted to a publicly accessible website, in whole or in part.

43. (a) $T(1, 0, 2, 3) = \begin{bmatrix} 0 & 1 & -2 & 1 \\ -1 & 4 & 5 & 0 \\ 0 & 1 & 3 & 1 \end{bmatrix} \cdot \begin{bmatrix} 1 \\ 0 \\ 2 \\ 3 \end{bmatrix} = \begin{bmatrix} -1 \\ 9 \\ 9 \end{bmatrix}$

(b) The preimage of $(0, 0, 0)$ is determined by solving the equation $T(w, x, y, z) = \begin{bmatrix} 0 & 1 & -2 & 1 \\ -1 & 4 & 5 & 0 \\ 0 & 1 & 3 & 1 \end{bmatrix}\begin{bmatrix} w \\ x \\ y \\ z \end{bmatrix} = \begin{bmatrix} 0 \\ 0 \\ 0 \end{bmatrix}$.

The equivalent system of linear equations has the solution $w = -4t$, $x = -t$, $y = 0$, $z = t$, where t is any real number. So, the preimage is given by the set of vectors $\{(-4t, -t, 0, t) : t \text{ is any real number}\}$.

45. (a) When $\theta = 45°$, $\cos\theta = \sin\theta = \frac{1}{\sqrt{2}}$, so $T(4, 4) = \left(4\left(\frac{1}{\sqrt{2}}\right) - 4\left(\frac{1}{\sqrt{2}}\right), 4\left(\frac{1}{\sqrt{2}}\right) + 4\left(\frac{1}{\sqrt{2}}\right)\right) = (0, 4\sqrt{2})$.

(b) When $\theta = 30°$, $\cos\theta = \frac{\sqrt{3}}{2}$ and $\sin\theta = \frac{1}{2}$, so $T(4, 4) = \left(4\left(\frac{\sqrt{3}}{2}\right) - 4\left(\frac{1}{2}\right), 4\left(\frac{1}{2}\right) + 4\left(\frac{\sqrt{3}}{2}\right)\right) = (2\sqrt{3} - 2, 2\sqrt{3} + 2)$.

(c) When $\theta = 120°$, $\cos\theta = -\frac{1}{2}$ and $\sin\theta = \frac{\sqrt{3}}{2}$, so $T(5, 0) = \left(5\left(-\frac{1}{2}\right) - 0\left(\frac{\sqrt{3}}{2}\right), 5\left(\frac{\sqrt{3}}{2}\right) + 0\left(-\frac{1}{2}\right)\right) = \left(-\frac{5}{2}, \frac{5\sqrt{3}}{2}\right)$.

47. $A = \begin{bmatrix} \cos\theta & -\sin\theta \\ \sin\theta & \cos\theta \end{bmatrix}$

$A^{-1} = \frac{1}{\cos^2\theta + \sin^2\theta}\begin{bmatrix} \cos\theta & \sin\theta \\ -\sin\theta & \cos\theta \end{bmatrix}$

$= \begin{bmatrix} \cos\theta & \sin\theta \\ -\sin\theta & \cos\theta \end{bmatrix}$

This represents a rotation in the clockwise direction.

49. If $\mathbf{v} = (x, y, z)$ is a vector in R^3, then $T(\mathbf{v}) = (x, 0, z)$. In other words, T maps every vector in R^3 to its orthogonal projection in the xz-plane.

51. T is *not* a linear transformation. Consider
$A = \begin{bmatrix} 8 & 2 \\ 3 & 6 \end{bmatrix}$ and $B = \begin{bmatrix} -6 & 9 \\ -1 & 0 \end{bmatrix}$.

Then $T(A + B) = (A + B)^{-1} \neq A^{-1} + B^{-1}$.

53. T is a linear transformation.

T preserves addition.

$T(A + C) = (A + C)B$
$= AB + CB$
$= T(A) + T(C)$

T preserves scalar multiplication.

$T(kA) = (kA)B$
$= k(AB)$
$= kT(A)$

55. $T(2 - 6x + x^2) = 2T(1) - 6T(x) + T(x^2)$
$= 2x - 6(1 + x) + (1 + x + x^2)$
$= -5 - 3x + x^2$

57. True. D_x is a linear transformation and therefore preserves addition and scalar multiplication.

59. False. $\sin 2x \neq 2\sin x$ for all x.

61. If $D_x(g(x)) = 2x + 1$, then $g(x) = x^2 + x + C$.

63. If $D_x(g(x)) = \sin x$, then $g(x) = -\cos x + C$.

65. (a) $T(3x^2 - 2) = \int_0^1 (3x^2 - 2)\,dx$
$= \left[x^3 - 2x\right]_0^1$
$= -1$

(b) $T(x^3 - x^5) = \int_0^1 (x^3 - x^5)\,dx$
$= \left[\frac{1}{4}x^4 - \frac{1}{6}x^6\right]_0^1$
$= \frac{1}{12}$

(c) $T(4x - 6) = \int_0^1 (4x - 6)\,dx = \left[2x^2 - 6x\right]_0^1 = -4$

67. (a) False. $\cos(x_1 + x_2) \neq \cos x_1 + \cos x_2$

(b) True. See Example 10, page 299.

© 2013 Cengage Learning. All Rights Reserved. May not be scanned, copied or duplicated, or posted to a publicly accessible website, in whole or in part.

69. (a) $T(x, y) = T[x(1, 0) + y(0, 1)] = xT(1, 0) + yT(0, 1) = x(1, 0) + y(0, 0) = (x, 0)$

(b) T is the projection onto the x-axis.

71. (a) Because

$$\text{proj}_{\mathbf{v}}\mathbf{u} = \frac{\mathbf{u} \cdot \mathbf{v}}{\mathbf{v} \cdot \mathbf{v}}\mathbf{v}$$

and $T(\mathbf{u}) = \text{proj}_{\mathbf{v}}\mathbf{u}$, you have $T(x, y) = \dfrac{x(1) + y(1)}{1^2 + 1^2}(1, 1) = \left(\dfrac{x + y}{2}, \dfrac{x + y}{2}\right)$.

(b) From the result of part (a), where $(x, y) = (5, 0)$, $T(x, y) = \left(\dfrac{x + y}{2}, \dfrac{x + y}{2}\right) = \left(\dfrac{5}{2}, \dfrac{5}{2}\right)$.

(c) From the result of part (a),

$$\begin{aligned} T(\mathbf{u} + \mathbf{w}) = T[(x_1, y_1) + (x_2, y_2)] &= T(x_1 + x_2, y_1 + y_2) \\ &= \left(\frac{x_1 + x_2 + y_1 + y_2}{2}, \frac{x_1 + x_2 + y_1 + y_2}{2}\right) \\ &= \left(\frac{x_1 + y_1}{2}, \frac{x_1 + y_1}{2}\right) + \left(\frac{x_2 + y_2}{2}, \frac{x_2 + y_2}{2}\right) \\ &= T(x_1, y_1) + T(x_2, y_2) \\ &= T(\mathbf{u}) + T(\mathbf{w}). \end{aligned}$$

From the result of part (a),

$$T(c\mathbf{u}) = T[c(x, y)] = T(cx, cy) = \left(\frac{cx + cy}{2}, \frac{cx + cy}{2}\right) = c\left(\frac{x + y}{2}, \frac{x + y}{2}\right) = cT(x, y) = cT(\mathbf{u}).$$

73. Observe that

$$A\mathbf{u} = \begin{bmatrix} \frac{1}{2} & \frac{1}{2} \\ \frac{1}{2} & \frac{1}{2} \end{bmatrix}\begin{bmatrix} x \\ y \end{bmatrix} = \begin{bmatrix} \frac{1}{2}x + \frac{1}{2}y \\ \frac{1}{2}x + \frac{1}{2}y \end{bmatrix} = T(\mathbf{u}).$$

75. (a) Because $T(\mathbf{0}) = \mathbf{0}$ for any linear transformation T, $\mathbf{0}$ is a fixed point of T.

(b) Let F be the set of fixed points of $T : V \to V$. F is nonempty because $\mathbf{0} \in F$. Furthermore, if $\mathbf{u}, \mathbf{v} \in F$, then $T(\mathbf{u} + \mathbf{v}) = T(\mathbf{u}) + T(\mathbf{v}) = \mathbf{u} + \mathbf{v}$ and $T(c\mathbf{u}) = cT(\mathbf{u}) = c\mathbf{u}$, which shows that F is closed under addition and scalar multiplication.

(c) A vector $\mathbf{u}$ is a fixed point if $T(\mathbf{u}) = \mathbf{u}$. Because $T(x, y) = (x, 2y) = (x, y)$ has solutions $x = t$ and $y = 0$, the set of all fixed points of T is $\{(t, 0) : t \text{ is any real number}\}$.

(d) A vector $\mathbf{u}$ is a fixed point if $T(\mathbf{u}) = \mathbf{u}$. Because $T(x, y) = (y, x)$ has solutions $x = y$, the set of fixed points of T is $\{(x, x) : x \text{ is any real number}\}$.

77. (a) Let $T(\mathbf{v}) = \mathbf{0}$ be the zero transformation. Because $T(\mathbf{u} + \mathbf{v}) = \mathbf{0} = T(\mathbf{u}) + T(\mathbf{v})$ and $T(c\mathbf{u}) = \mathbf{0} = cT(\mathbf{u})$, T is a linear transformation.

(b) Let $T(\mathbf{v}) = \mathbf{v}$ be the identity transformation. Because $T(\mathbf{u} + \mathbf{v}) = \mathbf{u} + \mathbf{v} = T(\mathbf{u}) + T(\mathbf{v})$ and $T(c\mathbf{u}) = c\mathbf{u} = cT(\mathbf{u})$, T is a linear transformation.

79. Because $\{\mathbf{v}_1, \ldots, \mathbf{v}_n\}$ are linearly dependent, there exist constants $c_1, \ldots, c_n$, not all zero, such that $c_1\mathbf{v}_1 + \cdots + c_n\mathbf{v}_n = \mathbf{0}$. So,

$$T(c_1\mathbf{v}_1 + \cdots + c_n\mathbf{v}_n) = c_1T(\mathbf{v}_1) + \cdots + c_nT(\mathbf{v}_n) = \mathbf{0},$$

which shows that the set $\{T(\mathbf{v}_1), \ldots, T(\mathbf{v}_n)\}$ is linearly dependent.

81. T is a linear transformation because

$$\begin{aligned} T(A + B) &= (a_{11} + b_{11}) + \cdots + (a_{nn} + b_{nn}) \\ &= (a_{11} + \cdots + a_{nn}) + (b_{11} + \cdots + b_{nn}) \\ &= T(A) + T(B) \end{aligned}$$

and

$$\begin{aligned} T(cA) &= ca_{11} + \cdots + ca_{nn} \\ &= c(a_{11} + \cdots + a_{nn}) \\ &= cT(A). \end{aligned}$$

© 2013 Cengage Learning. All Rights Reserved. May not be scanned, copied or duplicated, or posted to a publicly accessible website, in whole or in part.

83. Let $\mathbf{v} = c_1\mathbf{v}_1 + \cdots + c_n\mathbf{v}_n$ be an arbitrary vector in $\mathbf{v}$. Then,

$$\begin{aligned} T(\mathbf{v}) &= T(c_1\mathbf{v}_1 + \cdots + c_n\mathbf{v}_n) \\ &= c_1T(\mathbf{v}_1) + \cdots + c_nT(\mathbf{v}_n) \\ &= \mathbf{0} + \cdots + \mathbf{0} \\ &= \mathbf{0}. \end{aligned}$$

Section 6.2 The Kernel and Range of a Linear Transformation

1. Because T sends every vector in R^3 to the zero vector, the kernel is R^3.

3. Solving the equation $T(x, y, z, w) = (y, x, w, z) = (0, 0, 0, 0)$ yields the trivial solution $x = y = z = w = 0$.
So, $\ker(T) = \{(0, 0, 0, 0)\}$.

5. Solving the equation $T(a_0 + a_1x + a_2x^2 + a_3x^3) = a_0 = 0$ yields solutions of the form $a_0 = 0$ and a_1, a_2, and a_3 are any real numbers. So,
$\ker(T) = \{a_1x + a_2x^2 + a_3x^3 : a_1, a_2, a_3 \in R\}$.

7. Solving the equation $T(a_0 + a_1x + a_2x^2) = a_1 + 2a_2x = 0$ yields solutions of the form $a_1 = a_2 = 0$, and a_0 is any real number. So, $\ker(T) = \{a_0 : a_0 \in R\}$.

9. Solving the equation $T(x, y) = (x + 2y, y - x) = (0, 0)$ yields the trivial solution $x = y = 0$. So,
$\ker(T) = \{(0, 0)\}$.

11. (a) Because

$$T(\mathbf{v}) = \begin{bmatrix} 1 & 2 \\ 3 & 4 \end{bmatrix}\begin{bmatrix} v_1 \\ v_2 \end{bmatrix} = \begin{bmatrix} 0 \\ 0 \end{bmatrix}$$

has only the trivial solution $v_1 = v_2 = 0$, the kernel is $\{(0, 0)\}$.

(b) Transpose A and find the equivalent reduced row-echelon form.

$$A^T = \begin{bmatrix} 1 & 3 \\ 2 & 4 \end{bmatrix} \Rightarrow \begin{bmatrix} 1 & 0 \\ 0 & 1 \end{bmatrix}$$

So, a basis for the range of A is $\{(1, 0), (0, 1)\}$.

13. (a) Because

$$T(\mathbf{v}) = \begin{bmatrix} 1 & -1 & 2 \\ 0 & 1 & 2 \end{bmatrix}\begin{bmatrix} v_1 \\ v_2 \\ v_3 \end{bmatrix} = \begin{bmatrix} 0 \\ 0 \end{bmatrix}$$

has solutions of the form $(-4t, -2t, t)$, where t is any real number, a basis for $\ker(T)$ is $\{(-4, -2, 1)\}$.

(b) Transpose A and find the equivalent reduced row-echelon form.

$$A^T = \begin{bmatrix} 1 & 0 \\ -1 & 1 \\ 2 & 2 \end{bmatrix} \Rightarrow \begin{bmatrix} 1 & 0 \\ 0 & 1 \\ 0 & 0 \end{bmatrix}$$

So, a basis for the range of A is $\{(1, 0), (0, 1)\}$.

15. (a) Because

$$T(\mathbf{v}) = \begin{bmatrix} 1 & 2 \\ -1 & -2 \\ 1 & 1 \end{bmatrix}\begin{bmatrix} v_1 \\ v_2 \end{bmatrix} = \begin{bmatrix} 0 \\ 0 \\ 0 \end{bmatrix}$$

has only the trivial solution $v_1 = v_2 = 0$, the kernel is $\{(0, 0)\}$.

(b) Transpose A and find the equivalent reduced row-echelon form.

$$A^T = \begin{bmatrix} 1 & -1 & 1 \\ 2 & -2 & 1 \end{bmatrix} \Rightarrow \begin{bmatrix} 1 & -1 & 0 \\ 0 & 0 & 1 \end{bmatrix}$$

So, a basis for the range of A is $\{(1, -1, 0), (0, 0, 1)\}$.

17. (a) Because

$$T(\mathbf{v}) = \begin{bmatrix} 1 & 2 & -1 & 4 \\ 3 & 1 & 2 & -1 \\ -4 & -3 & -1 & -3 \\ -1 & -2 & 1 & 1 \end{bmatrix}\begin{bmatrix} v_1 \\ v_2 \\ v_3 \\ v_4 \end{bmatrix} = \begin{bmatrix} 0 \\ 0 \\ 0 \\ 0 \end{bmatrix}$$

has solutions of the form $(-t, t, t, 0)$, where t is any real number, a basis for $\ker(T)$ is $\{(-1, 1, 1, 0)\}$.

© 2013 Cengage Learning. All Rights Reserved. May not be scanned, copied or duplicated, or posted to a publicly accessible website, in whole or in part.

(b) Transpose A and find the equivalent reduced row-echelon form.

$$A^T = \begin{bmatrix} 1 & 3 & -4 & -1 \\ 2 & 1 & -3 & -2 \\ -1 & 2 & -1 & 1 \\ 4 & -1 & -3 & 1 \end{bmatrix} \Rightarrow \begin{bmatrix} 1 & 0 & -1 & 0 \\ 0 & 1 & -1 & 0 \\ 0 & 0 & 0 & 1 \\ 0 & 0 & 0 & 0 \end{bmatrix}$$

So, a basis for the range of A is

$\{(1, 0, -1, 0), (0, 1, -1, 0), (0, 0, 0, 1)\}$.

Equivalently, you could use columns 1, 2 and 4 of the original matrix A.

19. (a) Because $T(\mathbf{x}) = \mathbf{0}$ has only the trivial solution $\mathbf{x} = (0, 0)$, the kernel of T is $\{(0, 0)\}$.

(b) $\text{nullity}(T) = \dim(\ker(T)) = 0$

(c) Transpose A and find the equivalent reduced row-echelon form.

$$A^T = \begin{bmatrix} -1 & 1 \\ 1 & 1 \end{bmatrix} \Rightarrow \begin{bmatrix} 1 & 0 \\ 0 & 1 \end{bmatrix}$$

So, $\text{range}(T) = R^2$.

(d) $\text{rank}(T) = \dim(\text{range}(T)) = 2$

21. (a) Because $T(\mathbf{x}) = \mathbf{0}$ has only the trivial solution $\mathbf{x} = (0, 0)$, the kernel of T is $\{(0, 0)\}$.

(b) $\text{nullity}(T) = \dim(\ker(T)) = 0$

(c) Transpose A and find the equivalent reduced row-echelon form.

$$A^T = \begin{bmatrix} 5 & 1 & 1 \\ -3 & 1 & -1 \end{bmatrix} \Rightarrow \begin{bmatrix} 1 & 0 & \frac{1}{4} \\ 0 & 1 & -\frac{1}{4} \end{bmatrix}$$

So, $\text{range}(T) = \{(4s, 4t, s - t) : s, t \in R\}$.

(d) $\text{rank}(T) = \dim(\text{range}(T)) = 2$

23. (a) The kernel of T is given by the solution to the equation $T(\mathbf{x}) = \mathbf{0}$. So, $\ker(T) = \{(t, -3t) : t \in R\}$.

(b) $\text{nullity}(T) = \dim(\ker(T)) = 1$

(c) Transpose A and find its equivalent row-echelon form.

$$A^T = \begin{bmatrix} \frac{9}{10} & \frac{3}{10} \\ \frac{3}{10} & \frac{1}{10} \end{bmatrix} \Rightarrow \begin{bmatrix} 3 & 1 \\ 0 & 0 \end{bmatrix}$$

So, $\text{range}(T) = \{(3t, t) : t \in R\}$.

(d) $\text{rank}(T) = \dim(\text{range}(T)) = 1$

25. (a) The kernel of T is given by the solution to the equation $T(\mathbf{x}) = \mathbf{0}$.

So,

$\ker(T) = \{(s + t, s, -2t) : s \text{ and } t \text{ are real numbers}\}$.

(b) $\text{nullity}(T) = \dim(\ker(T)) = 2$

(c) Transpose A and find the equivalent reduced row-echelon form.

$$A^T = \begin{bmatrix} \frac{4}{9} & -\frac{4}{9} & \frac{2}{9} \\ -\frac{4}{9} & \frac{4}{9} & -\frac{2}{9} \\ \frac{2}{9} & -\frac{2}{9} & \frac{1}{9} \end{bmatrix} \Rightarrow \begin{bmatrix} 1 & -1 & \frac{1}{2} \\ 0 & 0 & 0 \\ 0 & 0 & 0 \end{bmatrix}$$

So,

$\text{range}(T) = \{(2t, -2t, t) : t \text{ is any real number}\}$.

(d) $\text{rank}(T) = \dim(\text{range}(T)) = 1$

27. (a) The kernel of T is given by the solution to the equation $T(\mathbf{x}) = \mathbf{0}$. So,

$\ker(T) = \{(-11t, 6t, 4t) : t \text{ is any real number}\}$.

(b) $\text{nullity}(T) = \dim(\ker(T)) = 1$

(c) Transpose A and find the equivalent reduced row-echelon form.

$$A^T = \begin{bmatrix} 0 & 4 \\ -2 & 0 \\ 3 & 11 \end{bmatrix} \Rightarrow \begin{bmatrix} 1 & 0 \\ 0 & 1 \\ 0 & 0 \end{bmatrix}$$

So, $\text{range}(T) = R^2$.

(d) $\text{rank}(T) = \dim(\text{range}(T)) = 2$

29. (a) The kernel of T is given by the solution to the equation $T(\mathbf{x}) = \mathbf{0}$.

So, $\ker(T) = \{(2s - t, t, 4s, -5s, s) : s, t \in R\}$.

(b) $\text{nullity}(T) = \dim(\ker(T)) = 2$

(c) Transpose A and find its equivalent row-echelon form.

$$A^T = \begin{bmatrix} 2 & 1 & 3 & 6 \\ 2 & 1 & 3 & 6 \\ -3 & 1 & -5 & -2 \\ 1 & 1 & 0 & 4 \\ 13 & -1 & 14 & 16 \end{bmatrix} \Rightarrow \begin{bmatrix} 7 & 0 & 0 & 8 \\ 0 & 7 & 0 & 20 \\ 0 & 0 & 7 & 2 \\ 0 & 0 & 0 & 0 \\ 0 & 0 & 0 & 0 \end{bmatrix}$$

So,

$\text{range}(T) = \{7r, 7s, 7t, 8r + 20s + 2t) : r, s, t \in R\}$.

Equivalently, the range of T is spanned by columns 1, 3 and 4 of A.

(d) $\text{rank}(T) = \dim(\text{range}(T)) = 3$

© 2013 Cengage Learning. All Rights Reserved. May not be scanned, copied or duplicated, or posted to a publicly accessible website, in whole or in part.

31. Use Theorem 6.5 to find nullity(T).

$$\text{rank}(T) + \text{nullity}(T) = \dim(R^3)$$
$$\text{nullity}(T) = 3 - 2 = 1$$

Because $\text{nullity}(T) = \dim(\ker(T)) = 1$, the kernel of T is a line in space. Furthermore, because $\text{rank}(T) = \dim(\text{range}(T)) = 2$, the range of T is a plane in space.

33. Because $\text{rank}(T) + \text{nullity}(T) = 3$ and you are given $\text{rank}(T) = 0$, it follows that $\text{nullity}(T) = 3$. So, the kernel of T is all of R^3, and the range is the single point $\{(0, 0, 0)\}$.

35. The preimage of $(0, 0, 0)$ is $\{(0, 0, 0)\}$.

So, $\text{nullity}(T) = 0$, and the rank of T is determined as follows.

$$\text{rank}(T) + \text{nullity}(T) = \dim(R^3)$$
$$\text{rank}(T) = 3 - 0 = 3$$

The kernel of T is the single point $(0, 0, 0)$. Because $\text{rank}(T) = \dim(\text{range}(T)) = 3$, the range of T is R^3.

37. The kernel of T is determined by solving $T(x, y, z) = \dfrac{x + 2y + 2z}{9}(1, 2, 2) = (0, 0, 0)$, which implies that $x + 2y + 2z = 0$. So, the nullity of T is 2, and the kernel is a plane. The range of T is found by observing that $\text{rank}(T) + \text{nullity}(T) = 3$. That is, the range of T is 1-dimensional, a line in R^3 and $\text{range}(T) = \{(t, 2t, 2t) : t \in R\}$.

39. $\text{rank}(T) + \text{nullity}(T) = \dim R^4 \Rightarrow \text{nullity}(T) = 4 - 2 = 2$

41. $\text{rank}(T) + \text{nullity}(T) = \dim R^4 \Rightarrow \text{nullity}(T) = 4 - 0 = 4$

43. Because $|A| = -1 \neq 0$, the homogeneous equation $A\mathbf{x} = \mathbf{0}$ has only the trivial solution. So, $\ker(T) = \{(0, 0)\}$ and T is one-to-one (by Theorem 6.6). Furthermore, because $\text{rank}(T) = \dim(R^2) - \text{nullity}(T) = 2 - 0 = 2 = \dim(R^2)$, T is onto (by Theorem 6.7).

45. Because $|A| = -1 \neq 0$, the homogeneous equation $A\mathbf{x} = \mathbf{0}$ has only the trivial solution. So, $\ker(T) = \{(0, 0, 0)\}$ and T is one-to-one (by Theorem 6.6). Furthermore, because $\text{rank}(T) = \dim R^3 - \text{nullity}(T) = 3 - 0 = 3 = \dim(R^3)$, T is onto (by Theorem 6.7).

47. The matrix representation of $T : R^4 \to R^4$ is given by

$$A = \begin{bmatrix} 0 & 1 & 0 & 0 \\ 1 & 0 & 0 & 0 \\ 0 & 0 & 0 & 1 \\ 0 & 0 & 1 & 0 \end{bmatrix}.$$

Since $|A| = 1 \neq 0$, the equation $A\mathbf{x} = \mathbf{0}$ has only the trivial solution. So, $\ker(T) = \{(0, 0, 0, 0)\}$ and T is one-to-one by Theorem 6.6.
Furthermore, since

$$\begin{aligned} \text{rank}(T) &= \dim(R^4) - \text{nullity}(T) \\ &= 4 - 0 \\ &= 4 \\ &= \dim(R^4) \end{aligned}$$

T is onto by Theorem 6.7.

49. $A = \begin{bmatrix} 5 & -3 \\ 1 & 1 \\ 1 & -1 \end{bmatrix}$

$T: R^2 \to R^3$

$\dim(\text{domain}) = 2$, $\text{rank}(T) = 2$, $\text{nullity}(T) = 0$

Because the rank of T is not equal to the dimension of R^3, T is not onto. Because $\ker(T) = \{0\}$, T is one-to-one.

© 2013 Cengage Learning. All Rights Reserved. May not be scanned, copied or duplicated, or posted to a publicly accessible website, in whole or in part.

51. **Zero** **Standard Basis**

(a) $(0, 0, 0, 0)$ $\{(1, 0, 0, 0), (0, 1, 0, 0), (0, 0, 1, 0), (0, 0, 0, 1)\}$

(b) $\begin{bmatrix} 0 \\ 0 \\ 0 \\ 0 \end{bmatrix}$ $\left\{\begin{bmatrix} 1 \\ 0 \\ 0 \\ 0 \end{bmatrix}, \begin{bmatrix} 0 \\ 1 \\ 0 \\ 0 \end{bmatrix}, \begin{bmatrix} 0 \\ 0 \\ 1 \\ 0 \end{bmatrix}, \begin{bmatrix} 0 \\ 0 \\ 0 \\ 1 \end{bmatrix}\right\}$

(c) $\begin{bmatrix} 0 & 0 \\ 0 & 0 \end{bmatrix}$ $\left\{\begin{bmatrix} 1 & 0 \\ 0 & 0 \end{bmatrix}, \begin{bmatrix} 0 & 1 \\ 0 & 0 \end{bmatrix}, \begin{bmatrix} 0 & 0 \\ 1 & 0 \end{bmatrix}, \begin{bmatrix} 0 & 0 \\ 0 & 1 \end{bmatrix}\right\}$

(d) $p(x) = 0$ $\{1, x, x^2, x^3\}$

(e) $(0, 0, 0, 0, 0)$ $\{(1, 0, 0, 0, 0), (0, 1, 0, 0, 0), (0, 0, 1, 0, 0), (0, 0, 0, 1, 0)\}$

53. Solve the equation $T(p) = \dfrac{d}{dx}\left(a_0 + a_1x + a_2x^2 + a_3x^3 + a_4x^4\right) = 0$ yielding $p = a_0$.

So, $\ker(T) = \{p(x) = a_0 : a_0 \text{ is a real number}\}$. (The constant polynomials)

55. First compute $T(\mathbf{u}) = \text{proj}_{\mathbf{v}}\mathbf{u}$, for $\mathbf{u} = (x, y, z)$.

$$T(\mathbf{u}) = \text{proj}_{\mathbf{v}}\mathbf{u} = \frac{(x, y, z) \cdot (2, -1, 1)}{(2, -1, 1) \cdot (2, -1, 1)}(2, -1, 1) = \frac{2x - y + z}{6}(2, -1, 1)$$

(a) Setting $T(\mathbf{u}) = \mathbf{0}$, you have $2x - y + z = 0$, so $\text{nullity}(T) = 2$, and $\text{rank}(T) = 3 - 2 = 1$.

(b) A basis for the kernel of T is obtained by solving $2x - y + z = 0$.

Letting $t = z$ and $s = y$, you have $x = \dfrac{1}{2}(y - z) = \dfrac{1}{2}s - \dfrac{1}{2}t$.

So, a basis for $\ker(T)$ is $\left\{\left(\dfrac{1}{2}, 1, 0\right), \left(-\dfrac{1}{2}, 0, 1\right)\right\}$, or $\{(1, 2, 0), (1, 0, -2)\}$.

57. (a) A is an $n \times n$ matrix and $\det(A) = \det(A^T) \neq 0$. So, the reduced row-echelon matrix equivalent to A^T has n nonzero rows and you can conclude that $\text{rank}(T) = n$.

(b) A is an $n \times n$ matrix and $\det(A) = \det(A^T) = 0$. So, the reduced row-echelon matrix equivalent to A^T has at least one row of zeros and you can conclude that $\text{rank}(T) < n$.

59. The kernel of T is given by $T(A) = \mathbf{0}$. So, $T(A) = A - A^T = \mathbf{0} \Rightarrow A = A^T$

and $\ker(T) = \{A : A = A^T\}$, the set of $n \times n$ symmetric matrices.

61. (a) False. See "Definition of Kernel of a Linear Transformation," page 374.

(b) False. See Theorem 6.4, page 306.

(c) True, $\dim(R^3) = \dim(M_{3,1}) = 3$ and any vector spaces of equal finite dimension are isomorphic. (See Theorem 6.9.)

63. Because T is a linear transformation of vector spaces with same dimension, you only need to show that T is one-to-one. It is sufficient to show that $\ker(T) = \{\mathbf{0}\}$. Let $A \in \ker(T)$. Then $T(A) = AB = \mathbf{0}$. Use the fact that B is invertible to obtain

$AB = \mathbf{0} \Rightarrow AB(B^{-1}) = \mathbf{0} \Rightarrow A = \mathbf{0}$. So, $\ker(T) = \{\mathbf{0}\}$ and T is one-to-one.

© 2013 Cengage Learning. All Rights Reserved. May not be scanned, copied or duplicated, or posted to a publicly accessible website, in whole or in part.

65. Let $T : V \to W$ be a linear transformation, where $\dim(W) = n$.

If T is onto, then W is equal to the range of T. So, W and the range of T have the same dimension, and the rank of T is n.

If the rank of T is n, then there are n linearly independent vectors $T(\mathbf{v}_1), T(\mathbf{v}_2), \ldots, T(\mathbf{v}_n)$ in the range of T. Since the range of T is a subspace of W, the vectors $T(\mathbf{v}_1), T(\mathbf{v}_2), \ldots T(\mathbf{v}_n)$ are linearly independent in W. By Theorem 4.12, they form a basis for W. So, any vector $\mathbf{w} \in W$ can be written as a linear combination

$$\begin{aligned}\mathbf{w} &= c_1T(\mathbf{v}_1) + c_2T(\mathbf{v}_2) + \cdots + c_nT(\mathbf{v}_n)\\ &= T(c_1\mathbf{v}_1 + c_2\mathbf{v}_2 + \cdots + c_n\mathbf{v}_n).\end{aligned}$$

So, any $\mathbf{w} \in W$ is in the range of T, and T is onto.

67. If T is onto, then $m \geq n$.

If T is one-to-one, then $m \leq n$.

Section 6.3 Matrices for Linear Transformations

1. Because

$$T\left(\begin{bmatrix}1\\0\end{bmatrix}\right) = \begin{bmatrix}1\\1\end{bmatrix} \text{ and } T\left(\begin{bmatrix}0\\1\end{bmatrix}\right) = \begin{bmatrix}2\\-2\end{bmatrix},$$

the standard matrix for T is $A = \begin{bmatrix}1 & 2\\1 & -2\end{bmatrix}$.

3. Because

$$T\left(\begin{bmatrix}1\\0\\0\end{bmatrix}\right) = \begin{bmatrix}1\\1\\-1\end{bmatrix}, T\left(\begin{bmatrix}0\\1\\0\end{bmatrix}\right) = \begin{bmatrix}1\\-1\\0\end{bmatrix}, \text{ and } T\left(\begin{bmatrix}0\\0\\1\end{bmatrix}\right) = \begin{bmatrix}0\\0\\1\end{bmatrix},$$

the standard matrix for T is

$$A = \begin{bmatrix}1 & 1 & 0\\1 & -1 & 0\\-1 & 0 & 1\end{bmatrix}.$$

5. Because

$$T\left(\begin{bmatrix}1\\0\\0\end{bmatrix}\right) = \begin{bmatrix}3\\0\end{bmatrix}, T\left(\begin{bmatrix}0\\1\\0\end{bmatrix}\right) = \begin{bmatrix}0\\2\end{bmatrix}, \text{ and } T\left(\begin{bmatrix}0\\0\\1\end{bmatrix}\right) = \begin{bmatrix}-2\\-1\end{bmatrix},$$

the standard matrix for T is

$$\begin{bmatrix}3 & 0 & -2\\0 & 2 & -1\end{bmatrix}.$$

7. Because

$$T\left(\begin{bmatrix}1\\0\\0\end{bmatrix}\right) = \begin{bmatrix}2\\0\end{bmatrix}, T\left(\begin{bmatrix}0\\1\\0\end{bmatrix}\right) = \begin{bmatrix}1\\3\end{bmatrix}, \text{ and } T\left(\begin{bmatrix}0\\0\\1\end{bmatrix}\right) = \begin{bmatrix}0\\-1\end{bmatrix},$$

the standard matrix for T is

$$\begin{bmatrix}2 & 1 & 0\\0 & 3 & -1\end{bmatrix}.$$

So,

$$T(\mathbf{v}) = \begin{bmatrix}2 & 1 & 0\\0 & 3 & -1\end{bmatrix}\begin{bmatrix}0\\1\\-1\end{bmatrix} = \begin{bmatrix}1\\4\end{bmatrix} \text{ and } T(0, 1, -1) = (1, 4).$$

9. Because

$$T\left(\begin{bmatrix}1\\0\end{bmatrix}\right) = \begin{bmatrix}1\\1\\0\end{bmatrix} \text{ and } T\left(\begin{bmatrix}0\\1\end{bmatrix}\right) = \begin{bmatrix}-1\\2\\1\end{bmatrix},$$

the standard matrix for T is

$$\begin{bmatrix}1 & -1\\1 & 2\\0 & 1\end{bmatrix}.$$

So,

$$T(\mathbf{v}) = \begin{bmatrix}1 & -1\\1 & 2\\0 & 1\end{bmatrix}\begin{bmatrix}2\\-2\end{bmatrix} = \begin{bmatrix}4\\-2\\-2\end{bmatrix} \text{ and } T(2, -2) = (4, -2, -2).$$

© 2013 Cengage Learning. All Rights Reserved. May not be scanned, copied or duplicated, or posted to a publicly accessible website, in whole or in part.

11. (a) The matrix of the reflection through the origin, $T(x, y) = (-x, -y)$, is given by

$$A = [T(1, 0) \vdots T(0, 1)] = \begin{bmatrix} -1 & 0 \\ 0 & -1 \end{bmatrix}.$$

(b) The image of $\mathbf{v} = (3, 4)$ is given by

$$A\mathbf{v} = \begin{bmatrix} -1 & 0 \\ 0 & -1 \end{bmatrix}\begin{bmatrix} 3 \\ 4 \end{bmatrix} = \begin{bmatrix} -3 \\ -4 \end{bmatrix}.$$

So, $T(3, 4) = (-3, -4)$.

(c)

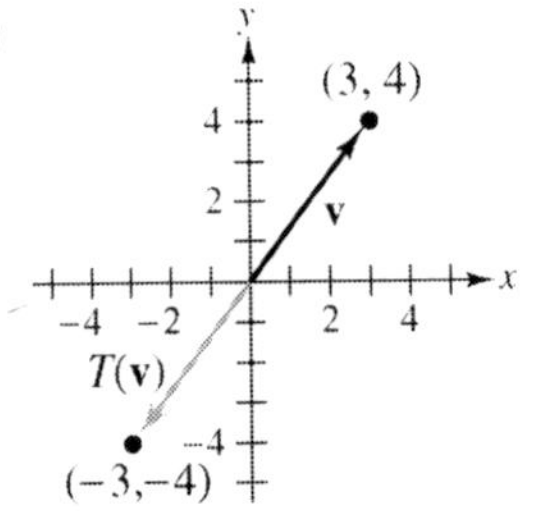

13. (a) The matrix of the reflection in the y-axis, $T(x, y) = (-x, y)$, is given

by $A = [T(1, 0) \vdots T(0, 1)] = \begin{bmatrix} -1 & 0 \\ 0 & 1 \end{bmatrix}.$

(b) The image of $\mathbf{v} = (2, -3)$ is given

by $A\mathbf{v} = \begin{bmatrix} -1 & 0 \\ 0 & 1 \end{bmatrix}\begin{bmatrix} 2 \\ -3 \end{bmatrix} = \begin{bmatrix} -2 \\ -3 \end{bmatrix}$. So,

$T(2, -3) = (-2, -3)$.

(c)

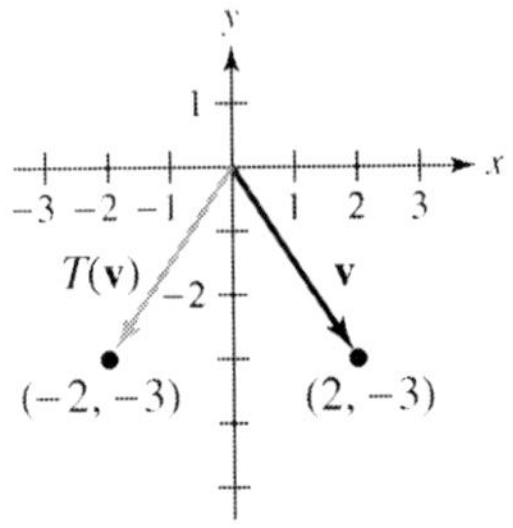

15. (a) The counterclockwise rotation of 45° is given by

$$T(x, y) = (\cos(45)x - \sin(45)y, \sin(45)x + \cos(45)y)$$

$$= \left(\frac{\sqrt{2}}{2}x - \frac{\sqrt{2}}{2}y, \frac{\sqrt{2}}{2}x + \frac{\sqrt{2}}{2}y\right).$$

So, the matrix is

$$A = [T(1, 0) \vdots T(0, 1)] = \begin{bmatrix} \frac{\sqrt{2}}{2} & -\frac{\sqrt{2}}{2} \\ \frac{\sqrt{2}}{2} & \frac{\sqrt{2}}{2} \end{bmatrix}.$$

(b) The image of $\mathbf{v} = (2, 2)$ is given by

$$A\mathbf{v} = \begin{bmatrix} \frac{\sqrt{2}}{2} & -\frac{\sqrt{2}}{2} \\ \frac{\sqrt{2}}{2} & \frac{\sqrt{2}}{2} \end{bmatrix}\begin{bmatrix} 2 \\ 2 \end{bmatrix} = \begin{bmatrix} 0 \\ 2\sqrt{2} \end{bmatrix}.$$

So, $T(2, 2) = (0, 2\sqrt{2})$.

(c)

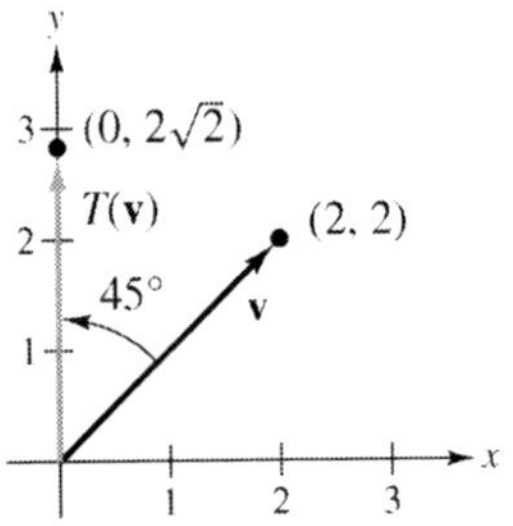

© 2013 Cengage Learning. All Rights Reserved. May not be scanned, copied or duplicated, or posted to a publicly accessible website, in whole or in part.

17. (a) The clockwise rotation of 60° is given by

$$T(x, y) = \left(\cos(-60)x - \sin(-60)y,\ \sin(-60)x + \cos(-60)y\right) = \left(\frac{1}{2}x + \frac{\sqrt{3}}{2}y,\ -\frac{\sqrt{3}}{2} + \frac{1}{2}y\right).$$

So, the matrix is $A = \left[T(1, 0) \vdots T(0, 1)\right] = \begin{bmatrix} \frac{1}{2} & \frac{\sqrt{3}}{2} \\ -\frac{\sqrt{3}}{2} & \frac{1}{2} \end{bmatrix}.$

(b) The image of $\mathbf{v} = (1, 2)$ is given by $A\mathbf{v} = \begin{bmatrix} \frac{1}{2} & \frac{\sqrt{3}}{2} \\ -\frac{\sqrt{3}}{2} & \frac{1}{2} \end{bmatrix}\begin{bmatrix} 1 \\ 2 \end{bmatrix} = \begin{bmatrix} \frac{1}{2} + \sqrt{3} \\ 1 - \frac{\sqrt{3}}{2} \end{bmatrix}$. So, $T(1, 2) = \left(\frac{1}{2} + \sqrt{3}, 1 - \frac{\sqrt{3}}{2}\right)$.

(c)

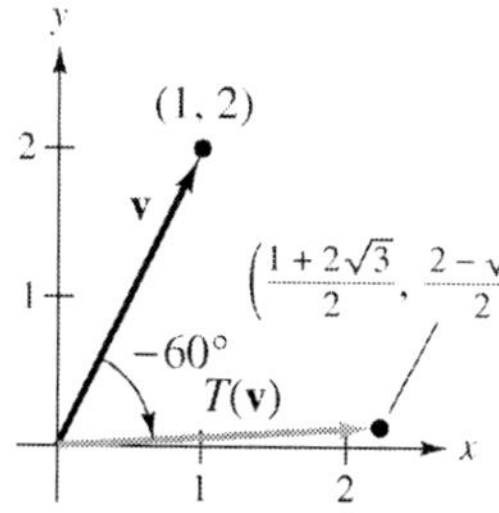

19. (a) The standard matrix for T is

$$A = \left[T(1, 0, 0) \vdots T(0, 1, 0) \vdots T(0, 0, 1)\right] = \begin{bmatrix} 1 & 0 & 0 \\ 0 & 1 & 0 \\ 0 & 0 & -1 \end{bmatrix}.$$

(b) The image of $\mathbf{v} = (3, 2, 2)$ is

$$A\mathbf{v} = \begin{bmatrix} 1 & 0 & 0 \\ 0 & 1 & 0 \\ 0 & 0 & -1 \end{bmatrix}\begin{bmatrix} 3 \\ 2 \\ 2 \end{bmatrix} = \begin{bmatrix} 3 \\ 2 \\ -2 \end{bmatrix}.$$

So, $T(3, 2, 2) = (3, 2, -2)$.

(c)

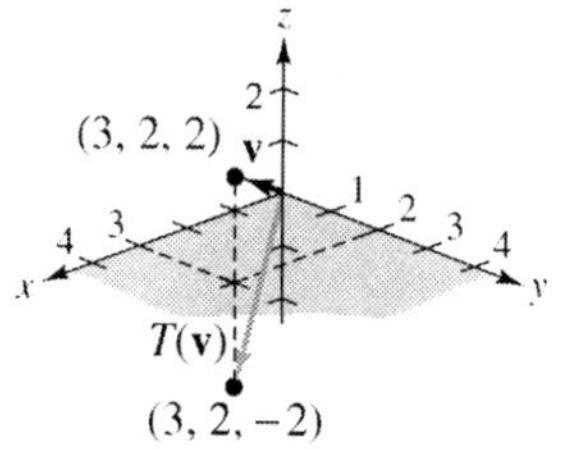

© 2013 Cengage Learning. All Rights Reserved. May not be scanned, copied or duplicated, or posted to a publicly accessible website, in whole or in part.

21. (a) The projection onto the vector $\mathbf{w} = (3, 1)$ is given by $T(\mathbf{v}) = \text{proj}_{\mathbf{w}}\mathbf{v} = \dfrac{3x + y}{10}(3, 1) = \left(\dfrac{3}{10}(3x + y), \dfrac{1}{10}(3x + y)\right)$.

So, the matrix is $A = [T(1, 0) \vdots T(0, 1)] = \begin{bmatrix} \frac{9}{10} & \frac{3}{10} \\ \frac{3}{10} & \frac{1}{10} \end{bmatrix}$.

(b) The image of $\mathbf{v} = (1, 4)$ is given by $A\mathbf{v} = \begin{bmatrix} \frac{9}{10} & \frac{3}{10} \\ \frac{3}{10} & \frac{1}{10} \end{bmatrix}\begin{bmatrix} 1 \\ 4 \end{bmatrix} = \begin{bmatrix} \frac{21}{10} \\ \frac{7}{10} \end{bmatrix}$. So, $T(1, 4) = \left(\dfrac{21}{10}, \dfrac{7}{10}\right)$.

(c)

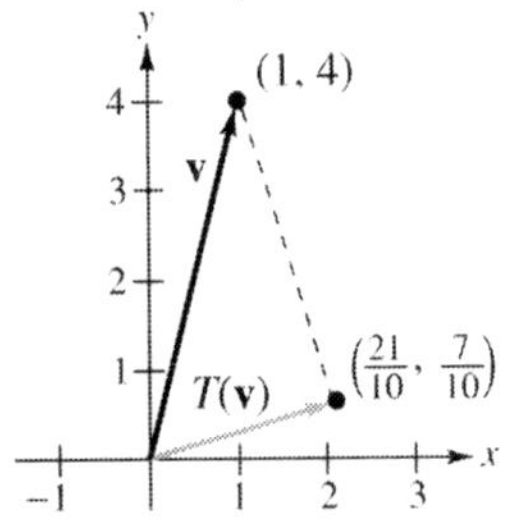

23. (a) The standard matrix for T is

$$A = \begin{bmatrix} 2 & 3 & -1 \\ 3 & 0 & -2 \\ 2 & -1 & 1 \end{bmatrix}$$

(b) The image of $\mathbf{v} = (1, 2, -1)$ is

$$A\mathbf{v} = \begin{bmatrix} 2 & 3 & -1 \\ 3 & 0 & -2 \\ 2 & -1 & 1 \end{bmatrix}\begin{bmatrix} 1 \\ 2 \\ -1 \end{bmatrix} = \begin{bmatrix} 9 \\ 5 \\ -1 \end{bmatrix}.$$

So, $T(1, 2, -1) = (9, 5, -1)$.

(c) Using a graphing utility or a computer software program to perform the multiplication in part (b) gives the same result.

25. (a) The standard matrix for T is

$$A = \begin{bmatrix} 1 & -1 & 0 & 0 \\ 0 & 0 & 1 & 0 \\ 1 & 2 & 0 & -1 \\ 0 & 0 & 0 & 1 \end{bmatrix}.$$

(b) The image of $\mathbf{v} = (1, 0, 1, -1)$ is

$$A\mathbf{v} = \begin{bmatrix} 1 & -1 & 0 & 0 \\ 0 & 0 & 1 & 0 \\ 1 & 2 & 0 & -1 \\ 0 & 0 & 0 & 1 \end{bmatrix}\begin{bmatrix} 1 \\ 0 \\ 1 \\ -1 \end{bmatrix} = \begin{bmatrix} 1 \\ 1 \\ 2 \\ -1 \end{bmatrix}.$$

So, $T(1, 0, 1, -1) = (1, 1, 2, -1)$.

(c) Using a graphing utility or a computer software program to perform the multiplication in part (b) gives the same result.

27. The standard matrices for T_1 and T_2 are

$$A_1 = \begin{bmatrix} 1 & -2 \\ 2 & 3 \end{bmatrix} \text{ and } A_2 = \begin{bmatrix} 0 & 1 \\ 0 & 0 \end{bmatrix}.$$

The standard matrix for $T = T_2 \circ T_1$ is

$$A_2A_1 = \begin{bmatrix} 0 & 1 \\ 0 & 0 \end{bmatrix}\begin{bmatrix} 1 & -2 \\ 2 & 3 \end{bmatrix} = \begin{bmatrix} 2 & 3 \\ 0 & 0 \end{bmatrix}$$

and the standard matrix for $T' = T_1 \circ T_2$ is

$$A_1A_2 = \begin{bmatrix} 1 & -2 \\ 2 & 3 \end{bmatrix}\begin{bmatrix} 0 & 1 \\ 0 & 0 \end{bmatrix} = \begin{bmatrix} 0 & 1 \\ 0 & 2 \end{bmatrix}.$$

29. The standard matrices for T_1 and T_2 are

$$A_1 = \begin{bmatrix} -1 & 2 \\ 1 & 1 \\ 1 & -1 \end{bmatrix} \text{ and } A_2 = \begin{bmatrix} 1 & -3 & 0 \\ 3 & 0 & 1 \end{bmatrix}.$$

The standard matrix for $T = T_2 \circ T_1$ is

$$A = A_2A_1 = \begin{bmatrix} 1 & -3 & 0 \\ 3 & 0 & 1 \end{bmatrix}\begin{bmatrix} -1 & 2 \\ 1 & 1 \\ 1 & -1 \end{bmatrix} = \begin{bmatrix} -4 & -1 \\ -2 & 5 \end{bmatrix}$$

and the standard matrix for $T' = T_1 \circ T_2$ is

$$A' = A_1A_2 = \begin{bmatrix} -1 & 2 \\ 1 & 1 \\ 1 & -1 \end{bmatrix}\begin{bmatrix} 1 & -3 & 0 \\ 3 & 0 & 1 \end{bmatrix} = \begin{bmatrix} 5 & 3 & 2 \\ 4 & -3 & 1 \\ -2 & -3 & -1 \end{bmatrix}.$$

© 2013 Cengage Learning. All Rights Reserved. May not be scanned, copied or duplicated, or posted to a publicly accessible website, in whole or in part.

31. The standard matrix for T is

$$A = \begin{bmatrix} -2 & 0 \\ 0 & 2 \end{bmatrix}.$$

Because $|A| = -4 \neq 0$, A is invertible

$$A^{-1} = \begin{bmatrix} -\frac{1}{2} & 0 \\ 0 & \frac{1}{2} \end{bmatrix}$$

and conclude that $T^{-1}(x, y) = \left(-\frac{1}{2}x, \frac{1}{2}y\right)$.

33. The standard matrix for T is

$$A = \begin{bmatrix} 1 & 1 \\ 3 & 3 \end{bmatrix}.$$

Because $|A| = 0$, A is not invertible, and so T is not invertible.

35. The standard matrix for T is

$$A = \begin{bmatrix} 1 & 0 & 0 \\ 1 & 1 & 0 \\ 1 & 1 & 1 \end{bmatrix}.$$

Because $|A| = 1 \neq 0$, A is invertible. Calculate A^{-1} by Gauss-Jordan elimination

$$A^{-1} = \begin{bmatrix} 1 & 0 & 0 \\ -1 & 1 & 0 \\ 0 & -1 & 1 \end{bmatrix}$$

and conclude that

$$T^{-1}(x_1, x_2, x_3) = (x_1, -x_1 + x_2, -x_2 + x_3).$$

37. (a) The standard matrix for T is $A' = \begin{bmatrix} 1 & 1 \\ 1 & 0 \\ 0 & 1 \end{bmatrix}$

and the image of $\mathbf{v}$ under T is

$$A'\mathbf{v} = \begin{bmatrix} 1 & 1 \\ 1 & 0 \\ 0 & 1 \end{bmatrix}\begin{bmatrix} 5 \\ 4 \end{bmatrix} = \begin{bmatrix} 9 \\ 5 \\ 4 \end{bmatrix}. \text{ So, } T(\mathbf{v}) = (9, 5, 4).$$

(b) The image of each vector in B is as follows.

$$T(1, -1) = (0, 1, -1) = (1, 1, 0) + 0(0, 1, 1) - (1, 0, 1)$$
$$T(0, 1) = (1, 0, 1) = 0(1, 1, 0) + 0(0, 1, 1) + (1, 0, 1)$$

So, $[T(1, -1)]_{B'} = \begin{bmatrix} 1 \\ 0 \\ -1 \end{bmatrix}$ and $[T(0, 1)]_{B'} = \begin{bmatrix} 0 \\ 0 \\ 1 \end{bmatrix}$

which implies that $A = \begin{bmatrix} 1 & 0 \\ 0 & 0 \\ -1 & 1 \end{bmatrix}$. Then, because $[\mathbf{v}]_B = \begin{bmatrix} 5 \\ 9 \end{bmatrix}$, you have $[T(\mathbf{v})]_{B'} = A[\mathbf{v}]_B = \begin{bmatrix} 1 & 0 \\ 0 & 0 \\ -1 & 1 \end{bmatrix}\begin{bmatrix} 5 \\ 9 \end{bmatrix} = \begin{bmatrix} 5 \\ 0 \\ 4 \end{bmatrix}$.

So, $T(\mathbf{v}) = 5(1, 1, 0) + 4(1, 0, 1) = (9, 5, 4)$.

© 2013 Cengage Learning. All Rights Reserved. May not be scanned, copied or duplicated, or posted to a publicly accessible website, in whole or in part.

39. (a) The standard matrix for T is $A' = \begin{bmatrix} 2 & 0 & 0 \\ 1 & 1 & 0 \\ 0 & 1 & 1 \\ 1 & 0 & 1 \end{bmatrix}$ and the image of $\mathbf{v} = (1, -5, 2)$ under T is $A'\mathbf{v} = \begin{bmatrix} 2 & 0 & 0 \\ 1 & 1 & 0 \\ 0 & 1 & 1 \\ 1 & 0 & 1 \end{bmatrix}\begin{bmatrix} 1 \\ -5 \\ 2 \end{bmatrix} = \begin{bmatrix} 2 \\ -4 \\ -3 \\ 3 \end{bmatrix}$.

So, $T(\mathbf{v}) = (2, -4, -3, 3)$.

(b) Because

$T(2, 0, 1) = (4, 2, 1, 3) = 2(1, 0, 0, 1) + (0, 1, 0, 1) + (1, 0, 1, 0) + (1, 1, 0, 0)$

$T(0, 2, 1) = (0, 2, 3, 1) = -2(1, 0, 0, 1) + 3(0, 1, 0, 1) + 3(1, 0, 1, 0) - (1, 1, 0, 0)$

$T(1, 2, 1) = (2, 3, 3, 2) = -(1, 0, 0, 1) + 3(0, 1, 0, 1) + 3(1, 0, 1, 0)$

the matrix for T relative to B and B' is

$$A = \begin{bmatrix} 2 & -2 & -1 \\ 1 & 3 & 3 \\ 1 & 3 & 3 \\ 1 & -1 & 0 \end{bmatrix}.$$

Because $\mathbf{v} = (1, -5, 2) = \frac{9}{2}(2, 0, 1) + \frac{11}{2}(0, 2, 1) - 8(1, 2, 1)$,

$$[T(\mathbf{v})]_{B'} = A[\mathbf{v}]_B = \begin{bmatrix} 2 & -2 & -1 \\ 1 & 3 & 3 \\ 1 & 3 & 3 \\ 1 & -1 & 0 \end{bmatrix}\begin{bmatrix} \frac{9}{2} \\ \frac{11}{2} \\ -8 \end{bmatrix} = \begin{bmatrix} 6 \\ -3 \\ -3 \\ -1 \end{bmatrix}.$$

So, $T(1, -5, 2) = 6(1, 0, 0, 1) - 3(0, 1, 0, 1) - 3(1, 0, 1, 0) - (1, 1, 0, 0) = (2, -4, -3, 3)$.

41. (a) The standard matrix for T is $A' = \begin{bmatrix} 1 & 1 & 1 \\ -1 & 0 & 2 \\ 0 & 2 & -1 \end{bmatrix}$ and the image of $\mathbf{v} = (4, -5, 10)$ under T is

$$A'\mathbf{v} = \begin{bmatrix} 1 & 1 & 1 \\ -1 & 0 & 2 \\ 0 & 2 & -1 \end{bmatrix}\begin{bmatrix} 4 \\ -5 \\ 10 \end{bmatrix} = \begin{bmatrix} 9 \\ 16 \\ -20 \end{bmatrix}.$$

So, $T(\mathbf{v}) = (9, 16, -20)$.

(b) Because

$T(2, 0, 1) = (3, 0, -1) = 2(1, 1, 1) + 1(1, 1, 0) - 3(0, 1, 1)$

$T(0, 2, 1) = (3, 2, 3) = 4(1, 1, 1) - (1, 1, 0) - (0, 1, 1)$

$T(1, 2, 1) = (4, 1, 3) = 6(1, 1, 1) - 2(1, 1, 0) - 3(0, 1, 1)$,

the matrix for T relative to B and B' is $A = \begin{bmatrix} 2 & 4 & 6 \\ 1 & -1 & -2 \\ -3 & -1 & -3 \end{bmatrix}$. Because $\mathbf{v} = (4, -5, 10) = \frac{25}{2}(2, 0, 1) + \frac{37}{2}(0, 2, 1) - 21(1, 2, 1)$,

$[T(\mathbf{v})]_{B'} = A[\mathbf{v}]_B = \begin{bmatrix} 2 & 4 & 6 \\ 1 & -1 & -2 \\ -3 & -1 & -3 \end{bmatrix}\begin{bmatrix} \frac{25}{2} \\ \frac{37}{2} \\ -21 \end{bmatrix} = \begin{bmatrix} -27 \\ 36 \\ 7 \end{bmatrix}$. So, $T(\mathbf{v}) = -27(1, 1, 1) + 36(1, 1, 0) + 7(0, 1, 1) = (9, 16, -20)$.

© 2013 Cengage Learning. All Rights Reserved. May not be scanned, copied or duplicated, or posted to a publicly accessible website, in whole or in part.

43. The image of each vector in B is as follows.

$T(1) = x, \quad T(x) = x^2, \quad T(x^2) = x^3.$

So, the matrix of T relative to B and B' is

$$A = \begin{bmatrix} 0 & 0 & 0 \\ 1 & 0 & 0 \\ 0 & 1 & 0 \\ 0 & 0 & 1 \end{bmatrix}.$$

45. The image of each vector in B is as follows.

$$D_x(1) = 0 = 0(1) + 0x + 0e^x + 0xe^x$$
$$D_x(x) = 1 = 1 + 0x + 0e^x + 0xe^x$$
$$D_x(e^x) = e^x = 0(1) + 0x + e^x + 0xe^x$$
$$D_x(xe^x) = e^x + xe^x = 0(1) + 0x + e^x + xe^x$$

So,

$$[D_x(1)]_B = \begin{bmatrix} 0 \\ 0 \\ 0 \\ 0 \end{bmatrix}, \quad [D_x(x)]_B = \begin{bmatrix} 1 \\ 0 \\ 0 \\ 0 \end{bmatrix},$$

$$[D_x(e^x)]_B = \begin{bmatrix} 0 \\ 0 \\ 1 \\ 0 \end{bmatrix}, \text{ and } [D_x(xe^x)]_B = \begin{bmatrix} 0 \\ 0 \\ 1 \\ 1 \end{bmatrix}$$

which implies that

$$A = \begin{bmatrix} 0 & 1 & 0 & 0 \\ 0 & 0 & 0 & 0 \\ 0 & 0 & 1 & 1 \\ 0 & 0 & 0 & 1 \end{bmatrix}.$$

47. Because $3x - 2xe^x = 0(1) + 3(x) + 0(e^x) - 2(xe^x)$,

$$A[\mathbf{v}]_B = \begin{bmatrix} 0 & 1 & 0 & 0 \\ 0 & 0 & 0 & 0 \\ 0 & 0 & 1 & 1 \\ 0 & 0 & 0 & 1 \end{bmatrix} \begin{bmatrix} 0 \\ 3 \\ 0 \\ -2 \end{bmatrix} = \begin{bmatrix} 3 \\ 0 \\ -2 \\ -2 \end{bmatrix}.$$

So, $D_x(3x - 2xe^x) = 3 - 2e^x - 2xe^x$.

49. (a) The image of each vector in B is as follows.

$$T(1) = \int_0^x dt$$
$$= x = 0(1) + x + 0x^2 + 0x^3 + 0x^4$$
$$T(x) = \int_0^x t\, dt$$
$$= \tfrac{1}{2}x^2 = 0(1) + 0x + \tfrac{1}{2}x^2 + 0x^3 + 0x^4$$
$$T(x^2) = \int_0^x t^2\, dt$$
$$= \tfrac{1}{3}x^3 = 0(1) + 0x + 0x^2 + \tfrac{1}{3}x^3 + 0x^4$$
$$T(x^3) = \int_0^x t^3\, dt$$
$$= \tfrac{1}{4}x^4 = 0(1) + 0(x) + 0x^2 + 0x^3 + \tfrac{1}{4}x^4$$

So,

$$A = \begin{bmatrix} 0 & 0 & 0 & 0 \\ 1 & 0 & 0 & 0 \\ 0 & \frac{1}{2} & 0 & 0 \\ 0 & 0 & \frac{1}{3} & 0 \\ 0 & 0 & 0 & \frac{1}{4} \end{bmatrix}.$$

(b) The image of $p(x) = 6 - 2x + 3x^3$ under T relative to the basis of B' is given by

$$A[p]_B = \begin{bmatrix} 0 & 0 & 0 & 0 \\ 1 & 0 & 0 & 0 \\ 0 & \frac{1}{2} & 0 & 0 \\ 0 & 0 & \frac{1}{3} & 0 \\ 0 & 0 & 0 & \frac{1}{4} \end{bmatrix} \begin{bmatrix} 6 \\ -2 \\ 0 \\ 3 \end{bmatrix} = \begin{bmatrix} 0 \\ 6 \\ -1 \\ 0 \\ \frac{3}{4} \end{bmatrix}.$$

So, $T(p) = 0(1) + 6x - x^2 + 0x^3 + \tfrac{3}{4}x^4$

$$= 6x - x^2 + \tfrac{3}{4}x^4 = \int_0^x p(t)\, dt.$$

© 2013 Cengage Learning. All Rights Reserved. May not be scanned, copied or duplicated, or posted to a publicly accessible website, in whole or in part.

51. (a) The standard basis for $M_{2,3}$ is

$$B = \left\{\begin{bmatrix}1&0&0\\0&0&0\end{bmatrix}, \begin{bmatrix}0&1&0\\0&0&0\end{bmatrix}, \begin{bmatrix}0&0&1\\0&0&0\end{bmatrix}, \begin{bmatrix}0&0&0\\1&0&0\end{bmatrix}, \begin{bmatrix}0&0&0\\0&1&0\end{bmatrix}, \begin{bmatrix}0&0&0\\0&0&1\end{bmatrix}\right\}$$

and the standard basis for $M_{3,2}$ is

$$B' = \left\{\begin{bmatrix}1&0\\0&0\\0&0\end{bmatrix}, \begin{bmatrix}0&1\\0&0\\0&0\end{bmatrix}, \begin{bmatrix}0&0\\1&0\\0&0\end{bmatrix}, \begin{bmatrix}0&0\\0&1\\0&0\end{bmatrix}, \begin{bmatrix}0&0\\0&0\\1&0\end{bmatrix}, \begin{bmatrix}0&0\\0&0\\0&1\end{bmatrix}\right\}.$$

By finding the image of each vector in B, you can find A.

$$T\left(\begin{bmatrix}1&0&0\\0&0&0\end{bmatrix}\right) = \begin{bmatrix}1&0\\0&0\\0&0\end{bmatrix}, \quad T\left(\begin{bmatrix}0&1&0\\0&0&0\end{bmatrix}\right) = \begin{bmatrix}0&0\\1&0\\0&0\end{bmatrix},$$

$$T\left(\begin{bmatrix}0&0&1\\0&0&0\end{bmatrix}\right) = \begin{bmatrix}0&0\\0&0\\1&0\end{bmatrix}, \quad T\left(\begin{bmatrix}0&0&0\\1&0&0\end{bmatrix}\right) = \begin{bmatrix}0&1\\0&0\\0&0\end{bmatrix},$$

$$T\left(\begin{bmatrix}0&0&0\\0&1&0\end{bmatrix}\right) = \begin{bmatrix}0&0\\0&1\\0&0\end{bmatrix}, \quad T\left(\begin{bmatrix}0&0&0\\0&0&1\end{bmatrix}\right) = \begin{bmatrix}0&0\\0&0\\0&1\end{bmatrix}$$

So,

$$A = \begin{bmatrix}1&0&0&0&0&0\\0&0&0&1&0&0\\0&1&0&0&0&0\\0&0&0&0&1&0\\0&0&1&0&0&0\\0&0&0&0&0&1\end{bmatrix}.$$

(b) Because $|T| \neq 0$, T is invertible and T is an isomorphism.

(c) The inverse of T is computed by Gauss-Jordan elimination, applied to A and $A^{-1} = A^T$, so the inverse of T exists and T^{-1} has the standard matrix A^T, and you find that $T^{-1} = T^T$, the transpose of T.

53. (a) True. See Theorem 6.10 on page 314.

(b) False. Let linear transformation $T: R^2 \to R^2$ be given by $T(x, y) = (x - y, -2x + 2y)$. Then the standard matrix for T is

$$\begin{bmatrix}1&-1\\-2&2\end{bmatrix},$$

which is not invertible. So, by Theorem 6.12 on page 393, T is not invertible.

55. Let

$$(T_2 \circ T_1)(\mathbf{u}) = (T_2 \circ T_1)(\mathbf{v})$$
$$T_2(T_1(\mathbf{u})) = T_2(T_1(\mathbf{v}))$$
$$T_1(\mathbf{u}) = T_1(\mathbf{v}) \qquad \text{because } T_2 \text{ one-to-one}$$
$$\mathbf{u} = \mathbf{v} \qquad \text{because } T_1 \text{ one-to-one}$$

Because $T_2 \circ T_1$ is one-to-one from V to V, it is also onto. The inverse is $T_1^{-1} \circ T_2^{-1}$ because

$$(T_2 \circ T_1) \circ (T_1^{-1} \circ T_2^{-1}) = T_2 \circ I \circ T_2^{-1} = I.$$

57. Sometimes it is preferable to use a nonstandard basis. If A_1 and A_2 are the standard matrices for T_1 and T_2 respectively, then the standard matrix for $T_2 \circ T_1$ is A_2A_1 and the standard matrix for $T_1^{-1} \circ T_2^{-1}$ is $A_1^{-1}A_2^{-1}$. Because $(A_2A_1)(A_1^{-1}A_2^{-1}) = A_2(I) = A_2^{-1} = I$, you have that the inverse of $T_2 \circ T_1$ is $T_1^{-1} \circ T_2^{-1}$.

© 2013 Cengage Learning. All Rights Reserved. May not be scanned, copied or duplicated, or posted to a publicly accessible website, in whole or in part.

Section 6.4 Transition Matrices and Similarity

1. The standard matrix for T is

$$A = \begin{bmatrix} 2 & -1 \\ -1 & 1 \end{bmatrix}.$$

Furthermore, the transition matrix P from B' to the standard basis B, and its inverse, are

$$P = \begin{bmatrix} 1 & 0 \\ -2 & 3 \end{bmatrix} \quad \text{and} \quad P^{-1} = \begin{bmatrix} 1 & 0 \\ \frac{2}{3} & \frac{1}{3} \end{bmatrix}.$$

Therefore, the matrix for T relative to B' is

$$A' = P^{-1}AP = \begin{bmatrix} 1 & 0 \\ \frac{2}{3} & \frac{1}{3} \end{bmatrix}\begin{bmatrix} 2 & -1 \\ -1 & 1 \end{bmatrix}\begin{bmatrix} 1 & 0 \\ -2 & 3 \end{bmatrix} = \begin{bmatrix} 4 & -3 \\ \frac{5}{3} & -1 \end{bmatrix}.$$

3. The standard matrix for T is

$$A = \begin{bmatrix} 1 & 1 \\ 0 & 4 \end{bmatrix}.$$

Furthermore, the transition matrix P from B' to the standard basis B, and its inverse, are

$$P = \begin{bmatrix} -4 & 1 \\ 1 & -1 \end{bmatrix} \quad \text{and} \quad P^{-1} = \begin{bmatrix} -\frac{1}{3} & -\frac{1}{3} \\ -\frac{1}{3} & -\frac{4}{3} \end{bmatrix}.$$

Therefore, the matrix for T relative to B' is

$$A' = P^{-1}AP = \begin{bmatrix} -\frac{1}{3} & -\frac{1}{3} \\ -\frac{1}{3} & -\frac{4}{3} \end{bmatrix}\begin{bmatrix} 1 & 1 \\ 0 & 4 \end{bmatrix}\begin{bmatrix} -4 & 1 \\ 1 & -1 \end{bmatrix} = \begin{bmatrix} -\frac{1}{3} & \frac{4}{3} \\ -\frac{13}{3} & \frac{16}{3} \end{bmatrix}.$$

5. The standard matrix for T is

$$A = \begin{bmatrix} 1 & 0 & 0 \\ 0 & 1 & 0 \\ 0 & 0 & 1 \end{bmatrix}.$$

Furthermore, the transition matrix B' to the standard basis B, and its inverse, are

$$P = \begin{bmatrix} 1 & 1 & 0 \\ 1 & 0 & 1 \\ 0 & 1 & 1 \end{bmatrix} \quad \text{and} \quad P^{-1} = \begin{bmatrix} \frac{1}{2} & \frac{1}{2} & -\frac{1}{2} \\ \frac{1}{2} & -\frac{1}{2} & \frac{1}{2} \\ -\frac{1}{2} & \frac{1}{2} & \frac{1}{2} \end{bmatrix}.$$

Therefore, the matrix for T relative to B' is

$$A' = P^{-1}AP = \begin{bmatrix} \frac{1}{2} & \frac{1}{2} & -\frac{1}{2} \\ \frac{1}{2} & -\frac{1}{2} & \frac{1}{2} \\ -\frac{1}{2} & \frac{1}{2} & \frac{1}{2} \end{bmatrix}\begin{bmatrix} 1 & 0 & 0 \\ 0 & 1 & 0 \\ 0 & 0 & 1 \end{bmatrix}\begin{bmatrix} 1 & 1 & 0 \\ 1 & 0 & 1 \\ 0 & 1 & 1 \end{bmatrix} = \begin{bmatrix} 1 & 0 & 0 \\ 0 & 1 & 0 \\ 0 & 0 & 1 \end{bmatrix}.$$

7. The standard matrix for T is

$$A = \begin{bmatrix} 1 & -1 & 2 \\ 2 & 1 & -1 \\ 1 & 2 & 1 \end{bmatrix}.$$

Furthermore, the transition matrix P from B' to the standard basis B, and its inverse, are

$$P = \begin{bmatrix} 1 & 0 & 1 \\ 0 & 2 & 2 \\ 1 & 2 & 0 \end{bmatrix} \quad \text{and} \quad P^{-1} = \begin{bmatrix} \frac{2}{3} & -\frac{1}{3} & \frac{1}{3} \\ -\frac{1}{3} & \frac{1}{6} & \frac{1}{3} \\ \frac{1}{3} & \frac{1}{3} & -\frac{1}{3} \end{bmatrix}.$$

Therefore, the matrix for T relative to B' is

$$A' = P^{-1}AP = \begin{bmatrix} \frac{2}{3} & -\frac{1}{3} & \frac{1}{3} \\ -\frac{1}{3} & \frac{1}{6} & \frac{1}{3} \\ \frac{1}{3} & \frac{1}{3} & -\frac{1}{3} \end{bmatrix}\begin{bmatrix} 1 & -1 & 2 \\ 2 & 1 & -1 \\ 1 & 2 & 1 \end{bmatrix}\begin{bmatrix} 1 & 0 & 1 \\ 0 & 2 & 2 \\ 1 & 2 & 0 \end{bmatrix} = \begin{bmatrix} \frac{7}{3} & \frac{10}{3} & -\frac{1}{3} \\ -\frac{1}{6} & \frac{4}{3} & \frac{8}{3} \\ \frac{2}{3} & -\frac{4}{3} & -\frac{2}{3} \end{bmatrix}.$$

© 2013 Cengage Learning. All Rights Reserved. May not be scanned, copied or duplicated, or posted to a publicly accessible website, in whole or in part.

9. (a) The transition matrix P from B' to B is found by row-reducing $[B \vdots B']$ to $[I \vdots P]$.

$$[B \vdots B'] = \begin{bmatrix} 1 & -2 & \vdots & -12 & -4 \\ 3 & -2 & \vdots & 0 & 4 \end{bmatrix} \Rightarrow [I \vdots P] = \begin{bmatrix} 1 & 0 & \vdots & 6 & 4 \\ 0 & 1 & \vdots & 9 & 4 \end{bmatrix}$$

So,

$$P = \begin{bmatrix} 6 & 4 \\ 9 & 4 \end{bmatrix}.$$

(b) The coordinate matrix for **v** relative to B is $[\mathbf{v}]_B = P[\mathbf{v}]_{B'} = \begin{bmatrix} 6 & 4 \\ 9 & 4 \end{bmatrix}\begin{bmatrix} -1 \\ 2 \end{bmatrix} = \begin{bmatrix} 2 \\ -1 \end{bmatrix}.$

Furthermore, the image of **v** under T relative to B is $[T(\mathbf{v})]_B = A[\mathbf{v}]_B = \begin{bmatrix} 3 & 2 \\ 0 & 4 \end{bmatrix}\begin{bmatrix} 2 \\ -1 \end{bmatrix} = \begin{bmatrix} 4 \\ -4 \end{bmatrix}.$

(c) The inverse of P is $P^{-1} = \begin{bmatrix} -\frac{1}{3} & \frac{1}{3} \\ \frac{3}{4} & -\frac{1}{2} \end{bmatrix}.$

The matrix of T relative to B' is then $A' = P^{-1}AP = \begin{bmatrix} -\frac{1}{3} & \frac{1}{3} \\ \frac{3}{4} & -\frac{1}{2} \end{bmatrix}\begin{bmatrix} 3 & 2 \\ 0 & 4 \end{bmatrix}\begin{bmatrix} 6 & 4 \\ 9 & 4 \end{bmatrix} = \begin{bmatrix} 0 & -\frac{4}{3} \\ 9 & 7 \end{bmatrix}.$

(d) The image of **v** under T relative to B' is $P^{-1}[T(\mathbf{v})]_B = \begin{bmatrix} -\frac{1}{3} & \frac{1}{3} \\ \frac{3}{4} & -\frac{1}{2} \end{bmatrix}\begin{bmatrix} 4 \\ -4 \end{bmatrix} = \begin{bmatrix} -\frac{8}{3} \\ 5 \end{bmatrix}.$

You can also find the image of **v** under T relative to B' by $A'[\mathbf{v}]_{B'} = \begin{bmatrix} 0 & -\frac{4}{3} \\ 9 & 7 \end{bmatrix}\begin{bmatrix} -1 \\ 2 \end{bmatrix} = \begin{bmatrix} -\frac{8}{3} \\ 5 \end{bmatrix}.$

11. (a) The transition matrix P from B' to B is found by row-reducing $[B \vdots B']$ to $[I \vdots P]$.

$$[B \vdots B'] = \begin{bmatrix} 1 & -1 & \vdots & -4 & 0 \\ 2 & -1 & \vdots & 1 & 2 \end{bmatrix} \Rightarrow [I \vdots P] = \begin{bmatrix} 1 & 0 & \vdots & 5 & 2 \\ 0 & 1 & \vdots & 9 & 2 \end{bmatrix}$$

So,

$$P = \begin{bmatrix} 5 & 2 \\ 9 & 2 \end{bmatrix}.$$

(b) The coordinate matrix for **v** relative to B is $[\mathbf{v}]_B = P[\mathbf{v}]_{B'} = \begin{bmatrix} 5 & 2 \\ 9 & 2 \end{bmatrix}\begin{bmatrix} -1 \\ 4 \end{bmatrix} = \begin{bmatrix} 3 \\ -1 \end{bmatrix}.$

Furthermore, the image of **v** under T relative to B is $[T(\mathbf{v})]_B = A[\mathbf{v}]_B = \begin{bmatrix} 2 & 1 \\ 0 & -1 \end{bmatrix}\begin{bmatrix} 3 \\ -1 \end{bmatrix} = \begin{bmatrix} 5 \\ 1 \end{bmatrix}.$

(c) The matrix of T relative to B' is $A' = P^{-1}AP = \begin{bmatrix} -\frac{1}{4} & \frac{1}{4} \\ \frac{9}{8} & -\frac{5}{8} \end{bmatrix}\begin{bmatrix} 2 & 1 \\ 0 & -1 \end{bmatrix}\begin{bmatrix} 5 & 2 \\ 9 & 2 \end{bmatrix} = \begin{bmatrix} -7 & -2 \\ 27 & 8 \end{bmatrix}.$

(d) The image of **v** under T relative to B' is $P^{-1}[T(\mathbf{v})]_B = \begin{bmatrix} -\frac{1}{4} & \frac{1}{4} \\ \frac{9}{8} & -\frac{5}{8} \end{bmatrix}\begin{bmatrix} 5 \\ 1 \end{bmatrix} = \begin{bmatrix} -1 \\ 5 \end{bmatrix}.$

You can also find the image of **v** under T relative to B' by $A'[\mathbf{v}]_{B'} = \begin{bmatrix} -7 & -2 \\ 27 & 8 \end{bmatrix}\begin{bmatrix} -1 \\ 4 \end{bmatrix} = \begin{bmatrix} -1 \\ 5 \end{bmatrix}.$

© 2013 Cengage Learning. All Rights Reserved. May not be scanned, copied or duplicated, or posted to a publicly accessible website, in whole or in part.

13. (a) The transition matrix P from B' to B is found by row-reducing $[B \vdots B']$ to $[I \vdots P]$.

$$[B \vdots B'] = \begin{bmatrix} 1 & 1 & 0 & \vdots & 1 & 0 & 0 \\ 1 & 0 & 1 & \vdots & 0 & 1 & 0 \\ 0 & 1 & 1 & \vdots & 0 & 0 & 1 \end{bmatrix} \Rightarrow [I \vdots P] = \begin{bmatrix} 1 & 0 & 0 & \vdots & \frac{1}{2} & \frac{1}{2} & -\frac{1}{2} \\ 0 & 1 & 0 & \vdots & \frac{1}{2} & -\frac{1}{2} & \frac{1}{2} \\ 0 & 0 & 1 & \vdots & -\frac{1}{2} & \frac{1}{2} & \frac{1}{2} \end{bmatrix}$$

So,

$$P = \frac{1}{2}\begin{bmatrix} 1 & 1 & -1 \\ 1 & -1 & 1 \\ -1 & 1 & 1 \end{bmatrix}.$$

(b) The coordinate matrix for $\mathbf{v}$ relative to B is $[\mathbf{v}]_B = P[\mathbf{v}]_{B'} = \frac{1}{2}\begin{bmatrix} 1 & 1 & -1 \\ 1 & -1 & 1 \\ -1 & 1 & 1 \end{bmatrix}\begin{bmatrix} 1 \\ 0 \\ -1 \end{bmatrix} = \begin{bmatrix} 1 \\ 0 \\ -1 \end{bmatrix}.$

Furthermore, the image of $\mathbf{v}$ under T relative to B is $[T(\mathbf{v})]_B = A[\mathbf{v}]_B = \begin{bmatrix} \frac{3}{2} & -1 & -\frac{1}{2} \\ -\frac{1}{2} & 2 & \frac{1}{2} \\ \frac{1}{2} & 1 & \frac{5}{2} \end{bmatrix}\begin{bmatrix} 1 \\ 0 \\ -1 \end{bmatrix} = \begin{bmatrix} 2 \\ -1 \\ -2 \end{bmatrix}.$

(c) The matrix of T relative to B' is $A' = P^{-1}AP = \begin{bmatrix} 1 & 1 & 0 \\ 1 & 0 & 1 \\ 0 & 1 & 1 \end{bmatrix}\begin{bmatrix} \frac{3}{2} & -1 & -\frac{1}{2} \\ -\frac{1}{2} & 2 & \frac{1}{2} \\ \frac{1}{2} & 1 & \frac{5}{2} \end{bmatrix}\begin{bmatrix} \frac{1}{2} & \frac{1}{2} & -\frac{1}{2} \\ \frac{1}{2} & -\frac{1}{2} & \frac{1}{2} \\ -\frac{1}{2} & \frac{1}{2} & \frac{1}{2} \end{bmatrix} = \begin{bmatrix} 1 & 0 & 0 \\ 0 & 2 & 0 \\ 0 & 0 & 3 \end{bmatrix}.$

(d) The image of $\mathbf{v}$ under T relative to B' is $P^{-1}[T(\mathbf{v})]_B = \begin{bmatrix} 1 & 1 & 0 \\ 1 & 0 & 1 \\ 0 & 1 & 1 \end{bmatrix}\begin{bmatrix} 2 \\ -1 \\ -2 \end{bmatrix} = \begin{bmatrix} 1 \\ 0 \\ -3 \end{bmatrix}.$

You can also find the image of $\mathbf{v}$ under T relative to B' by $A'[\mathbf{v}]_{B'} = \begin{bmatrix} 1 & 0 & 0 \\ 0 & 2 & 0 \\ 0 & 0 & 3 \end{bmatrix}\begin{bmatrix} 1 \\ 0 \\ -1 \end{bmatrix} = \begin{bmatrix} 1 \\ 0 \\ -3 \end{bmatrix}.$

15. A and A' are similar since

$$A' = P^{-1}AP = \begin{bmatrix} -2 & -1 \\ 1 & 1 \end{bmatrix}\begin{bmatrix} 12 & 7 \\ -20 & -11 \end{bmatrix}\begin{bmatrix} -1 & -1 \\ 1 & 2 \end{bmatrix} = \begin{bmatrix} 1 & -2 \\ 4 & 0 \end{bmatrix}.$$

17. A and A' are similar since

$$A' = P^{-1}AP = \begin{bmatrix} \frac{1}{5} & 0 & 0 \\ 0 & \frac{1}{4} & 0 \\ 0 & 0 & \frac{1}{3} \end{bmatrix}\begin{bmatrix} 5 & 10 & 0 \\ 8 & 4 & 0 \\ 0 & 9 & 6 \end{bmatrix}\begin{bmatrix} 5 & 0 & 0 \\ 0 & 4 & 0 \\ 0 & 0 & 3 \end{bmatrix} = \begin{bmatrix} 5 & 8 & 0 \\ 10 & 4 & 0 \\ 0 & 12 & 6 \end{bmatrix}.$$

19. The transition matrix from B' to the standard matrix has columns consisting of the vectors in B'.

$$P = \begin{bmatrix} -1 & 2 & 0 \\ 1 & 1 & 0 \\ 0 & 0 & 1 \end{bmatrix}$$

and it follows that

$$P^{-1} = \begin{bmatrix} -\frac{1}{3} & \frac{2}{3} & 0 \\ \frac{1}{3} & \frac{1}{3} & 0 \\ 0 & 0 & 1 \end{bmatrix}.$$

So, the matrix for T relative to B' is

$$A' = \rho^{-1}AP = \begin{bmatrix} -\frac{1}{3} & \frac{2}{3} & 0 \\ \frac{1}{3} & \frac{1}{3} & 0 \\ 0 & 0 & 1 \end{bmatrix}\begin{bmatrix} 0 & 2 & 0 \\ 1 & -1 & 0 \\ 0 & 0 & 1 \end{bmatrix}\begin{bmatrix} -1 & 2 & 0 \\ 1 & 1 & 0 \\ 0 & 0 & 1 \end{bmatrix} = \begin{bmatrix} -2 & 0 & 0 \\ 0 & 1 & 0 \\ 0 & 0 & 1 \end{bmatrix}.$$

21. If A and B are similar, then $B = P^{-1}AP$, for some nonsingular matrix P. So,

$|B| = |P^{-1}AP| = |P^{-1}||A||P| = |A|\frac{1}{|P|}|P| = |A|$. No, the converse is not true. For example, $\begin{vmatrix} 1 & 1 \\ 0 & 1 \end{vmatrix} = \begin{vmatrix} 1 & 0 \\ 0 & 1 \end{vmatrix}$, but these matrices are not similar.

© 2013 Cengage Learning. All Rights Reserved. May not be scanned, copied or duplicated, or posted to a publicly accessible website, in whole or in part.

23. $B = P^{-1}AP \Rightarrow$

$$B^k = \left(P^{-1}AP\right)^k = \left(P^{-1}AP\right)\left(P^{-1}AP\right)\cdots\left(P^{-1}AP\right) \qquad (k \text{ times})$$
$$= P^{-1}A^kP.$$

25. Let A be an $n \times n$ matrix similar to I_n. Then there exists an invertible matrix P such that

$A = P^{-1}I_nP = P^{-1}P = I_n$. So, I_n is similar only to itself.

27. Let $A^2 = O$ and $B = P^{-1}AP$, then

$$B^2 = \left(P^{-1}AP\right)^2 = \left(P^{-1}AP\right)\left(P^{-1}AP\right) = P^{-1}A^2P = P^{-1}OP = O.$$

29. If A is similar to B, then $B = P^{-1}AP$. If B is similar to C, then $C = Q^{-1}BQ$.

So, $C = Q^{-1}BQ = Q^{-1}\left(P^{-1}AP\right)Q = (PQ)^{-1}A(PQ)$, which shows that A is similar to C.

31. $B = P^{-1}AP \Rightarrow B^T = \left(P^{-1}AP\right)^T = P^TA^T\left(P^{-1}\right)^T = P^TA^T\left(P^T\right)^{-1}$, which shows that A^T and B^T are similar.

33. If $A = CD$ and C is nonsingular, then $C^{-1}A = D \Rightarrow C^{-1}AC = DC$, which shows that DC is similar to A.

35. The matrix for I relative to B and B' is the square matrix whose columns are the coordinates of $\mathbf{v}_1, \ldots \mathbf{v}_n$ relative to the standard basis. The matrix for I relative to B, or relative to B', is the identity matrix.

37. (a) True. See page 400.

(b) False. If T is a linear transformation with matrices A and A' relative to bases B and B' respectively, then $A' = P^{-1}AP$, where P is the transition matrix from B' to B. Therefore, two matrices representing the *same* linear transformation must be similar.

Section 6.5 Applications of Linear Transformations

1. The standard matrix for T is

$$A = \begin{bmatrix} 1 & 0 \\ 0 & -1 \end{bmatrix}.$$

(a) $\begin{bmatrix} 1 & 0 \\ 0 & -1 \end{bmatrix}\begin{bmatrix} 3 \\ 5 \end{bmatrix} = \begin{bmatrix} 3 \\ -5 \end{bmatrix} \Rightarrow T(3, 5) = (3, -5)$

(b) $\begin{bmatrix} 1 & 0 \\ 0 & -1 \end{bmatrix}\begin{bmatrix} 2 \\ -1 \end{bmatrix} = \begin{bmatrix} 2 \\ 1 \end{bmatrix} \Rightarrow T(2, -1) = (2, 1)$

(c) $\begin{bmatrix} 1 & 0 \\ 0 & -1 \end{bmatrix}\begin{bmatrix} a \\ 0 \end{bmatrix} = \begin{bmatrix} a \\ 0 \end{bmatrix} \Rightarrow T(a, 0) = (a, 0)$

(d) $\begin{bmatrix} 1 & 0 \\ 0 & -1 \end{bmatrix}\begin{bmatrix} 0 \\ b \end{bmatrix} = \begin{bmatrix} 0 \\ -b \end{bmatrix} \Rightarrow T(0, b) = (0, -b)$

(e) $\begin{bmatrix} 1 & 0 \\ 0 & -1 \end{bmatrix}\begin{bmatrix} -c \\ d \end{bmatrix} = \begin{bmatrix} -c \\ -d \end{bmatrix} \Rightarrow T(-c, d) = (-c, -d)$

(f) $\begin{bmatrix} 1 & 0 \\ 0 & -1 \end{bmatrix}\begin{bmatrix} f \\ -g \end{bmatrix} = \begin{bmatrix} f \\ g \end{bmatrix} \Rightarrow T(f, -g) = (f, g)$

3. The standard matrix for T is

$$A = \begin{bmatrix} 0 & 1 \\ 1 & 0 \end{bmatrix}.$$

(a) $\begin{bmatrix} 0 & 1 \\ 1 & 0 \end{bmatrix}\begin{bmatrix} 0 \\ 1 \end{bmatrix} = \begin{bmatrix} 1 \\ 0 \end{bmatrix} \Rightarrow T(0, 1) = (1, 0)$

(b) $\begin{bmatrix} 0 & 1 \\ 1 & 0 \end{bmatrix}\begin{bmatrix} -1 \\ 3 \end{bmatrix} = \begin{bmatrix} 3 \\ -1 \end{bmatrix} \Rightarrow T(-1, 3) = (3, -1)$

(c) $\begin{bmatrix} 0 & 1 \\ 1 & 0 \end{bmatrix}\begin{bmatrix} a \\ 0 \end{bmatrix} = \begin{bmatrix} 0 \\ a \end{bmatrix} \Rightarrow T(a, 0) = (0, a)$

(d) $\begin{bmatrix} 0 & 1 \\ 1 & 0 \end{bmatrix}\begin{bmatrix} 0 \\ b \end{bmatrix} = \begin{bmatrix} b \\ 0 \end{bmatrix} \Rightarrow T(0, b) = (b, 0)$

(e) $\begin{bmatrix} 0 & 1 \\ 1 & 0 \end{bmatrix}\begin{bmatrix} -c \\ d \end{bmatrix} = \begin{bmatrix} d \\ -c \end{bmatrix} \Rightarrow T(-c, d) = (d, -c)$

(f) $\begin{bmatrix} 0 & 1 \\ 1 & 0 \end{bmatrix}\begin{bmatrix} f \\ -g \end{bmatrix} = \begin{bmatrix} -g \\ f \end{bmatrix} \Rightarrow T(f, -g) = (-g, f)$

© 2013 Cengage Learning. All Rights Reserved. May not be scanned, copied or duplicated, or posted to a publicly accessible website, in whole or in part.

5. (a) $T(x, y) = xT(1, 0) + yT(0, 1)$

$= x(2, 0) + y(0, 1) = (2x, y).$

(b) T is a horizontal expansion.

7. (a) Identify T as a vertical contraction from its standard matrix.

$$A = \begin{bmatrix} 1 & 0 \\ 0 & \frac{1}{2} \end{bmatrix}$$

(b)

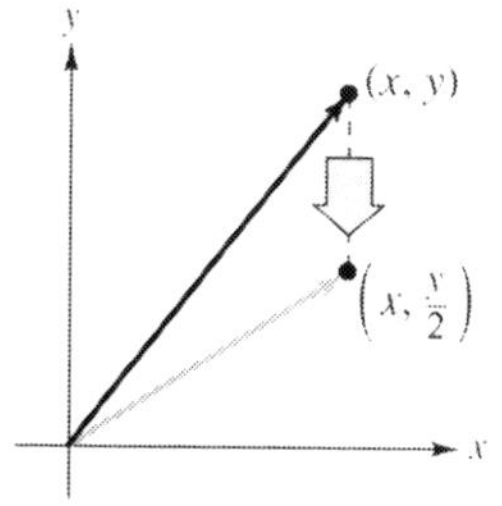

9. (a) Identify T as a horizontal expansion from its standard matrix

$$A = \begin{bmatrix} 4 & 0 \\ 0 & 1 \end{bmatrix}.$$

(b)

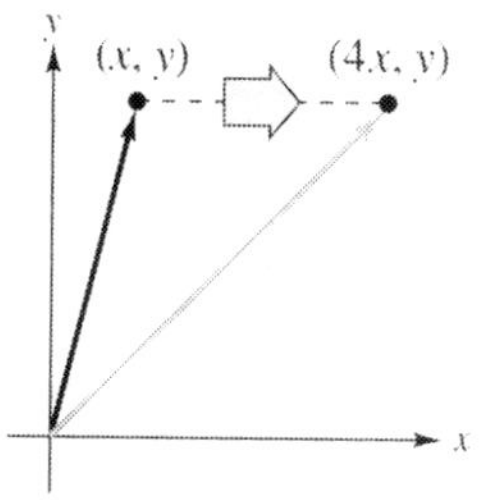

11. (a) Identify T as a horizontal shear from its matrix.

$$A = \begin{bmatrix} 1 & 3 \\ 0 & 1 \end{bmatrix}.$$

(b)

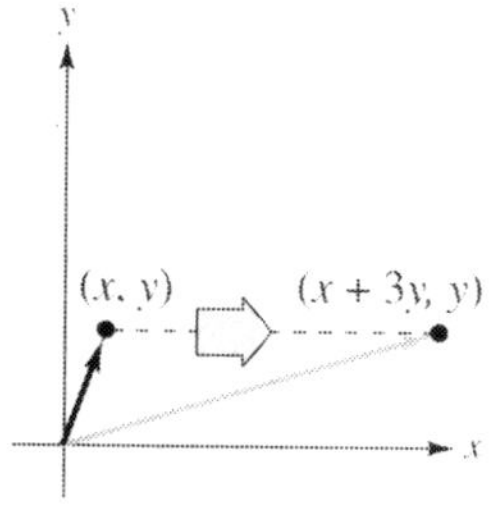

13. (a) Identify T is a vertical shear from its matrix

$$A = \begin{bmatrix} 1 & 0 \\ 5 & 1 \end{bmatrix}.$$

(b)

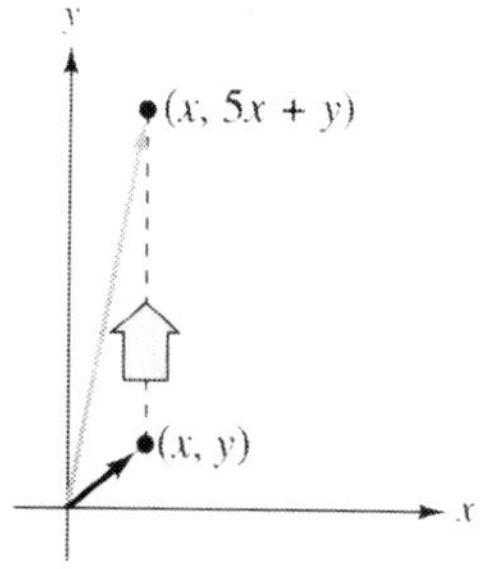

15. The reflection in the y-axis is given by $T(x, y) = (-x, y)$. If (x, y) is a fixed point, then $T(x, y) = (x, y) = (-x, y)$ which implies that $x = 0$. So the set of fixed points is $\{(0, t) : t \in R\}$.

17. The reflection in the line $y = x$ is given by $T(x, y) = (y, x)$. If (x, y) is a fixed point, then $T(x, y) = (x, y) = (y, x)$ which implies that $x = y$. So, the set of fixed points is $\{(t, t) : t \in R\}$.

19. A vertical contraction has the standard matrix $(k < 1)$

$$\begin{bmatrix} 1 & 0 \\ 0 & k \end{bmatrix}.$$

A fixed point of T satisfies the equation

$$T(\mathbf{v}) = \begin{bmatrix} 1 & 0 \\ 0 & k \end{bmatrix}\begin{bmatrix} v_1 \\ v_2 \end{bmatrix} = \begin{bmatrix} v_1 \\ kv_2 \end{bmatrix} = \begin{bmatrix} v_1 \\ v_2 \end{bmatrix} = \mathbf{v}.$$

So, the set of fixed points is $\{(t, 0) : t \text{ is a real number}\}$.

21. A horizontal shear has the form $T(x, y) = (x + ky, y)$. If (x, y) is a fixed point, then $T(x, y) = (x, y) = (x + ky, y)$ which implies that $y = 0$. So, the set of fixed points is $\{(t, 0) : t \in R\}$.

23. Find the image of each vertex under $T(x, y) = (x, -y)$.

$T(0, 0) = (0, 0)$, $\quad T(1, 0) = (1, 0)$,

$T(1, 1) = (1, -1)$, $\quad T(0, 1) = (0, -1)$

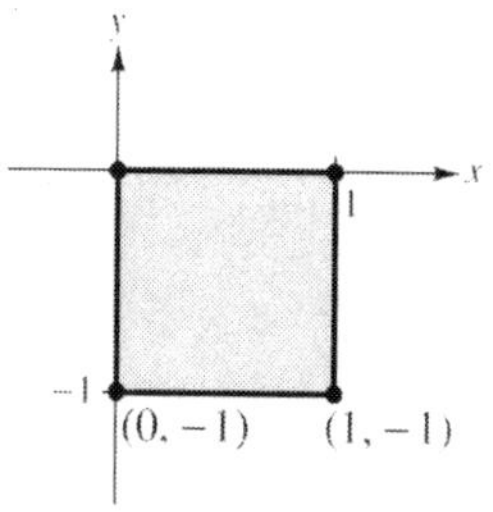

© 2013 Cengage Learning. All Rights Reserved. May not be scanned, copied or duplicated, or posted to a publicly accessible website, in whole or in part.

25. Find the image of each vertex under $T(x, y) = \left(\frac{x}{2}, y\right)$.

$T(0, 0) = (0, 0)$, $\quad T(1, 0) = \left(\frac{1}{2}, 0\right)$,

$T(1, 1) = \left(\frac{1}{2}, 1\right)$, $\quad T(0, 1) = (0, 1)$

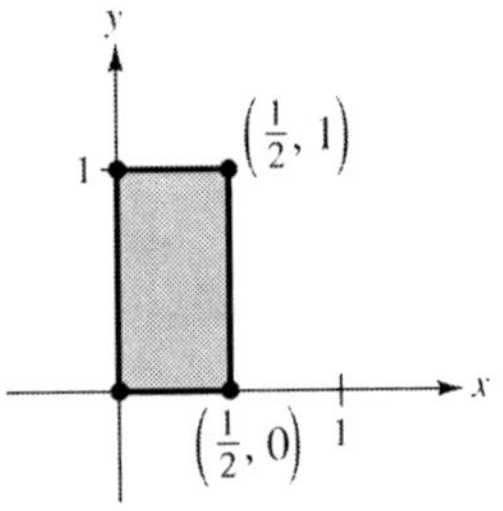

27. Find the image of each vertex under $T(x, y) = (x + 2y, y)$.

$T(0, 0) = (0, 0)$, $\quad T(1, 0) = (1, 0)$,
$T(1, 1) = (3, 1)$, $\quad T(0, 1) = (2, 1)$

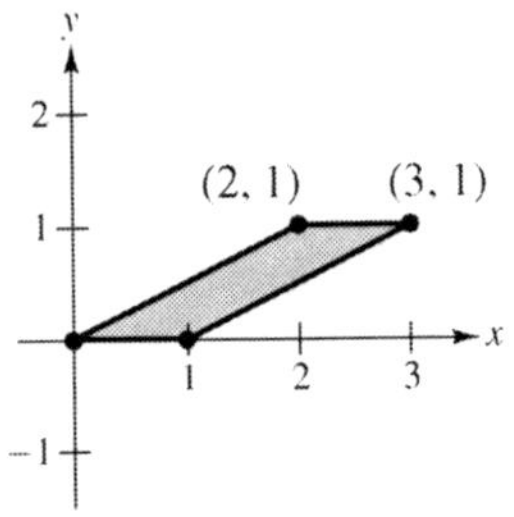

29. Find the image of each vertex under $T(x, y) = (-x, y)$.

$T(0, 0) = (0, 0)$, $\quad T(0, 2) = (0, 2)$,
$T(1, 2) = (-1, 2)$, $\quad T(1, 0) = (-1, 0)$

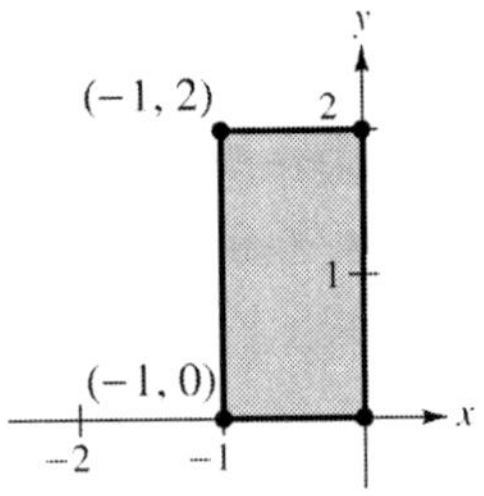

31. Find the image of each vertex under $T(x, y) = \left(x, \frac{1}{2}y\right)$.

$T(0, 0) = (0, 0)$, $\quad T(0, 2) = (0, 1)$,
$T(1, 2) = (1, 1)$, $\quad T(1, 0) = (1, 0)$

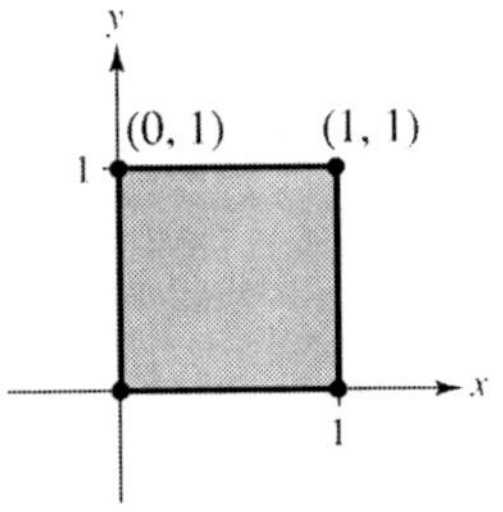

33. Find the image of each vertex under
$T(x, y) = (x + y, y)$.

$T(0, 0) = (0, 0)$, $\quad T(0, 2) = (2, 2)$,
$T(1, 2) = (3, 2)$, $\quad T(1, 0) = (1, 0)$

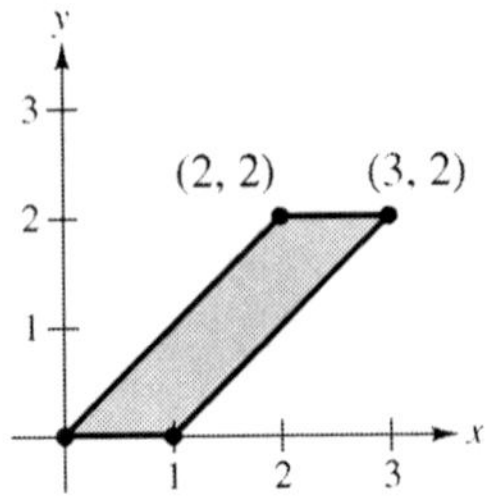

35. Find the image of each vertex under
$T(x, y) = (x + y, y)$.

(a) $T(0, 0) = (0, 0)$, $\quad T(1, 2) = (3, 2)$,
$T(3, 6) = (9, 6)$, $\quad T(5, 2) = (7, 2)$,
$T(6, 0) = (6, 0)$

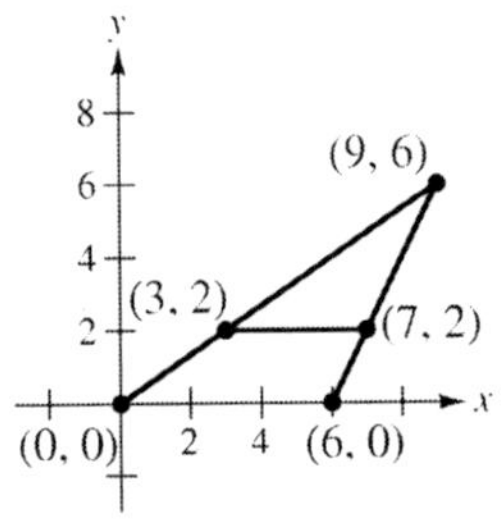

(b) $T(0, 0) = (0, 0)$ $\quad T(0, 6) = (6, 6)$
$T(6, 6) = (12, 6)$ $\quad T(6, 0) = (6, 0)$

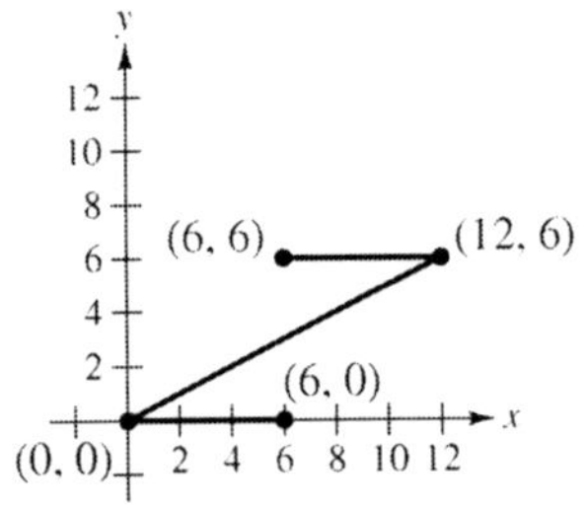

37. Find the image of each vertex under
$T(x, y) = \left(2x, \frac{1}{2}y\right)$.

(a) $T(0, 0) = (0, 0)$ $\quad T(1, 2) = (2, 1)$
$T(3, 6) = (6, 3)$ $\quad T(5, 2) = (10, 1)$
$T(6, 0) = (12, 0)$

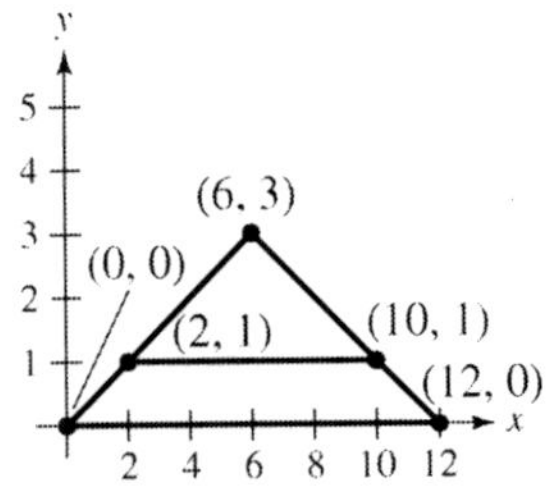

© 2013 Cengage Learning. All Rights Reserved. May not be scanned, copied or duplicated, or posted to a publicly accessible website, in whole or in part.

(b) $T(0, 0) = (0, 0)$ $\quad T(0, 6) = (0, 3)$

$T(6, 6) = (12, 3)$ $\quad T(6, 0) = (12, 0)$

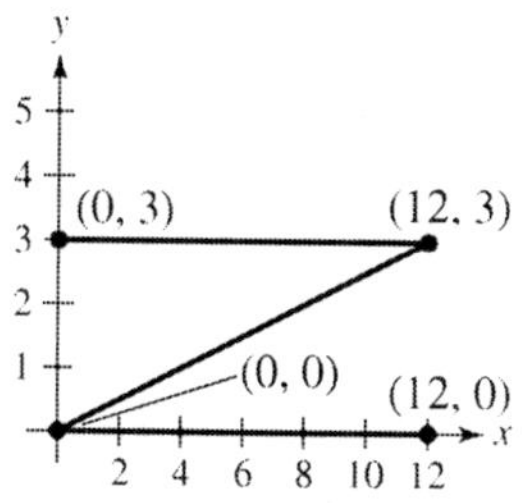

39. The images of the given vectors are as follows.

$$T(1, 0) = \begin{bmatrix} 2 & 0 \\ 0 & 3 \end{bmatrix}\begin{bmatrix} 1 \\ 0 \end{bmatrix} = \begin{bmatrix} 2 \\ 0 \end{bmatrix} = (2, 0)$$

$$T(0, 1) = \begin{bmatrix} 2 & 0 \\ 0 & 3 \end{bmatrix}\begin{bmatrix} 0 \\ 1 \end{bmatrix} = \begin{bmatrix} 0 \\ 3 \end{bmatrix} = (0, 3)$$

$$T(2, 2) = \begin{bmatrix} 2 & 0 \\ 0 & 3 \end{bmatrix}\begin{bmatrix} 2 \\ 2 \end{bmatrix} = \begin{bmatrix} 4 \\ 6 \end{bmatrix} = (4, 6)$$

The two triangles are shown in the following figure.

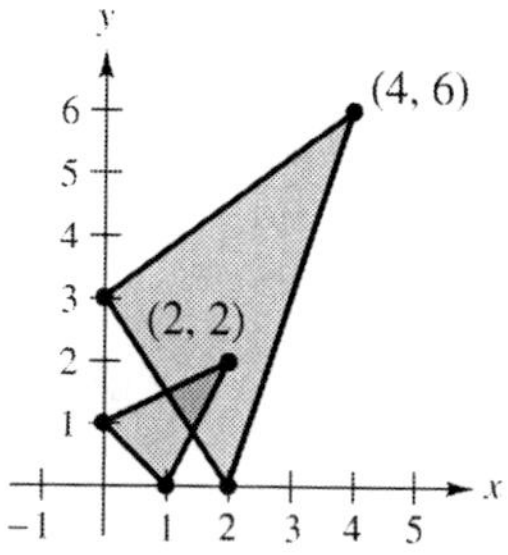

41. The linear transformation defined by A is a horizontal expansion.

43. The linear transformation defined by A is a horizontal shear.

45. The linear transformation defined by A is a reflection in the x-axis followed by a vertical expansion.

47. Because

$$\begin{bmatrix} 1 & 0 \\ 2 & 1 \end{bmatrix}$$

represents a vertical shear and

$$\begin{bmatrix} 2 & 0 \\ 0 & 1 \end{bmatrix}$$

represents a horizontal expansion, A is a vertical shear *followed* by a horizontal expansion.

49. A rotation of 30° about the z-axis is given by the matrix

$$A = \begin{bmatrix} \cos 30° & -\sin 30° & 0 \\ \sin 30° & \cos 30° & 0 \\ 0 & 0 & 1 \end{bmatrix} = \begin{bmatrix} \frac{\sqrt{3}}{2} & -\frac{1}{2} & 0 \\ \frac{1}{2} & \frac{\sqrt{3}}{2} & 0 \\ 0 & 0 & 1 \end{bmatrix}.$$

51. A rotation of 60° about the y-axis is given by the matrix

$$A = \begin{bmatrix} \cos 60° & 0 & \sin 60° \\ 0 & 1 & 0 \\ -\sin 60° & 0 & \cos 60° \end{bmatrix} = \begin{bmatrix} \frac{1}{2} & 0 & \frac{\sqrt{3}}{2} \\ 0 & 1 & 0 \\ -\frac{\sqrt{3}}{2} & 0 & \frac{1}{2} \end{bmatrix}.$$

53. Using the matrix obtained in Exercise 49,

$$T(1, 1, 1) = \begin{bmatrix} \frac{\sqrt{3}}{2} & -\frac{1}{2} & 0 \\ \frac{1}{2} & \frac{\sqrt{3}}{2} & 0 \\ 0 & 0 & 1 \end{bmatrix}\begin{bmatrix} 1 \\ 1 \\ 1 \end{bmatrix} = \begin{bmatrix} \frac{\left(\sqrt{3} - 1\right)}{2} \\ \frac{\left(1 + \sqrt{3}\right)}{2} \\ 1 \end{bmatrix}.$$

55. Using the matrix obtained in Exercise 51,

$$T(1, 1, 1) = \begin{bmatrix} \frac{1}{2} & 0 & \frac{\sqrt{3}}{2} \\ 0 & 1 & 0 \\ -\frac{\sqrt{3}}{2} & 0 & \frac{1}{2} \end{bmatrix}\begin{bmatrix} 1 \\ 1 \\ 1 \end{bmatrix} = \begin{bmatrix} \frac{1 + \sqrt{3}}{2} \\ 1 \\ \frac{1 - \sqrt{3}}{2} \end{bmatrix}.$$

57. The indicated tetrahedron is produced by a 90° rotation about the x-axis.

59. The indicated tetrahedron is produced by a 180° rotation about the y-axis.

61. The indicated tetrahedron is produced by a 90° rotation about the z-axis.

63. The matrix is $\begin{bmatrix} 0 & 0 & 1 \\ 0 & 1 & 0 \\ -1 & 0 & 0 \end{bmatrix}\begin{bmatrix} 1 & 0 & 0 \\ 0 & 0 & -1 \\ 0 & 1 & 0 \end{bmatrix} = \begin{bmatrix} 0 & 1 & 0 \\ 0 & 0 & -1 \\ -1 & 0 & 0 \end{bmatrix}.$

$$T(1, 1, 1) = (1, -1, -1)$$

© 2013 Cengage Learning. All Rights Reserved. May not be scanned, copied or duplicated, or posted to a publicly accessible website, in whole or in part.

65. The matrix is $\begin{bmatrix} \cos 60° & 0 & \sin 60° \\ 0 & 1 & 0 \\ -\sin 60° & 0 & \cos 60° \end{bmatrix}\begin{bmatrix} \cos 30° & -\sin 30° & 0 \\ \sin 30° & \cos 30° & 0 \\ 0 & 0 & 1 \end{bmatrix} = \begin{bmatrix} \frac{1}{2} & 0 & \frac{\sqrt{3}}{2} \\ 0 & 1 & 0 \\ -\frac{\sqrt{3}}{2} & 0 & \frac{1}{2} \end{bmatrix}\begin{bmatrix} \frac{\sqrt{3}}{2} & -\frac{1}{2} & 0 \\ \frac{1}{2} & \frac{\sqrt{3}}{2} & 0 \\ 0 & 0 & 1 \end{bmatrix} = \begin{bmatrix} \frac{\sqrt{3}}{4} & -\frac{1}{4} & \frac{\sqrt{3}}{2} \\ \frac{1}{2} & \frac{\sqrt{3}}{2} & 0 \\ -\frac{3}{4} & \frac{\sqrt{3}}{4} & \frac{1}{2} \end{bmatrix}.$

$$T(1, 1, 1) = \left(\frac{3\sqrt{3} - 1}{4}, \frac{\sqrt{3} + 1}{2}, \frac{\sqrt{3} - 1}{4}\right)$$

67. The matrix is

$$\begin{bmatrix} \cos 135° & -\sin 135° & 0 \\ \sin 135° & \cos 135° & 0 \\ 0 & 0 & 1 \end{bmatrix}\begin{bmatrix} 1 & 0 & 0 \\ 0 & \cos 120° & -\sin 120° \\ 0 & \sin 120° & \cos 120° \end{bmatrix} = \begin{bmatrix} -\frac{\sqrt{2}}{2} & -\frac{\sqrt{2}}{2} & 0 \\ \frac{\sqrt{2}}{2} & -\frac{\sqrt{2}}{2} & 0 \\ 0 & 0 & 1 \end{bmatrix}\begin{bmatrix} 1 & 0 & 0 \\ 0 & -\frac{1}{2} & -\frac{\sqrt{3}}{2} \\ 0 & \frac{\sqrt{3}}{2} & -\frac{1}{2} \end{bmatrix} = \begin{bmatrix} -\frac{\sqrt{2}}{2} & \frac{\sqrt{2}}{4} & \frac{\sqrt{6}}{4} \\ \frac{\sqrt{2}}{2} & \frac{\sqrt{2}}{4} & \frac{\sqrt{6}}{4} \\ 0 & \frac{\sqrt{3}}{2} & -\frac{1}{2} \end{bmatrix}.$$

$$T(1, 1, 1) = \left(\frac{\sqrt{6} - \sqrt{2}}{4}, \frac{\sqrt{6} + 3\sqrt{2}}{4}, \frac{\sqrt{3} - 1}{2}\right)$$

Review Exercises for Chapter 6

1. (a) $T(\mathbf{v}) = T(2, -3) = (2, -4)$

(b) The preimage of **w** is given by solving the equation

$T(v_1, v_2) = (v_1, v_1 + 2v_2) = (4, 12).$

The resulting system of linear equations

$$\begin{aligned} v_1 \qquad &= 4 \\ v_1 + 2v_2 &= 12 \end{aligned}$$

has the solution $v_1 = v_2 = 4$. So, the preimage of **w** is $(4, 4)$.

3. (a) $T(\mathbf{v}) = T(-3, 2, 5) = (0, -1, 7).$

(b) The preimage of **w** is given by solving the equation $T(v_1, v_2, v_3) = (0, v_1 + v_2, v_2 + v_3) = (0, 2, 5).$

The resulting system of linear equations has the solution of the form $v_1 = t - 3, v_2 = 5 - t, v_3 = t$, where t is any real number. So, the preimage of **w** is $\{(t - 3, 5 - t, t) : t \in R\}$.

5. T preserves addition.

$$\begin{aligned} T(x_1, x_2) + T(y_1, y_2) &= (x_1 + 2x_2, -x_1 - x_2) + (y_1 + 2y_2, -y_1 - y_2) \\ &= (x_1 + 2x_2 + y_1 + 2y_2, -x_1 - x_2 - y_1 - y_2) \\ &= ((x_1 + y_1) + 2(x_2 + y_2), -(x_1 + y_1) - (x_2 + y_2)) \\ &= T(x_1 + y_1, x_2 + y_2) \end{aligned}$$

T preserves scalar multiplication.

$$cT(x_1, x_2) = c(x_1 + 2x_2, -x_1 - x_2) = (cx_1 + 2(cx_2), -cx_1 - cx_2) = T(cx_1, cx_2)$$

So, T is a linear transformation with standard matrix $A = \begin{bmatrix} 1 & 2 \\ -1 & -1 \end{bmatrix}$.

© 2013 Cengage Learning. All Rights Reserved. May not be scanned, copied or duplicated, or posted to a publicly accessible website, in whole or in part.

7. T preserves addition.

$$\begin{aligned}T(x_1, y_1) + T(x_2, y_2) &= (x_1 - 2y_1, 2y_1 - x_1) + (x_2 - 2y_2, 2y_2 - x_2)\\ &= x_1 - 2y_1 + x_2 - 2y_2, 2y_1 - x_1 + 2y_2 - x_2\\ &= (x_1 + x_2) - 2(y_1 + y_2), 2(y_1 + y_2) - (x_1 + x_2)\\ &= T(x_1 + y_1, x_2 + y_2)\end{aligned}$$

T preserves scalar multiplication.

$$cT(x, y) = c(x - 2y, 2y - x) = cx - 2cy, 2cy - cx = T(cx, cy)$$

So, T is a linear transformation with standard matrix $A = \begin{bmatrix} 1 & -2 \\ -1 & 2 \end{bmatrix}$.

9. T does not preserve addition or scalar multiplication, so, T is *not* a linear transformation. A counterexample is $T(1, 0) + T(0, 1) = (1 + h, k) + (h, 1 + k) = (1 + 2h, 1 + 2k) \neq T(1 + h, 1 + k) = T(1, 1)$.

11. T preserves addition.

$$\begin{aligned}&T(x_1, x_2, x_3) + T(y_1, y_2, y_3)\\ &= (x_1 - x_2, x_2 - x_3, x_3 - x_1) + (y_1 - y_2, y_2 - y_3, y_3 - y_1)\\ &= (x_1 - x_2 + y_1 - y_2, x_2 - x_3 + y_2 - y_3, x_3 - x_1 + y_3 - y_1)\\ &= ((x_1 + y_1) - (x_2 + y_2), (x_2 + y_2) - (x_3 + y_3), (x_3 + y_3) - (x_1 + y_1))\\ &= T(x_1 + y_1, x_2 + y_2, x_3 + y_3)\end{aligned}$$

T preserves scalar multiplication.

$$\begin{aligned}cT(x_1, x_2, x_3) &= c(x_1 - x_2, x_2 - x_3, x_3 - x_1)\\ &= (c(x_1 - x_2), c(x_2 - x_3), c(x_3 - x_1))\\ &= (cx_1 - cx_2, cx_2 - cx_3, cx_3 - cx_1)\\ &= T(cx_1, cx_2, cx_3)\end{aligned}$$

So, T is a linear transformation with standard matrix $A = \begin{bmatrix} 1 & -1 & 0 \\ 0 & 1 & -1 \\ -1 & 0 & 1 \end{bmatrix}$.

13. Because $(1, 1) = \frac{1}{2}(2, 0) + \frac{1}{3}(0, 3)$,

$$\begin{aligned}T(1, 1) &= \tfrac{1}{2}T(2, 0) + \tfrac{1}{3}T(0, 3)\\ &= \tfrac{1}{2}(1, 1) + \tfrac{1}{3}(3, 3) = \left(\tfrac{3}{2}, \tfrac{3}{2}\right).\end{aligned}$$

Because $(0, 1) = \frac{1}{3}(0, 3)$,

$$\begin{aligned}T(0, 1) &= \tfrac{1}{3}T(0, 3)\\ &= \tfrac{1}{3}(3, 3) = (1, 1).\end{aligned}$$

15. Because $(0, -1) = -\frac{2}{3}(1, 1) - \frac{1}{3}(2, -1)$

$$\begin{aligned}T(0, -1) &= -\tfrac{2}{3}T(1, 1) + \tfrac{1}{3}T(2, -1)\\ &= -\tfrac{2}{3}(2, 3) + \tfrac{1}{3}(1, 0)\\ &= \left(-\tfrac{4}{3}, -2\right) + \left(\tfrac{1}{3}, 0\right) = (-1, -2).\end{aligned}$$

17. (a) Because A is a 2×3 matrix, it maps R^3 into R^2, $(n = 3, m = 2)$.

(b) Because $T(\mathbf{v}) = A\mathbf{v}$ and

$$A\mathbf{v} = \begin{bmatrix} 0 & 1 & 2 \\ -2 & 0 & 0 \end{bmatrix}\begin{bmatrix} 6 \\ 1 \\ 1 \end{bmatrix} = \begin{bmatrix} 3 \\ -12 \end{bmatrix},$$

it follows that $T(6, 1, 1) = (3, -12)$.

(c) The preimage of $\mathbf{w}$ is given by the solution to the equation

$T(v_1, v_2, v_3) = \mathbf{w} = (3, 5)$.

The equivalent system of linear equations

$$\begin{aligned}v_2 + 2v_3 &= 3\\ -2v_1 \qquad &= 5\end{aligned}$$

has the solution

$\left\{\left(-\frac{5}{2}, 3 - 2t, t\right) : t \text{ is a real number}\right\}$.

© 2013 Cengage Learning. All Rights Reserved. May not be scanned, copied or duplicated, or posted to a publicly accessible website, in whole or in part.

19. (a) Because A is a 1×2 matrix, it maps R^2 into R^1, $(n = 2, m = 1)$.

(b) Because $T(\mathbf{v}) = A\mathbf{v}$ and $A\mathbf{v} = \begin{bmatrix} 1 & 1 \end{bmatrix}\begin{bmatrix} 2 \\ 3 \end{bmatrix} = 5,$

it follows that $T(2, 3) = 5$.

(c) The preimage of $\mathbf{w} = (4)$ is given by the solution to this equation $T(v_1, v_2) = \mathbf{w} = (4)$.

The equivalent system of linear equations is $v_1 + v_2 = 4$, which has the solution $\{(4 - t, t) : t \in R\}$.

21. (a) Because A is a 3×3 matrix, it maps R^3 into R^3, $(n = 3, m = 3)$.

(b) Because $T(\mathbf{v}) = A\mathbf{v}$ and

$$A\mathbf{v} = \begin{bmatrix} 1 & 1 & 1 \\ 0 & 1 & 1 \\ 0 & 0 & 1 \end{bmatrix}\begin{bmatrix} 2 \\ 1 \\ -5 \end{bmatrix} = \begin{bmatrix} -2 \\ -4 \\ -5 \end{bmatrix},$$

it follows that $T(2, 1, -5) = (-2, -4, -5)$.

(c) The preimage of $\mathbf{w} = (6, 4, 2)$ is given by the solution to the equation

$T(v_1, v_2, v_3) = (6, 4, 2) = \mathbf{w}$.

The equivalent system of linear equations has the solution $v_1 = v_2 = v_3 = 2$. So, the preimage is $(2, 2, 2)$.

23. (a) Because A is a 3×2 matrix, it maps R^2 into R^3, $(n = 2, m = 3)$.

(b) Because $T(\mathbf{v}) = A\mathbf{v}$ and

$$A\mathbf{v} = \begin{bmatrix} 4 & 0 \\ 0 & 5 \\ 1 & 1 \end{bmatrix}\begin{bmatrix} 2 \\ 2 \end{bmatrix} = \begin{bmatrix} 8 \\ 10 \\ 4 \end{bmatrix},$$

it follows that $T(2, 2) = (8, 10, 4)$.

(c) The preimage of $\mathbf{w} = (4, -5, 0)$ is given by the solution to the equation

$T(v_1, v_2) = (4, -5, 0) = \mathbf{w}$.

The equivalent system of linear equations has the solution $v_1 = 1$ and $v_2 = -1$. So, the preimage is $(1, -1)$.

25. The standard matrix for the 90° counterclockwise notation is

$$A = \begin{bmatrix} \cos 90° & -\sin 90° \\ \sin 90° & \cos 90° \end{bmatrix} = \begin{bmatrix} 0 & -1 \\ 1 & 0 \end{bmatrix}.$$

Calculating the image of the three vertices,

$$\begin{bmatrix} 0 & -1 \\ 1 & 0 \end{bmatrix}\begin{bmatrix} 3 \\ 5 \end{bmatrix} = \begin{bmatrix} -5 \\ 3 \end{bmatrix},$$

$$\begin{bmatrix} 0 & -1 \\ 1 & 0 \end{bmatrix}\begin{bmatrix} 5 \\ 3 \end{bmatrix} = \begin{bmatrix} -3 \\ 5 \end{bmatrix},$$

$$\begin{bmatrix} 0 & -1 \\ 1 & 0 \end{bmatrix}\begin{bmatrix} 3 \\ 0 \end{bmatrix} = \begin{bmatrix} 0 \\ 3 \end{bmatrix},$$

you have the following graph.

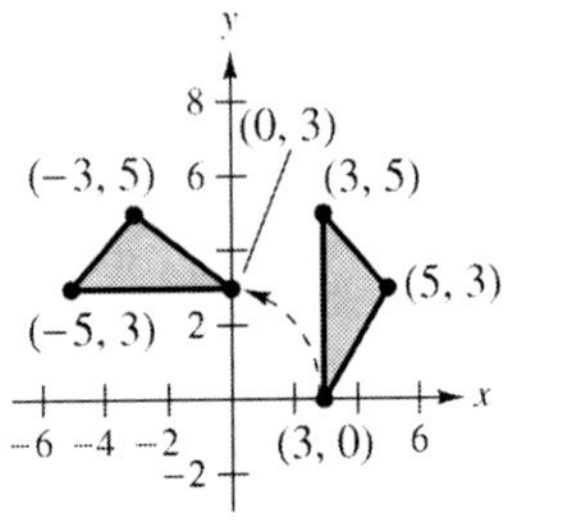

27. (a) The standard matrix for T is

$$A = \begin{bmatrix} 2 & 4 & 6 & 5 \\ -1 & -2 & 2 & 0 \\ 0 & 0 & 8 & 4 \end{bmatrix}.$$

Solving $A\mathbf{v} = \mathbf{0}$ yields the solution $\{(-2s + 2t, s, t, -2t) : s \text{ and } t \text{ are real numbers}\}$.

So, a basis for $\ker(T)$ is $\{(-2, 1, 0, 0), (2, 0, 1, -2)\}$.

(b) Use Gauss-Jordan elimination to reduce A^T as follows.

$$A^T = \begin{bmatrix} 2 & -1 & 0 \\ 4 & -2 & 0 \\ 6 & 2 & 8 \\ 5 & 0 & 4 \end{bmatrix} \Rightarrow \begin{bmatrix} 1 & 0 & \frac{4}{5} \\ 0 & 1 & \frac{8}{5} \\ 0 & 0 & 0 \\ 0 & 0 & 0 \end{bmatrix}$$

The nonzero row vectors form a basis for the range of T, $\left\{\left(1, 0, \frac{4}{5}\right), \left(0, 1, \frac{8}{5}\right)\right\}$.

29. (a) The standard matrix for T is

$$A = \begin{bmatrix} 1 & 0 & 0 \\ 0 & 1 & 2 \\ 0 & 0 & 1 \end{bmatrix}.$$

Solving $A\mathbf{v} = \mathbf{0}$ yields the solution $\mathbf{v} = \mathbf{0}$. So, $\ker(T)$ consists of the zero vector, $\ker(T) = \{(0, 0, 0)\}$.

(b) Because $\ker(T)$ is dimension 0, the range(T) must be all of R^3. So, a basis for the range is $\{(1, 0, 0), (0, 1, 0), (0, 0, 1)\}$.

© 2013 Cengage Learning. All Rights Reserved. May not be scanned, copied or duplicated, or posted to a publicly accessible website, in whole or in part.

31. (a) To find the kernel of T, row-reduce A,

$$A = \begin{bmatrix} 1 & 2 \\ -1 & 0 \\ 1 & 1 \end{bmatrix} \Rightarrow \begin{bmatrix} 1 & 0 \\ 0 & 1 \\ 0 & 0 \end{bmatrix}$$

which shows that $\ker(T) = \{(0, 0)\}$.

(b) $\dim(\ker(T)) = \text{nullity}(T) = 0$

(c) The range of T can be found by row-reducing the transpose of A.

$$A^T = \begin{bmatrix} 1 & -1 & 1 \\ 2 & 0 & 1 \end{bmatrix} \Rightarrow \begin{bmatrix} 1 & 0 & \frac{1}{2} \\ 0 & 1 & -\frac{1}{2} \end{bmatrix}$$

range(T) is span $\left\{\left(1, 0, \frac{1}{2}\right), \left(0, 1, -\frac{1}{2}\right)\right\}$

(d) $\dim(\text{range}(T)) = \text{rank}(T) = 2$

33. (a) To find the kernel of T, row-reduce A,

$$A = \begin{bmatrix} 2 & 1 & 3 \\ 1 & 1 & 0 \\ 0 & 1 & -3 \end{bmatrix} \Rightarrow \begin{bmatrix} 1 & 0 & 3 \\ 0 & 1 & -3 \\ 0 & 0 & 0 \end{bmatrix}$$

which shows that kernel $(T) = \{(-3t, 3t, t) : t \in R\}$.

(b) $\dim(\ker(T)) = \text{nullity}(T) = 1$

(c) The range of T can be found by row-reducing the transpose of A.

$$A^T = \begin{bmatrix} 2 & 1 & 0 \\ 1 & 1 & 1 \\ 3 & 0 & -3 \end{bmatrix} \Rightarrow \begin{bmatrix} 1 & 0 & -1 \\ 0 & 1 & 2 \\ 0 & 0 & 0 \end{bmatrix}$$

range(T) is span $\{(1, 0, -1), (0, 1, 2)\}$

(d) $\dim(\text{range}(T)) = \text{rank}(T) = 2$

35. $\text{rank}(T) = \dim R^5 - \text{nullity}(T) = 5 - 2 = 3$

37. $\text{nullity}(T) = \dim(P_4) - \text{rank}(T) = 5 - 3 = 2$

39. The standard matrix for T is

$$A = \begin{bmatrix} 1 & 0 & 0 \\ 0 & 1 & 0 \\ 0 & 0 & -1 \end{bmatrix}.$$

Therefore,

$$A^2 = \begin{bmatrix} 1^2 & 0 & 0 \\ 0 & 1^2 & 0 \\ 0 & 0 & (-1)^2 \end{bmatrix} = \begin{bmatrix} 1 & 0 & 0 \\ 0 & 1 & 0 \\ 0 & 0 & 1 \end{bmatrix} = I_3.$$

41. The standard matrix for T is

$$A = \begin{bmatrix} \cos\theta & -\sin\theta \\ \sin\theta & \cos\theta \end{bmatrix}.$$ Therefore,

$$A^3 = \begin{bmatrix} \cos 3\theta & -\sin 3\theta \\ \sin 3\theta & \cos 3\theta \end{bmatrix}.$$

43. The standard matrices for T_1 and T_2 are

$$A_1 = \begin{bmatrix} 1 & 0 \\ 1 & 1 \\ 0 & 1 \end{bmatrix} \text{ and } A_2 = \begin{bmatrix} 0 & 0 & 0 \\ 0 & 1 & 0 \end{bmatrix}.$$

The standard matrix for $T = T_1 \circ T_2$ is

$$A = A_1A_2 = \begin{bmatrix} 1 & 0 \\ 1 & 1 \\ 0 & 1 \end{bmatrix}\begin{bmatrix} 0 & 0 & 0 \\ 0 & 1 & 0 \end{bmatrix} = \begin{bmatrix} 0 & 0 & 0 \\ 0 & 1 & 0 \\ 0 & 1 & 0 \end{bmatrix}$$

and the standard matrix for $T' = T_2 \circ T_1$ is

$$A' = A_2A_1 = \begin{bmatrix} 0 & 0 & 0 \\ 0 & 1 & 0 \end{bmatrix}\begin{bmatrix} 1 & 0 \\ 1 & 1 \\ 0 & 1 \end{bmatrix} = \begin{bmatrix} 0 & 0 \\ 1 & 1 \end{bmatrix}.$$

45. The standard matrix for T is

$$A = \begin{bmatrix} 0 & 0 \\ 0 & 1 \end{bmatrix}.$$

Because A is *not* invertible, T has no inverse.

47. The standard matrix for T is

$$A = \begin{bmatrix} 1 & 0 \\ 0 & -1 \end{bmatrix}.$$

A is invertible and its inverse is given by

$$A^{-1} = \begin{bmatrix} 1 & 0 \\ 0 & -1 \end{bmatrix}.$$

49. (a) Because $|A| = 6 \neq 0$, $\ker(T) = \{(0, 0)\}$ and the transformation is one-to-one.

(b) Because the nullity of T is 0, the rank of T equals the dimension of the domain and the transformation is onto.

(c) The transformation is one-to-one and onto (an isomorphism) and is, therefore, invertible.

51. (a) Because $T: R^3 \rightarrow R^2$ and $\text{rank}(T) \leq 2$, $\text{nullity}(T) \geq 1$. So, T is not one-to-one.

(b) Because $\text{rank}(A) = 2$, T is onto.

(c) T is not invertible (A is not square).

© 2013 Cengage Learning. All Rights Reserved. May not be scanned, copied or duplicated, or posted to a publicly accessible website, in whole or in part.

53. (a) The standard matrix for T is

$$A = \begin{bmatrix} -1 & 0 \\ 0 & 1 \\ 1 & 1 \end{bmatrix}$$

so it follows that

$$T(\mathbf{v}) = A(\mathbf{v}) = \begin{bmatrix} -1 & 0 \\ 0 & 1 \\ 1 & 1 \end{bmatrix}\begin{bmatrix} 0 \\ 1 \end{bmatrix} = \begin{bmatrix} 0 \\ 1 \\ 1 \end{bmatrix} = (0, 1, 1).$$

(b) The image of each vector in B is as follows.

$$T(1, 1) = (-1, 1, 2) = (0, 1, 0) + 2(0, 0, 1) - (1, 0, 0)$$

$$T(1, -1) = (-1, -1, 0) = -(0, 1, 0) + 0(0, 0, 1) - (1, 0, 0)$$

Therefore,

$$[T(1, 1)]_{B'} = [1, 2, -1]^T$$

and

$$[T(1, -1)]_{B'} = [-1, 0, -1]^T$$

and

$$A' = \begin{bmatrix} 1 & -1 \\ 2 & 0 \\ -1 & -1 \end{bmatrix}.$$

Because

$$[\mathbf{v}]_B = \begin{bmatrix} \frac{1}{2} \\ -\frac{1}{2} \end{bmatrix},$$

the image of $\mathbf{v}$ under T relative to B' is

$$[T(\mathbf{v})]_{B'} = A'[\mathbf{v}]_B = \begin{bmatrix} 1 & -1 \\ 2 & 0 \\ -1 & -1 \end{bmatrix}\begin{bmatrix} \frac{1}{2} \\ -\frac{1}{2} \end{bmatrix} = \begin{bmatrix} 1 \\ 1 \\ 0 \end{bmatrix}.$$

So,

$$T(\mathbf{v}) = (0, 1, 0) + (0, 0, 1) + 0(1, 0, 0) = (0, 1, 1).$$

55. The standard matrix for T is

$$A = \begin{bmatrix} 1 & -3 \\ -1 & 1 \end{bmatrix}.$$

The transformation matrix from B' to the standard basis $B = \{(1, 0), (0, 1)\}$ is

$$P = \begin{bmatrix} 1 & 1 \\ -1 & 1 \end{bmatrix}.$$

The matrix A' for T relative to B' is

$$A' = P^{-1}AP = \begin{bmatrix} \frac{1}{2} & -\frac{1}{2} \\ \frac{1}{2} & \frac{1}{2} \end{bmatrix}\begin{bmatrix} 1 & -3 \\ -1 & 1 \end{bmatrix}\begin{bmatrix} 1 & 1 \\ -1 & 1 \end{bmatrix} = \begin{bmatrix} 3 & -1 \\ 1 & -1 \end{bmatrix}.$$

Because $A' = P^{-1}AP$, it follows that A and A' are similar.

57. Since

$$A' = P^{-1}AP$$

$$= \begin{bmatrix} \frac{4}{7} & -\frac{5}{7} \\ \frac{1}{7} & -\frac{3}{7} \end{bmatrix}\begin{bmatrix} 18 & -19 \\ 11 & -12 \end{bmatrix}\begin{bmatrix} 3 & -5 \\ 1 & -4 \end{bmatrix}$$

$$= \begin{bmatrix} 5 & -3 \\ -4 & 1 \end{bmatrix}$$

A and A' are similar.

59. (a) Because $T(\mathbf{v}) = T(x, y, z) = \text{proj}_{\mathbf{u}}\mathbf{v}$ where $\mathbf{u} = (0, 1, 2)$,

$$T(\mathbf{v}) = \frac{y + 2z}{5}(0, 1, 2).$$

So,

$$T(1, 0, 0) = (0, 0, 0),\ T(0, 1, 0) = \left(0, \tfrac{1}{5}, \tfrac{2}{5}\right),\ T(0, 0, 1) = \left(0, \tfrac{2}{5}, \tfrac{4}{5}\right)$$

and the standard matrix for T is

$$A = \begin{bmatrix} 0 & 0 & 0 \\ 0 & \frac{1}{5} & \frac{2}{5} \\ 0 & \frac{2}{5} & \frac{4}{5} \end{bmatrix}.$$

© 2013 Cengage Learning. All Rights Reserved. May not be scanned, copied or duplicated, or posted to a publicly accessible website, in whole or in part.

(b) $S = I - A$ satisfies $S(\mathbf{u}) = \mathbf{0}$. Letting $\mathbf{w}_1 = (1, 0, 0)$ and $\mathbf{w}_2 = (0, 2, -1)$ be two vectors orthogonal to $\mathbf{u}$,

$$\text{proj}_{\mathbf{w}_1}\mathbf{v} = \frac{x}{1}(1, 0, 0) \quad \Rightarrow \quad P_1 = \begin{bmatrix} 1 & 0 & 0 \\ 0 & 0 & 0 \\ 0 & 0 & 0 \end{bmatrix}$$

$$\text{proj}_{\mathbf{w}_2}\mathbf{v} = \frac{2y - z}{5}(0, 2, -1) \quad \Rightarrow \quad P_2 = \begin{bmatrix} 0 & 0 & 0 \\ 0 & \frac{4}{5} & -\frac{2}{5} \\ 0 & -\frac{2}{5} & \frac{1}{5} \end{bmatrix}.$$

So,

$$S = I - A = \begin{bmatrix} 1 & 0 & 0 \\ 0 & \frac{4}{5} & -\frac{2}{5} \\ 0 & -\frac{2}{5} & \frac{1}{5} \end{bmatrix} = P_1 + P_2$$

verifying that $S(\mathbf{v}) = \text{proj}_{\mathbf{w}_1}\mathbf{v} + \text{proj}_{\mathbf{w}_2}\mathbf{v}$.

(c) The kernel of T has basis $\{(1, 0, 0), (0, 2, -1)\}$, which is precisely the column space of S.

61. $S + T$ preserves addition.

$$\begin{aligned}(S + T)(\mathbf{v} + \mathbf{w}) &= S(\mathbf{v} + \mathbf{w}) + T(\mathbf{v} + \mathbf{w}) \\ &= S(\mathbf{v}) + S(\mathbf{w}) + T(\mathbf{v}) + T(\mathbf{w}) \\ &= S(\mathbf{v}) + T(\mathbf{v}) + S(\mathbf{w}) + T(\mathbf{w}) \\ &= (S + T)\mathbf{v} + (S + T)(\mathbf{w})\end{aligned}$$

$S + T$ preserves scalar multiplication.

$$\begin{aligned}(S + T)(c\mathbf{v}) &= S(c\mathbf{v}) + T(c\mathbf{v}) \\ &= cS(\mathbf{v}) + cT(\mathbf{v}) \\ &= c(S(\mathbf{v}) + T(\mathbf{v})) \\ &= c(S + T)(\mathbf{v})\end{aligned}$$

kT preserves addition.

$$\begin{aligned}(kT)(\mathbf{v} + \mathbf{w}) &= kT(\mathbf{v} + \mathbf{w}) = k(T(\mathbf{v}) + T(\mathbf{w})) \\ &= kT(\mathbf{v}) + kT(\mathbf{w}) \\ &= (kT)(\mathbf{v}) + (kT)(\mathbf{w})\end{aligned}$$

kT preserves scalar multiplication.

$$(kT)(c\mathbf{v}) = kT(c\mathbf{v}) = kcT(\mathbf{v}) = ckT(\mathbf{v}) = c(kT)(\mathbf{v}).$$

63. If S, T and $S + T$ are written as matrices, the number of linearly independent columns in $S + T$ cannot exceed the number of linearly independent columns in S plus the number of linearly independent columns in T, because the columns in $S + T$ are created by summing columns in S and T.

65. (a) T preserves addition.

$$\begin{aligned}&T\left[\left(a_0 + a_1x + a_2x^2 + a_3x^3\right) + \left(b_0 + b_1x + b_2x^2 + b_3x^3\right)\right] \\ &= T\left[(a_0 + b_0) + (a_1 + b_1)x + (a_2 + b_2)x^2 + (a_3 + b_3)x^3\right] \\ &= (a_0 + b_0) + (a_1 + b_1) + (a_2 + b_2) + (a_3 + b_3) \\ &= (a_0 + a_1 + a_2 + a_3) + (b_0 + b_1 + b_2 + b_3) \\ &= T\left(a_0 + a_1x + a_2x^2 + a_3x^3\right) + T\left(b_0 + b_1x + b_2x^2 + b_3x^3\right)\end{aligned}$$

T preserves scalar multiplication.

$$\begin{aligned}T\left(c\left(a_0 + a_1x + a_2x^2 + a_3x^3\right)\right) &= T\left(ca_0 + ca_1x + ca_2x^2 + ca_3x^3\right) \\ &= ca_0 + ca_1 + ca_2 + ca_3 = c(a_0 + a_1 + a_2 + a_3) = cT\left(a_0 + a_1x + a_2x^2 + a_3x^3\right)\end{aligned}$$

© 2013 Cengage Learning. All Rights Reserved. May not be scanned, copied or duplicated, or posted to a publicly accessible website, in whole or in part.

(b) Because the range of T is R, $\text{rank}(T) = 1$. So, $\text{nullity}(T) = 4 - 1 = 3$.

(c) A basis for the kernel of T is obtained by solving $T(a_0 + a_1x + a_2x^2 + a_3x^3) = a_0 + a_1 + a_2 + a_3 = 0$.

Letting $a_3 = t$, $a_2 = s$, $a_1 = r$ be the free variables, $a_0 = -t - s - r$ and a basis is $\{-1 + x^3, -1 + x^2, -1 + x\}$.

67. Let B be a basis for V and let $[\mathbf{v}_0]_B = [a_1, a_2, \ldots, a_n]^T$, where at least one $a_i \neq 0$ for $i = 1, \ldots, n$. Then for $[\mathbf{v}]_B = [v_1, v_2, \ldots, v_n]^T$ you have

$[T(\mathbf{v})]_B = \langle \mathbf{v}, \mathbf{v}_0 \rangle = a_1v_1 + a_2v_2 + \cdots + a_nv_n.$

The matrix for T relative to B is then

$A = [a_1 \;\; a_2 \;\; \cdots \;\; a_n].$

Because A^T row-reduces to one nonzero row, the range of T is $\{t : t \in R\} = R$. So, the rank of T is 1 and $\text{nullity}(T) = n - 1$. Finally, $\ker(T) = \{\mathbf{v} : \langle \mathbf{v}, \mathbf{v}_0 \rangle = 0\}$.

69. Although they are not the same, they have the same dimension (4) and are isomorphic.

71. (a) T is a vertical expansion.

(b)

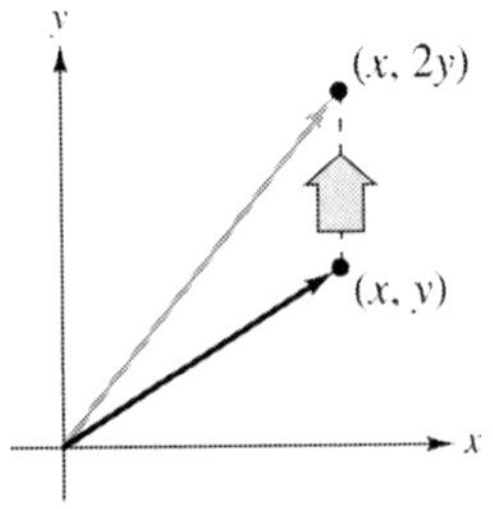

73. (a) T is a vertical shear.

(b)

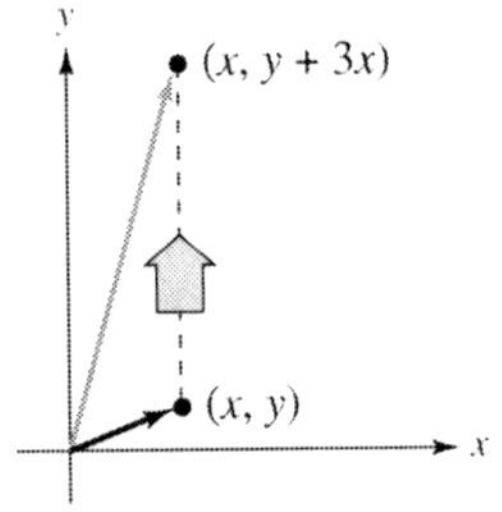

75. (a) T is a horizontal shear.

(b)

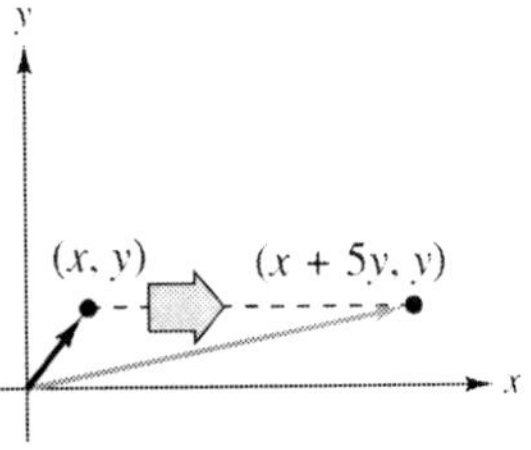

77. The image of each vertex is
$T(0, 0) = (0, 0)$, $T(1, 0) = (1, 0)$, $T(0, 1) = (0, -1)$.

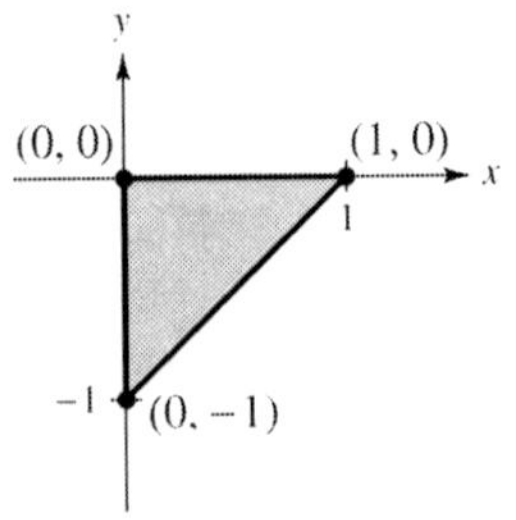

79. The image of each vertex is
$T(0, 0) = (0, 0)$, $T(1, 0) = (1, 0)$, and $T(0, 1) = (3, 1)$.

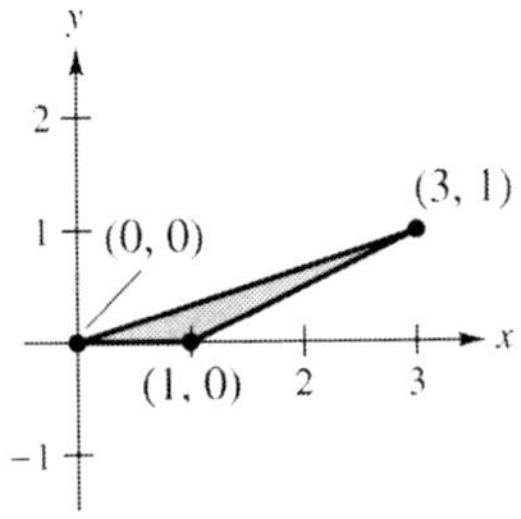

81. The transformation is a reflection in the line $y = x$

$$\begin{bmatrix} 0 & 1 \\ 1 & 0 \end{bmatrix}$$

followed by a horizontal expansion $\begin{bmatrix} 2 & 0 \\ 0 & 1 \end{bmatrix}$.

83. A rotation of 45° about the z-axis is given by

$$A = \begin{bmatrix} \cos 45^\circ & -\sin 45^\circ & 0 \\ \sin 45^\circ & \cos 45^\circ & 0 \\ 0 & 0 & 1 \end{bmatrix} = \begin{bmatrix} \frac{\sqrt{2}}{2} & -\frac{\sqrt{2}}{2} & 0 \\ \frac{\sqrt{2}}{2} & \frac{\sqrt{2}}{2} & 0 \\ 0 & 0 & 1 \end{bmatrix}.$$

Because

$$A\mathbf{v} = \begin{bmatrix} \frac{\sqrt{2}}{2} & -\frac{\sqrt{2}}{2} & 0 \\ \frac{\sqrt{2}}{2} & \frac{\sqrt{2}}{2} & 0 \\ 0 & 0 & 1 \end{bmatrix} \begin{bmatrix} 1 \\ -1 \\ 1 \end{bmatrix} = \begin{bmatrix} \sqrt{2} \\ 0 \\ 1 \end{bmatrix}$$

the image of $(1, -1, 1)$ is $(\sqrt{2}, 0, 1)$.

© 2013 Cengage Learning. All Rights Reserved. May not be scanned, copied or duplicated, or posted to a publicly accessible website, in whole or in part.

85. A rotation of 60° about the x-axis is given by

$$A = \begin{bmatrix} 1 & 0 & 0 \\ 0 & \cos 60° & -\sin 60° \\ 0 & \sin 60° & \cos 60° \end{bmatrix} = \begin{bmatrix} 1 & 0 & 0 \\ 0 & \frac{1}{2} & -\frac{\sqrt{3}}{2} \\ 0 & \frac{\sqrt{3}}{2} & \frac{1}{2} \end{bmatrix}.$$

Because

$$A\mathbf{v} = \begin{bmatrix} 1 & 0 & 0 \\ 0 & \frac{1}{2} & -\frac{\sqrt{3}}{2} \\ 0 & \frac{\sqrt{3}}{2} & \frac{1}{2} \end{bmatrix}\begin{bmatrix} 1 \\ -1 \\ 1 \end{bmatrix} = \begin{bmatrix} 1 \\ -\frac{1}{2} - \frac{\sqrt{3}}{2} \\ \frac{1}{2} - \frac{\sqrt{3}}{2} \end{bmatrix},$$

the image of $(1, -1, 1)$ is $\left(1, -\frac{1}{2} - \frac{\sqrt{3}}{2}, \frac{1}{2} - \frac{\sqrt{3}}{2}\right)$.

87. A rotation of 60° about the x-axis has the standard matrix

$$\begin{bmatrix} 1 & 0 & 0 \\ 0 & \cos 60° & -\sin 60° \\ 0 & \sin 60° & \cos 60° \end{bmatrix} = \begin{bmatrix} 1 & 0 & 0 \\ 0 & \frac{1}{2} & -\frac{\sqrt{3}}{2} \\ 0 & \frac{\sqrt{3}}{2} & \frac{1}{2} \end{bmatrix},$$

while a rotation of 30° about the z-axis has the standard matrix

$$\begin{bmatrix} \cos 30° & -\sin 30° & 0 \\ \sin 30° & \cos 30° & 0 \\ 0 & 0 & 1 \end{bmatrix} = \begin{bmatrix} \frac{\sqrt{3}}{2} & -\frac{1}{2} & 0 \\ \frac{1}{2} & \frac{\sqrt{3}}{2} & 0 \\ 0 & 0 & 1 \end{bmatrix}.$$

So, the pair of rotations is given by

$$\begin{bmatrix} \frac{\sqrt{3}}{2} & -\frac{1}{2} & 0 \\ \frac{1}{2} & \frac{\sqrt{3}}{2} & 0 \\ 0 & 0 & 1 \end{bmatrix}\begin{bmatrix} 1 & 0 & 0 \\ 0 & \frac{1}{2} & -\frac{\sqrt{3}}{2} \\ 0 & \frac{\sqrt{3}}{2} & \frac{1}{2} \end{bmatrix} = \begin{bmatrix} \frac{\sqrt{3}}{2} & -\frac{1}{4} & \frac{\sqrt{3}}{4} \\ \frac{1}{2} & \frac{\sqrt{3}}{4} & -\frac{3}{4} \\ 0 & \frac{\sqrt{3}}{2} & \frac{1}{2} \end{bmatrix}.$$

89. A rotation of 30° about the y-axis has the standard matrix

$$\begin{bmatrix} \cos 30° & 0 & \sin 30° \\ 0 & 1 & 0 \\ -\sin 30° & 0 & \cos 30° \end{bmatrix} = \begin{bmatrix} \frac{\sqrt{3}}{2} & 0 & \frac{1}{2} \\ 0 & 1 & 0 \\ -\frac{1}{2} & 0 & \frac{\sqrt{3}}{2} \end{bmatrix},$$

while a rotation of 45° about the z-axis has the standard matrix

$$\begin{bmatrix} \cos 45° & -\sin 45° & 0 \\ \sin 45° & \cos 45° & 0 \\ 0 & 0 & 1 \end{bmatrix} = \begin{bmatrix} \frac{\sqrt{2}}{2} & -\frac{\sqrt{2}}{2} & 0 \\ \frac{\sqrt{2}}{2} & \frac{\sqrt{2}}{2} & 0 \\ 0 & 0 & 1 \end{bmatrix}.$$

So, the pair of rotations is given by

$$\begin{bmatrix} \frac{\sqrt{2}}{2} & -\frac{\sqrt{2}}{2} & 0 \\ \frac{\sqrt{2}}{2} & \frac{\sqrt{2}}{2} & 0 \\ 0 & 0 & 1 \end{bmatrix}\begin{bmatrix} \frac{\sqrt{3}}{2} & 0 & \frac{1}{2} \\ 0 & 1 & 0 \\ -\frac{1}{2} & 0 & \frac{\sqrt{3}}{2} \end{bmatrix} = \begin{bmatrix} \frac{\sqrt{6}}{4} & -\frac{\sqrt{2}}{2} & \frac{\sqrt{2}}{4} \\ \frac{\sqrt{6}}{4} & \frac{\sqrt{2}}{2} & \frac{\sqrt{2}}{4} \\ -\frac{1}{2} & 0 & \frac{\sqrt{3}}{2} \end{bmatrix}.$$

91. The standard matrix for T is

$$\begin{bmatrix} \cos 45° & -\sin 45° & 0 \\ \sin 45° & \cos 45° & 0 \\ 0 & 0 & 1 \end{bmatrix} = \begin{bmatrix} \frac{\sqrt{2}}{2} & -\frac{\sqrt{2}}{2} & 0 \\ \frac{\sqrt{2}}{2} & \frac{\sqrt{2}}{2} & 0 \\ 0 & 0 & 1 \end{bmatrix}.$$

Therefore, T is given by

$$T(x, y, z) = \left(\frac{\sqrt{2}}{2}x - \frac{\sqrt{2}}{2}y, \frac{\sqrt{2}}{2}x + \frac{\sqrt{2}}{2}y, z\right).$$

The image of each vertex is as follows.

$T(0, 0, 0) = (0, 0, 0)$

$T(1, 0, 0) = \left(\frac{\sqrt{2}}{2}, \frac{\sqrt{2}}{2}, 0\right)$

$T(1, 1, 0) = \left(0, \sqrt{2}, 0\right)$

$T(0, 1, 0) = \left(-\frac{\sqrt{2}}{2}, \frac{\sqrt{2}}{2}, 0\right)$

$T(0, 0, 1) = (0, 0, 1)$

$T(1, 0, 1) = \left(\frac{\sqrt{2}}{2}, \frac{\sqrt{2}}{2}, 1\right)$

$T(1, 1, 1) = \left(0, \sqrt{2}, 1\right)$

$T(0, 1, 1) = \left(-\frac{\sqrt{2}}{2}, \frac{\sqrt{2}}{2}, 1\right)$

© 2013 Cengage Learning. All Rights Reserved. May not be scanned, copied or duplicated, or posted to a publicly accessible website, in whole or in part.

93. The standard matrix for T is

$$\begin{bmatrix} 1 & 0 & 0 \\ 0 & \cos 30° & -\sin 30° \\ 0 & \sin 30° & \cos 30° \end{bmatrix} = \begin{bmatrix} 1 & 0 & 0 \\ 0 & \frac{\sqrt{3}}{2} & -\frac{1}{2} \\ 0 & \frac{1}{2} & \frac{\sqrt{3}}{2} \end{bmatrix}.$$

Therefore, T is given by

$$T(x, y, z) = \left(x, \frac{\sqrt{3}}{2}y - \frac{1}{2}z, \frac{1}{2}y + \frac{\sqrt{3}}{2}z\right).$$

The image of each vertex is as follows.

$$T(0, 0, 0) = (0, 0, 0)$$

$$T(1, 0, 0) = (1, 0, 0)$$

$$T(1, 1, 0) = \left(1, \frac{\sqrt{3}}{2}, \frac{1}{2}\right)$$

$$T(0, 1, 0) = \left(0, \frac{\sqrt{3}}{2}, \frac{1}{2}\right)$$

$$T(0, 0, 1) = \left(0, -\frac{1}{2}, \frac{\sqrt{3}}{2}\right)$$

$$T(1, 0, 1) = \left(1, -\frac{1}{2}, \frac{\sqrt{3}}{2}\right)$$

$$T(1, 1, 1) = \left(1, \frac{\sqrt{3}}{2} - \frac{1}{2}, \frac{1}{2} + \frac{\sqrt{3}}{2}\right)$$

$$T(0, 1, 1) = \left(0, \frac{\sqrt{3}}{2} - \frac{1}{2}, \frac{1}{2} + \frac{\sqrt{3}}{2}\right)$$

95. (a) False. See "Elementary Matrices for Linear Transformations in the Plane," page 330.

(b) True. See "Elementary Matrices for Linear Transformations in the Plane," page 330.

(c) True. See discussion following Example 4, page 334.

97. (a) False. See "Remark," page 294.

(b) False. See Theorem 6.7, page 310.

(c) True. See discussion following Example 5, page 327.

© 2013 Cengage Learning. All Rights Reserved. May not be scanned, copied or duplicated, or posted to a publicly accessible website, in whole or in part.

CHAPTER 7
Eigenvalues and Eigenvectors

© 2013 Cengage Learning. All Rights Reserved. May not be scanned, copied or duplicated, or posted to a publicly accessible website, in whole or in part.

CHAPTER 7
Eigenvalues and Eigenvectors

Section 7.1 Eigenvalues and Eigenvectors

1. $A\mathbf{x}_1 = \begin{bmatrix} 2 & 0 \\ 0 & -2 \end{bmatrix}\begin{bmatrix} 1 \\ 0 \end{bmatrix} = \begin{bmatrix} 2 \\ 0 \end{bmatrix} = 2\begin{bmatrix} 1 \\ 0 \end{bmatrix} = \lambda_1\mathbf{x}_1$

$A\mathbf{x}_2 = \begin{bmatrix} 2 & 0 \\ 0 & -2 \end{bmatrix}\begin{bmatrix} 0 \\ -1 \end{bmatrix} = \begin{bmatrix} 0 \\ 2 \end{bmatrix} = -2\begin{bmatrix} 0 \\ -1 \end{bmatrix} = \lambda_2\mathbf{x}_2$

3. $A\mathbf{x}_1 = \begin{bmatrix} 1 & 1 \\ 1 & 1 \end{bmatrix}\begin{bmatrix} 1 \\ -1 \end{bmatrix} = \begin{bmatrix} 0 \\ 0 \end{bmatrix} = 0\begin{bmatrix} 1 \\ -1 \end{bmatrix} = \lambda_1\mathbf{x}_1$

$A\mathbf{x}_2 = \begin{bmatrix} 1 & 1 \\ 1 & 1 \end{bmatrix}\begin{bmatrix} 1 \\ 1 \end{bmatrix} = \begin{bmatrix} 2 \\ 2 \end{bmatrix} = 2\begin{bmatrix} 1 \\ 1 \end{bmatrix} = \lambda_2\mathbf{x}_2$

5. $A\mathbf{x}_1 = \begin{bmatrix} 2 & 3 & 1 \\ 0 & -1 & 2 \\ 0 & 0 & 3 \end{bmatrix}\begin{bmatrix} 1 \\ 0 \\ 0 \end{bmatrix} = \begin{bmatrix} 2 \\ 0 \\ 0 \end{bmatrix} = 2\begin{bmatrix} 1 \\ 0 \\ 0 \end{bmatrix} = \lambda_1\mathbf{x}_1$

$A\mathbf{x}_2 = \begin{bmatrix} 2 & 3 & 1 \\ 0 & -1 & 2 \\ 0 & 0 & 3 \end{bmatrix}\begin{bmatrix} 1 \\ -1 \\ 0 \end{bmatrix} = \begin{bmatrix} -1 \\ 1 \\ 0 \end{bmatrix} = -1\begin{bmatrix} 1 \\ -1 \\ 0 \end{bmatrix} = \lambda_2\mathbf{x}_2$

$A\mathbf{x}_3 = \begin{bmatrix} 2 & 3 & 1 \\ 0 & -1 & 2 \\ 0 & 0 & 3 \end{bmatrix}\begin{bmatrix} 5 \\ 1 \\ 2 \end{bmatrix} = \begin{bmatrix} 15 \\ 3 \\ 6 \end{bmatrix} = 3\begin{bmatrix} 5 \\ 1 \\ 2 \end{bmatrix} = \lambda_3\mathbf{x}_3$

7. $A\mathbf{x}_1 = \begin{bmatrix} 0 & 1 & 0 \\ 0 & 0 & 1 \\ 1 & 0 & 0 \end{bmatrix}\begin{bmatrix} 1 \\ 1 \\ 1 \end{bmatrix} = \begin{bmatrix} 1 \\ 1 \\ 1 \end{bmatrix} = 1\begin{bmatrix} 1 \\ 1 \\ 1 \end{bmatrix} = \lambda_1\mathbf{x}_1$

9. (a) $A(c\mathbf{x}_1) = \begin{bmatrix} 1 & 1 \\ 1 & 1 \end{bmatrix}\begin{bmatrix} c \\ -c \end{bmatrix} = \begin{bmatrix} 0 \\ 0 \end{bmatrix} = 0\begin{bmatrix} c \\ -c \end{bmatrix} = 0(c\mathbf{x}_1)$

(b) $A(c\mathbf{x}_2) = \begin{bmatrix} 1 & 1 \\ 1 & 1 \end{bmatrix}\begin{bmatrix} c \\ c \end{bmatrix} = \begin{bmatrix} 2c \\ 2c \end{bmatrix} = 2\begin{bmatrix} c \\ c \end{bmatrix} = 2(c\mathbf{x}_2)$

11. (a) Because

$A\mathbf{x} = \begin{bmatrix} 7 & 2 \\ 2 & 4 \end{bmatrix}\begin{bmatrix} 1 \\ 2 \end{bmatrix} = \begin{bmatrix} 11 \\ 10 \end{bmatrix} \neq \lambda\begin{bmatrix} 1 \\ 2 \end{bmatrix}$

x is *not* an eigenvector of A.

(b) Because

$A\mathbf{x} = \begin{bmatrix} 7 & 2 \\ 2 & 4 \end{bmatrix}\begin{bmatrix} 2 \\ 1 \end{bmatrix} = \begin{bmatrix} 16 \\ 8 \end{bmatrix} = 8\begin{bmatrix} 2 \\ 1 \end{bmatrix}$

x *is* an eigenvector of A (with a corresponding eigenvalue 8).

(c) Because

$A\mathbf{x} = \begin{bmatrix} 7 & 2 \\ 2 & 4 \end{bmatrix}\begin{bmatrix} 1 \\ -2 \end{bmatrix} = \begin{bmatrix} 3 \\ -6 \end{bmatrix} = 3\begin{bmatrix} 1 \\ -2 \end{bmatrix}$

x *is* an eigenvector of A (with a corresponding eigenvalue 3).

(d) Because

$A\mathbf{x} = \begin{bmatrix} 7 & 2 \\ 2 & 4 \end{bmatrix}\begin{bmatrix} -1 \\ 0 \end{bmatrix} = \begin{bmatrix} -7 \\ -2 \end{bmatrix} \neq \lambda\begin{bmatrix} -1 \\ 0 \end{bmatrix}$

x is *not* an eigenvector of A.

13. (a) Because

$A\mathbf{x} = \begin{bmatrix} -1 & -1 & 1 \\ -2 & 0 & -2 \\ 3 & -3 & 1 \end{bmatrix}\begin{bmatrix} 2 \\ -4 \\ 6 \end{bmatrix} = \begin{bmatrix} 8 \\ -16 \\ 24 \end{bmatrix} = 4\begin{bmatrix} 2 \\ -4 \\ 6 \end{bmatrix}$

x *is* an eigenvector of A (with a corresponding eigenvalue 4).

(b) Because

$A\mathbf{x} = \begin{bmatrix} -1 & -1 & 1 \\ -2 & 0 & -2 \\ 3 & -3 & 1 \end{bmatrix}\begin{bmatrix} 2 \\ 0 \\ 6 \end{bmatrix} = \begin{bmatrix} 4 \\ -16 \\ 12 \end{bmatrix} \neq \lambda\begin{bmatrix} 2 \\ 0 \\ 6 \end{bmatrix}$

x is *not* an eigenvector of A.

(c) Because

$A\mathbf{x} = \begin{bmatrix} -1 & -1 & 1 \\ -2 & 0 & -2 \\ 3 & -3 & 1 \end{bmatrix}\begin{bmatrix} 2 \\ 2 \\ 0 \end{bmatrix} = \begin{bmatrix} -4 \\ -4 \\ 0 \end{bmatrix} = -2\begin{bmatrix} 2 \\ 2 \\ 0 \end{bmatrix}$

x *is* an eigenvector of A (with a corresponding eigenvalue −2).

(d) Because

$A\mathbf{x} = \begin{bmatrix} -1 & -1 & 1 \\ -2 & 0 & -2 \\ 3 & -3 & 1 \end{bmatrix}\begin{bmatrix} -1 \\ 0 \\ 1 \end{bmatrix} = \begin{bmatrix} 2 \\ 0 \\ -2 \end{bmatrix} = -2\begin{bmatrix} -1 \\ 0 \\ 1 \end{bmatrix}$

x *is* an eigenvector of A (with a corresponding eigenvalue −2).

 © 2013 Cengage Learning. All Rights Reserved. May not be scanned, copied or duplicated, or posted to a publicly accessible website, in whole or in part.

15. Geometrically, multiplying a vector in R^2 by A corresponds to a reflection in the x-axis.

$$\begin{bmatrix}1 & 0\\0 & -1\end{bmatrix}\begin{bmatrix}x\\y\end{bmatrix} = \begin{bmatrix}x\\-y\end{bmatrix}$$

The only vectors reflected onto scalar multiples of themselves are those lying on the x-axis or the y-axis.

$$\begin{bmatrix}1 & 0\\0 & -1\end{bmatrix}\begin{bmatrix}x\\0\end{bmatrix} = \begin{bmatrix}x\\0\end{bmatrix} = 1\begin{bmatrix}x\\0\end{bmatrix} \qquad \begin{bmatrix}1 & 0\\0 & -1\end{bmatrix}\begin{bmatrix}0\\y\end{bmatrix} = \begin{bmatrix}0\\-y\end{bmatrix} = -1\begin{bmatrix}0\\y\end{bmatrix}$$

So, the eigenvalues are $\lambda_1 = 1$ and $\lambda_2 = -1$, and the corresponding eigenspaces are the x-axis and the y-axis.

17. (a) The characteristic equation is $|\lambda I - A| = \begin{vmatrix}\lambda - 6 & 3\\2 & \lambda - 1\end{vmatrix} = \lambda^2 - 7\lambda = \lambda(\lambda - 7) = 0.$

(b) The eigenvalues are $\lambda_1 = 0$ and $\lambda_2 = 7$.

For $\lambda_1 = 0$, $\begin{bmatrix}\lambda_1 - 6 & 3\\2 & \lambda_1 - 1\end{bmatrix}\begin{bmatrix}x_1\\x_2\end{bmatrix} = \begin{bmatrix}0\\0\end{bmatrix} \Rightarrow \begin{bmatrix}2 & -1\\0 & 0\end{bmatrix}\begin{bmatrix}x_1\\x_2\end{bmatrix} = \begin{bmatrix}0\\0\end{bmatrix}.$

The solution is $\{(t, 2t) : t \in R\}$. So, an eigenvector corresponding to $\lambda_1 = 0$ is $(1, 2)$.

For $\lambda_2 = 7$, $\begin{bmatrix}\lambda_2 - 6 & 3\\2 & \lambda_2 - 1\end{bmatrix}\begin{bmatrix}x_1\\x_2\end{bmatrix} = \begin{bmatrix}0\\0\end{bmatrix} \Rightarrow \begin{bmatrix}1 & 3\\0 & 0\end{bmatrix}\begin{bmatrix}x_1\\x_2\end{bmatrix} = \begin{bmatrix}0\\0\end{bmatrix}.$

The solution is $\{(-3t, t) : t \in R\}$. So, an eigenvector corresponding to $\lambda_2 = 7$ is $(-3, 1)$.

19. (a) The characteristic equation is $|\lambda I - A| = \begin{vmatrix}\lambda - 1 & \frac{3}{2}\\-\frac{1}{2} & \lambda + 1\end{vmatrix} = \lambda^2 - \frac{1}{4} = 0.$

(b) The eigenvalues are $\lambda_1 = \frac{1}{2}$ and $\lambda_2 = -\frac{1}{2}$.

For $\lambda_1 = \frac{1}{2}$, $\begin{vmatrix}\lambda_1 - 1 & \frac{3}{2}\\-\frac{1}{2} & \lambda_1 + 1\end{vmatrix}\begin{bmatrix}x_1\\x_2\end{bmatrix} = \begin{bmatrix}0\\0\end{bmatrix} \Rightarrow \begin{bmatrix}1 & -3\\0 & 0\end{bmatrix}\begin{bmatrix}x_1\\x_2\end{bmatrix} = \begin{bmatrix}0\\0\end{bmatrix}.$

The solution is $\{(-3t, t) : t \in R\}$. So, an eigenvector corresponding to $\lambda_1 = \frac{1}{2}$ is $(3, 1)$.

For $\lambda_2 = -\frac{1}{2}$, $\begin{vmatrix}\lambda_2 - 1 & \frac{3}{2}\\-\frac{1}{2} & \lambda_2 + 1\end{vmatrix}\begin{bmatrix}x_1\\x_2\end{bmatrix} = \begin{bmatrix}0\\0\end{bmatrix} \Rightarrow \begin{bmatrix}1 & -1\\0 & 0\end{bmatrix}\begin{bmatrix}x_1\\x_2\end{bmatrix} = \begin{bmatrix}0\\0\end{bmatrix}.$

The solution is $\{(t, t) : t \in R\}$. So, an eigenvector corresponding to $\lambda_2 = -\frac{1}{2}$ is $(1, 1)$.

© 2013 Cengage Learning. All Rights Reserved. May not be scanned, copied or duplicated, or posted to a publicly accessible website, in whole or in part.

21. (a) The characteristic equation is $|\lambda I - A| = \begin{vmatrix} \lambda - 2 & 2 & -3 \\ 0 & \lambda - 3 & 2 \\ 0 & 1 & \lambda - 2 \end{vmatrix} = (\lambda - 2)(\lambda - 4)(\lambda - 1) = 0.$

(b) The eigenvalues are $\lambda_1 = 1, \lambda_2 = 2$, and $\lambda_3 = 4$.

For $\lambda_1 = 1$, $\begin{bmatrix} \lambda_1 - 2 & 2 & -3 \\ 0 & \lambda_1 - 3 & 2 \\ 0 & 1 & \lambda_1 - 2 \end{bmatrix}\begin{bmatrix} x_1 \\ x_2 \\ x_3 \end{bmatrix} = \begin{bmatrix} 0 \\ 0 \\ 0 \end{bmatrix} \Rightarrow \begin{bmatrix} -1 & 2 & -3 \\ 0 & -2 & 2 \\ 0 & 1 & -1 \end{bmatrix}\begin{bmatrix} x_1 \\ x_2 \\ x_3 \end{bmatrix} = \begin{bmatrix} 0 \\ 0 \\ 0 \end{bmatrix}.$

The solution is $\{(-t, t, t) : t \in R\}$. So, an eigenvector corresponding to $\lambda_1 = 1$ is $(-1, 1, 1)$.

For $\lambda_2 = 2$, $\begin{bmatrix} \lambda_2 - 2 & 2 & -3 \\ 0 & \lambda_2 - 3 & 2 \\ 0 & 1 & \lambda_2 - 2 \end{bmatrix}\begin{bmatrix} x_1 \\ x_2 \\ x_3 \end{bmatrix} = \begin{bmatrix} 0 \\ 0 \\ 0 \end{bmatrix} \Rightarrow \begin{bmatrix} 0 & 2 & -3 \\ 0 & -1 & 2 \\ 0 & 1 & 0 \end{bmatrix}\begin{bmatrix} x_1 \\ x_2 \\ x_3 \end{bmatrix} = \begin{bmatrix} 0 \\ 0 \\ 0 \end{bmatrix}.$

The solution is $\{(t, 0, 0) : t \in R\}$. So, an eigenvector corresponding to $\lambda_2 = 2$ is $(1, 0, 0)$.

For $\lambda_3 = 4$, $\begin{bmatrix} \lambda_3 - 2 & 2 & -3 \\ 0 & \lambda_3 - 3 & 2 \\ 0 & 1 & \lambda_3 - 2 \end{bmatrix}\begin{bmatrix} x_1 \\ x_2 \\ x_3 \end{bmatrix} = \begin{bmatrix} 0 \\ 0 \\ 0 \end{bmatrix} \Rightarrow \begin{bmatrix} 2 & 2 & -3 \\ 0 & 1 & 2 \\ 0 & 1 & 2 \end{bmatrix}\begin{bmatrix} x_1 \\ x_2 \\ x_3 \end{bmatrix} = \begin{bmatrix} 0 \\ 0 \\ 0 \end{bmatrix}.$

The solution is $\{(7t, -4t, 2t) : t \in R\}$. So, an eigenvector corresponding to $\lambda_3 = 4$ is $(7, -4, 2)$.

23. (a) The characteristic equation is $|\lambda I - A| = \begin{vmatrix} \lambda - 1 & -2 & 2 \\ 2 & \lambda - 5 & 2 \\ 6 & -6 & \lambda + 3 \end{vmatrix} = \lambda^3 - 3\lambda^2 - 9\lambda + 27 = (\lambda + 3)(\lambda - 3)^2 = 0.$

(b) The eigenvalues are $\lambda_1 = -3$ and $\lambda_2 = 3$ (repeated).

For $\lambda_1 = -3$, $\begin{bmatrix} \lambda_1 - 1 & -2 & 2 \\ 2 & \lambda_1 - 5 & 2 \\ 6 & -6 & \lambda_1 + 3 \end{bmatrix}\begin{bmatrix} x_1 \\ x_2 \\ x_3 \end{bmatrix} = \begin{bmatrix} 0 \\ 0 \\ 0 \end{bmatrix} \Rightarrow \begin{bmatrix} -4 & -2 & 2 \\ 2 & -8 & 2 \\ 6 & -6 & 0 \end{bmatrix}\begin{bmatrix} x_1 \\ x_2 \\ x_3 \end{bmatrix} = \begin{bmatrix} 0 \\ 0 \\ 0 \end{bmatrix}.$

The solution is $\{(t, t, 3t) : t \in R\}$. So, an eigenvector corresponding to $\lambda_1 = -3$ is $(1, 1, 3)$.

For $\lambda_2 = 3$, $\begin{bmatrix} \lambda_2 - 1 & -2 & 2 \\ 2 & \lambda_2 - 5 & 2 \\ 6 & -6 & \lambda_2 + 3 \end{bmatrix}\begin{bmatrix} x_1 \\ x_2 \\ x_3 \end{bmatrix} = \begin{bmatrix} 0 \\ 0 \\ 0 \end{bmatrix} \Rightarrow \begin{bmatrix} 2 & -2 & 2 \\ 2 & -2 & 2 \\ 6 & -6 & 6 \end{bmatrix}\begin{bmatrix} x_1 \\ x_2 \\ x_3 \end{bmatrix} = \begin{bmatrix} 0 \\ 0 \\ 0 \end{bmatrix}.$

The solution is $\{(s - t, s, t) : s, t \in R\}$. So, two eigenvector corresponding to $\lambda_2 = 3$ are $(1, 1, 0)$ and $(1, 0, -1)$.

© 2013 Cengage Learning. All Rights Reserved. May not be scanned, copied or duplicated, or posted to a publicly accessible website, in whole or in part.

25. (a) The characteristic equation is

$$|\lambda I - A| = \begin{vmatrix} \lambda & 3 & -5 \\ 4 & \lambda - 4 & 10 \\ 0 & 0 & \lambda - 4 \end{vmatrix} = (\lambda - 4)(\lambda^2 - 4\lambda - 12) = (\lambda - 4)(\lambda - 6)(\lambda + 2) = 0.$$

(b) The eigenvalues are $\lambda_1 = 4$, $\lambda_2 = 6$ and $\lambda_3 = -2$.

For $\lambda_1 = 4$, $\begin{bmatrix} \lambda_1 & 3 & -5 \\ 4 & \lambda_1 - 4 & 10 \\ 0 & 0 & \lambda_1 - 4 \end{bmatrix}\begin{bmatrix} x_1 \\ x_2 \\ x_3 \end{bmatrix} = \begin{bmatrix} 0 \\ 0 \\ 0 \end{bmatrix} \Rightarrow \begin{bmatrix} 2 & 0 & 5 \\ 0 & 1 & -5 \\ 0 & 0 & 0 \end{bmatrix}\begin{bmatrix} x_1 \\ x_2 \\ x_3 \end{bmatrix} = \begin{bmatrix} 0 \\ 0 \\ 0 \end{bmatrix}$.

The solution is $\{(-5t, 10t, 2t) : t \in R\}$. So, an eigenvector corresponding to $\lambda_1 = 4$ is $(-5, 10, 2)$.

For $\lambda_2 = 6$, $\begin{bmatrix} \lambda_2 & 3 & -5 \\ 4 & \lambda_2 - 4 & 10 \\ 0 & 0 & \lambda_2 - 4 \end{bmatrix}\begin{bmatrix} x_1 \\ x_2 \\ x_3 \end{bmatrix} = \begin{bmatrix} 0 \\ 0 \\ 0 \end{bmatrix} \Rightarrow \begin{bmatrix} 2 & 1 & 0 \\ 0 & 0 & 1 \\ 0 & 0 & 0 \end{bmatrix}\begin{bmatrix} x_1 \\ x_2 \\ x_3 \end{bmatrix} = \begin{bmatrix} 0 \\ 0 \\ 0 \end{bmatrix}$.

The solution is $\{(t, -2t, 0) : t \in R\}$. So, an eigenvector corresponding to $\lambda_2 = 6$ is $(1, -2, 0)$.

For $\lambda_3 = -2$, $\begin{bmatrix} \lambda_3 & 3 & -5 \\ 4 & \lambda_3 - 4 & 10 \\ 0 & 0 & \lambda_3 - 4 \end{bmatrix}\begin{bmatrix} x_1 \\ x_2 \\ x_3 \end{bmatrix} = \begin{bmatrix} 0 \\ 0 \\ 0 \end{bmatrix} \Rightarrow \begin{bmatrix} 2 & -3 & 0 \\ 0 & 0 & 1 \\ 0 & 0 & 0 \end{bmatrix}\begin{bmatrix} x_1 \\ x_2 \\ x_3 \end{bmatrix} = \begin{bmatrix} 0 \\ 0 \\ 0 \end{bmatrix}$.

The solution is $\{(3t, 2t, 0) : t \in R\}$. So, an eigenvector corresponding to $\lambda_3 = -2$ is $(3, 2, 0)$.

27. (a) The characteristic equation is

$$|\lambda I - A| = \begin{vmatrix} \lambda - 2 & 0 & 0 & 0 \\ 0 & \lambda - 2 & 0 & 0 \\ 0 & 0 & \lambda - 3 & -1 \\ 0 & 0 & -4 & \lambda \end{vmatrix} = (\lambda - 2)^2(\lambda^2 - 3\lambda - 4)$$

$$= (\lambda - 4)(\lambda + 1)(\lambda - 2)^2.$$

(b) The eigenvalues are $\lambda_1 = 4$, $\lambda_2 = -1$, and $\lambda_3 = 2$ (repeated).

For $\lambda_1 = 4$, $\begin{bmatrix} 2 & 0 & 0 & 0 \\ 0 & 2 & 0 & 0 \\ 0 & 0 & 1 & -1 \\ 0 & 0 & -4 & 4 \end{bmatrix}\begin{bmatrix} x_1 \\ x_2 \\ x_3 \\ x_4 \end{bmatrix} = \begin{bmatrix} 0 \\ 0 \\ 0 \\ 0 \end{bmatrix} \Rightarrow \begin{bmatrix} 1 & 0 & 0 & 0 \\ 0 & 1 & 0 & 0 \\ 0 & 0 & 1 & -1 \\ 0 & 0 & 0 & 0 \end{bmatrix}\begin{bmatrix} x_1 \\ x_2 \\ x_3 \\ x_4 \end{bmatrix} = \begin{bmatrix} 0 \\ 0 \\ 0 \\ 0 \end{bmatrix}$.

The solution is $\{(0, 0, t, t) : t \in R\}$. So, an eigenvector corresponding to $\lambda_1 = 4$ is $(0, 0, 1, 1)$.

For $\lambda_2 = -1$, $\begin{bmatrix} -3 & 0 & 0 & 0 \\ 0 & -3 & 0 & 0 \\ 0 & 0 & -4 & -1 \\ 0 & 0 & -4 & -1 \end{bmatrix}\begin{bmatrix} x_1 \\ x_2 \\ x_3 \\ x_4 \end{bmatrix} = \begin{bmatrix} 0 \\ 0 \\ 0 \\ 0 \end{bmatrix} \Rightarrow \begin{bmatrix} 1 & 0 & 0 & 0 \\ 0 & 1 & 0 & 0 \\ 0 & 0 & 1 & \frac{1}{4} \\ 0 & 0 & 0 & 0 \end{bmatrix}\begin{bmatrix} x_1 \\ x_2 \\ x_3 \\ x_4 \end{bmatrix} = \begin{bmatrix} 0 \\ 0 \\ 0 \\ 0 \end{bmatrix}$.

The solution is $\{(0, 0, -t, 4t) : t \in R\}$. So, an eigenvector corresponding to $\lambda_2 = -1$ is $(0, 0, -1, 4)$.

For $\lambda_3 = 2$, $\begin{bmatrix} 0 & 0 & 0 & 0 \\ 0 & 0 & 0 & 0 \\ 0 & 0 & -1 & -1 \\ 0 & 0 & -4 & 2 \end{bmatrix}\begin{bmatrix} x_1 \\ x_2 \\ x_3 \\ x_4 \end{bmatrix} = \begin{bmatrix} 0 \\ 0 \\ 0 \\ 0 \end{bmatrix} \Rightarrow \begin{bmatrix} 0 & 0 & 1 & 0 \\ 0 & 0 & 0 & 1 \\ 0 & 0 & 0 & 0 \\ 0 & 0 & 0 & 0 \end{bmatrix}\begin{bmatrix} x_1 \\ x_2 \\ x_3 \\ x_4 \end{bmatrix} = \begin{bmatrix} 0 \\ 0 \\ 0 \\ 0 \end{bmatrix}$.

The solution is $\{(s, t, 0, 0) : s, t \in R\}$. So, two eigenvectors corresponding to $\lambda_3 = 2$ are $(1, 0, 0, 0)$ and $(0, 1, 0, 0)$.

© 2013 Cengage Learning. All Rights Reserved. May not be scanned, copied or duplicated, or posted to a publicly accessible website, in whole or in part.

29. Using a graphing utility: $\lambda = -2, 1$

31. Using a graphing utility: $\lambda = \frac{1}{3}, -\frac{1}{2}, 4$

33. Using a graphing utility: $\lambda = -1, 4, 4$

35. Using a graphing utility: $\lambda = 0, 0, 0, 21$

37. Using a graphing utility: $\lambda = 0, 3$

39. The eigenvalues are the entries on the main diagonal, 2, 3, and 1.

41. The eigenvalues are the entries on the main diagonal, $-2, 4, -3$, and -3.

43. (a) The characteristic equation is $|\lambda I - A| = \begin{vmatrix} \lambda - 2 & 2 \\ -1 & \lambda - 5 \end{vmatrix} = \lambda^2 - 7\lambda + 12 = (\lambda - 3)(\lambda - 4).$

The eigenvalues are $\lambda_1 = 3$ and $\lambda_2 = 4$.

(b) For $\lambda_1 = 3$, $\begin{bmatrix} 1 & 2 \\ -1 & -2 \end{bmatrix}\begin{bmatrix} x_1 \\ x_2 \end{bmatrix} = \begin{bmatrix} 0 \\ 0 \end{bmatrix} \Rightarrow \begin{bmatrix} 1 & 2 \\ 0 & 0 \end{bmatrix}\begin{bmatrix} x_1 \\ x_2 \end{bmatrix} = \begin{bmatrix} 0 \\ 0 \end{bmatrix}$.

The solution is $\{(-2t, t) : t \in R\}$. So, a basis for the eigenspace is $B_1 = \{(-2, 1)\}$.

For $\lambda_2 = 4$, $\begin{bmatrix} 2 & 2 \\ -1 & -1 \end{bmatrix}\begin{bmatrix} x_1 \\ x_2 \end{bmatrix} = \begin{bmatrix} 0 \\ 0 \end{bmatrix} \Rightarrow \begin{bmatrix} 1 & 1 \\ 0 & 0 \end{bmatrix}\begin{bmatrix} x_1 \\ x_2 \end{bmatrix} = \begin{bmatrix} 0 \\ 0 \end{bmatrix}$.

The solution is $\{(t, -t) : t \in R\}$. So, a basis for the eigenspace is $B_2 = \{(1, -1)\}$.

(c) $A' = \begin{bmatrix} 3 & 0 \\ 0 & 4 \end{bmatrix}$

45. (a) The characteristic equation is $|\lambda I - A| = \begin{vmatrix} \lambda & -2 & 1 \\ 1 & \lambda - 3 & -1 \\ 0 & 0 & \lambda + 1 \end{vmatrix} = (\lambda + 1)(\lambda^2 - 3\lambda + 2)$

$= (\lambda + 1)(\lambda - 1)(\lambda - 2).$

The eigenvalues are $\lambda_1 = -1$, $\lambda_2 = 1$, and $\lambda_3 = 2$.

(b) For $\lambda_1 = -1$, $\begin{bmatrix} -1 & -2 & 1 \\ 1 & -4 & -1 \\ 0 & 0 & 0 \end{bmatrix}\begin{bmatrix} x_1 \\ x_2 \\ x_3 \end{bmatrix} = \begin{bmatrix} 0 \\ 0 \\ 0 \end{bmatrix} \Rightarrow \begin{bmatrix} 1 & 0 & -1 \\ 0 & 1 & 0 \\ 0 & 0 & 0 \end{bmatrix}\begin{bmatrix} x_1 \\ x_2 \\ x_3 \end{bmatrix} = \begin{bmatrix} 0 \\ 0 \\ 0 \end{bmatrix}$.

The solution is $\{(t, 0, t) : t \in R\}$. So, a basis for the eigenspace is $B_1 = \{(1, 0, 1)\}$.

For $\lambda_2 = 1$, $\begin{bmatrix} 1 & -2 & 1 \\ 1 & -2 & -1 \\ 0 & 0 & 2 \end{bmatrix}\begin{bmatrix} x_1 \\ x_2 \\ x_3 \end{bmatrix} = \begin{bmatrix} 0 \\ 0 \\ 0 \end{bmatrix} \Rightarrow \begin{bmatrix} 1 & -2 & 0 \\ 0 & 0 & 1 \\ 0 & 0 & 0 \end{bmatrix}\begin{bmatrix} x_1 \\ x_2 \\ x_3 \end{bmatrix} = \begin{bmatrix} 0 \\ 0 \\ 0 \end{bmatrix}$.

The solution is $\{(2t, t, 0) : t \in R\}$. So, a basis for the eigenspace is $B_2 = \{(2, 1, 0)\}$.

For $\lambda_3 = 2$, $\begin{bmatrix} 2 & -2 & 1 \\ 1 & -1 & -1 \\ 0 & 0 & 3 \end{bmatrix}\begin{bmatrix} x_1 \\ x_2 \\ x_3 \end{bmatrix} = \begin{bmatrix} 0 \\ 0 \\ 0 \end{bmatrix} \Rightarrow \begin{bmatrix} 1 & -1 & 0 \\ 0 & 0 & 1 \\ 0 & 0 & 0 \end{bmatrix}\begin{bmatrix} x_1 \\ x_2 \\ x_3 \end{bmatrix} = \begin{bmatrix} 0 \\ 0 \\ 0 \end{bmatrix}$.

The solution is $\{(t, t, 0) : t \in R\}$. So, a basis for the eigenspace is $B_3 = \{(1, 1, 0)\}$.

(c) $A' = \begin{bmatrix} -1 & 0 & 0 \\ 0 & 1 & 0 \\ 0 & 0 & 2 \end{bmatrix}$

© 2013 Cengage Learning. All Rights Reserved. May not be scanned, copied or duplicated, or posted to a publicly accessible website, in whole or in part.

47. The characteristic equation is

$$|\lambda I - A| = \begin{vmatrix} \lambda - 4 & 0 \\ 3 & \lambda - 2 \end{vmatrix} = \lambda^2 - 6\lambda + 8 = 0.$$

Because

$$A^2 - 6A + 8I = \begin{bmatrix} 4 & 0 \\ -3 & 2 \end{bmatrix}^2 - 6\begin{bmatrix} 4 & 0 \\ -3 & 2 \end{bmatrix} + 8\begin{bmatrix} 1 & 0 \\ 0 & 1 \end{bmatrix} = \begin{bmatrix} 16 & 0 \\ -18 & 4 \end{bmatrix} - \begin{bmatrix} 24 & 0 \\ -18 & 12 \end{bmatrix} + \begin{bmatrix} 8 & 0 \\ 0 & 8 \end{bmatrix} = \begin{bmatrix} 0 & 0 \\ 0 & 0 \end{bmatrix}$$

the theorem holds for this matrix.

49. The characteristic equation is

$$|\lambda I - A| = \begin{vmatrix} \lambda - 1 & 0 & 4 \\ 0 & \lambda - 3 & -1 \\ -2 & 0 & \lambda - 1 \end{vmatrix} = \lambda^3 - 5\lambda^2 + 15\lambda - 27 = 0.$$

Because

$$A^3 - 5A^2 + 15A - 27I = \begin{bmatrix} 1 & 0 & -4 \\ 0 & 3 & 1 \\ 2 & 0 & 1 \end{bmatrix}^3 - 5\begin{bmatrix} 1 & 0 & -4 \\ 0 & 3 & 1 \\ 2 & 0 & 1 \end{bmatrix}^2 + 15\begin{bmatrix} 1 & 0 & -4 \\ 0 & 3 & 1 \\ 2 & 0 & 1 \end{bmatrix} - 27\begin{bmatrix} 1 & 0 & 0 \\ 0 & 1 & 0 \\ 0 & 0 & 1 \end{bmatrix}$$

$$= \begin{bmatrix} -23 & 0 & 20 \\ 10 & 27 & 5 \\ -10 & 0 & -23 \end{bmatrix} - 5\begin{bmatrix} -7 & 0 & -8 \\ 2 & 9 & 4 \\ 4 & 0 & -7 \end{bmatrix} + 15\begin{bmatrix} 1 & 0 & -4 \\ 0 & 3 & 1 \\ 2 & 0 & 1 \end{bmatrix} - \begin{bmatrix} 27 & 0 & 0 \\ 0 & 27 & 0 \\ 0 & 0 & 27 \end{bmatrix}$$

$$= \begin{bmatrix} 0 & 0 & 0 \\ 0 & 0 & 0 \\ 0 & 0 & 0 \end{bmatrix}$$

the theorem holds for this matrix.

51. For the $n \times n$ matrix $A = [a_{ij}]$, the sum of the diagonal entries, or the trace, of A is given by $\sum_{i=1}^{n} a_{ii}$.

Exercise 15: $\lambda_1 = 0, \lambda_2 = 7$

(a) $\sum_{i=1}^{2} \lambda_i = 7 = \sum_{i=1}^{2} a_{ii}$

(b) $|A| = \begin{vmatrix} 6 & -3 \\ -2 & 1 \end{vmatrix} = 0 = 0 \cdot 7 = \lambda_1 \cdot \lambda_2$

Exercise 17: $\lambda_1 = \frac{1}{2}, \lambda_2 = -\frac{1}{2}$

(a) $\sum_{i=1}^{2} \lambda_i = 0 = \sum_{i=1}^{2} a_{ii}$

(b) $|A| = \begin{vmatrix} 1 & -\frac{3}{2} \\ \frac{1}{2} & -1 \end{vmatrix} = -\frac{1}{4} = \frac{1}{2} \cdot \left(-\frac{1}{2}\right) = \lambda_1 \cdot \lambda_2$

Exercise 19: $\lambda_1 = 2, \lambda_2 = 3, \lambda_3 = 1$

(a) $\sum_{i=1}^{3} \lambda_i = 6 = \sum_{i=1}^{3} a_{ii}$

(b) $|A| = \begin{vmatrix} 2 & 0 & 1 \\ 0 & 3 & 4 \\ 0 & 0 & 1 \end{vmatrix} = 6 = 2 \cdot 3 \cdot 1 = \lambda_1 \cdot \lambda_2 \cdot \lambda_3$

© 2013 Cengage Learning. All Rights Reserved. May not be scanned, copied or duplicated, or posted to a publicly accessible website, in whole or in part.

Exercise 21: $\lambda_1 = 1, \lambda_2 = 2, \lambda_3 = 4$

(a) $\sum_{i=1}^{3} \lambda_i = 7 = \sum_{i=1}^{3} a_{ii}$

(b) $|A| = \begin{vmatrix} 2 & -2 & 3 \\ 0 & 3 & -2 \\ 0 & -1 & 2 \end{vmatrix} = 8 = 1 \cdot 2 \cdot 4 = \lambda_1 \cdot \lambda_2 \cdot \lambda_3$

Exercise 23: $\lambda_1 = -3, \lambda_2 = 3, \lambda_3 = 3$

(a) $\sum_{i=1}^{3} \lambda_i = 3 = \sum_{i=1}^{3} a_{ii}$

(b) $|A| = \begin{vmatrix} 1 & 2 & -2 \\ -2 & 5 & -2 \\ -6 & 6 & -3 \end{vmatrix} = -27 = -3 \cdot 3 \cdot 3 = \lambda_1 \cdot \lambda_2 \cdot \lambda_3$

Exercise 25: $\lambda_1 = 4, \lambda_2 = 6, \lambda_3 = -2$

(a) $\sum_{i=1}^{3} \lambda_i = 8 = \sum_{i=1}^{3} a_{ii}$

(b) $|A| = \begin{vmatrix} 0 & -3 & 5 \\ -4 & 4 & -10 \\ 0 & 0 & 4 \end{vmatrix} = -48 = 4 \cdot 6 \cdot (-2) = \lambda_1 \cdot \lambda_2 \cdot \lambda_3$

Exercise 27: $\lambda_1 = 4, \lambda_2 = -1, \lambda_3 = 2, \lambda_4 = 2$

(a) $\sum_{i=1}^{4} \lambda_i = 7 = \sum_{i=1}^{4} a_{ii}$

(b) $|A| = \begin{bmatrix} 2 & 0 & 0 & 0 \\ 0 & 2 & 0 & 0 \\ 0 & 0 & 3 & 1 \\ 0 & 0 & 4 & 0 \end{bmatrix} = -16 = 4 \cdot -1 \cdot 2 \cdot 2 = \lambda_1 \cdot \lambda_2 \cdot \lambda_3 \cdot \lambda_4$

53. Because the ith row of A is identical to the ith row of I, the ith row of A consists of zeros except for the main diagonal entry, which is 1. The ith row of $\lambda I - A$ then consists of zeros except for the main diagonal entry, which is $\lambda - 1$. Because $\det(A) = \sum_{j=1}^{n} a_{ij}C_{ij} = a_{i1}C_{i1} + a_{i2}C_{i2} + \cdots + a_{in}C_{in}$ and because each a_{ij} equals zero except for the main diagonal entry, $\det(\lambda I - A) = (\lambda - 1)c_{im} = 0$ where C_{im} is the cofactor determined for the main diagonal entry for this row. So, 1 is an eigenvalue of A.

55. If $A\mathbf{x} = \lambda\mathbf{x}$, then $A^{-1}A\mathbf{x} = A^{-1}\lambda\mathbf{x}$ and $\mathbf{x} = \lambda A^{-1}\mathbf{x}$. So, $A^{-1}\mathbf{x} = \frac{1}{\lambda}\mathbf{x}$, which shows that $\mathbf{x}$ is an eigenvector of A^{-1} with eigenvalue $\frac{1}{\lambda}$. The eigenvectors of A and A^{-1} are the same.

57. The characteristic polynomial of A is $|\lambda I - A|$. The constant term of this polynomial (in λ) is obtained by setting $\lambda = 0$. So, the constant term is $|0I - A| = |-A| = \pm|A|$.

59. Assume that A is a real matrix with eigenvalues $\lambda_1, \lambda_2, \ldots, \lambda_n$ as its main diagonal entries. From Theorem 7.3, the eigenvalues of A are $\lambda_1, \lambda_2, \ldots, \lambda_n$. So, A has real eigenvalues. Because the determinant of A is $|A| = \lambda_1, \lambda_2, \ldots, \lambda_n$, it follows that A is nonsingular if and only if each λ is nonzero.

© 2013 Cengage Learning. All Rights Reserved. May not be scanned, copied or duplicated, or posted to a publicly accessible website, in whole or in part.

61. Let λ_i be an eigenvalue of an $n \times n$ matrix A, and the dimension of the corresponding eigenspace be k. Then there are k linearly independent eigenvectors $u_1, u_2, \ldots, u_k$ corresponding to λ_i. This set can be extended to form a basis for R^n, $B' = \{u_1, u_2, \ldots, u_k, u_{k+1}, \ldots, u_n\}$. Since A represents a linear transformation from R^n to R^n, the matrix A' can be found relative to the basis B'. This matrix has the form

$$A' = \begin{bmatrix} \lambda_i & 0 & 0 & \cdots & 0 & a_{1(k+1)} & \cdots & a_{1n} \\ 0 & \lambda_i & 0 & \cdots & 0 & a_{2(k+1)} & \cdots & a_{2n} \\ 0 & 0 & \lambda_i & \cdots & 0 & a_{3(k+1)} & \cdots & a_{3n} \\ \vdots & \vdots & \vdots & & \vdots & \vdots & & \vdots \\ 0 & 0 & 0 & \cdots & \lambda_i & a_{k(k+1)} & \cdots & a_{kn} \\ 0 & 0 & 0 & \cdots & 0 & a_{(k+1)(k+1)} & \cdots & a_{(k+1)n} \\ \vdots & \vdots & \vdots & & \vdots & \vdots & & \vdots \\ 0 & 0 & 0 & \cdots & 0 & a_{n(k+1)} & \cdots & a_{nn} \end{bmatrix}$$

because A' maps the eigenvectors $u_1, u_2, \ldots, u_k$ onto $\lambda_i u_1, \lambda_i u_2, \ldots, \lambda_i u_k$. Since A and A' represent the same transformation, they have the same eigenvalues, and therefore the same characteristic equation. The polynomial $|\lambda I - A'|$ has a factor of $(\lambda - \lambda_i)^k$ from the first k columns that have λ_i as the main diagonal entries and zeros elsewhere. So, the multiplicity of λ_i is at least k.

63. The characteristic equation of A is

$$\begin{vmatrix} \lambda - a & -b \\ 0 & \lambda - d \end{vmatrix} = (\lambda - a)(\lambda - d)$$
$$= \lambda^2 - (a + d)\lambda + ad = 0.$$

Because the given eigenvalues indicate a characteristic equation of $\lambda(\lambda - 1) = \lambda^2 - \lambda$,

$\lambda^2 - (a + d)\lambda + ad = \lambda^2 - \lambda.$

So, $a = 0$ and $d = 1$, or $a = 1$ and $d = 0$.

65. (a) False. See "Definition of Eigenvalue and Eigenvector," page 342.

(b) True. See discussion before Theorem 7.1, page 344.

(c) True. See Theorem 7.2, page 345.

67. Substituting the value $\lambda = 3$ yields the system

$$\begin{vmatrix} \lambda - 3 & 0 & 0 \\ 0 & \lambda - 3 & 0 \\ 0 & 0 & \lambda - 3 \end{vmatrix} \begin{bmatrix} x_1 \\ x_2 \\ x_3 \end{bmatrix} = \begin{bmatrix} 0 \\ 0 \\ 0 \end{bmatrix} \Rightarrow \begin{bmatrix} 0 & 0 & 0 \\ 0 & 0 & 0 \\ 0 & 0 & 0 \end{bmatrix} \begin{bmatrix} x_1 \\ x_2 \\ x_3 \end{bmatrix} = \begin{bmatrix} 0 \\ 0 \\ 0 \end{bmatrix}.$$

So, 3 has three linearly independent eigenvectors and the dimension of the eigenspace is 3.

69. Substituting the value $\lambda = 3$ yields the system

$$\begin{vmatrix} \lambda - 3 & 1 & 0 \\ 0 & \lambda - 3 & 1 \\ 0 & 0 & \lambda - 3 \end{vmatrix} \begin{bmatrix} x_1 \\ x_2 \\ x_3 \end{bmatrix} = \begin{bmatrix} 0 \\ 0 \\ 0 \end{bmatrix} \Rightarrow \begin{bmatrix} 0 & -1 & 0 \\ 0 & 0 & -1 \\ 0 & 0 & 0 \end{bmatrix} \begin{bmatrix} x_1 \\ x_2 \\ x_3 \end{bmatrix} = \begin{bmatrix} 0 \\ 0 \\ 0 \end{bmatrix}.$$

So, 3 has one linearly independent eigenvector, and the dimension of the eigenspace is 1.

71. $T(e^x) = \dfrac{d}{dx}[e^x] = e^x = 1(e^x)$

Therefore, $\lambda = 1$ is an eigenvalue.

© 2013 Cengage Learning. All Rights Reserved. May not be scanned, copied or duplicated, or posted to a publicly accessible website, in whole or in part.

73. The standard matrix for T is

$$A = \begin{bmatrix} 0 & -3 & 5 \\ -4 & 4 & -10 \\ 0 & 0 & 4 \end{bmatrix}.$$

The characteristic equation of A is

$$\begin{vmatrix} \lambda & 3 & -5 \\ 4 & \lambda - 4 & 10 \\ 0 & 0 & \lambda - 4 \end{vmatrix} = (\lambda + 2)(\lambda - 4)(\lambda - 6) = 0.$$

The eigenvalues of A are $\lambda_1 = -2$, $\lambda_2 = 4$, and $\lambda_3 = 6$. The corresponding eigenvectors are found by solving

$$\begin{bmatrix} \lambda_i & 3 & -5 \\ 4 & \lambda_i - 4 & 10 \\ 0 & 0 & \lambda_i - 4 \end{bmatrix}\begin{bmatrix} a_0 \\ a_1 \\ a_2 \end{bmatrix} = \begin{bmatrix} 0 \\ 0 \\ 0 \end{bmatrix}$$

for each λ_i. Thus,

$p_1(x) = 3 + 2x$, $p_2(x) = -5 + 10x + 2x^2$ and $p_3(x) = -1 + 2x$ are eigenvectors corresponding to λ_1, λ_2, and λ_3.

75. The standard matrix for T is

$$A = \begin{bmatrix} 1 & 0 & -1 & 1 \\ 0 & 1 & 0 & 1 \\ -2 & 0 & 2 & -2 \\ 0 & 2 & 0 & 2 \end{bmatrix}.$$

Because the standard matrix is the same as that in Exercise 37, you know the eigenvalues are $\lambda_1 = 0$ and $\lambda_2 = 3$. So, two eigenvectors corresponding to $\lambda_1 = 0$ are

$$\mathbf{x}_1 = \begin{bmatrix} 1 \\ 0 \\ 1 \\ 0 \end{bmatrix} \text{ and } \mathbf{x}_2 = \begin{bmatrix} 1 \\ 1 \\ 0 \\ -1 \end{bmatrix}$$

and an eigenvector corresponding to $\lambda_2 = 3$ is

$$\mathbf{x}_3 = \begin{bmatrix} 1 \\ 0 \\ -2 \\ 0 \end{bmatrix}.$$

77. The possible eigenvalues of an idempotent matrix are 0 and 1. Suppose $A\mathbf{x} = \lambda\mathbf{x}$, where $A^2 = A$. Then

$$\lambda\mathbf{x} = A\mathbf{x} = A^2\mathbf{x} = A(A\mathbf{x}) = A(\lambda\mathbf{x}) = \lambda^2\mathbf{x} \Rightarrow (\lambda^2 - \lambda)\mathbf{x} = \mathbf{0}.$$

Because $\mathbf{x} \neq \mathbf{0}$, $\lambda^2 - \lambda = 0 \Rightarrow \lambda = 0, 1$.

79. Let $\mathbf{x} = \begin{bmatrix} 1 \\ 1 \\ \vdots \\ 1 \end{bmatrix}$. Then $A\mathbf{x} = \begin{bmatrix} r \\ r \\ \vdots \\ r \end{bmatrix} = r\mathbf{x}$,

which shows that r is an eigenvalue of A with eigenvector $\mathbf{x}$.

For example, let $A = \begin{bmatrix} 1 & 2 \\ 3 & 0 \end{bmatrix}$.

Then $\begin{bmatrix} 1 & 2 \\ 3 & 0 \end{bmatrix}\begin{bmatrix} 1 \\ 1 \end{bmatrix} = \begin{bmatrix} 3 \\ 3 \end{bmatrix} = 3\begin{bmatrix} 1 \\ 1 \end{bmatrix}$.

Section 7.2 Diagonalization

1. (b) $\lambda = 1, -2$

3. (b) $\lambda = 2, -3$

5. (b) $\lambda = 5, 3, -1$

7. The eigenvalues of A are $\lambda_1 = 0$, $\lambda_2 = 7$ (see Exercise 17, Section 7.1). The corresponding eigenvectors $(1, 2)$ and $(-3, 1)$ are used to form the columns of P. So,

$$P = \begin{bmatrix} 1 & -3 \\ 2 & 1 \end{bmatrix} \Rightarrow P^{-1} = \begin{bmatrix} \frac{1}{7} & \frac{3}{7} \\ -\frac{2}{7} & \frac{1}{7} \end{bmatrix}$$

and

$$P^{-1}AP = \begin{bmatrix} \frac{1}{7} & \frac{3}{7} \\ -\frac{2}{7} & \frac{1}{7} \end{bmatrix}\begin{bmatrix} 6 & -3 \\ -2 & 1 \end{bmatrix}\begin{bmatrix} 1 & -3 \\ 2 & 1 \end{bmatrix} = \begin{bmatrix} 0 & 0 \\ 0 & 7 \end{bmatrix}.$$

© 2013 Cengage Learning. All Rights Reserved. May not be scanned, copied or duplicated, or posted to a publicly accessible website, in whole or in part.

9. The eigenvalues of A are $\lambda_1 = 1$, $\lambda_2 = 2$, $\lambda_3 = 4$ (see Exercise 21, Section 7.1). The corresponding eigenvectors $(-1, 1, 1)$, $(1, 0, 0)$, $(7, -4, 2)$ are used to form the columns of P. So,

$$P = \begin{bmatrix} -1 & 1 & 7 \\ 1 & 0 & -4 \\ 1 & 0 & 2 \end{bmatrix} \Rightarrow P^{-1} = \begin{bmatrix} 0 & \frac{1}{3} & \frac{2}{3} \\ 1 & \frac{3}{2} & -\frac{1}{2} \\ 0 & -\frac{1}{6} & \frac{1}{6} \end{bmatrix}$$

and

$$P^{-1}AP = \begin{bmatrix} 0 & \frac{1}{3} & \frac{2}{3} \\ 1 & \frac{3}{2} & -\frac{1}{2} \\ 0 & -\frac{1}{6} & \frac{1}{6} \end{bmatrix}\begin{bmatrix} 2 & -2 & 3 \\ 0 & 3 & -2 \\ 0 & -1 & 2 \end{bmatrix}\begin{bmatrix} -1 & 1 & 7 \\ 1 & 0 & -4 \\ 1 & 0 & 2 \end{bmatrix} = \begin{bmatrix} 1 & 0 & 0 \\ 0 & 2 & 0 \\ 0 & 0 & 4 \end{bmatrix}.$$

11. The eigenvalues of A are $\lambda_1 = -3$ and $\lambda_2 = 3$ (repeated) (see Exercise 23, Section 7.1).

The corresponding eigenvectors $(1, 1, 3)$, $(1, 1, 0)$, and $(1, 0, -1)$ are used to form the columns of P. So,

$$P = \begin{bmatrix} 1 & 1 & 1 \\ 1 & 1 & 0 \\ 3 & 0 & -1 \end{bmatrix} \Rightarrow P^{-1} = \begin{bmatrix} \frac{1}{3} & -\frac{1}{3} & \frac{1}{3} \\ -\frac{1}{3} & \frac{4}{3} & -\frac{1}{3} \\ 1 & -1 & 0 \end{bmatrix}$$

and

$$P^{-1}AP = \begin{bmatrix} \frac{1}{3} & -\frac{1}{3} & \frac{1}{3} \\ -\frac{1}{3} & \frac{4}{3} & -\frac{1}{3} \\ 1 & -1 & 0 \end{bmatrix}\begin{bmatrix} 1 & 2 & -2 \\ -2 & 5 & -2 \\ -6 & 6 & -3 \end{bmatrix}\begin{bmatrix} 1 & 1 & 1 \\ 1 & 1 & 0 \\ 3 & 0 & -1 \end{bmatrix} = \begin{bmatrix} -3 & 0 & 0 \\ 0 & 3 & 0 \\ 0 & 0 & 3 \end{bmatrix}.$$

13. The eigenvalues of A are $\lambda_1 = 1$ and $\lambda_2 = 2$. Furthermore, there are just two linearly independent eigenvectors of A, $\mathbf{x}_1 = (-1, 0, 1)$ and $\mathbf{x}_2 = (0, 1, 0)$. So, A is not diagonalizable.

15. A has only one eigenvalue, $\lambda = 0$, and a basis for the eigenspace is $\{(0, 1)\}$. So, A does not satisfy Theorem 7.5 (it does not have two linearly independent eigenvectors) and is not diagonalizable.

17. The matrix has eigenvalues $\lambda = 1$ (repeated), and a basis for the eigenspace is $\{(1, 0)\}$. So, A does not satisfy Theorem 7.5 (it does not have two linearly independent eigenvectors) and it is not diagonalizable.

19. The matrix has eigenvalues $\lambda = 1$ (repeated) and $\lambda_2 = 2$. A basis for the eigenspace associated with $\lambda = 1$ is $\{(1, 0, 0)\}$. So, the matrix has only two linearly independent eigenvectors and, by Theorem 7.5 it is not diagonalizable.

21. From Exercise 37, Section 7.1, A has only three linearly independent eigenvectors. So, A does not satisfy Theorem 7.5 and is not diagonalizable.

23. The eigenvalues of A are $\lambda_1 = 0$ and $\lambda_2 = 2$. Because A has two distinct eigenvalues, it is diagonalizable (by Theorem 7.6).

25. The eigenvalues of A are $\lambda = 0$ and $\lambda = -2$ (repeated). Because A does not have three distinct eigenvalues, Theorem 7.6 does not guarantee that A is diagonalizable.

27. The standard matrix for T is

$$A = \begin{bmatrix} 1 & 1 \\ 1 & 1 \end{bmatrix}$$

which has eigenvalues $\lambda_1 = 0$ and $\lambda_2 = 2$ and corresponding eigenvectors $(1, -1)$ and $(1, 1)$. Let $B = \{(1, -1), (1, 1)\}$ and find the image of each vector in B.

$$[T(1, -1)]_B = [(0, 0)]_B = (0, 0)$$
$$[T(1, 1)]_B = [(2, 2)]_B = (0, 2)$$

The matrix of T relative to B is then

$$A' = \begin{bmatrix} 0 & 0 \\ 0 & 2 \end{bmatrix}.$$

© 2013 Cengage Learning. All Rights Reserved. May not be scanned, copied or duplicated, or posted to a publicly accessible website, in whole or in part.

29. The standard matrix for T is

$$A = \begin{bmatrix} 1 & 0 \\ 1 & 2 \end{bmatrix}$$

which has eigenvalues $\lambda_1 = 1$ and $\lambda_2 = 2$ and corresponding eigenvectors $-1 + x$ and x. Let $B = \{-1 + x, x\}$ and find the image of each vector in B.

$$[T(-1 + x)]_B = [-1 + x]_B = (1, 0)$$

$$[T(x)]_B = [2x]_B = (0, 2)$$

The matrix of T relative to B is then

$$A' = \begin{bmatrix} 1 & 0 \\ 0 & 2 \end{bmatrix}.$$

31. $B = P^{-1}AP \Rightarrow A = PBP^{-1} \Rightarrow A^k = PB^kP^{-1}$

33. The eigenvalues and corresponding eigenvectors of A are $\lambda_1 = -2$, $\lambda_2 = 1$, $\mathbf{x}_1 = \left(-\frac{3}{2}, 1\right)$, and $\mathbf{x}_2 = (-2, 1)$. Construct a nonsingular matrix P from the eigenvectors of A,

$$P = \begin{bmatrix} -\frac{3}{2} & -2 \\ 1 & 1 \end{bmatrix}$$

and find a diagonal matrix B similar to A.

$$B = P^{-1}AP = \begin{bmatrix} 2 & 4 \\ -2 & -3 \end{bmatrix}\begin{bmatrix} 10 & 18 \\ -6 & -11 \end{bmatrix}\begin{bmatrix} -\frac{3}{2} & -2 \\ 1 & 1 \end{bmatrix} = \begin{bmatrix} -2 & 0 \\ 0 & 1 \end{bmatrix}$$

Then

$$A^6 = PB^6P^{-1} = \begin{bmatrix} -\frac{3}{2} & -2 \\ 1 & 1 \end{bmatrix}\begin{bmatrix} 64 & 0 \\ 0 & 1 \end{bmatrix}\begin{bmatrix} 2 & 4 \\ -2 & -3 \end{bmatrix} = \begin{bmatrix} -188 & -378 \\ 126 & 253 \end{bmatrix}.$$

35. The eigenvalues and corresponding eigenvectors of A are $\lambda_1 = 0$, $\lambda_2 = 2$ (repeated), $\mathbf{x}_1 = (-1, 3, 1)$, $\mathbf{x}_2 = (3, 0, 1)$ and $\mathbf{x}_3 = (-2, 1, 0)$. Construct a nonsingular matrix P from the eigenvectors of A.

$$P = \begin{bmatrix} -1 & 3 & -2 \\ 3 & 0 & 1 \\ 1 & 1 & 0 \end{bmatrix}$$

and find a diagonal matrix B similar to A.

$$B = P^{-1}AP = \begin{bmatrix} \frac{1}{2} & 1 & -\frac{3}{2} \\ -\frac{1}{2} & -1 & \frac{5}{2} \\ -\frac{3}{2} & -2 & \frac{9}{2} \end{bmatrix}\begin{bmatrix} 3 & 2 & -3 \\ -3 & -4 & 9 \\ -1 & -2 & 5 \end{bmatrix}\begin{bmatrix} -1 & 3 & -2 \\ 3 & 0 & 1 \\ 1 & 1 & 0 \end{bmatrix} = \begin{bmatrix} 0 & 0 & 0 \\ 0 & 2 & 0 \\ 0 & 0 & 2 \end{bmatrix}$$

Then,

$$A^8 = PB^8P^{-1} = P\begin{bmatrix} 0 & 0 & 0 \\ 0 & 256 & 0 \\ 0 & 0 & 256 \end{bmatrix}P^{-1} = \begin{bmatrix} 384 & 256 & -384 \\ -384 & -512 & 1152 \\ -128 & -256 & 640 \end{bmatrix}.$$

37. (a) True. See the proof of Theorem 7.4, pages 354.

(b) False. See Theorem 7.6, page 358.

© 2013 Cengage Learning. All Rights Reserved. May not be scanned, copied or duplicated, or posted to a publicly accessible website, in whole or in part.

39. Yes, the matrices are similar. Let $P = \begin{bmatrix} 0 & 0 & 1 \\ 0 & 1 & 0 \\ 1 & 0 & 0 \end{bmatrix} \Rightarrow P^{-1} = \begin{bmatrix} 0 & 0 & 1 \\ 0 & 1 & 0 \\ 1 & 0 & 0 \end{bmatrix}$ and observe that

$$P^{-1}AP = \begin{bmatrix} 0 & 0 & 1 \\ 0 & 1 & 0 \\ 1 & 0 & 0 \end{bmatrix}\begin{bmatrix} 1 & 0 & 0 \\ 0 & 2 & 0 \\ 0 & 0 & 3 \end{bmatrix}\begin{bmatrix} 0 & 0 & 1 \\ 0 & 1 & 0 \\ 1 & 0 & 0 \end{bmatrix} = \begin{bmatrix} 3 & 0 & 0 \\ 0 & 2 & 0 \\ 0 & 0 & 1 \end{bmatrix} = B.$$

41. Yes. The matrix $\begin{bmatrix} 2 & 0 & 0 \\ 0 & 5 & 0 \\ 0 & 0 & 7 \end{bmatrix}$ is similar to $\begin{bmatrix} 5 & 0 & 0 \\ 0 & 7 & 0 \\ 0 & 0 & 2 \end{bmatrix}$ and $\begin{bmatrix} 7 & 0 & 0 \\ 0 & 2 & 0 \\ 0 & 0 & 5 \end{bmatrix}$.

43. Assume that A is diagonalizable with n real eigenvalues $\lambda_1, \ldots, \lambda_n$. Then if $PAP^{-1} = D$, D is diagonal,

$$|A| = |P^{-1}AP| = \begin{vmatrix} \lambda_1 & 0 & \ldots & 0 \\ 0 & \lambda_2 & & \vdots \\ \vdots & & & 0 \\ 0 & \ldots & 0 & \lambda_n \end{vmatrix} = \lambda_1\lambda_2 \ldots \lambda_n.$$

45. Let the eigenvalues of the diagonalizable matrix A be all ± 1. Then there exists an invertible matrix P such that

$$P^{-1}AP = D$$

where D is diagonal with ± 1 along the main diagonal. So $A = PDP^{-1}$ and because $D^{-1} = D$,

$$A^{-1} = \left(PDP^{-1}\right)^{-1} = \left(P^{-1}\right)^{-1} D^{-1}P^{-1} = PDP^{-1} = A.$$

47. Given that $P^{-1}AP = D$, where D is diagonal,

$$A = PDP^{-1} \text{ and } A^{-1} = \left(PDP^{-1}\right)^{-1} = \left(P^{-1}\right)^{-1} D^{-1}P^{-1} = PD^{-1}P^{-1} \quad \Rightarrow \quad P^{-1}A^{-1}P = D^{-1},$$

which shows that A^{-1} is diagonalizable.

49. A is triangular so, the eigenvalues are simply the entries on the main diagonal. So, the only eigenvalue is $\lambda = 3$, and a basis for the eigenspace is $\{(1, 0)\}$.

Because matrix A does not have two linearly independent eigenvectors, it does not satisfy Theorem 7.5 and it is not diagonalizable.

Section 7.3 Symmetric Matrices and Orthogonal Diagonalization

1. Because $\begin{bmatrix} 1 & 3 \\ 3 & -1 \end{bmatrix}^T = \begin{bmatrix} 1 & 3 \\ 3 & -1 \end{bmatrix}$

the matrix *is* symmetric.

3. Because

$$\begin{bmatrix} 4 & -2 & 1 \\ 3 & 1 & 2 \\ 1 & 2 & 1 \end{bmatrix}^T \neq \begin{bmatrix} 4 & 3 & 1 \\ -2 & 1 & 2 \\ 1 & 2 & 1 \end{bmatrix}$$

the matrix is *not* symmetric.

5. Because

$$\begin{bmatrix} 0 & 1 & 2 & -1 \\ 1 & 0 & -3 & 2 \\ 2 & -3 & 0 & 1 \\ -1 & 2 & 1 & -2 \end{bmatrix}^T = \begin{bmatrix} 0 & 1 & 2 & -1 \\ 1 & 0 & -3 & 2 \\ 2 & -3 & 0 & 1 \\ -1 & 2 & 1 & -2 \end{bmatrix}$$

the matrix *is* symmetric.

© 2013 Cengage Learning. All Rights Reserved. May not be scanned, copied or duplicated, or posted to a publicly accessible website, in whole or in part.

7. The characteristic equation of A is

$$|\lambda I - A| = \begin{vmatrix} \lambda & 0 & -a \\ 0 & \lambda - a & 0 \\ -a & 0 & \lambda \end{vmatrix} = \lambda^2(\lambda - a) - a^2(\lambda - a) = (\lambda + a)(\lambda - a)^2 = 0.$$

The eigenvalues are $\lambda_1 = -a$ and $\lambda_2 = a$. Since the eigenvalues are real, A is diagonalizable. The corresponding eigenvectors are $(1, 0, -1)$ for λ_1 and $(1, 0, 1)$ and $(0, 1, 0)$ for λ_2. So, $P = \begin{bmatrix} 1 & 1 & 0 \\ 0 & 0 & 1 \\ -1 & 1 & 0 \end{bmatrix}$ and

$$P^{-1}AP = \begin{bmatrix} \frac{1}{2} & 0 & -\frac{1}{2} \\ \frac{1}{2} & 0 & \frac{1}{2} \\ 0 & 1 & 0 \end{bmatrix}\begin{bmatrix} 0 & 0 & a \\ 0 & a & 0 \\ a & 0 & 0 \end{bmatrix}\begin{bmatrix} 1 & 1 & 0 \\ 0 & 0 & 1 \\ -1 & 1 & 0 \end{bmatrix} = \begin{bmatrix} -a & 0 & 0 \\ 0 & a & 0 \\ 0 & 0 & a \end{bmatrix}.$$

9. The characteristic equation of A is

$$|\lambda I - A| = \begin{vmatrix} \lambda - a & 0 & -a \\ 0 & \lambda - a & 0 \\ -a & 0 & \lambda - a \end{vmatrix} = \lambda(\lambda - a)(\lambda - 2a).$$

The eigenvalues are $\lambda_1 = 0$, $\lambda_2 = a$, and $\lambda_3 = 2a$. Since the eigenvalues are real, A is diagonalizable. The corresponding eigenvectors are $(1, 0, -1)$, $(0, 1, 0)$, and $(1, 0, 1)$, respectively. So,

$$P = \begin{bmatrix} 1 & 0 & 1 \\ 0 & 1 & 0 \\ -1 & 0 & 1 \end{bmatrix} \text{ and}$$

$$P^{-1}AP = \begin{bmatrix} \frac{1}{2} & 0 & -\frac{1}{2} \\ 0 & 1 & 0 \\ \frac{1}{2} & 0 & \frac{1}{2} \end{bmatrix}\begin{bmatrix} a & 0 & a \\ 0 & a & 0 \\ a & 0 & a \end{bmatrix}\begin{bmatrix} 1 & 0 & 1 \\ 0 & 1 & 0 \\ -1 & 0 & 1 \end{bmatrix} = \begin{bmatrix} 0 & 0 & 0 \\ 0 & a & 0 \\ 0 & 0 & 2a \end{bmatrix}.$$

11. The characteristic equation of A is

$$|\lambda I - A| = \begin{vmatrix} \lambda - 2 & -1 \\ -1 & \lambda - 2 \end{vmatrix} = (\lambda - 3)(\lambda - 1).$$

The eigenvalues are $\lambda_1 = 3$ and $\lambda_2 = 1$. The dimension of the corresponding eigenspace of each eigenvalue is 1 (by Theorem 7.7).

13. The characteristic equation of A is

$$|\lambda I - A| = \begin{vmatrix} \lambda - 3 & 0 & 0 \\ 0 & \lambda - 2 & 0 \\ 0 & 0 & \lambda - 2 \end{vmatrix}$$

$$= (\lambda - 3)(\lambda - 2)^2 = 0.$$

Therefore, the eigenvalues of A are $\lambda_1 = 3$ and $\lambda_2 = 2$. The dimension of the eigenspace corresponding $\lambda_1 = 3$ is 1. The multiplicity of $\lambda_2 = 2$ is 2, so the dimension of the corresponding eigenspace is 2 (by Theorem 7.7).

15. The characteristic equation of A is

$$|\lambda I - A| = \begin{vmatrix} \lambda & -2 & -2 \\ -2 & \lambda & -2 \\ -2 & -2 & \lambda \end{vmatrix} = (\lambda + 2)^2(\lambda - 4) = 0.$$

The eigenvalues of A are $\lambda_1 = -2$ and $\lambda_2 = 4$. The multiplicity of $\lambda_1 = -2$ is 2, so the dimension of the corresponding eigenspace is 2 (by Theorem 7.7). The dimension for the eigenspace corresponding to $\lambda_2 = 4$ is 1.

17. The characteristic equation of A is

$$|\lambda I - A| = \begin{vmatrix} \lambda & -1 & -1 \\ -1 & \lambda & -1 \\ -1 & -1 & \lambda - 1 \end{vmatrix}$$

$$= (\lambda + 1)(\lambda^2 - 2\lambda - 1) = 0.$$

Therefore, the eigenvalues are $\lambda_1 = -1$, $\lambda_2 = 1 - \sqrt{2}$, and $\lambda_3 = 1 + \sqrt{2}$. The dimension of the eigenspace corresponding to each eigenvalue is 1.

© 2013 Cengage Learning. All Rights Reserved. May not be scanned, copied or duplicated, or posted to a publicly accessible website, in whole or in part.

19. The characteristic equation of A is

$$|\lambda I - A| = \begin{vmatrix} \lambda - 3 & 0 & 0 & 0 \\ 0 & \lambda - 3 & 0 & 0 \\ 0 & 0 & \lambda - 3 & -5 \\ 0 & 0 & -5 & \lambda - 3 \end{vmatrix} = (\lambda + 2)(\lambda - 3)^2(\lambda - 8).$$

The eigenvalues are $\lambda_1 = -2$, $\lambda_2 = 3$, and $\lambda_3 = 8$. The dimensions of the corresponding eigenspaces are 1, 2, and 1, respectively (by Theorem 7.7).

21. The characteristic equation of A is

$$|\lambda I - A| = \begin{vmatrix} \lambda - 2 & 1 & 0 & 0 & 0 \\ 1 & \lambda - 2 & 0 & 0 & 0 \\ 0 & 0 & \lambda - 2 & 0 & 0 \\ 0 & 0 & 0 & \lambda - 2 & 0 \\ 0 & 0 & 0 & 0 & \lambda - 2 \end{vmatrix} = (\lambda - 1)(\lambda - 3)(\lambda - 2)^3.$$

The eigenvalues are $\lambda_1 = 1$, $\lambda_2 = 3$, and $\lambda_3 = 2$. The dimensions of the corresponding eigenspaces are 1, 1, and 3, respectively (by Theorem 7.7).

23. Because $PP^T = \begin{bmatrix} \frac{\sqrt{2}}{2} & \frac{\sqrt{2}}{2} \\ -\frac{\sqrt{6}}{2} & \frac{\sqrt{2}}{2} \end{bmatrix}\begin{bmatrix} \frac{\sqrt{2}}{2} & -\frac{\sqrt{2}}{2} \\ \frac{\sqrt{2}}{2} & \frac{\sqrt{2}}{2} \end{bmatrix} = I_2$, $P^T = P^{-1}$ and P is orthogonal.

Letting $P_1 = \begin{bmatrix} \frac{\sqrt{2}}{2} \\ -\frac{\sqrt{2}}{2} \end{bmatrix}$ and $P_2 = \begin{bmatrix} \frac{\sqrt{2}}{2} \\ \frac{\sqrt{2}}{2} \end{bmatrix}$ produces $p_1 \cdot p_2 = 0$ and $\|p_1\| = \|p_2\| = 1$. So, $\{p_1, p_2\}$ is an orthonormal set.

25. Because $PP^T = \begin{bmatrix} \frac{2}{3} & -\frac{2}{3} & \frac{1}{3} \\ \frac{2}{3} & \frac{1}{3} & -\frac{2}{3} \\ \frac{1}{3} & \frac{2}{3} & \frac{2}{3} \end{bmatrix}\begin{bmatrix} \frac{2}{3} & \frac{2}{3} & \frac{1}{3} \\ -\frac{2}{3} & \frac{1}{3} & \frac{2}{3} \\ \frac{1}{3} & -\frac{2}{3} & \frac{2}{3} \end{bmatrix} = I_3$, $P^T = P^{-1}$ and P is orthogonal.

Letting $p_1 = \begin{bmatrix} \frac{2}{3} \\ \frac{2}{3} \\ \frac{1}{3} \end{bmatrix}$, $p_2 = \begin{bmatrix} -\frac{2}{3} \\ \frac{1}{3} \\ \frac{2}{3} \end{bmatrix}$, and $p_3 = \begin{bmatrix} \frac{1}{3} \\ -\frac{2}{3} \\ \frac{2}{3} \end{bmatrix}$ produces

$p_1 \cdot p_2 = p_1 \cdot p_3 = p_2 \cdot p_3 = 0$ and $\|p_1\| = \|p_2\| = \|p_3\| = 1$. So, $\{p_1, p_2, p_3\}$ is an orthonormal set.

27. Because $PP^T = \begin{bmatrix} -4 & 0 & 3 \\ 0 & 1 & 0 \\ 3 & 0 & 4 \end{bmatrix}\begin{bmatrix} -4 & 0 & 3 \\ 0 & 1 & 0 \\ 3 & 0 & 4 \end{bmatrix} = \begin{bmatrix} 25 & 0 & 0 \\ 0 & 1 & 0 \\ 0 & 0 & 25 \end{bmatrix} \neq I_3$,

P is not orthogonal.

© 2013 Cengage Learning. All Rights Reserved. May not be scanned, copied or duplicated, or posted to a publicly accessible website, in whole or in part.

29. Because $PP^T = \begin{bmatrix} \frac{\sqrt{2}}{2} & -\frac{\sqrt{6}}{6} & \frac{\sqrt{3}}{3} \\ 0 & \frac{\sqrt{6}}{3} & \frac{\sqrt{3}}{3} \\ \frac{\sqrt{2}}{2} & \frac{\sqrt{6}}{6} & -\frac{\sqrt{3}}{3} \end{bmatrix} \begin{bmatrix} \frac{\sqrt{2}}{2} & 0 & \frac{\sqrt{2}}{2} \\ -\frac{\sqrt{6}}{6} & \frac{\sqrt{6}}{3} & \frac{\sqrt{6}}{6} \\ \frac{\sqrt{3}}{3} & \frac{\sqrt{3}}{3} & -\frac{\sqrt{3}}{3} \end{bmatrix} = I_3$, $P^T = P^{-1}$ and P is orthogonal.

Letting $p_1 = \begin{bmatrix} \frac{\sqrt{2}}{2} \\ 0 \\ \frac{\sqrt{2}}{2} \end{bmatrix}$, $p_2 = \begin{bmatrix} -\frac{\sqrt{6}}{6} \\ \frac{\sqrt{6}}{3} \\ \frac{\sqrt{6}}{6} \end{bmatrix}$, and $p_3 = \begin{bmatrix} \frac{\sqrt{3}}{3} \\ \frac{\sqrt{3}}{3} \\ -\frac{\sqrt{3}}{3} \end{bmatrix}$ produces $p_1 \cdot p_2 = p_1 \cdot p_3 = p_2 \cdot p_3 = 0$ and

$\|p_1\| = \|p_2\| = \|p_3\| = 1$. So, $\{p_1, p_2, p_3\}$ is an orthonormal set.

31. Because $PP^T = \begin{bmatrix} \frac{1}{8} & 0 & 0 & \frac{3\sqrt{7}}{8} \\ 0 & 1 & 0 & 0 \\ 0 & 0 & 1 & 0 \\ \frac{3\sqrt{7}}{8} & 0 & 0 & \frac{1}{8} \end{bmatrix} \begin{bmatrix} \frac{1}{8} & 0 & 0 & \frac{3\sqrt{7}}{8} \\ 0 & 1 & 0 & 0 \\ 0 & 0 & 1 & 0 \\ \frac{3\sqrt{7}}{8} & 0 & 0 & \frac{1}{8} \end{bmatrix} = \begin{bmatrix} 1 & 0 & 0 & \frac{3\sqrt{7}}{32} \\ 0 & 1 & 0 & 0 \\ 0 & 0 & 1 & 0 \\ \frac{3\sqrt{7}}{32} & 0 & 0 & 1 \end{bmatrix} \neq I_4$,

P is not orthogonal.

33. The characteristic polynomial of A is

$$|\lambda I - A| = \begin{vmatrix} \lambda - 3 & -3 \\ -3 & \lambda - 3 \end{vmatrix} = \lambda(\lambda - 6).$$

The eigenvalues are $\lambda_1 = 0$ and $\lambda_2 = 6$. Every eigenvector corresponding to $\lambda_1 = 0$ is of the form $x_1 = (t, -t)$, and every eigenvector corresponding to $\lambda_2 = 6$ is of the form $x_2 = (s, s)$.

$x_1 \cdot x_2 = st - st = 0$

So, x_1 and x_2 are orthogonal.

35. The matrix is diagonal, so the eigenvalues are $\lambda_1 = 1$ and $\lambda_2 = 3$. Every eigenvector corresponding to $\lambda_1 = 1$ is of the form $x_1 = (t_1, t_2, 0)$, and every eigenvector corresponding to $\lambda_2 = 3$ is of the form $x_2 = (0, 0, s)$.

$x_1 \cdot x_2 = 0$

So, x_1 and x_2 are orthogonal.

37. The characteristic polynomial of A is

$$|\lambda I - A| = \begin{vmatrix} \lambda & -\sqrt{3} & 0 \\ -\sqrt{3} & \lambda & 1 \\ 0 & 1 & \lambda \end{vmatrix} = \lambda(\lambda + 2)(\lambda - 2).$$

The eigenvalues are $\lambda_1 = 0$, $\lambda_2 = -2$, and $\lambda_3 = 2$. Every eigenvector corresponding to $\lambda_1 = 0$ is of the form $x_1 = (t, 0, \sqrt{3}t)$, every eigenvector corresponding to $\lambda_2 = -2$ is of the form $x_2 = (-\sqrt{3}s, 2s, s)$, and every eigenvector corresponding to $\lambda_3 = 2$ is of the form $x_3 = (-\sqrt{3}u, -2u, u)$.

$x_1 \cdot x_2 = x_1 \cdot x_3 = x_2 \cdot x_3 = 0$

So, $\{x_1, x_2, x_3\}$ is an orthogonal set.

39. The matrix is not symmetric, so it is not orthogonally diagonalizable.

41. The matrix is symmetric, so it is orthogonally diagonalizable.

© 2013 Cengage Learning. All Rights Reserved. May not be scanned, copied or duplicated, or posted to a publicly accessible website, in whole or in part.

43. The eigenvalues of A are $\lambda_1 = 0$ and $\lambda_2 = 2$, with corresponding eigenvectors $(1, -1)$ and $(1, 1)$, respectively.

Normalize each eigenvector to form the columns of P. Then

$$P = \begin{bmatrix} \frac{\sqrt{2}}{2} & \frac{\sqrt{2}}{2} \\ -\frac{\sqrt{2}}{2} & \frac{\sqrt{2}}{2} \end{bmatrix}$$

and

$$P^T AP = \begin{bmatrix} \frac{\sqrt{2}}{2} & -\frac{\sqrt{2}}{2} \\ \frac{\sqrt{2}}{2} & \frac{\sqrt{2}}{2} \end{bmatrix} \begin{bmatrix} 1 & 1 \\ 1 & 1 \end{bmatrix} \begin{bmatrix} \frac{\sqrt{2}}{2} & \frac{\sqrt{2}}{2} \\ -\frac{\sqrt{2}}{2} & \frac{\sqrt{2}}{2} \end{bmatrix} = \begin{bmatrix} 0 & 0 \\ 0 & 2 \end{bmatrix}.$$

45. The eigenvalues of A are $\lambda_1 = 0$ and $\lambda_2 = 3$, with corresponding eigenvectors $\left(\frac{\sqrt{2}}{2}, -1\right)$ and $\left(\sqrt{2}, 1\right)$, respectively. Normalize each eigenvector to form the columns of P. Then

$$P = \begin{bmatrix} \frac{\sqrt{3}}{3} & \frac{\sqrt{6}}{3} \\ -\frac{\sqrt{6}}{3} & \frac{\sqrt{3}}{3} \end{bmatrix}$$

and

$$P^T AP = \begin{bmatrix} \frac{\sqrt{3}}{3} & -\frac{\sqrt{6}}{3} \\ \frac{\sqrt{6}}{3} & \frac{\sqrt{3}}{3} \end{bmatrix} \begin{bmatrix} 2 & \sqrt{2} \\ \sqrt{2} & 1 \end{bmatrix} \begin{bmatrix} \frac{\sqrt{3}}{3} & \frac{\sqrt{6}}{3} \\ -\frac{\sqrt{6}}{3} & \frac{\sqrt{3}}{3} \end{bmatrix} = \begin{bmatrix} 0 & 0 \\ 0 & 3 \end{bmatrix}.$$

47. The eigenvalues of A are $\lambda_1 = -15$ and $\lambda_2 = 0$, and $\lambda_3 = 15$, with corresponding eigenvectors $(-2, 1, 2)$, $(-1, 2, -2)$ and $(2, 2, 1)$, respectively. Normalize each eigenvector to form the columns of P. Then

$$P = \begin{bmatrix} -\frac{2}{3} & -\frac{1}{3} & \frac{2}{3} \\ \frac{1}{3} & \frac{2}{3} & \frac{2}{3} \\ \frac{2}{3} & -\frac{2}{3} & \frac{1}{3} \end{bmatrix}$$

and

$$P^T AP = \begin{bmatrix} -\frac{2}{3} & \frac{1}{3} & \frac{2}{3} \\ -\frac{1}{3} & \frac{2}{3} & -\frac{2}{3} \\ \frac{2}{3} & \frac{2}{3} & \frac{1}{3} \end{bmatrix} \begin{bmatrix} 0 & 10 & 10 \\ 10 & 5 & 0 \\ 10 & 0 & -5 \end{bmatrix} \begin{bmatrix} -\frac{2}{3} & -\frac{1}{3} & \frac{2}{3} \\ \frac{1}{3} & \frac{2}{3} & \frac{2}{3} \\ \frac{2}{3} & -\frac{2}{3} & \frac{1}{3} \end{bmatrix} = \begin{bmatrix} -15 & 0 & 0 \\ 0 & 0 & 0 \\ 0 & 0 & 15 \end{bmatrix}.$$

49. The eigenvalues of A are $\lambda_1 = -2$, $\lambda_2 = 2$, and $\lambda_3 = 4$, with corresponding eigenvectors $(-1, -1, 1)$, $(-1, 1, 0)$, and $(1, 1, 2)$, respectively. Normalize each eigenvector to form the columns of P. Then

$$P = \begin{bmatrix} -\frac{\sqrt{3}}{3} & -\frac{\sqrt{2}}{2} & \frac{\sqrt{6}}{6} \\ -\frac{\sqrt{3}}{2} & \frac{\sqrt{2}}{2} & \frac{\sqrt{6}}{6} \\ \frac{\sqrt{3}}{3} & 0 & \frac{\sqrt{6}}{3} \end{bmatrix}$$

and

$$P^T AP = \begin{bmatrix} -\frac{\sqrt{3}}{3} & -\frac{\sqrt{3}}{3} & \frac{\sqrt{3}}{3} \\ -\frac{\sqrt{2}}{2} & \frac{\sqrt{2}}{2} & 0 \\ \frac{\sqrt{6}}{6} & \frac{\sqrt{6}}{6} & \frac{\sqrt{6}}{3} \end{bmatrix} \begin{bmatrix} 1 & -1 & 2 \\ -1 & 1 & 2 \\ 2 & 2 & 2 \end{bmatrix} \begin{bmatrix} -\frac{\sqrt{3}}{3} & -\frac{\sqrt{2}}{2} & \frac{\sqrt{6}}{6} \\ -\frac{\sqrt{3}}{3} & \frac{\sqrt{2}}{2} & \frac{\sqrt{6}}{6} \\ \frac{\sqrt{3}}{2} & 0 & \frac{\sqrt{6}}{3} \end{bmatrix} = \begin{bmatrix} -2 & 0 & 0 \\ 0 & 2 & 0 \\ 0 & 0 & 4 \end{bmatrix}.$$

© 2013 Cengage Learning. All Rights Reserved. May not be scanned, copied or duplicated, or posted to a publicly accessible website, in whole or in part.

51. The eigenvalues of A are $\lambda_1 = 2$ and $\lambda_2 = 6$, with corresponding eigenvectors $(1, -1, 0, 0)$ and $(0, 0, 1, -1)$ for λ_1 and $(1, 1, 0, 0)$ and $(0, 0, 1, 1)$ for λ_2. Normalize each eigenvector to form the columns of P. Then

$$P = \begin{bmatrix} \frac{\sqrt{2}}{2} & 0 & \frac{\sqrt{2}}{2} & 0 \\ -\frac{\sqrt{2}}{2} & 0 & \frac{\sqrt{2}}{2} & 0 \\ 0 & \frac{\sqrt{2}}{2} & 0 & \frac{\sqrt{2}}{2} \\ 0 & -\frac{\sqrt{2}}{2} & 0 & \frac{\sqrt{2}}{2} \end{bmatrix}$$

and

$$P^T AP = \begin{bmatrix} \frac{\sqrt{2}}{2} & -\frac{\sqrt{2}}{2} & 0 & 0 \\ 0 & 0 & \frac{\sqrt{2}}{2} & -\frac{\sqrt{2}}{2} \\ \frac{\sqrt{2}}{2} & \frac{\sqrt{2}}{2} & 0 & 0 \\ 0 & 0 & \frac{\sqrt{2}}{2} & \frac{\sqrt{2}}{2} \end{bmatrix} \begin{bmatrix} 4 & 2 & 0 & 0 \\ 2 & 4 & 0 & 0 \\ 0 & 0 & 4 & 2 \\ 0 & 0 & 2 & 4 \end{bmatrix} \begin{bmatrix} \frac{\sqrt{2}}{2} & 0 & \frac{\sqrt{2}}{2} & 0 \\ -\frac{\sqrt{2}}{2} & 0 & \frac{\sqrt{2}}{2} & 0 \\ 0 & \frac{\sqrt{2}}{2} & 0 & \frac{\sqrt{2}}{2} \\ 0 & -\frac{\sqrt{2}}{2} & 0 & \frac{\sqrt{2}}{2} \end{bmatrix}$$

$$= \begin{bmatrix} 2 & 0 & 0 & 0 \\ 0 & 2 & 0 & 0 \\ 0 & 0 & 6 & 0 \\ 0 & 0 & 0 & 6 \end{bmatrix}.$$

53. (a) True. See Theorem 7.10, page 367.

(b) True. See Theorem 7.9, page 366.

55. $(AB)^{-1} = B^{-1}A^{-1} = B^T A^T = (AB)^T \Rightarrow AB$ is orthogonal

$(BA)^{-1} = A^{-1}B^{-1} = A^T B^T = (BA)^T \Rightarrow BA$ is orthogonal

57. Let A be orthogonal, $A^{-1} = A^T$.

Then $(A^T)^{-1} = (A^{-1})^{-1} = A = (A^T)^T \Rightarrow A^T$ is orthogonal.

Furthermore,

$(A^{-1})^{-1} = (A^T)^{-1} = (A^{-1})^T \Rightarrow A^{-1}$ is orthogonal.

59. $A^T A = \begin{bmatrix} 1 & 4 \\ -3 & -6 \\ 2 & 1 \end{bmatrix} \begin{bmatrix} 1 & -3 & 2 \\ 4 & -6 & 1 \end{bmatrix} = \begin{bmatrix} 17 & -27 & 6 \\ -27 & 45 & -12 \\ 6 & -12 & 5 \end{bmatrix}$

$AA^T = \begin{bmatrix} 1 & -3 & 2 \\ 4 & -6 & 1 \end{bmatrix} \begin{bmatrix} 1 & 4 \\ -3 & -6 \\ 2 & 1 \end{bmatrix} = \begin{bmatrix} 14 & 24 \\ 24 & 53 \end{bmatrix}$

Both products are symmetric.

© 2013 Cengage Learning. All Rights Reserved. May not be scanned, copied or duplicated, or posted to a publicly accessible website, in whole or in part.

Section 7.4 Applications of Eigenvalues and Eigenvectors

1. $\mathbf{x}_2 = A\mathbf{x}_1 = \begin{bmatrix} 0 & 2 \\ \frac{1}{2} & 0 \end{bmatrix}\begin{bmatrix} 10 \\ 10 \end{bmatrix} = \begin{bmatrix} 20 \\ 5 \end{bmatrix}$

$\mathbf{x}_3 = A\mathbf{x}_2 = \begin{bmatrix} 0 & 2 \\ \frac{1}{2} & 0 \end{bmatrix}\begin{bmatrix} 20 \\ 5 \end{bmatrix} = \begin{bmatrix} 10 \\ 10 \end{bmatrix}$

3. $\mathbf{x}_2 = A\mathbf{x}_1 = \begin{bmatrix} 0 & 3 & 4 \\ 1 & 0 & 0 \\ 0 & \frac{1}{2} & 0 \end{bmatrix}\begin{bmatrix} 12 \\ 12 \\ 12 \end{bmatrix} = \begin{bmatrix} 84 \\ 12 \\ 6 \end{bmatrix}$

$\mathbf{x}_3 = A\mathbf{x}_2 = \begin{bmatrix} 0 & 3 & 4 \\ 1 & 0 & 0 \\ 0 & \frac{1}{2} & 0 \end{bmatrix}\begin{bmatrix} 84 \\ 12 \\ 6 \end{bmatrix} = \begin{bmatrix} 60 \\ 84 \\ 6 \end{bmatrix}$

5. $\mathbf{x}_2 = \begin{bmatrix} 0 & 2 & 2 & 0 \\ \frac{1}{4} & 0 & 0 & 0 \\ 0 & 1 & 0 & 0 \\ 0 & 0 & \frac{1}{2} & 0 \end{bmatrix}\begin{bmatrix} 100 \\ 100 \\ 100 \\ 100 \end{bmatrix} = \begin{bmatrix} 400 \\ 25 \\ 100 \\ 50 \end{bmatrix}$

$\mathbf{x}_3 = \begin{bmatrix} 0 & 2 & 2 & 0 \\ \frac{1}{4} & 0 & 0 & 0 \\ 0 & 1 & 0 & 0 \\ 0 & 0 & \frac{1}{2} & 0 \end{bmatrix}\begin{bmatrix} 400 \\ 25 \\ 100 \\ 50 \end{bmatrix} = \begin{bmatrix} 250 \\ 100 \\ 25 \\ 50 \end{bmatrix}$

7. The eigenvalues are 1 and -1. Choosing the positive eigenvalue, $\lambda = 1$, the corresponding eigenvector is found by row-reducing $\lambda I - A = I - A$.

$\begin{bmatrix} 1 & -2 \\ -\frac{1}{2} & 1 \end{bmatrix} \Rightarrow \begin{bmatrix} 1 & -2 \\ 0 & 0 \end{bmatrix}$

So, an eigenvector is $(2, 1)$, and the stable age distribution vector is $\mathbf{x} = t\begin{bmatrix} 2 \\ 1 \end{bmatrix}$.

9. The eigenvalues of A are -1 and 2. Choosing the positive eigenvalue, let $\lambda = 2$.

An eigenvector corresponding to $\lambda = 2$ is found by row-reducing $2I - A$.

$\begin{bmatrix} 2 & -3 & -4 \\ -1 & 2 & 0 \\ 0 & -\frac{1}{2} & 2 \end{bmatrix} \Rightarrow \begin{bmatrix} 1 & 0 & -8 \\ 0 & 1 & -4 \\ 0 & 0 & 0 \end{bmatrix}$

So, an eigenvector is $(8, 4, 1)$ and stable age distribution vector is $\mathbf{x} = t\begin{bmatrix} 8 \\ 4 \\ 1 \end{bmatrix}$.

11. The characteristic equation of A is

$$|\lambda I - A| = \lambda^4 - \tfrac{1}{2}\lambda^2 - \tfrac{1}{2}\lambda$$
$$= \lambda(\lambda - 1)\left(\lambda^2 + \lambda + \tfrac{1}{2}\right) = 0.$$

Choosing the positive eigenvalue $\lambda = 1$, you find its corresponding eigenvector by row-reducing $\lambda I - A = I - A$. So, an eigenvector is (8, 2, 2, 1) and the stable age distribution vector is $\mathbf{x} = t\begin{bmatrix} 8 \\ 2 \\ 2 \\ 1 \end{bmatrix}$.

13. Construct the age transition matrix.

$$A = \begin{bmatrix} 2 & 4 & 2 \\ 0.75 & 0 & 0 \\ 0 & 0.25 & 0 \end{bmatrix}$$

The current age distribution vector is

$$x_1 = \begin{bmatrix} 160 \\ 160 \\ 160 \end{bmatrix}.$$

In 1 year, the age distribution vector will be

$$x_2 = Ax_1 = \begin{bmatrix} 2 & 4 & 2 \\ 0.75 & 0 & 0 \\ 0 & 0.25 & 0 \end{bmatrix}\begin{bmatrix} 160 \\ 160 \\ 160 \end{bmatrix} = \begin{bmatrix} 1280 \\ 120 \\ 40 \end{bmatrix}.$$

In 2 years, the age distribution will be

$$x_3 = Ax_2 = \begin{bmatrix} 2 & 4 & 2 \\ 0.75 & 0 & 0 \\ 0 & 0.25 & 0 \end{bmatrix}\begin{bmatrix} 1280 \\ 120 \\ 40 \end{bmatrix} = \begin{bmatrix} 3120 \\ 960 \\ 30 \end{bmatrix}.$$

© 2013 Cengage Learning. All Rights Reserved. May not be scanned, copied or duplicated, or posted to a publicly accessible website, in whole or in part.

15. Construct the age transition matrix.

$$A = \begin{bmatrix} 2 & 5 & 2 \\ 0.6 & 0 & 0 \\ 0 & 0.5 & 0 \end{bmatrix}$$

The current age distribution vector is

$$\mathbf{x}_1 = \begin{bmatrix} 100 \\ 100 \\ 100 \end{bmatrix}.$$

In one year, the age distribution vector will be

$$\mathbf{x}_2 = A\mathbf{x}_1 = \begin{bmatrix} 2 & 5 & 2 \\ 0.6 & 0 & 0 \\ 0 & 0.5 & 0 \end{bmatrix}\begin{bmatrix} 100 \\ 100 \\ 100 \end{bmatrix} = \begin{bmatrix} 900 \\ 60 \\ 50 \end{bmatrix}.$$

In two years, the age distribution vector will be

$$\mathbf{x}_3 = A\mathbf{x}_2 = \begin{bmatrix} 2 & 5 & 2 \\ 0.6 & 0 & 0 \\ 0 & 0.5 & 0 \end{bmatrix}\begin{bmatrix} 900 \\ 60 \\ 50 \end{bmatrix} = \begin{bmatrix} 2200 \\ 540 \\ 30 \end{bmatrix}.$$

17. The solution to the differential equation $y' = ky$ is $y = Ce^{kt}$. So, $y_1 = C_1e^{2t}$ and $y_2 = C_2e^{t}$.

19. The solution to the differential equation $y' = ky$ is $y = Ce^{kt}$. So, $y_1 = C_1e^{-4t}$ and $y_2 = C_2e^{-1/2t}$.

21. The solution to the differential equation $y' = ky$ is $y = Ce^{kt}$. So, $y_1 = C_1e^{-t}$, $y_2 = C_2e^{6t}$ and $y_3 = C_3e^{t}$.

23. The solution to the differential equation $y' = ky$ is $y = Ce^{kt}$. So, $y_1 = C_1e^{-12t}$, $y_2 = C_2e^{-6t}$, and $y_3 = C_3e^{7t}$.

25. The solution to the differential equation $y' = ky$ is $y = Ce^{kt}$. So, $y_1 = C_1e^{-0.3t}$, $y_2 = C_2e^{0.4t}$, and $y_3 = C_3e^{-0.6t}$.

27. The solution to the differential equation $y' = ky$ is $y = Ce^{kt}$. So, $y_1 = C_1e^{7t}$, $y_2 = C_2e^{9t}$, $y_3 = C_3e^{-7t}$, and $y_4 = C_4e^{-9t}$.

29. This system has the matrix form

$$\mathbf{y}' = \begin{bmatrix} y_1' \\ y_2' \end{bmatrix} = \begin{bmatrix} 1 & -4 \\ 0 & 2 \end{bmatrix}\begin{bmatrix} y_1 \\ y_2 \end{bmatrix} = A\mathbf{y}.$$

The eigenvalues of A are $\lambda_1 = 1$ and $\lambda_2 = 2$, with corresponding eigenvectors $(1, 0)$ and $(-4, 1)$, respectively. So, diagonalize A using a matrix P whose columns are the eigenvectors of A.

$$P = \begin{bmatrix} 1 & -4 \\ 0 & 1 \end{bmatrix} \text{ and } P^{-1}AP = \begin{bmatrix} 1 & 0 \\ 0 & 2 \end{bmatrix}$$

The solution of the system $\mathbf{w}' = P^{-1}AP\mathbf{w}$ is $w_1 = C_1e^{t}$ and $w_2 = C_2e^{2t}$. Return to the original system by applying the substitution $\mathbf{y} = P\mathbf{w}$.

$$\mathbf{y} = \begin{bmatrix} y_1 \\ y_2 \end{bmatrix} = \begin{bmatrix} 1 & -4 \\ 0 & 1 \end{bmatrix}\begin{bmatrix} w_1 \\ w_2 \end{bmatrix} = \begin{bmatrix} w_1 - 4w_2 \\ w_2 \end{bmatrix}.$$

So, the solution is

$$y_1 = C_1e^{t} - 4C_2e^{2t}$$
$$y_2 = C_2e^{2t}.$$

31. This system has the matrix form

$$\mathbf{y}' = \begin{bmatrix} y_1' \\ y_2' \end{bmatrix} = \begin{bmatrix} 1 & 2 \\ 2 & 1 \end{bmatrix}\begin{bmatrix} y_1 \\ y_2 \end{bmatrix} = A\mathbf{y}.$$

The eigenvalues of A are $\lambda_1 = -1$ and $\lambda_2 = 3$ with corresponding eigenvectors $\mathbf{x}_1 = (1, -1)$ and $\mathbf{x}_2 = (1, 1)$, respectively. So, diagonalize A using a matrix P whose columns vectors are the eigenvectors of A.

$$P = \begin{bmatrix} 1 & 1 \\ -1 & 1 \end{bmatrix} \text{ and } P^{-1}AP = \begin{bmatrix} -1 & 0 \\ 0 & 3 \end{bmatrix}$$

The solution of the system $\mathbf{w}' = P^{-1}AP\mathbf{w}$ is $w_1 = C_1e^{-t}$ and $w_2 = C_2e^{3t}$. Return to the original system by applying the substitution $\mathbf{y} = P\mathbf{w}$.

$$\mathbf{y} = \begin{bmatrix} y_1 \\ y_2 \end{bmatrix} = \begin{bmatrix} 1 & 1 \\ -1 & 1 \end{bmatrix}\begin{bmatrix} w_1 \\ w_2 \end{bmatrix} = \begin{bmatrix} w_1 + w_2 \\ -w_1 + w_2 \end{bmatrix}$$

So, the solution is

$$y_1 = C_1e^{-t} + C_2e^{3t}$$
$$y_2 = -C_1e^{-t} + C_2e^{3t}.$$

© 2013 Cengage Learning. All Rights Reserved. May not be scanned, copied or duplicated, or posted to a publicly accessible website, in whole or in part.

33. This system has the matrix form

$$\mathbf{y}' = \begin{bmatrix} y_1' \\ y_2' \\ y_3' \end{bmatrix} = \begin{bmatrix} 0 & -3 & 5 \\ -4 & 4 & -10 \\ 0 & 0 & 4 \end{bmatrix}\begin{bmatrix} y_1 \\ y_2 \\ y_3 \end{bmatrix} = A\mathbf{y}.$$

The eigenvalues of A are $\lambda_1 = -2$, $\lambda_2 = 6$ and $\lambda_3 = 4$, with corresponding eigenvectors $(3, 2, 0)$, $(-1, 2, 0)$, and $(-5, 10, 2)$, respectively. So, diagonalize A using a matrix P whose column vectors are the eigenvectors of A.

$$P = \begin{bmatrix} 3 & -1 & -5 \\ 2 & 2 & 10 \\ 0 & 0 & 2 \end{bmatrix} \text{ and } P^{-1}AP = \begin{bmatrix} -2 & 0 & 0 \\ 0 & 6 & 0 \\ 0 & 0 & 4 \end{bmatrix}$$

The solution of the system $\mathbf{w}' = P^{-1}AP\mathbf{w}$ is $w_1 = C_1e^{-2t}$, $w_2 = C_2e^{6t}$ and $w_3 = C_3e^{4t}$.

Return to the original system by applying the substitution $\mathbf{y} = P\mathbf{w}$.

$$\mathbf{y} = \begin{bmatrix} y_1 \\ y_2 \\ y_3 \end{bmatrix} = \begin{bmatrix} 3 & -1 & -5 \\ 2 & 2 & 10 \\ 0 & 0 & 2 \end{bmatrix}\begin{bmatrix} w_1 \\ w_2 \\ w_3 \end{bmatrix} = \begin{bmatrix} 3w_1 - w_2 - 5w_3 \\ 2w_1 + 2w_2 + 10w_3 \\ 2w_3 \end{bmatrix}$$

So, the solution is

$$y_1 = 3C_1e^{-2t} - C_2e^{6t} - 5C_3e^{4t}$$
$$y_2 = 2C_1e^{-2t} + 2C_2e^{6t} + 10C_3e^{4t}$$
$$y_3 = 2C_3e^{4t}.$$

35. This system has the matrix form

$$\mathbf{y}' = \begin{bmatrix} y_1' \\ y_2' \\ y_3' \end{bmatrix} = \begin{bmatrix} 1 & -2 & 1 \\ 0 & 2 & 4 \\ 0 & 0 & 3 \end{bmatrix}\begin{bmatrix} y_1 \\ y_2 \\ y_3 \end{bmatrix} = A\mathbf{y}.$$

The eigenvalues of A are $\lambda_1 = 1$, $\lambda_2 = 2$, and $\lambda_3 = 3$ with corresponding eigenvectors $\mathbf{x}_1 = (1, 0, 0)$, $\mathbf{x}_2 = (-2, 1, 0)$, and $\mathbf{x}_3 = (-7, 8, 2)$. So, diagonalize A using a matrix P whose column vectors are the eigenvectors of A.

$$P = \begin{bmatrix} 1 & -2 & -7 \\ 0 & 1 & 8 \\ 0 & 0 & 2 \end{bmatrix} \text{ and } P^{-1}AP = \begin{bmatrix} 1 & 0 & 0 \\ 0 & 2 & 0 \\ 0 & 0 & 3 \end{bmatrix}$$

The solution of the system $\mathbf{w}' = P^{-1}AP\mathbf{w}$ is $w_1 = C_1e^{t}$, $w_2 = C_2e^{2t}$, and $w_3 = C_3e^{3t}$.

Return to the original system by applying the substitution $\mathbf{y} = P\mathbf{w}$.

$$\mathbf{y} = \begin{bmatrix} y_1 \\ y_2 \\ y_3 \end{bmatrix} = \begin{bmatrix} 1 & -2 & -7 \\ 0 & 1 & 8 \\ 0 & 0 & 2 \end{bmatrix}\begin{bmatrix} w_1 \\ w_2 \\ w_3 \end{bmatrix} = \begin{bmatrix} w_1 - 2w_2 - 7w_3 \\ w_2 + 8w_3 \\ 2w_3 \end{bmatrix}$$

So, the solution is

$$y_1 = C_1e^{t} - 2C_2e^{2t} - 7C_3e^{3t}$$
$$y_2 = C_2e^{2t} + 8C_3e^{3t}$$
$$y_3 = 2C_3e^{3t}.$$

© 2013 Cengage Learning. All Rights Reserved. May not be scanned, copied or duplicated, or posted to a publicly accessible website, in whole or in part.

37. Because

$$\mathbf{y}' = \begin{bmatrix} y_1' \\ y_2' \end{bmatrix} = \begin{bmatrix} 1 & 1 \\ 0 & 1 \end{bmatrix}\begin{bmatrix} y_1 \\ y_2 \end{bmatrix} = A\mathbf{y},$$

the system represented by $\mathbf{y}' = A\mathbf{y}$ is

$$\begin{aligned} y_1' &= y_1 + y_2 \\ y_2' &= y_2. \end{aligned}$$

Note that

$$\begin{aligned} y_1' &= C_1e^t + C_2te^t + C_2e^t = y_1 + y_2 \\ y_2' &= C_2e^t = y_2. \end{aligned}$$

39. Because

$$\mathbf{y}' = \begin{bmatrix} y_1' \\ y_2' \\ y_3' \end{bmatrix} = \begin{bmatrix} 0 & 1 & 0 \\ 0 & 0 & 1 \\ 0 & -4 & 0 \end{bmatrix}\begin{bmatrix} y_1 \\ y_2 \\ y_3 \end{bmatrix} = A\mathbf{y},$$

the system represented by $\mathbf{y}' = A\mathbf{y}$ is

$$\begin{aligned} y_1' &= y_2 \\ y_2' &= y_3 \\ y_3' &= -4y_2. \end{aligned}$$

Note that

$$\begin{aligned} y_1' &= -2C_2 \sin 2t + 2C_3 \cos 2t = y_2 \\ y_2' &= -4C_3 \sin 2t - 4C_2 \cos 2t = y_3 \\ y_3' &= 8C_2 \sin 2t - 8C_3 \cos 2t = -4y_2. \end{aligned}$$

41. The matrix of the quadratic form is

$$A = \begin{bmatrix} a & \frac{b}{2} \\ \frac{b}{2} & c \end{bmatrix} = \begin{bmatrix} 1 & 0 \\ 0 & 1 \end{bmatrix}.$$

43. The matrix of the quadratic form is

$$A = \begin{bmatrix} a & \frac{b}{2} \\ \frac{b}{2} & c \end{bmatrix} = \begin{bmatrix} 9 & 5 \\ 5 & -4 \end{bmatrix}.$$

45. The matrix of the quadratic form is

$$A = \begin{bmatrix} a & \frac{b}{2} \\ \frac{b}{2} & c \end{bmatrix} = \begin{bmatrix} 0 & 5 \\ 5 & -10 \end{bmatrix}.$$

47. The matrix of the quadratic form is

$$A = \begin{bmatrix} a & \frac{b}{2} \\ \frac{b}{2} & c \end{bmatrix} = \begin{bmatrix} 2 & -\frac{3}{2} \\ -\frac{3}{2} & -2 \end{bmatrix}.$$

The eigenvalues of A are $\lambda_1 = -\frac{5}{2}$ and $\lambda_2 = \frac{5}{2}$ with corresponding eigenvectors $\mathbf{x}_1 = (1, 3)$ and $\mathbf{x}_2 = (-3, 1)$ respectively.

Using unit vectors in the direction of $\mathbf{x}_1$ and $\mathbf{x}_2$ to form the columns of P yields

$$P = \begin{bmatrix} \frac{1}{\sqrt{10}} & -\frac{3}{\sqrt{10}} \\ \frac{3}{\sqrt{10}} & \frac{1}{\sqrt{10}} \end{bmatrix}.$$

Note that

$$P^TAP = \begin{bmatrix} \frac{1}{\sqrt{10}} & \frac{3}{\sqrt{10}} \\ -\frac{3}{\sqrt{10}} & \frac{1}{\sqrt{10}} \end{bmatrix}\begin{bmatrix} 2 & -\frac{3}{2} \\ -\frac{3}{2} & -2 \end{bmatrix}\begin{bmatrix} \frac{1}{\sqrt{10}} & -\frac{3}{\sqrt{10}} \\ \frac{3}{\sqrt{10}} & \frac{1}{\sqrt{10}} \end{bmatrix} = \begin{bmatrix} -\frac{5}{2} & 0 \\ 0 & \frac{5}{2} \end{bmatrix}.$$

© 2013 Cengage Learning. All Rights Reserved. May not be scanned, copied or duplicated, or posted to a publicly accessible website, in whole or in part.

49. The matrix of the quadratic form is

$$A = \begin{bmatrix} a & \frac{b}{2} \\ \frac{b}{2} & c \end{bmatrix} = \begin{bmatrix} 13 & 3\sqrt{3} \\ 3\sqrt{3} & 7 \end{bmatrix}.$$

The eigenvalues of A are $\lambda_1 = 4$ and $\lambda_2 = 16$, with corresponding eigenvectors $\mathbf{x}_1 = (1, -\sqrt{3})$ and $\mathbf{x}_2 = (\sqrt{3}, 1)$, respectively. Using unit vectors in the direction of $\mathbf{x}_1$ and $\mathbf{x}_2$ to form the columns of P,

$$P = \begin{bmatrix} \frac{1}{2} & \frac{\sqrt{3}}{2} \\ -\frac{\sqrt{3}}{2} & \frac{1}{2} \end{bmatrix} \quad \text{and} \quad P^T AP = \begin{bmatrix} 4 & 0 \\ 0 & 16 \end{bmatrix}.$$

51. The matrix of the quadratic form is

$$A = \begin{bmatrix} a & \frac{b}{2} \\ \frac{b}{2} & c \end{bmatrix} = \begin{bmatrix} 16 & -12 \\ -12 & 9 \end{bmatrix}.$$

The eigenvalues of A are $\lambda_1 = 0$ and $\lambda_2 = 25$, with corresponding eigenvectors $\mathbf{x}_1 = (3, 4)$ and $\mathbf{x}_2 = (-4, 3)$ respectively. Using unit vectors in the direction of $\mathbf{x}_1$ and $\mathbf{x}_2$ to form the columns of P,

$$P = \begin{bmatrix} \frac{3}{5} & -\frac{4}{5} \\ \frac{4}{5} & \frac{3}{5} \end{bmatrix} \quad \text{and} \quad P^T AP = \begin{bmatrix} 0 & 0 \\ 0 & 25 \end{bmatrix}.$$

53. The matrix of the quadratic form is

$$A = \begin{bmatrix} a & \frac{b}{2} \\ \frac{b}{2} & c \end{bmatrix} = \begin{bmatrix} 13 & -4 \\ -4 & 7 \end{bmatrix}.$$

This matrix has eigenvalues of 5 and 15 with corresponding unit eigenvectors $\left(\frac{1}{\sqrt{5}}, \frac{2}{\sqrt{5}}\right)$ and $\left(-\frac{2}{\sqrt{5}}, \frac{1}{\sqrt{5}}\right)$ respectively. Let

$$P = \begin{bmatrix} \frac{1}{\sqrt{5}} & -\frac{2}{\sqrt{5}} \\ \frac{2}{\sqrt{5}} & \frac{1}{\sqrt{5}} \end{bmatrix} \quad \text{and} \quad P^T AP = \begin{bmatrix} 5 & 0 \\ 0 & 15 \end{bmatrix}.$$

This implies that the rotated conic is an ellipse with equation $5(x')^2 + 15(y')^2 = 45$.

55. The matrix of the quadratic form is

$$A = \begin{bmatrix} a & \frac{b}{2} \\ \frac{b}{2} & c \end{bmatrix} = \begin{bmatrix} 2 & -2 \\ -2 & 5 \end{bmatrix}.$$

This matrix has eigenvalues of 1 and 6, and corresponding unit eigenvectors and $\left(\frac{2}{\sqrt{5}}, \frac{1}{\sqrt{5}}\right)$ and $\left(-\frac{1}{\sqrt{5}}, \frac{2}{\sqrt{5}}\right)$, respectively.

So, let

$$P = \begin{bmatrix} \frac{2}{\sqrt{5}} & -\frac{1}{\sqrt{5}} \\ \frac{1}{\sqrt{5}} & \frac{2}{\sqrt{5}} \end{bmatrix} \quad \text{and} \quad P^T AP = \begin{bmatrix} 1 & 0 \\ 0 & 6 \end{bmatrix}.$$

This implies that the rotated conic is an ellipse with equation $(x')^2 + 6(y')^2 - 36 = 0$.

© 2013 Cengage Learning. All Rights Reserved. May not be scanned, copied or duplicated, or posted to a publicly accessible website, in whole or in part.

57. The matrix of the quadratic form is $A = \begin{bmatrix} a & \frac{b}{2} \\ \frac{b}{2} & c \end{bmatrix} = \begin{bmatrix} 2 & 2 \\ 2 & 2 \end{bmatrix}$.

This matrix has eigenvalues of 0 and 4, with corresponding unit eigenvectors $\left(\frac{1}{\sqrt{2}}, -\frac{1}{\sqrt{2}}\right)$ and $\left(\frac{1}{\sqrt{2}}, \frac{1}{\sqrt{2}}\right)$ respectively. Let $P = \begin{bmatrix} \frac{1}{\sqrt{2}} & \frac{1}{\sqrt{2}} \\ -\frac{1}{\sqrt{2}} & \frac{1}{\sqrt{2}} \end{bmatrix}$ and $P^T AP = \begin{bmatrix} 0 & 0 \\ 0 & 4 \end{bmatrix}$.

This implies that the rotated conic is a parabola. Furthermore, $\begin{bmatrix} d & e \end{bmatrix}P = \begin{bmatrix} 6\sqrt{2} & 2\sqrt{2} \end{bmatrix}\begin{bmatrix} \frac{1}{\sqrt{2}} & \frac{1}{\sqrt{2}} \\ -\frac{1}{\sqrt{2}} & \frac{1}{\sqrt{2}} \end{bmatrix} = \begin{bmatrix} 4 & 8 \end{bmatrix} = \begin{bmatrix} d' & e' \end{bmatrix}$.

So, the equation in the $x'y'$-coordinate system is $4(y')^2 + 4x' + 8y' + 4 = 0$.

59. The matrix of the quadratic form is $A = \begin{bmatrix} 0 & \frac{1}{2} \\ \frac{1}{2} & 0 \end{bmatrix}$.

This matrix has eigenvalues of $-\frac{1}{2}$ and $\frac{1}{2}$ with corresponding unit eigenvectors $\left(-\frac{1}{\sqrt{2}}, \frac{1}{\sqrt{2}}\right)$ and $\left(\frac{1}{\sqrt{2}}, \frac{1}{\sqrt{2}}\right)$ respectively. Let $P = \begin{bmatrix} -\frac{1}{\sqrt{2}} & \frac{1}{\sqrt{2}} \\ \frac{1}{\sqrt{2}} & \frac{1}{\sqrt{2}} \end{bmatrix}$ and $P^T AP = \begin{bmatrix} -\frac{1}{2} & 0 \\ 0 & \frac{1}{2} \end{bmatrix}$.

This implies that the rotated conic is a hyperbola. Furthermore,

$$\begin{bmatrix} d & e \end{bmatrix}P = \begin{bmatrix} 1 & -2 \end{bmatrix}\begin{bmatrix} -\frac{1}{\sqrt{2}} & \frac{1}{\sqrt{2}} \\ \frac{1}{\sqrt{2}} & \frac{1}{\sqrt{2}} \end{bmatrix} = \begin{bmatrix} -\frac{3}{\sqrt{2}} & -\frac{1}{\sqrt{2}} \end{bmatrix} = \begin{bmatrix} d' & e' \end{bmatrix},$$

so, the equation in the $x'y'$-coordinate system is $-\frac{1}{2}(x')^2 + \frac{1}{2}(y')^2 - \frac{3}{\sqrt{2}}x' - \frac{1}{\sqrt{2}}y' + 3 = 0$.

© 2013 Cengage Learning. All Rights Reserved. May not be scanned, copied or duplicated, or posted to a publicly accessible website, in whole or in part.

61. The matrix of the quadratic form is $A = \begin{bmatrix} 3 & -1 & 0 \\ -1 & 3 & 0 \\ 0 & 0 & 8 \end{bmatrix}$.

The eigenvalues of A are 2, 4, and 8 with corresponding unit eigenvectors $\left(\frac{1}{\sqrt{2}}, \frac{1}{\sqrt{2}}, 0\right), \left(-\frac{1}{\sqrt{2}}, \frac{1}{\sqrt{2}}, 0\right)$ and

$(0, 0, 1)$ respectively. Then let $P = \begin{bmatrix} \frac{1}{\sqrt{2}} & -\frac{1}{\sqrt{2}} & 0 \\ \frac{1}{\sqrt{2}} & \frac{1}{\sqrt{2}} & 0 \\ 0 & 0 & 1 \end{bmatrix}$ and $P^T AP = \begin{bmatrix} 2 & 0 & 0 \\ 0 & 4 & 0 \\ 0 & 0 & 8 \end{bmatrix}$.

Furthermore,

$$[g \quad h \quad i]P = [0 \quad 0 \quad 0]\begin{bmatrix} \frac{1}{\sqrt{2}} & -\frac{1}{\sqrt{2}} & 0 \\ \frac{1}{\sqrt{2}} & \frac{1}{\sqrt{2}} & 0 \\ 0 & 0 & 1 \end{bmatrix} = [0 \quad 0 \quad 0] = [g' \quad h' \quad i'].$$

So, the equation of the rotated quadratic surface is $2(x')^2 + 4(y')^2 + 8(z')^2 - 16 = 0$.

63. The matrix of the quadratic form is $A = \begin{bmatrix} 1 & 0 & 0 \\ 0 & 2 & 1 \\ 0 & 1 & 2 \end{bmatrix}$.

The eigenvalues of A are 1, 1, and 3 with corresponding unit eigenvectors $(1, 0, 0)$, $(0, -1, 1)$, and $(0, 1, 1)$ respectively.

Then let $P = \begin{bmatrix} 1 & 0 & 0 \\ 0 & -\frac{\sqrt{2}}{2} & \frac{\sqrt{2}}{2} \\ 0 & \frac{\sqrt{2}}{2} & \frac{\sqrt{2}}{\sqrt{2}} \end{bmatrix}$ and $P^T AP = \begin{bmatrix} 1 & 0 & 0 \\ 0 & 1 & 0 \\ 0 & 0 & 3 \end{bmatrix}$.

Furthermore,

$$[g \quad h \quad i]P = [0 \quad 0 \quad 0]\begin{bmatrix} 1 & 0 & 0 \\ 0 & -\frac{\sqrt{2}}{2} & \frac{\sqrt{2}}{2} \\ 0 & \frac{\sqrt{2}}{2} & \frac{\sqrt{2}}{2} \end{bmatrix} = [0 \quad 0 \quad 0] = [g' \quad h' \quad i'].$$

So, the equation of the rotated quadratic surface is $(x')^2 + (y')^2 + 3(z')^2 - 1 = 0$.

65. Let $P = \begin{bmatrix} a & c \\ c & d \end{bmatrix}$ be a 2×2 orthogonal matrix such that $|P| = 1$. Define $\theta \in [0, 2\pi]$ as follows.

(i) If $a = 1$, then $c = 0, b = 0$ and $d = 1$, so let $\theta = 0$.

(ii) If $a = -1$, then $c = 0, b = 0$ and $d = -1$, so let $\theta = \pi$.

(iii) If $a \geq 0$ and $c > 0$, let $\theta = \arccos(a), 0 < \theta \leq \pi/2$.

(iv) If $a \geq 0$ and $c < 0$, let $\theta = 2\pi - \arccos(a), 3\pi/2 \leq \theta < 2\pi$.

(v) If $a \leq 0$ and $c > 0$, let $\theta = \arccos(a), \pi/2 \leq \theta < \pi$.

(vi) If $a \leq 0$ and $c < 0$, let $\theta = 2\pi - \arccos(a), \pi < \theta \leq 3\pi/2$.

In each of these cases, you can confirm that $P = \begin{bmatrix} a & b \\ c & d \end{bmatrix} = \begin{bmatrix} \cos\theta & -\sin\theta \\ \sin\theta & \cos\theta \end{bmatrix}$.

© 2013 Cengage Learning. All Rights Reserved. May not be scanned, copied or duplicated, or posted to a publicly accessible website, in whole or in part.

Review Exercises for Chapter 7

1. (a) The characteristic equation of A is given by

$$|\lambda I - A| = \begin{vmatrix} \lambda - 2 & -1 \\ -5 & \lambda + 2 \end{vmatrix} = \lambda^2 - 9 = 0.$$

(b) The eigenvalues of A are $\lambda_1 = -3$ and $\lambda_2 = 3$.

(c) To find the eigenvectors corresponding to $\lambda_1 = -3$, solve the matrix equation $(\lambda_1 I - A)\mathbf{x} = \mathbf{0}$. Row-reduce the augmented matrix to yield

$$\begin{bmatrix} -5 & -1 & \vdots & 0 \\ -5 & -1 & \vdots & 0 \end{bmatrix} \Rightarrow \begin{bmatrix} 1 & \frac{1}{5} & \vdots & 0 \\ 0 & 0 & \vdots & 0 \end{bmatrix}.$$

So, $\mathbf{x}_1 = (1, -5)$ is an eigenvector and $\{(1, -5)\}$ is a basis for the eigenspace corresponding to $\lambda_1 = -3$. Similarly, solve $(\lambda_2 I - A)\mathbf{x} = \mathbf{0}$ for $\lambda_2 = 3$. So, $\mathbf{x}_2 = (1, 1)$ is an eigenvector and $\{(1, 1)\}$ is a basis for the eigenspace corresponding to $\lambda_2 = 3$.

3. (a) The characteristic equation of A is given by

$$|\lambda I - A| = \begin{vmatrix} \lambda - 9 & -4 & 3 \\ 2 & \lambda & -6 \\ 1 & 4 & \lambda - 11 \end{vmatrix} = \lambda^3 - 20\lambda^2 + 128\lambda - 256 = (\lambda - 4)(\lambda - 8)^2.$$

(b) The eigenvalues of A are $\lambda_1 = 4$ and $\lambda_2 = 8$ (repeated).

(c) To find the eigenvectors corresponding to $\lambda_1 = 4$, solve the matrix equation $(\lambda_1 I - A)\mathbf{x} = \mathbf{0}$. Row-reducing the augmented matrix,

$$\begin{bmatrix} -5 & -4 & 3 & \vdots & 0 \\ 2 & 4 & -6 & \vdots & 0 \\ 1 & 4 & -7 & \vdots & 0 \end{bmatrix} \Rightarrow \begin{bmatrix} 1 & 0 & 1 & \vdots & 0 \\ 0 & 1 & -2 & \vdots & 0 \\ 0 & 0 & 0 & \vdots & 0 \end{bmatrix}$$

you can see that a basis for the eigenspace of $\lambda_1 = 4$ is $\{(-1, 2, 1)\}$. Similarly, solve $(\lambda_2 I - A)\mathbf{x} = \mathbf{0}$ for $\lambda_2 = 8$. So, a basis for the eigenspace of $\lambda_2 = 8$ is $\{(3, 0, 1), (-4, 1, 0)\}$.

5. (a) The characteristic equation of A is given by

$$|\lambda I - A| = \begin{vmatrix} \lambda - 2 & 0 & -1 \\ 0 & \lambda - 3 & -4 \\ 0 & 0 & \lambda - 1 \end{vmatrix} = (\lambda - 2)(\lambda - 3)(\lambda - 1) = 0.$$

(b) The eigenvalues of A are $\lambda_1 = 1$, $\lambda_2 = 2$ and $\lambda_3 = 3$.

(c) To find the eigenvectors corresponding to $\lambda_1 = 1$, solve the matrix equation $(\lambda_1 I - A)\mathbf{x} = \mathbf{0}$. Row-reducing the augmented matrix,

$$\begin{bmatrix} -1 & 0 & -1 & \vdots & 0 \\ 0 & -2 & -4 & \vdots & 0 \\ 0 & 0 & 0 & \vdots & 0 \end{bmatrix} \Rightarrow \begin{bmatrix} 1 & 0 & 1 & \vdots & 0 \\ 0 & 1 & 2 & \vdots & 0 \\ 0 & 0 & 0 & \vdots & 0 \end{bmatrix}$$

you can see that a basis for the eigenspace of $\lambda_1 = 1$ is $\{(-1, -2, 1)\}$. Similarly, solve $(\lambda_2 I - A)\mathbf{x} = \mathbf{0}$ for $\lambda_2 = 2$, and you see that $\{(1, 0, 0)\}$ is a basis for the eigenspace of $\lambda_2 = 2$. Finally, solve $(\lambda_3 I - A)\mathbf{x} = \mathbf{0}$ for $\lambda_3 = 3$, and you discover that $\{(0, 1, 0)\}$ is a basis for its eigenspace.

© 2013 Cengage Learning. All Rights Reserved. May not be scanned, copied or duplicated, or posted to a publicly accessible website, in whole or in part.

7. (a) The characteristic equation of A is given by

$$|\lambda I - A| = \begin{vmatrix} \lambda - 2 & -1 & 0 & 0 \\ -1 & \lambda - 2 & 0 & 0 \\ 0 & 0 & \lambda - 2 & -1 \\ 0 & 0 & -1 & \lambda - 2 \end{vmatrix} = (\lambda - 1)^2(\lambda - 3)^2 = 0.$$

(b) The eigenvalues of A are $\lambda_1 = 1$ (repeated) and $\lambda_2 = 3$ (repeated).

(c) To find the eigenvectors corresponding to $\lambda_1 = 1$, solve the matrix equation $(\lambda_1 I - A)\mathbf{x} = \mathbf{0}$ for $\lambda_1 = 1$. Row reducing the augmented matrix,

$$\left[\begin{array}{cccc:c} -1 & -1 & 0 & 0 & 0 \\ -1 & -1 & 0 & 0 & 0 \\ 0 & 0 & -1 & -1 & 0 \\ 0 & 0 & -1 & -1 & 0 \end{array}\right] \Rightarrow \left[\begin{array}{cccc:c} 1 & 1 & 0 & 0 & 0 \\ 0 & 0 & 1 & 1 & 0 \\ 0 & 0 & 0 & 0 & 0 \\ 0 & 0 & 0 & 0 & 0 \end{array}\right]$$

you see that a basis for the eigenspace of $\lambda_1 = 1$ is $\{(1, -1, 0, 0), (0, 0, 1, -1)\}$. Similarly, solve $(\lambda_2 I - A)\mathbf{x} = \mathbf{0}$ for $\lambda_2 = 3$, and discover that a basis for the eigenspace of $\lambda_2 = 3$ is $\{(1, 1, 0, 0), (0, 0, 1, 1)\}$.

9. The eigenvalues of A are $\lambda_1 = 0$ and $\lambda_2 = 9$. From Exercise 18, Section 7.1, the corresponding eigenvectors $(4, 1)$ and $(-1, 2)$ are used to form the columns of P. So,

$$P = \begin{bmatrix} 4 & -1 \\ 1 & 2 \end{bmatrix} \Rightarrow P^{-1} = \begin{bmatrix} \frac{2}{9} & \frac{1}{9} \\ -\frac{1}{9} & \frac{4}{9} \end{bmatrix}$$

and

$$P^{-1}AP = \begin{bmatrix} \frac{2}{9} & \frac{1}{9} \\ -\frac{1}{9} & \frac{4}{9} \end{bmatrix}\begin{bmatrix} 1 & -4 \\ -2 & 8 \end{bmatrix}\begin{bmatrix} 4 & -1 \\ 1 & 2 \end{bmatrix} = \begin{bmatrix} 0 & 0 \\ 0 & 9 \end{bmatrix}.$$

11. The eigenvalues of A are the solutions of

$$|\lambda I - A| = \begin{vmatrix} \lambda + 2 & 1 & -3 \\ 0 & \lambda - 1 & -2 \\ 0 & 0 & \lambda - 1 \end{vmatrix} = (\lambda + 2)(\lambda - 1)^2 = 0.$$

The eigenspace corresponding to the repeated eigenvalue $\lambda = 1$ has dimension 1, and so A is *not* diagonalizable.

13. The eigenvalues of A are the solutions to

$$|\lambda I - A| = \begin{vmatrix} \lambda - 1 & 0 & -2 \\ 0 & \lambda - 1 & 0 \\ -2 & 0 & \lambda - 1 \end{vmatrix} = (\lambda - 3)(\lambda - 1)(\lambda + 1) = 0.$$

Therefore, the eigenvalues are 3, 1, and -1. The corresponding eigenvectors are the solutions of $(\lambda I - A)\mathbf{x} = \mathbf{0}$. So, an eigenvector corresponding to 3 is $(1, 0, 1)$, an eigenvector corresponding to 1 is $(0, 1, 0)$, an eigenvector corresponding to -1 is $(1, 0, -1)$. Now form P using the eigenvectors of A as column vectors.

$$P = \begin{vmatrix} 1 & 0 & 1 \\ 0 & 1 & 0 \\ 1 & 0 & -1 \end{vmatrix}$$

Note that

$$P^{-1}AP = \begin{bmatrix} \frac{1}{2} & 0 & \frac{1}{2} \\ 0 & 1 & 0 \\ \frac{1}{2} & 0 & -\frac{1}{2} \end{bmatrix}\begin{bmatrix} 1 & 0 & 2 \\ 0 & 1 & 0 \\ 2 & 0 & 1 \end{bmatrix}\begin{bmatrix} 1 & 0 & 1 \\ 0 & 1 & 0 \\ 1 & 0 & -1 \end{bmatrix} = \begin{bmatrix} 3 & 0 & 0 \\ 0 & 1 & 0 \\ 0 & 0 & -1 \end{bmatrix}.$$

15. The characteristic equation of A is given by

$$|\lambda I - A| = \begin{vmatrix} \lambda & -1 \\ -a & \lambda - 1 \end{vmatrix} = \lambda^2 - \lambda - a = 0.$$

The discriminant d of this quadratic equation in λ is $1 + 4a$.

(a) A has an eigenvalue of multiplicity 2 if and only if $d = 1 + 4a = 0$; that is, $a = -\frac{1}{4}$.

(b) A has -1 and 2 as eigenvalues if and only if $\lambda^2 - \lambda - a = (\lambda + 1)(\lambda - 2)$; that is, $a = 2$.

(c) A has real eigenvalues if and only if $d = 1 + 4a \geq 0$; that is, $a \geq -\frac{1}{4}$.

© 2013 Cengage Learning. All Rights Reserved. May not be scanned, copied or duplicated, or posted to a publicly accessible website, in whole or in part.

17. The eigenvalue is $\lambda = 0$ (repeated). To find its corresponding eigenspace, solve

$$\begin{bmatrix} \lambda & -2 & \vdots & 0 \\ 0 & \lambda & \vdots & 0 \end{bmatrix} = \begin{bmatrix} 0 & -2 & \vdots & 0 \\ 0 & 0 & \vdots & 0 \end{bmatrix} \Rightarrow \begin{bmatrix} 0 & 1 & \vdots & 0 \\ 0 & 0 & \vdots & 0 \end{bmatrix}.$$

Because the eigenspace is only one-dimensional, the matrix A is not diagonalizable.

19. The eigenvalue is $\lambda = 3$ (repeated). To find its corresponding eigenspace, solve $(\lambda I - A)\mathbf{x} = \mathbf{0}$ with $\lambda = 3$.

$$\begin{bmatrix} \lambda - 3 & 0 & 0 & \vdots & 0 \\ -1 & \lambda - 3 & 0 & \vdots & 0 \\ 0 & 0 & \lambda - 3 & \vdots & 0 \end{bmatrix} = \begin{bmatrix} 0 & 0 & 0 & \vdots & 0 \\ -1 & 0 & 0 & \vdots & 0 \\ 0 & 0 & 0 & \vdots & 0 \end{bmatrix} \Rightarrow \begin{bmatrix} 1 & 0 & 0 & \vdots & 0 \\ 0 & 0 & 0 & \vdots & 0 \\ 0 & 0 & 0 & \vdots & 0 \end{bmatrix}$$

Because the eigenspace is only two-dimensional, the matrix A is not diagonalizable.

21. The eigenvalues of B are 1 and 2 with corresponding eigenvectors $(0, 1)$ and $(1, 0)$, respectively. Form the columns of P from the eigenvectors of B. So,

$$P = \begin{bmatrix} 0 & 1 \\ 1 & 0 \end{bmatrix}$$

$$P^{-1}BP = \begin{bmatrix} 0 & 1 \\ 1 & 0 \end{bmatrix}\begin{bmatrix} 2 & 0 \\ 0 & 1 \end{bmatrix}\begin{bmatrix} 0 & 1 \\ 1 & 0 \end{bmatrix} = \begin{bmatrix} 1 & 0 \\ 0 & 2 \end{bmatrix} = A.$$

Therefore, A and B are similar.

23. Because the eigenspace corresponding to $\lambda = 1$ of matrix A has dimension 1, while that of matrix B has dimension 2, the matrices are not similar.

25. Because

$$A^T = \begin{bmatrix} -\frac{\sqrt{2}}{2} & \frac{\sqrt{2}}{2} \\ \frac{\sqrt{2}}{2} & \frac{\sqrt{2}}{2} \end{bmatrix} = A,$$

A is symmetric. Furthermore, the column vectors of A form an orthonormal set. So, A is both symmetric and orthogonal.

27. Because

$$A^T = \begin{bmatrix} 0 & 0 & 1 \\ 0 & 1 & 0 \\ 1 & 0 & 1 \end{bmatrix} = A,$$

A is symmetric. However, column 3 is not a unit vector, so A is *not* orthogonal.

29. Because

$$A^T = \begin{bmatrix} -\frac{2}{3} & \frac{2}{3} & \frac{1}{3} \\ \frac{1}{3} & \frac{2}{3} & -\frac{2}{3} \\ -\frac{2}{3} & -\frac{1}{3} & \frac{2}{3} \end{bmatrix} \neq A,$$

A is *not* symmetric. Because the column vectors of A do not form an orthonormal set (columns 2 and 3 are not orthogonal), A is *not* orthogonal.

31. The matrix is diagonal, so the eigenvalues are $\lambda_1 = 2$ and $\lambda_2 = -3$. Every eigenvector corresponding to $\lambda_1 = 2$ is of the form $x_1 = (t, 0)$, and every eigenvector corresponding to $\lambda_2 = -3$ is of the form $x_2 = (0, s)$.

$$x_1 \cdot x_2 = 0$$

So, x_1 and x_2 are orthogonal.

33. The characteristic polynomial of A is

$$|\lambda \mathbf{I} - A| = \begin{vmatrix} \lambda + 1 & 0 & 1 \\ 0 & \lambda + 1 & 0 \\ 1 & 0 & \lambda - 1 \end{vmatrix}$$

$$= (\lambda + 1)(\lambda - \sqrt{2})(\lambda + \sqrt{2}).$$

The eigenvalues are $\lambda_1 = -1$, $\lambda_2 = \sqrt{2}$, and $\lambda_3 = -\sqrt{2}$. Every eigenvector corresponding to $\lambda_1 = -1$ is of the form $x_1 = (0, t, 0)$, every eigenvector corresponding to $\lambda_2 = \sqrt{2}$ is of the form $x_2 = ((1 - \sqrt{2})s, 0, s)$, and every eigenvector corresponding to $\lambda_3 = -\sqrt{2}$ is of the form $x_3 = ((\sqrt{2} + 1)u, 0, u)$.

$$x_1 \cdot x_2 = x_1 \cdot x_3 = x_2 \cdot x_3 = 0$$

So, $\{x_1, x_2, x_3\}$ is an orthogonal set.

35. The matrix is symmetric, so it is orthogonally diagonalizable.

© 2013 Cengage Learning. All Rights Reserved. May not be scanned, copied or duplicated, or posted to a publicly accessible website, in whole or in part.

37. The eigenvalues of A are 5 and -5 with corresponding unit eigenvectors $\left(\frac{2}{\sqrt{5}}, \frac{1}{\sqrt{5}}\right)$ and $\left(-\frac{1}{\sqrt{5}}, \frac{2}{\sqrt{5}}\right)$, respectively. Form the columns of P with the eigenvectors of A.

$$P = \begin{bmatrix} \frac{2}{\sqrt{5}} & -\frac{1}{\sqrt{5}} \\ \frac{1}{\sqrt{5}} & \frac{2}{\sqrt{5}} \end{bmatrix}$$

$$P^T AP = \begin{bmatrix} \frac{2}{\sqrt{5}} & \frac{1}{\sqrt{5}} \\ -\frac{1}{\sqrt{5}} & \frac{2}{\sqrt{5}} \end{bmatrix}\begin{bmatrix} 3 & 4 \\ 4 & -3 \end{bmatrix}\begin{bmatrix} \frac{2}{\sqrt{5}} & -\frac{1}{\sqrt{5}} \\ \frac{1}{\sqrt{5}} & \frac{2}{\sqrt{5}} \end{bmatrix} = \begin{bmatrix} 5 & 0 \\ 0 & -5 \end{bmatrix}$$

39. The eigenvalues of A are 0 (repeated) and 2, with corresponding unit eigenvectors $(0, 0, 1)$, $\left(\frac{1}{\sqrt{2}}, -\frac{1}{\sqrt{2}}, 0\right)$, and $\left(\frac{1}{\sqrt{2}}, \frac{1}{\sqrt{2}}, 0\right)$. Form the columns of P from the eigenvectors of A.

$$P = \begin{bmatrix} 0 & \frac{1}{\sqrt{2}} & \frac{1}{\sqrt{2}} \\ 0 & -\frac{1}{\sqrt{2}} & \frac{1}{\sqrt{2}} \\ 1 & 0 & 0 \end{bmatrix}$$

$$P^T AP = \begin{bmatrix} 0 & 0 & 1 \\ \frac{1}{\sqrt{2}} & -\frac{1}{\sqrt{2}} & 0 \\ \frac{1}{\sqrt{2}} & \frac{1}{\sqrt{2}} & 0 \end{bmatrix}\begin{bmatrix} 1 & 1 & 0 \\ 1 & 1 & 0 \\ 0 & 0 & 1 \end{bmatrix}\begin{bmatrix} 0 & \frac{1}{\sqrt{2}} & \frac{1}{\sqrt{2}} \\ 0 & -\frac{1}{\sqrt{2}} & \frac{1}{\sqrt{2}} \\ 1 & 0 & 0 \end{bmatrix} = \begin{bmatrix} 0 & 0 & 0 \\ 0 & 0 & 0 \\ 0 & 0 & 2 \end{bmatrix}$$

41. The eigenvalues of A are 3 and 1 (repeated), with corresponding unit eigenvectors $\left(\frac{1}{\sqrt{2}}, 0, -\frac{1}{\sqrt{2}}\right)$, $\left(\frac{1}{\sqrt{2}}, 0, \frac{1}{\sqrt{2}}\right)$ and $(0, 1, 0)$. Form the columns of P from the eigenvectors of A.

$$P = \begin{bmatrix} \frac{1}{\sqrt{2}} & \frac{1}{\sqrt{2}} & 0 \\ 0 & 0 & 1 \\ -\frac{1}{\sqrt{2}} & \frac{1}{\sqrt{2}} & 0 \end{bmatrix}$$

$$P^T AP = \begin{bmatrix} \frac{1}{\sqrt{2}} & 0 & -\frac{1}{\sqrt{2}} \\ \frac{1}{\sqrt{2}} & 0 & \frac{1}{\sqrt{2}} \\ 0 & 1 & 0 \end{bmatrix}\begin{bmatrix} 2 & 0 & -1 \\ 0 & 1 & 0 \\ -1 & 0 & 2 \end{bmatrix}\begin{bmatrix} \frac{1}{\sqrt{2}} & \frac{1}{\sqrt{2}} & 0 \\ 0 & 0 & 1 \\ -\frac{1}{\sqrt{2}} & \frac{1}{\sqrt{2}} & 0 \end{bmatrix} = \begin{bmatrix} 3 & 0 & 0 \\ 0 & 1 & 0 \\ 0 & 0 & 1 \end{bmatrix}$$

43. The eigenvalues of A are $\frac{1}{6}$ and 1. The eigenvectors corresponding to $\lambda = 1$ are $\mathbf{x} = t(3, 2)$. By choosing $t = \frac{1}{5}$, you can find the steady state probability vector for A to be $\mathbf{v} = \left(\frac{3}{5}, \frac{2}{5}\right)$. Note that

$$A\mathbf{v} = \begin{bmatrix} \frac{2}{3} & \frac{1}{2} \\ \frac{1}{3} & \frac{1}{2} \end{bmatrix}\begin{bmatrix} \frac{3}{5} \\ \frac{2}{5} \end{bmatrix} = \begin{bmatrix} \frac{3}{5} \\ \frac{2}{5} \end{bmatrix} = \mathbf{v}.$$

© 2013 Cengage Learning. All Rights Reserved. May not be scanned, copied or duplicated, or posted to a publicly accessible website, in whole or in part.

45. The eigenvalues of A are $\frac{1}{2}$ and 1. The eigenvectors corresponding to $\lambda = 1$ are $\mathbf{x} = t(3, 2)$. By choosing $t = \frac{1}{5}$, you can find the steady state probability vector for A to be $\mathbf{v} = \left(\frac{3}{5}, \frac{2}{5}\right)$. Note that

$$A\mathbf{v} = \begin{bmatrix} 0.8 & 0.3 \\ 0.2 & 0.7 \end{bmatrix} \begin{bmatrix} \frac{3}{5} \\ \frac{2}{5} \end{bmatrix} = \begin{bmatrix} \frac{3}{5} \\ \frac{2}{5} \end{bmatrix} = \mathbf{v}.$$

47. The eigenvalues of A are $0, \frac{1}{2}$ and 1. The eigenvectors corresponding to $\lambda = 1$ are $\mathbf{x} = t(1, 2, 1)$. By choosing $t = \frac{1}{4}$, you can find the steady state probability vector for A to be $\mathbf{v} = \left(\frac{1}{4}, \frac{1}{2}, \frac{1}{4}\right)$. Note that

$$A\mathbf{v} = \begin{bmatrix} \frac{1}{2} & \frac{1}{4} & 0 \\ \frac{1}{2} & \frac{1}{2} & \frac{1}{2} \\ 0 & \frac{1}{4} & \frac{1}{2} \end{bmatrix} \begin{bmatrix} \frac{1}{4} \\ \frac{1}{2} \\ \frac{1}{4} \end{bmatrix} = \begin{bmatrix} \frac{1}{4} \\ \frac{1}{2} \\ \frac{1}{4} \end{bmatrix} = \mathbf{v}.$$

49. The eigenvalues of A are 0.6 and 1. The eigenvectors corresponding to $\lambda = 1$ are $\mathbf{x} = t(4, 5, 7)$. By choosing $t = \frac{1}{16}$, you can find the steady state probability vector for A to be $\mathbf{v} = \left(\frac{1}{4}, \frac{5}{16}, \frac{7}{16}\right)$.

Note that

$$A\mathbf{v} = \begin{bmatrix} 0.7 & 0.1 & 0.1 \\ 0.2 & 0.7 & 0.1 \\ 0.1 & 0.2 & 0.8 \end{bmatrix} \begin{bmatrix} 0.25 \\ 0.3125 \\ 0.4375 \end{bmatrix} = \begin{bmatrix} 0.25 \\ 0.3125 \\ 0.4375 \end{bmatrix} = \mathbf{v}.$$

51. $(P^TAP)^T = P^TA^T(P^T)^T = P^TAP$ (because A is symmetric), which shows that P^TAP is symmetric.

53. From the form $p(\lambda) = a_0 + a_1\lambda + a_2\lambda^2$, you have $a_0 = 0$, $a_1 = -9$, and $a_2 = 4$. This implies that the companion matrix of p is

$$A = \begin{bmatrix} 0 & 1 \\ -\frac{a_0}{a_2} & -\frac{a_1}{a_2} \end{bmatrix} \begin{bmatrix} 0 & 1 \\ 0 & \frac{9}{4} \end{bmatrix}.$$

The eigenvalues of A are 0 and $\frac{9}{4}$, the zeros of p.

55. $A^2 = 10A - 24I_2 = \begin{bmatrix} 80 & -40 \\ 20 & 20 \end{bmatrix} - \begin{bmatrix} 24 & 0 \\ 0 & 24 \end{bmatrix} = \begin{bmatrix} 56 & -40 \\ 20 & -4 \end{bmatrix}$

$A^3 = 10A^2 - 24A = \begin{bmatrix} 560 & -400 \\ 200 & -40 \end{bmatrix} - \begin{bmatrix} 192 & -96 \\ 48 & 48 \end{bmatrix} = \begin{bmatrix} 368 & -304 \\ 152 & -88 \end{bmatrix}$

57. (a) True. If $A\mathbf{x} = \lambda\mathbf{x}$, then $A^2\mathbf{x} = A(A\mathbf{x}) = A(\lambda\mathbf{x}) = \lambda(A\mathbf{x}) = \lambda^2\mathbf{x}$ showing that $\mathbf{x}$ is an eigenvector of A^2.

(b) False. For example, $(1, 0)$ is an eigenvector of $A^2 = \begin{bmatrix} 1 & 0 \\ 0 & 1 \end{bmatrix}$, but not of $A = \begin{bmatrix} 0 & 1 \\ 1 & 0 \end{bmatrix}$.

59. Because $A^{-1}(AB)A = BA$, you can see that AB and BA are similar.

61. The eigenvalues of A are $a + b$ and $a - b$, with corresponding unit eigenvectors $\left(\frac{1}{\sqrt{2}}, \frac{1}{\sqrt{2}}\right)$ and $\left(-\frac{1}{\sqrt{2}}, \frac{1}{\sqrt{2}}\right)$, respectively. So, $P = \begin{bmatrix} \frac{1}{\sqrt{2}} & -\frac{1}{\sqrt{2}} \\ \frac{1}{\sqrt{2}} & \frac{1}{\sqrt{2}} \end{bmatrix}$. Note that

$$P^{-1}AP = \begin{bmatrix} \frac{1}{\sqrt{2}} & \frac{1}{\sqrt{2}} \\ -\frac{1}{\sqrt{2}} & \frac{1}{\sqrt{2}} \end{bmatrix} \begin{bmatrix} a & b \\ b & a \end{bmatrix} \begin{bmatrix} \frac{1}{\sqrt{2}} & -\frac{1}{\sqrt{2}} \\ \frac{1}{\sqrt{2}} & \frac{1}{\sqrt{2}} \end{bmatrix} = \begin{bmatrix} a + b & 0 \\ 0 & a - b \end{bmatrix}.$$

63. (a) A is diagonalizable if and only if $a = b = c = 0$.

(b) If exactly two of a, b, and c are zero, then the eigenspace of 2 has dimension 3. If exactly one of a, b, c is zero, then the dimension of the eigenspace is 2. If none of a, b, c is zero, the eigenspace is dimension 1.

65. (a) True. See "Definitions of Eigenvalue and Eigenvector," page 342.

(b) False. See Theorem 7.4, page 354.

(c) True. See "Definition of a Diagonalizable Matrix," page 353.

© 2013 Cengage Learning. All Rights Reserved. May not be scanned, copied or duplicated, or posted to a publicly accessible website, in whole or in part.

67. The population after one transition is

$$\mathbf{x}_2 = A\mathbf{x}_1 = \begin{bmatrix} 0 & 1 \\ \frac{1}{4} & 0 \end{bmatrix}\begin{bmatrix} 100 \\ 100 \end{bmatrix} = \begin{bmatrix} 100 \\ 25 \end{bmatrix}$$

and after two transitions is

$$\mathbf{x}_3 = A\mathbf{x}_2 = \begin{bmatrix} 0 & 1 \\ \frac{1}{4} & 0 \end{bmatrix}\begin{bmatrix} 100 \\ 25 \end{bmatrix} = \begin{bmatrix} 25 \\ 25 \end{bmatrix}.$$

The eigenvalues of A are $-\frac{1}{2}$ and $\frac{1}{2}$. Choose the positive eigenvalue and find the corresponding eigenvectors to be multiples of (2, 1). So, the stable age distribution vector is $\mathbf{x} = t\begin{bmatrix} 2 \\ 1 \end{bmatrix}$.

69. The population after one transition is

$$\mathbf{x}_2 = \begin{bmatrix} 0 & 3 & 12 \\ 1 & 0 & 0 \\ 0 & \frac{1}{6} & 0 \end{bmatrix}\begin{bmatrix} 300 \\ 300 \\ 300 \end{bmatrix} = \begin{bmatrix} 4500 \\ 300 \\ 50 \end{bmatrix}$$

and after two transitions is

$$\mathbf{x}_3 = \begin{bmatrix} 0 & 3 & 12 \\ 1 & 0 & 0 \\ 0 & \frac{1}{6} & 0 \end{bmatrix}\begin{bmatrix} 4500 \\ 300 \\ 50 \end{bmatrix} = \begin{bmatrix} 1500 \\ 4500 \\ 50 \end{bmatrix}.$$

The positive eigenvalue 2 has corresponding eigenvector (24, 12, 1), which is a stable distribution.

So, the stable age distribution vector is $\mathbf{x} = t\begin{bmatrix} 24 \\ 12 \\ 1 \end{bmatrix}$.

71. Construct the age transition matrix.

$$A = \begin{bmatrix} 4 & 6 & 2 \\ 0.9 & 0 & 0 \\ 0 & 0.75 & 0 \end{bmatrix}$$

The current age distribution vector is

$$\mathbf{x}_1 = \begin{bmatrix} 120 \\ 120 \\ 120 \end{bmatrix}.$$

In one year, the age distribution vector will be

$$\mathbf{x}_2 = A\mathbf{x}_1 = \begin{bmatrix} 4 & 6 & 2 \\ 0.9 & 0 & 0 \\ 0 & 0.75 & 0 \end{bmatrix}\begin{bmatrix} 120 \\ 120 \\ 120 \end{bmatrix} = \begin{bmatrix} 1440 \\ 108 \\ 90 \end{bmatrix}.$$

In two years, the age distribution vector will be

$$\mathbf{x}_3 = A\mathbf{x}_2 = \begin{bmatrix} 4 & 6 & 2 \\ 0.9 & 0 & 0 \\ 0 & 0.75 & 0 \end{bmatrix}\begin{bmatrix} 1440 \\ 108 \\ 90 \end{bmatrix} = \begin{bmatrix} 6588 \\ 1296 \\ 81 \end{bmatrix}.$$

73. The solution to the differential equation $y' = ky$ is $y = Ce^{kt}$. So, $y_1 = C_1e^{3t}$, $y_2 = C_2e^{8t}$, and $y_3 = C_3e^{-8t}$.

75. The matrix corresponding to the system $\mathbf{y}' = A\mathbf{y}$ is

$$A = \begin{bmatrix} 1 & 2 \\ 0 & 0 \end{bmatrix}.$$

This matrix has eigenvalues of 0 and 1, with corresponding eigenvectors $(-2, 1)$ and $(1, 0)$, respectively. So, a matrix P that diagonalizes A is

$$P = \begin{bmatrix} -2 & 1 \\ 1 & 0 \end{bmatrix} \quad \text{and} \quad P^{-1}AP = \begin{bmatrix} 0 & 0 \\ 0 & 1 \end{bmatrix}.$$

The system represented by $\mathbf{w}' = P^{-1}AP\mathbf{w}$ yields the solution $w_1' = 0$ and $w_2' = w_2$.

So $w_1 = C_1$ and $w_2 = C_2e^t$. Substitute $\mathbf{y} = P\mathbf{w}$ and write

$$\begin{bmatrix} y_1 \\ y_2 \end{bmatrix} = \begin{bmatrix} -2 & 1 \\ 1 & 0 \end{bmatrix}\begin{bmatrix} w_1 \\ w_2 \end{bmatrix} = \begin{bmatrix} -2w_1 + w_2 \\ w_1 \end{bmatrix}.$$

This implies that the solution is

$$y_1 = -2C_1 + C_2e^t$$
$$y_2 = C_1.$$

77. The matrix corresponding to the system $\mathbf{y}' = A\mathbf{y}$ is

$$A = \begin{bmatrix} 0 & 1 & 0 \\ 1 & 0 & 0 \\ 0 & 0 & 0 \end{bmatrix}.$$

This matrix has eigenvalues 1, −1, and 0 with corresponding eigenvectors $(1, 1, 0)$, $(1, -1, 0)$, and $(0, 0, 1)$.

So, a matrix P that diagonalizes A is

$$P = \begin{bmatrix} 1 & 1 & 0 \\ 1 & -1 & 0 \\ 0 & 0 & 1 \end{bmatrix} \quad \text{and} \quad P^{-1}AP = \begin{bmatrix} 1 & 0 & 0 \\ 0 & -1 & 0 \\ 0 & 0 & 0 \end{bmatrix}.$$

The system represented by $\mathbf{w}' = P^{-1}AP\mathbf{w}$ has solutions $w_1 = C_1e^t$, $w_2 = C_2e^{-t}$, and $w_3 = C_3e^0 = C_3$

Substitute $\mathbf{y} = P\mathbf{w}$ and obtain

$$\begin{bmatrix} y_1 \\ y_2 \\ y_3 \end{bmatrix} = \begin{bmatrix} 1 & 1 & 0 \\ 1 & -1 & 0 \\ 0 & 0 & 1 \end{bmatrix}\begin{bmatrix} w_1 \\ w_2 \\ w_3 \end{bmatrix} = \begin{bmatrix} w_1 + w_2 \\ w_1 - w_2 \\ w_3 \end{bmatrix},$$

which yields the solution

$$y_1 = C_1e^t + C_2e^{-t}$$
$$y_2 = C_1e^t - C_2e^{-t}$$
$$y_3 = C_3.$$

© 2013 Cengage Learning. All Rights Reserved. May not be scanned, copied or duplicated, or posted to a publicly accessible website, in whole or in part.

79. (a) The matrix of the quadratic form is

$$A = \begin{bmatrix} a & \frac{b}{2} \\ \frac{b}{2} & c \end{bmatrix} = \begin{bmatrix} 1 & \frac{3}{2} \\ \frac{3}{2} & 1 \end{bmatrix}.$$

(b) The eigenvalues are $\frac{5}{2}$ and $-\frac{1}{2}$ with corresponding unit eigenvectors $\left(\frac{1}{\sqrt{2}}, \frac{1}{\sqrt{2}}\right)$ and $\left(-\frac{1}{\sqrt{2}}, \frac{1}{\sqrt{2}}\right)$, respectively.

Then form the columns of P from the eigenvectors of A.

$$P = \begin{bmatrix} \frac{1}{\sqrt{2}} & -\frac{1}{\sqrt{2}} \\ \frac{1}{\sqrt{2}} & \frac{1}{\sqrt{2}} \end{bmatrix} \quad \text{and} \quad P^T AP = \begin{bmatrix} \frac{5}{2} & 0 \\ 0 & -\frac{1}{2} \end{bmatrix}$$

(c) This implies that the equation of the rotated conic is $\frac{5}{2}(x')^2 - \frac{1}{2}(y')^2 - 3 = 0$.

(d)

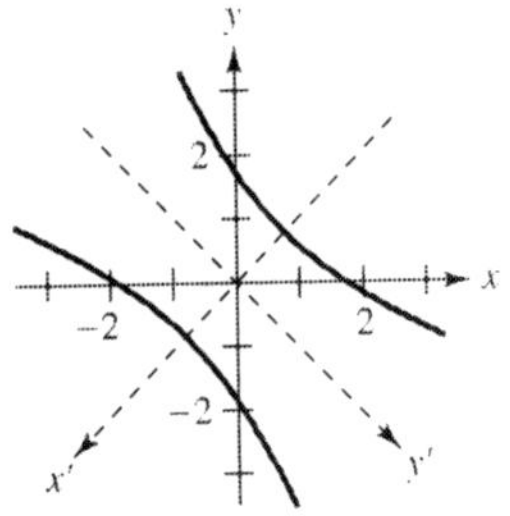

81. (a) The matrix of the quadratic form is

$$A = \begin{bmatrix} a & \frac{b}{2} \\ \frac{b}{2} & c \end{bmatrix} = \begin{bmatrix} 0 & \frac{1}{2} \\ \frac{1}{2} & 0 \end{bmatrix}.$$

(b) The eigenvalues are $\frac{1}{2}$ and $-\frac{1}{2}$, with corresponding unit eigenvectors $\left(\frac{1}{\sqrt{2}}, \frac{1}{\sqrt{2}}\right)$ and $\left(-\frac{1}{\sqrt{2}}, \frac{1}{\sqrt{2}}\right)$. Use these eigenvectors to form the columns of P.

$$P = \begin{bmatrix} \frac{1}{\sqrt{2}} & -\frac{1}{\sqrt{2}} \\ \frac{1}{\sqrt{2}} & \frac{1}{\sqrt{2}} \end{bmatrix} \quad \text{and} \quad P^T AP = \begin{bmatrix} \frac{1}{2} & 0 \\ 0 & -\frac{1}{2} \end{bmatrix}$$

(c) This implies that the equation of the rotated conic is $\frac{1}{2}(x')^2 - \frac{1}{2}(y')^2 - 2 = 0$, a hyperbola.

(d)

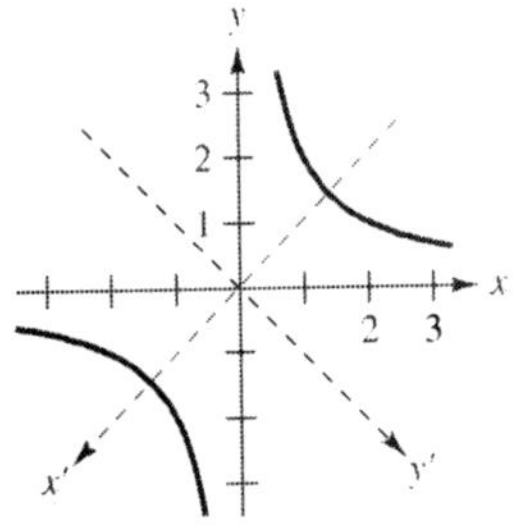

© 2013 Cengage Learning. All Rights Reserved. May not be scanned, copied or duplicated, or posted to a publicly accessible website, in whole or in part.

Cumulative Test for Chapters 6 and 7

1. T preserves addition.

$$\begin{aligned} T(x_1, y_1, z_1) + T(x_2, y_2, z_2) &= (2x_1, x_1 + y_1) + (2x_2, x_2 + y_2) \\ &= (2x_1 + 2x_2, x_1 + y_1 + x_2 + y_2) \\ &= (2(x_1 + x_2), (x_1 + x_2) + (y_1 + y_2)) \\ &= T(x_1 + x_2, y_1 + y_2, z_1 + z_2) \end{aligned}$$

T preserves scalar multiplication.

$$\begin{aligned} T(c(x, y, z)) &= T(cx, cy, cz) \\ &= (2cx, cx + cy) \\ &= c(2x, x + y) \\ &= cT(x, y, z) \end{aligned}$$

Therefore, T is a linear transformation.

2. No, T is not a linear transformation. For example,

$$T\left(2\begin{bmatrix} 1 & 0 \\ 0 & 1 \end{bmatrix}\right) = T\begin{bmatrix} 2 & 0 \\ 0 & 2 \end{bmatrix} = \left|\begin{bmatrix} 2 & 0 \\ 0 & 2 \end{bmatrix} + \begin{bmatrix} 2 & 0 \\ 0 & 2 \end{bmatrix}\right| = \begin{vmatrix} 4 & 0 \\ 0 & 4 \end{vmatrix} = 16$$

$$2T\begin{bmatrix} 1 & 0 \\ 0 & 1 \end{bmatrix} = 2\left|\begin{bmatrix} 1 & 0 \\ 0 & 1 \end{bmatrix} + \begin{bmatrix} 1 & 0 \\ 0 & 1 \end{bmatrix}\right| = 2\begin{vmatrix} 2 & 0 \\ 0 & 2 \end{vmatrix} = 2 \cdot 4 = 8.$$

3. Because the matrix has five columns, the dimension of R^n is 5. Because the matrix has two rows, the dimension of R^m is 2. So $T : R^5 \to R^2$.

4. (a) $T(1, -2) = \begin{bmatrix} 1 & 0 \\ -1 & 0 \\ 0 & 0 \end{bmatrix}\begin{bmatrix} 1 \\ -2 \end{bmatrix} = \begin{bmatrix} 1 \\ -1 \\ 0 \end{bmatrix}$

(b) $\begin{bmatrix} 1 & 0 \\ -1 & 0 \\ 0 & 0 \end{bmatrix}\begin{bmatrix} x \\ y \end{bmatrix} = \begin{bmatrix} x \\ -x \\ 0 \end{bmatrix} = \begin{bmatrix} 5 \\ -5 \\ 0 \end{bmatrix} \Rightarrow x = 5, y = t$

The preimage of $(5, -5, 0)$ is $(5, t)$, where t is any real number.

5. The kernel is the solution space of the homogeneous system

$$\begin{aligned} x_1 - x_2 \qquad\qquad &= 0 \\ -x_1 + x_2 \qquad\qquad &= 0 \\ x_3 + x_4 &= 0. \end{aligned}$$

$$\begin{bmatrix} 1 & -1 & 0 & 0 \\ -1 & 1 & 0 & 0 \\ 0 & 0 & 1 & 1 \end{bmatrix} \Rightarrow \begin{bmatrix} 1 & -1 & 0 & 0 \\ 0 & 0 & 0 & 0 \\ 0 & 0 & 1 & 1 \end{bmatrix} \Rightarrow \begin{aligned} x_1 &= x_2 \\ x_3 &= -x_4 \end{aligned}$$

So, $\ker(T) = \{(s, s, -t, t) : s, t \in R\}$.

6. $A = \begin{bmatrix} 1 & 0 & 1 & 0 \\ 0 & -1 & 0 & -1 \end{bmatrix} \Rightarrow \begin{bmatrix} 1 & 0 & 1 & 0 \\ 0 & 1 & 0 & 1 \end{bmatrix}$

(a) basis for kernel: $\left\{\begin{bmatrix} 0 \\ -1 \\ 0 \\ 1 \end{bmatrix}, \begin{bmatrix} 1 \\ 0 \\ -1 \\ 0 \end{bmatrix}\right\}$

(b) basis for range (column space of A): $\left\{\begin{pmatrix} 1 \\ 0 \end{pmatrix}, \begin{pmatrix} 0 \\ 1 \end{pmatrix}\right\}$

(c) rank $= 2$, nullity $= 2$

7. Because

$$T\left(\begin{bmatrix} 1 \\ 0 \end{bmatrix}\right) = \begin{bmatrix} 3 \\ -1 \end{bmatrix} \text{ and } T\left(\begin{bmatrix} 0 \\ 1 \end{bmatrix}\right) = \begin{bmatrix} 2 \\ 2 \end{bmatrix},$$

the standard matrix for T is

$$\begin{bmatrix} 3 & 2 \\ -1 & 2 \end{bmatrix}.$$

© 2013 Cengage Learning. All Rights Reserved. May not be scanned, copied or duplicated, or posted to a publicly accessible website, in whole or in part.

8. Because

$$T\left(\begin{bmatrix}1\\0\\0\end{bmatrix}\right) = \begin{bmatrix}1\\0\\1\end{bmatrix},\quad T\left(\begin{bmatrix}0\\1\\0\end{bmatrix}\right) = \begin{bmatrix}1\\1\\0\end{bmatrix} \quad\text{and}\quad T\left(\begin{bmatrix}0\\0\\1\end{bmatrix}\right) = \begin{bmatrix}0\\1\\-1\end{bmatrix},$$

the standard matrix for T is

$$A = \begin{bmatrix}1 & 1 & 0\\0 & 1 & 1\\1 & 0 & -1\end{bmatrix}.$$

9. Because

$$T\left(\begin{bmatrix}1\\0\\0\end{bmatrix}\right) = \begin{bmatrix}0\\4\end{bmatrix},\quad T\left(\begin{bmatrix}0\\1\\0\end{bmatrix}\right) = \begin{bmatrix}-2\\0\end{bmatrix},$$

$$\text{and } T\left(\begin{bmatrix}0\\0\\1\end{bmatrix}\right) = \begin{bmatrix}3\\11\end{bmatrix},$$

the standard matrix for T is $A = \begin{bmatrix}0 & -2 & 3\\4 & 0 & 11\end{bmatrix}.$

10. Because

$$T\left(\begin{bmatrix}1\\0\\0\end{bmatrix}\right) = \begin{bmatrix}0\\0\\0\end{bmatrix},\quad T\left(\begin{bmatrix}0\\1\\0\end{bmatrix}\right) = \begin{bmatrix}0\\0\\0\end{bmatrix},\quad\text{and}\quad T\left(\begin{bmatrix}0\\0\\1\end{bmatrix}\right) = \begin{bmatrix}0\\0\\0\end{bmatrix},$$

the standard matrix for T is

$$\begin{bmatrix}0 & 0 & 0\\0 & 0 & 0\\0 & 0 & 0\end{bmatrix}.$$

11. $T\left(\begin{bmatrix}1\\0\end{bmatrix}\right) = \begin{bmatrix}\frac{1}{2}\\-\frac{1}{2}\end{bmatrix}, T\left(\begin{bmatrix}0\\1\end{bmatrix}\right) = \begin{bmatrix}-\frac{1}{2}\\\frac{1}{2}\end{bmatrix}, A = \begin{bmatrix}\frac{1}{2} & -\frac{1}{2}\\-\frac{1}{2} & \frac{1}{2}\end{bmatrix}$

$$T\left(\begin{bmatrix}1\\1\end{bmatrix}\right) = \begin{bmatrix}\frac{1}{2} & -\frac{1}{2}\\-\frac{1}{2} & \frac{1}{2}\end{bmatrix}\begin{bmatrix}1\\1\end{bmatrix} = \begin{bmatrix}0\\0\end{bmatrix}$$

$$T\left(\begin{bmatrix}-2\\2\end{bmatrix}\right) = \begin{bmatrix}\frac{1}{2} & -\frac{1}{2}\\-\frac{1}{2} & \frac{1}{2}\end{bmatrix}\begin{bmatrix}-2\\2\end{bmatrix} = \begin{bmatrix}-2\\2\end{bmatrix}$$

12. (a) The counterclockwise rotation of 30° is given by

$$T(x, y) = \left(\cos(30)x - \sin(30)y, \sin(30)x + \cos(30)y\right) = \left(\frac{\sqrt{3}}{2}x - \frac{1}{2}y, \frac{1}{2}x + \frac{\sqrt{3}}{2}y\right).$$

So, the matrix is $A = \left[T(1, 0) \vdots T(0, 1)\right] = \begin{bmatrix}\frac{\sqrt{3}}{2} & -\frac{1}{2}\\\frac{1}{2} & \frac{\sqrt{3}}{2}\end{bmatrix}.$

(b) The image of $\mathbf{v} = (1, 2)$ is $A\mathbf{v} = \begin{bmatrix}\frac{\sqrt{3}}{2} & -\frac{1}{2}\\\frac{1}{2} & \frac{\sqrt{3}}{2}\end{bmatrix}\begin{bmatrix}1\\2\end{bmatrix} = \begin{bmatrix}\frac{\sqrt{3}}{2} - 1\\\frac{1}{2} + \sqrt{3}\end{bmatrix}$. So, $T(1, 2) = \left(\frac{\sqrt{3}}{2} - 1, \frac{1}{2} + \sqrt{3}\right).$

(c)

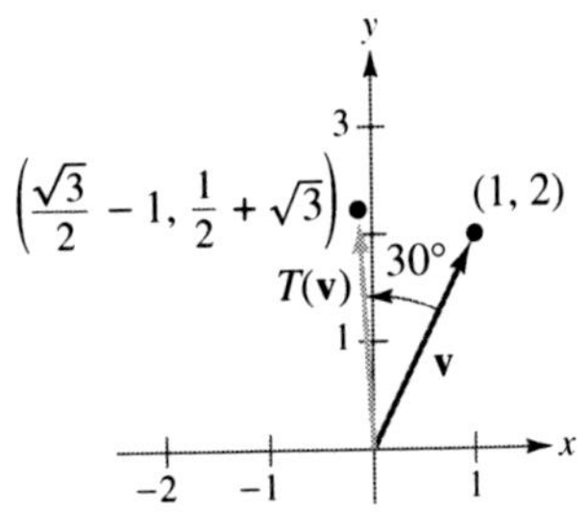

© 2013 Cengage Learning. All Rights Reserved. May not be scanned, copied or duplicated, or posted to a publicly accessible website, in whole or in part.

13. The standard matrices for T_1 and T_2 are

$$A_1 = \begin{bmatrix} 1 & -2 \\ 2 & 3 \end{bmatrix} \quad \text{and} \quad A_2 = \begin{bmatrix} 2 & 0 \\ 1 & -1 \end{bmatrix}.$$

The standard matrix for $T = T_2 \circ T_1$ is

$$A = A_2A_1 = \begin{bmatrix} 2 & 0 \\ 1 & -1 \end{bmatrix}\begin{bmatrix} 1 & -2 \\ 2 & 3 \end{bmatrix} = \begin{bmatrix} 2 & -4 \\ -1 & -5 \end{bmatrix}$$

and the standard matrix for $T' = T_1 \circ T_2$ is

$$A' = A_1A_2 = \begin{bmatrix} 1 & -2 \\ 2 & 3 \end{bmatrix}\begin{bmatrix} 2 & 0 \\ 1 & -1 \end{bmatrix} = \begin{bmatrix} 0 & 2 \\ 7 & -3 \end{bmatrix}.$$

14. The standard matrices for T_1 and T_2 are

$$A_1 = \begin{bmatrix} 1 & 2 & 0 \\ 0 & 1 & -1 \\ -2 & 1 & 2 \end{bmatrix} \quad \text{and} \quad A_2 = \begin{bmatrix} 0 & 1 & 1 \\ 1 & 0 & 1 \\ 0 & 2 & -2 \end{bmatrix}.$$

The standard matrix for $T = T_2 \circ T_1$ is

$$A_2A_1 = \begin{bmatrix} 0 & 1 & 1 \\ 1 & 0 & 1 \\ 0 & 2 & -2 \end{bmatrix}\begin{bmatrix} 1 & 2 & 0 \\ 0 & 1 & -1 \\ -2 & 1 & 2 \end{bmatrix} = \begin{bmatrix} -2 & 2 & 1 \\ -1 & 3 & 2 \\ 4 & 0 & -6 \end{bmatrix}$$

and the standard matrix for $T' = T_1 \circ T_2$ is

$$A_1A_2 = \begin{bmatrix} 1 & 2 & 0 \\ 0 & 1 & -1 \\ -2 & 1 & 2 \end{bmatrix}\begin{bmatrix} 0 & 1 & 1 \\ 1 & 0 & 1 \\ 0 & 2 & -2 \end{bmatrix} = \begin{bmatrix} 2 & 1 & 3 \\ 1 & -2 & 3 \\ 1 & 2 & -5 \end{bmatrix}.$$

15. Matrix of T is $A = \begin{bmatrix} 1 & -1 \\ 2 & 1 \end{bmatrix}$.

$$A^{-1} = \frac{1}{3}\begin{bmatrix} 1 & 1 \\ -2 & 1 \end{bmatrix} \Rightarrow T^{-1}(x, y) = \left(\tfrac{1}{3}x + \tfrac{1}{3}y, -\tfrac{2}{3}x + \tfrac{1}{3}y\right)$$

$$\left(T^{-1} \circ T\right)(3, -2) = T^{-1}\left(T(3, -2)\right) = T^{-1}(5, 4) = (3, -2)$$

16. The standard matrix for T is $A = \begin{bmatrix} 1 & 1 & 0 \\ 0 & 1 & 1 \\ 1 & 0 & 1 \end{bmatrix}$. Because $|A| = 2 \neq 0$, A is invertible.

$$A^{-1} = \begin{bmatrix} \frac{1}{2} & -\frac{1}{2} & \frac{1}{2} \\ \frac{1}{2} & \frac{1}{2} & -\frac{1}{2} \\ -\frac{1}{2} & \frac{1}{2} & \frac{1}{2} \end{bmatrix} \text{ and conclude that } T^{-1}(x_1, x_2, x_3) = \left(\frac{x_1 - x_2 + x_3}{2}, \frac{x_1 + x_2 - x_3}{2}, \frac{-x_1 + x_2 + x_3}{2}\right).$$

17. $T(1, 1) = (1, 2, 2) = -1(1, 0, 0) + 0(1, 1, 0) + 2(1, 1, 1)$

$T(1, 0) = (0, 2, 1) = -2(1, 0, 0) + 1(1, 1, 0) + 1(1, 1, 1)$

$$A = \begin{bmatrix} -1 & -2 \\ 0 & 1 \\ 2 & 1 \end{bmatrix}$$

$$T(0, 1) = A[\mathbf{v}]_B = \begin{bmatrix} -1 & -2 \\ 0 & 1 \\ 2 & 1 \end{bmatrix}\begin{bmatrix} 1 \\ -1 \end{bmatrix} = \begin{bmatrix} 1 \\ -1 \\ 1 \end{bmatrix} = (1, 0, 0) - (1, 1, 0) + (1, 1, 1) = (1, 0, 1)$$

18. (a) $A = \begin{bmatrix} 1 & -2 \\ 1 & 4 \end{bmatrix}$

(b) $[B : B'] \Rightarrow [I : P] \Rightarrow P = \begin{bmatrix} 1 & 1 \\ 1 & 2 \end{bmatrix}$

(c) $P^{-1} = \begin{bmatrix} 2 & -1 \\ -1 & 1 \end{bmatrix} \Rightarrow A' = P^{-1}AP = \begin{bmatrix} -7 & -15 \\ 6 & 12 \end{bmatrix}$

(d) $[T(\mathbf{v})]_{B'} = A'[\mathbf{v}]_{B'} = \begin{bmatrix} -7 & -15 \\ 6 & 12 \end{bmatrix}\begin{bmatrix} 3 \\ -2 \end{bmatrix} = \begin{bmatrix} 9 \\ -6 \end{bmatrix}$

© 2013 Cengage Learning. All Rights Reserved. May not be scanned, copied or duplicated, or posted to a publicly accessible website, in whole or in part.

(e) $[\mathbf{v}]_B = P[\mathbf{v}]_{B'} = \begin{bmatrix} 1 & 1 \\ 1 & 2 \end{bmatrix}\begin{bmatrix} 3 \\ -2 \end{bmatrix} = \begin{bmatrix} 1 \\ -1 \end{bmatrix}$

$T(\mathbf{v}) = P[T(\mathbf{v})]_{B'} = \begin{bmatrix} 1 & 1 \\ 1 & 2 \end{bmatrix}\begin{bmatrix} 9 \\ -6 \end{bmatrix} = \begin{bmatrix} 3 \\ -3 \end{bmatrix}$

$[T(\mathbf{v})]_B = A[\mathbf{v}]_B = \begin{bmatrix} 1 & -2 \\ 1 & 4 \end{bmatrix}\begin{bmatrix} 1 \\ -1 \end{bmatrix} = \begin{bmatrix} 3 \\ -3 \end{bmatrix}$

19. The characteristic equation is

$$|\lambda I - A| = \begin{vmatrix} \lambda - 7 & -2 \\ 2 & \lambda - 3 \end{vmatrix}$$
$$= \lambda^2 - 10\lambda + 25 = (\lambda - 5)^2 = 0.$$

So, the only eigenvalue is $\lambda = 5$. A corresponding eigenvector is $(1, -1)$.

20. The characteristic equation is

$$|\lambda I - A| = \begin{vmatrix} \lambda - 4 & -1 \\ 2 & \lambda - 1 \end{vmatrix}$$
$$= \lambda^2 - 5\lambda + 6 = (\lambda - 2)(\lambda - 3) = 0.$$

The eigenvalues are $\lambda_1 = 2$ and $\lambda_2 = 3$. Corresponding eigenvectors are $(1, -2)$ and $(1, -1)$, respectively.

21. The characteristic equation is

$$|\lambda I - A| = \begin{vmatrix} \lambda - 1 & -2 & -1 \\ 0 & \lambda - 3 & -1 \\ 0 & 3 & \lambda + 1 \end{vmatrix} = (\lambda - 1)(\lambda^2 - 2\lambda) = \lambda(\lambda - 1)(\lambda - 2) = 0.$$

For $\lambda_1 = 1$:

$$\begin{bmatrix} 0 & -2 & -1 \\ 0 & -2 & -1 \\ 0 & 3 & 2 \end{bmatrix}\begin{bmatrix} x_1 \\ x_2 \\ x_3 \end{bmatrix} = \begin{bmatrix} 0 \\ 0 \\ 0 \end{bmatrix}$$

The solution is $\{(t, 0, 0) : t \in R\}$ and an eigenvector is $(1, 0, 0)$.

For $\lambda_2 = 0$:

$$\begin{bmatrix} -1 & -2 & -1 \\ 0 & -3 & -1 \\ 0 & 3 & 1 \end{bmatrix}\begin{bmatrix} x_1 \\ x_2 \\ x_3 \end{bmatrix} = \begin{bmatrix} 0 \\ 0 \\ 0 \end{bmatrix}$$

The solution is $\{(-t, -t, 3t) : t \in R\}$ and an eigenvector is $(-1, -1, 3)$.

For $\lambda_3 = 2$:

$$\begin{bmatrix} 1 & -2 & -1 \\ 0 & -1 & -1 \\ 0 & 3 & 3 \end{bmatrix}\begin{bmatrix} x_1 \\ x_2 \\ x_3 \end{bmatrix} = \begin{bmatrix} 0 \\ 0 \\ 0 \end{bmatrix}$$

The solution is $\{(t, t, -t) : t \in R\}$ and an eigenvector is $(1, 1, -1)$.

22. Because A is a triangular matrix, $\lambda = 1$ (repeated).

For $\lambda = 1$:

$$\begin{bmatrix} 0 & 1 & -1 \\ 0 & 0 & -2 \\ 0 & 0 & 0 \end{bmatrix}\begin{bmatrix} x_1 \\ x_2 \\ x_3 \end{bmatrix} = \begin{bmatrix} 0 \\ 0 \\ 0 \end{bmatrix}$$

The solution is $\{(t, 0, 0) : t \in R\}$ and an eigenvector is $(1, 0, 0)$.

© 2013 Cengage Learning. All Rights Reserved. May not be scanned, copied or duplicated, or posted to a publicly accessible website, in whole or in part.

23. The eigenvalues of A are $\lambda_1 = 2$, $\lambda_2 = -1$, and $\lambda_3 = 3$. The corresponding eigenvectors $(1, 0, 0)$, $(1, -1, 0)$, and $(5, 1, 2)$ are used to form the columns of P. So,

$$P = \begin{bmatrix} 1 & 1 & 5 \\ 0 & -1 & 1 \\ 0 & 0 & 2 \end{bmatrix} \Rightarrow P^{-1} = \begin{bmatrix} 1 & 1 & -3 \\ 0 & -1 & \frac{1}{2} \\ 0 & 0 & \frac{1}{2} \end{bmatrix} \text{ and } P^{-1}AP = \begin{bmatrix} 1 & 1 & -3 \\ 0 & -1 & \frac{1}{2} \\ 0 & 0 & \frac{1}{2} \end{bmatrix}\begin{bmatrix} 2 & 3 & 1 \\ 0 & -1 & 2 \\ 0 & 0 & 3 \end{bmatrix}\begin{bmatrix} 1 & 1 & 5 \\ 0 & -1 & 1 \\ 0 & 0 & 2 \end{bmatrix} = \begin{bmatrix} 2 & 0 & 0 \\ 0 & -1 & 0 \\ 0 & 0 & 3 \end{bmatrix}.$$

24. The eigenvalues of A are $\lambda_1 = -2$, $\lambda_2 = 6$ and $\lambda_3 = 4$ (see Exercise 25, Section 7.1).

The corresponding eigenvectors $(3, 2, 0)$, $(-1, 2, 0)$ and $(-5, 10, 2)$ are used to form the columns of P. So,

$$P = \begin{bmatrix} 3 & -1 & -5 \\ 2 & 2 & 10 \\ 0 & 0 & 2 \end{bmatrix} \Rightarrow P^{-1} = \begin{bmatrix} \frac{1}{4} & \frac{1}{8} & 0 \\ -\frac{1}{4} & \frac{3}{8} & -\frac{5}{2} \\ 0 & 0 & \frac{1}{2} \end{bmatrix}$$

and

$$P^{-1}AP = \begin{bmatrix} \frac{1}{4} & \frac{1}{8} & 0 \\ -\frac{1}{4} & \frac{3}{8} & -\frac{5}{2} \\ 0 & 0 & \frac{1}{2} \end{bmatrix}\begin{bmatrix} 0 & -3 & 5 \\ -4 & 4 & -10 \\ 0 & 0 & 4 \end{bmatrix}\begin{bmatrix} 3 & -1 & -5 \\ 2 & 2 & 10 \\ 0 & 0 & 2 \end{bmatrix} = \begin{bmatrix} -2 & 0 & 0 \\ 0 & 6 & 0 \\ 0 & 0 & 4 \end{bmatrix}.$$

25. The standard matrix for T is

$$A = \begin{bmatrix} 2 & 0 & -2 \\ 0 & 2 & -2 \\ 3 & 0 & -3 \end{bmatrix}$$

which has eigenvalues $\lambda_1 = 2$, $\lambda_2 = 0$, and $\lambda_3 = -1$ and corresponding eigenvectors $(0, 1, 0)$, $(1, 1, 1)$, and $(2, 2, 3)$.

So, $B = \{(0, 1, 0), (1, 1, 1), (2, 2, 3)\}$ and

$$P = \begin{bmatrix} 0 & 1 & 2 \\ 1 & 1 & 2 \\ 0 & 1 & 3 \end{bmatrix} \Rightarrow P^{-1}\begin{bmatrix} -2 & 2 & 0 \\ 0 & 0 & 0 \\ 1 & 0 & -1 \end{bmatrix}$$

and

$$P^{-1}AP = \begin{bmatrix} 2 & 0 & 0 \\ 0 & 0 & 0 \\ 0 & 0 & -1 \end{bmatrix}.$$

26. The eigenvalues of A are $\lambda_1 = -2$ and $\lambda_2 = 4$, with corresponding eigenvectors $(1, -1)$ and $(1, 1)$, respectively. Normalize each eigenvector to form the columns of P. Then

$$P = \begin{bmatrix} \frac{1}{\sqrt{2}} & \frac{1}{\sqrt{2}} \\ -\frac{1}{\sqrt{2}} & \frac{1}{\sqrt{2}} \end{bmatrix}$$

and

$$P^T AP = \begin{bmatrix} \frac{1}{\sqrt{2}} & -\frac{1}{\sqrt{2}} \\ \frac{1}{\sqrt{2}} & \frac{1}{\sqrt{2}} \end{bmatrix}\begin{bmatrix} 1 & 3 \\ 3 & 1 \end{bmatrix}\begin{bmatrix} \frac{1}{\sqrt{2}} & \frac{1}{\sqrt{2}} \\ -\frac{1}{\sqrt{2}} & \frac{1}{\sqrt{2}} \end{bmatrix} = \begin{bmatrix} -2 & 0 \\ 0 & 4 \end{bmatrix}.$$

27. Eigenvalues and eigenvectors of A are

$$\lambda = 4, \begin{bmatrix} 1 \\ 1 \\ 1 \end{bmatrix}; \lambda = -2, \begin{bmatrix} 1 \\ 0 \\ -1 \end{bmatrix}, \begin{bmatrix} 1 \\ -1 \\ 0 \end{bmatrix}.$$

Using the Gram-Schmidt orthonormalization process,

$$P = \begin{bmatrix} \frac{1}{\sqrt{3}} & \frac{1}{\sqrt{2}} & \frac{1}{\sqrt{6}} \\ \frac{1}{\sqrt{3}} & 0 & -\frac{2}{\sqrt{6}} \\ \frac{1}{\sqrt{3}} & -\frac{1}{\sqrt{2}} & \frac{1}{\sqrt{6}} \end{bmatrix} \text{ and}$$

$$P^T AP = \begin{bmatrix} 4 & 0 & 0 \\ 0 & -2 & 0 \\ 0 & 0 & -2 \end{bmatrix}.$$

28. The solution to the differential equation $y' = Ky$ is $y = Ce^{Kt}$. So, $y_1 = C_1e^t$ and $y_2 = C_2e^{3t}$.

29. Because $a = 4$, $b = -8$, and $c = 4$, the matrix is

$$A = \begin{bmatrix} a & \frac{b}{2} \\ \frac{b}{2} & c \end{bmatrix} = \begin{bmatrix} 4 & -4 \\ -4 & 4 \end{bmatrix}.$$

© 2013 Cengage Learning. All Rights Reserved. May not be scanned, copied or duplicated, or posted to a publicly accessible website, in whole or in part.

30. Construct the age transition matrix.

$$A = \begin{bmatrix} 3 & 6 & 3 \\ 0.8 & 0 & 0 \\ 0 & 0.4 & 0 \end{bmatrix}$$

The current age distribution vector is

$$\mathbf{x}_1 = \begin{bmatrix} 150 \\ 150 \\ 150 \end{bmatrix}.$$

In 1 year, the age distribution vector will be

$$\mathbf{x}_2 = A\mathbf{x}_1 = \begin{bmatrix} 3 & 6 & 3 \\ 0.8 & 0 & 0 \\ 0 & 0.4 & 0 \end{bmatrix}\begin{bmatrix} 150 \\ 150 \\ 150 \end{bmatrix} = \begin{bmatrix} 1800 \\ 120 \\ 60 \end{bmatrix}.$$

In 2 years, the age distribution vector will be

$$\mathbf{x}_3 = A\mathbf{x}_2 = \begin{bmatrix} 3 & 6 & 3 \\ 0.8 & 0 & 0 \\ 0 & 0.4 & 0 \end{bmatrix}\begin{bmatrix} 1800 \\ 120 \\ 60 \end{bmatrix} = \begin{bmatrix} 6300 \\ 1440 \\ 48 \end{bmatrix}.$$

31. P is *orthogonal* if $P^{-1} = P^T$.

$$1 = \det\left(P \cdot P^{-1}\right)$$

$$= \det\left(PP^T\right) = (\det P)^2 \Rightarrow \det P = \pm 1$$

32. There exists P such that $P^{-1}AP = D$. A and B are similar implies that there exists Q such that

$A = Q^{-1}BQ$. Then

$D = P^{-1}AP = P^{-1}\left(Q^{-1}BQ\right)P = (QP)^{-1}B(QP)$.

© 2013 Cengage Learning. All Rights Reserved. May not be scanned, copied or duplicated, or posted to a publicly accessible website, in whole or in part.